*Advances in Numerical Mathematics*

Andreas Prohl

# Computational Micromagnetism

*Advances in Numerical Mathematics*

Andreas Prohl

# Computational Micromagnetism

Springer Fachmedien Wiesbaden GmbH

Die Deutsche Bibliothek – CIP-Einheitsaufnahme

**Dr. Andreas Prohl**

Geboren 1968 in Lübeck. Von 1988 bis 1993 Studium der Mathematik und Physik an der Ruprecht-Karls-Universität Heidelberg, 1993 Diplom in Mathematik, Promotion 1996. Einjähriger Aufenthalt am Institut for Mathematics and its Applications (IMA) der University of Minnesota, Minneapolis als DFG-Forschungsstipendiat. Seit 1997 Mitarbeiter des Mathematischen Seminars der Christian-Albrechts-Universität Kiel, seit 1999 gefördert durch ein DFG-Habilitandenstipendium, Habilitation 2001. Seit 2001 gefördert durch ein DFG-Heisenbergstipendium.

1. Auflage Dezember 2001


Originally published by B.G. Teubner GmbH, Stuttgart/Leipzig/Wiesbaden in 2001


www.teubner.de



Umschlaggestaltung: Ulrike Weigel, www.CorporateDesignGroup.de

Gedruckt auf säurefreiem und chlorfrei gebleichtem Papier.

ISBN 978-3-519-00358-8 ISBN 978-3-663-09498-2 (eBook)
DOI 10.1007/978-3-663-09498-2

*Dedicated to my parents*

*Numerus clavis rerum, res fontes numerorum*

# Introduction

In this work, we study numerical issues related to a common mathematical model which describes ferromagnetic materials, both in a stationary and nonstationary context. Electromagnetic effects are accounted for in an extended model to study nonstationary magneto-electronics. The last part deals with the numerical analysis of the commonly used Ericksen-Leslie model to study the fluid flow of nematic liquid crystals which find applications in display technologies, for example.

All these mathematical models to describe different microstructural phenomena share common features like (i) strong nonlinearities, and (ii) nonconvex side constraints (i.e., $|\mathbf{m}| = 1$, almost everywhere in $\omega \subset \mathbb{R}^d$, for the order parameter $\mathbf{m} : \omega \to \mathbb{R}^d$). One key issue in numerical modeling of such problems is to make sure that the non-convex constraint is fulfilled for computed solutions.

We present and analyze different solution strategies to deal with the variational problem of stationary micromagnetism, which builds part I of the book: *direct minimization*, *convexification*, and *relaxation using Young measure-valued solutions*. In particular, we address the following points:

- *Direct minimization:* A spatial triangulation 'generates an artificial exchange energy contribution' in the discretized minimizing problem which may pollute physically relevant exchange energy contributions; its minimizers exhibit multiple scales (with branching structures near the boundary of the ferromagnet) and are difficult to be computed efficiently. We exploit this observation to construct an adaptive scheme which better resolves multiple scale structures (cubic ferromagnets).

- *Convexification:* It is due to the degeneracy of the convex envelope and the mixed formulation that a stable finite element realization requires careful balancing of used trial spaces for magnetization and magnetic potential; moreover, a penalty approach to satisfy the convexified side constraint (i.e., $|\mathbf{m}| \leq 1$, a.e. in $\omega$) needs clarification how to choose an optimal penalty parameter in terms of the mesh-size of the underlying triangulation. We discuss three stable discretization strategies: the first is based on non-conforming finite element functions for the magnetic potential. The second approach uses an additional stabilization term which enables to use conforming piecewise affine functions; a second stabilization strategy allows for convergence results for the related Young measure $\nu_h = \nu_h(\mathbf{m}_h)$ in cases where the computed magnetization $\mathbf{m}_h$ is smooth (at least locally). In addition, adaptive concepts are studied.

- *Relaxation using Young measure-valued solutions:* Minimization of the original problem is extended to Young measure-valued functions, which are approximated element-wise by sums of Dirac distributions. This approach is attractive since it does not require manipulation of the energy functional, and the problem that we end up with has a solution. Its main drawback is the huge computational effort which is necessary to finally solve the problem. A significant reduction of work to make it a competitive method is by means of an active set strategy that singles out atoms locally which are relevant for minimizing sequences.

The subject of part II is the numerical analysis of nonstationary problems: *ferromagnetism*, *magneto-electronics*, and *nematic liquid crystals*. It is known that a straightforward application of the implicit Euler method leads to a violation of the non-convexity constraint in each specific model (i.e., in general $|\mathbf{m}^j| \neq 1$, versus $|\mathbf{m}(t_j)| = 1$ as an implication for the continuous problem, $0 \leq j \leq J$). To get rid of this problem, time discretization schemes have been developed which project computed vector fields back to the sphere at the end of each iteration step.

- *Ferromagnetism:* In order to understand the effect of the basic projection scheme [2, 43], we reformulate it as a semi-implicit penalty method. From the analysis of different penalization strategies, it will become clear how to modify the basic projection scheme, while keeping the same overall computational effort: the new splitting scheme involves algebraic manipulation of computed magnetizations which is no projection to the sphere any more.

- *Magneto-electronics:* The subtle interplay between 'electric' and 'magnetic' quantities is described by Maxwell's equations, together with the Landau-Lifshitz equation. The construction of an efficient splitting scheme requires understanding of (related) penalization and decoupling effects under the action of strong nonlinearities, with special attention given to the system character of the problem: further stabilizing terms and stretched time-grid structures will be employed to handle these difficulties.

- *Nematic liquid crystals:* The Ericksen-Leslie model involves an additional incompressibility constraint that needs to be accounted for in an efficient way: we combine the idea of Chorin's projection scheme [28] with the above [2, 43] (or its new modifications, see above). It is by reformulation of the scheme as a problem which involves two penalization terms that we can prove optimal convergence behavior of the constructed scheme.

All convergence analyses in part II are carried out for periodic boundary data and two spatial dimensions, using the notion of strong solutions which can be verified locally in time (at least). There is no doubt that corresponding numerical analyses which deal with weak solutions and/or a three-dimensional setting are of considerable interest; we hope that this work will stimulate further research in this direction, leading to efficient numerical schemes which reliably detect (dynamics of) microstructures, defects.

This monograph summarizes the research of the author over the last four years in this area, and most of the results presented here are new. However, in order to complete our picture of nowadays computational micromagnetism — with no claim of being exhaustive —, we add some further material that has been obtained jointly with other researchers and has already been published before: it is a pleasure for me to mention the joint work with C. Carstensen on stable non-conforming methods in stationary convexified ferromagnetism [20], S. A. Funken on stabilized conforming methods in convexified ferromagnetism [52], and M. Kružík on relaxed micromagnetism using Young measures [82]. This material is presented in Sections 2.1, 2.2, and Chapter 3, respectively. The rest of part I presents results that have not been published before. Part II covers [110].

I am very grateful to C. Carstensen (Technische Universität Wien) and S. Müller (MPI Leipzig) for introducing me to numerical micromagnetism

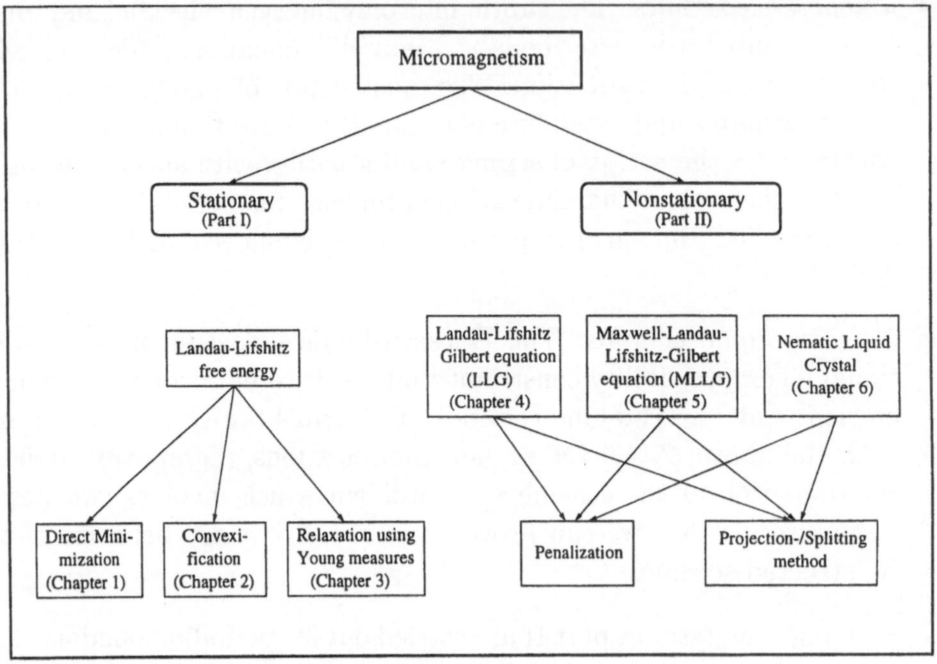


Figure 1: Organization of the book

when I joined the DFG-project: 'Numerische Behandlung des Mikromagnetismus' in 1997, continuing support and interest in my results. I am indebted to them and W. Hackbusch (MPI Leipzig) for many valuable suggestions, without which this work would not have been possible. It is also a pleasure for me to thank M. Kružík (Research Center Caesar, Bonn), C. Liu (Penn State University, USA), P. Plecháč (University of Warwick, UK), T. Roubíček (Charles University, Prague), and N. Walkington (Carnegie Mellon University, Pittsburgh) for many stimulating discussions and continuous support. Moreover, financial support by a DFG scholarship is gratefully acknowledged which enables this work, and fostering by the Graduiertenkolleg 357 "Effiziente Algorithmen und Mehrskalenmethoden" (Universität Kiel).

# Contents

# Notation

| | |
|---|---|
| $o, \mathcal{O}$ | Landau symbols |
| $C$ | generic constant |
| $C_\gamma$ | generic constant that depends on $\gamma$, in particular |
| $C(a,b,c)$ | generic constant that depends on $a, b, c$, in particular |
| $\mu_d(\omega)$ | $d$-dimensional Lebesgue measure of $\omega \subset \mathbb{R}^d$ |
| $\partial A$ | boundary of a set $A$ |
| $D^k v(a)$ | $k$-th (Fréchet) derivative of a function at a point $a$ |
| $\mathbf{v}$ | $= (v_1, ..., v_d)$ |
| $\partial_i v(a)$ | $= Dv(a)e_i$ |
| $\operatorname{div} \mathbf{v}$ | $= \sum_{i=1}^d \partial_i v_i$ |
| $\mathbf{curl}\, \mathbf{v}$ | $\equiv \nabla \times \mathbf{v} = (\partial_2 v_3 - \partial_3 v_2, \partial_3 v_1 - \partial_1 v_3, \partial_1 v_2 - \partial_2 v_1)^\top$ |
| $\mathbf{curl}\, v$ | $= (\partial_2 v, -\partial_1 v)^\top$ |
| $\nabla$ | $= (\partial_1, \partial_2, .., \partial_d)$ |
| $\Delta$ | $= \sum_{i=1}^d \partial_{ii}$ |
| $\nabla \mathbf{v} \odot \nabla \mathbf{w}$ | $= \nabla \mathbf{v}^\top \nabla \mathbf{w}$ |
| $\partial_{\mathbf{n}}$ | $= \sum_{i=1}^d n_i \partial_i$ (outer) normal derivative operator |
| $\lvert \alpha \rvert$ | $= \sum_{i=1}^d \alpha_i$, for $\alpha$ multi-index $\alpha = (\alpha_1, ..., \alpha_d) \in \mathbb{N}^d$ |
| $\partial^\alpha v(a)$ | $= D^{\lvert \alpha \rvert} v(a)(\underbrace{e_1, ..., e_1}_{\alpha_1 \text{ times}}, \underbrace{e_2, ..., e_2}_{\alpha_2 \text{ times}}, ..., \underbrace{e_d, ..., e_d}_{\alpha_d \text{ times}})$ |
| $B_r(a)$ | $= \{x \in X : \lVert x - a \rVert_X \le r\}$ |
| $\mathcal{S}^{d-1}$ | $\subset \mathbb{R}^d$ $(d-1)$-dimensional unit sphere |
| $X^*$ | dual of a Hilbert space $X$ |
| $\mathrm{Id}$ | identity mapping |
| $-\tilde{\Delta}$ | $= \mathrm{Id} - \Delta : \mathbf{W}^{1,2}(\omega, \mathbb{R}^3) \to \mathbf{W}^{-1,2}(\omega, \mathbb{R}^3)$ isomorphism |
| $\hookrightarrow$ | inclusion with continuous injection |
| $\Subset$ | inclusion with compact injection |

| | |
|---|---|
| $\mathbf{a} \times \mathbf{b}$ | vector product of vectors $\mathbf{a}, \mathbf{b} \in \mathbb{R}^3$ |
| $\tau(t)$ | $= \min\{1, t\}$ |
| $I$ | time interval $[0, T]$ |
| $\omega_T$ | $= \omega \times I$ |
| $\mathbf{m}, \mathbf{H}, \mathbf{E}$ | magnetization, magnetic and electric field, resp., with $\mathbf{m}, \mathbf{H}, \mathbf{E} : \mathbb{R}^2 \supset \omega \to \mathbb{R}^3$ |
| $\mathbf{e}, \mathbf{e}_i$ | easy axis (uniaxial) resp. axes (cubic, $i = 1, 2.3$) |
| $\mathbf{H}_{\text{eff}}$ | effective field |
| $u$ | magnetostatic potential, $u : \mathbb{R}^2 \supset \Omega \to \mathbb{R}$ |
| $\mathbf{u}, \mathbf{d}$ | velocity and director field, resp., $\mathbf{u}, \mathbf{d} : \mathbb{R}^2 \supset \omega \to \mathbb{R}^2$ |
| $p$ | pressure, $p : \mathbb{R}^2 \supset \omega \to \mathbb{R}$ |
| $(s)_+$ | $:= \max\{0, s\}$ |
| $(s)_\bullet$ | $:= \max\{1, s\}$ |
| $\mathbf{l}_\varepsilon(\mathbf{v})$ | $= \frac{1}{\varepsilon}\{\lvert \mathbf{v} \rvert^2 - 1\}\mathbf{v}$ |

## Specific vector spaces

| | |
|---|---|
| $\lVert \cdot \rVert_X$ | norm (in the space $X$) |
| $\langle \cdot, \cdot \rangle_{\mathbb{R}^d}$ | scalar product of $\mathbb{R}^d$ |
| $\lvert \cdot \rvert$ | Euclidean norm in $\mathbb{R}^d$ |
| $(\cdot, \cdot)_\omega$ | inner product in $L^2(\omega)$, $\omega \subset \mathbb{R}^2$ |
| $\mathcal{A}$ | $= \big\{\mathbf{m} \in \mathbf{L}^\infty(\omega, \mathbb{R}^d) : \lvert \mathbf{m} \rvert = 1, \text{ a.e. in } \omega\big\}$ |
| $\tilde{\mathcal{A}}$ | $= \big\{\mathbf{m} \in \mathbf{L}^\infty(\omega, \mathbb{R}^d) : \lvert \mathbf{m} \rvert \leq 1, \text{ a.e. in } \omega\big\}$ |
| $\overline{\mathcal{A}}$ | $= \big\{\nu = \{\nu_{\mathbf{x}}\}_{\mathbf{x} \in \omega}$ weakly measurable; $\operatorname{supp} \nu_{\mathbf{x}} \subset \mathcal{S}^{d-1}$, f.a.a. $\mathbf{x} \in \omega\big\}$ |
| $\mathbf{H}(\mathbf{curl}, \omega)$ | $= \big\{\mathbf{v} \in \mathbf{W}^{0,2}_{\text{per}}(\omega, \mathbb{R}^d) : \mathbf{curl}\, \mathbf{v} = 0\big\}$, for $\omega = (0, 2D)^2$ |
| $\mathbf{H}(\operatorname{div}, \omega)$ | $= \big\{\mathbf{v} \in \mathbf{W}^{0,2}_{\text{per}}(\omega, \mathbb{R}^d) : \operatorname{div} \mathbf{v} = 0\big\}$, for $\omega = (0, 2D)^2$ |
| $L^2_0(\omega)$ | $= \big\{v \in L^2(\omega) : \int_\omega v \,\mathrm{d}\mathbf{x} = 0\big\}$, isomorphic to $L^2(\omega)/\mathbb{R}$ |

| | |
|---|---|
| $\mathbf{W}^{m,p}(\omega,\mathbb{R}^d)$ | $= \{\mathbf{v} \in \mathbf{L}^p(\omega,\mathbb{R}^d) : \partial^\alpha v_i \in L^p$ for all $\alpha$, with $\lvert\alpha\rvert \leq m$ , and all $1 \leq i \leq d\}$ |
| $\lVert \mathbf{v} \rVert_{\mathbf{W}^{m,p}}$ | $= \left(\sum_{\lvert\alpha\rvert\leq m} \int_\omega \lvert \partial^\alpha \mathbf{v} \rvert^p \,\mathrm{d}\mathbf{x}\right)^{1/p}, \quad 1 \leq p < \infty$ |
| $\lVert \mathbf{v} \rVert_{\mathbf{W}^{m,\infty}}$ | $= \max_{\lvert\alpha\rvert\leq m}\{\operatorname{ess\,sup}_{\mathbf{x}\in\omega} \lvert \partial^\alpha \mathbf{v}(\mathbf{x}) \rvert\}$ |
| $W_0^{1,2}(\omega)$ | $= \{v \in W^{1,2}(\omega) : v\big\rvert_{\partial\omega} = 0\}$ |
| $W^{-1,2}(\omega)$ | $= [W^{1,2}(\omega)]^*$ |
| $\mathbf{W}^{1,2}(\omega,\mathcal{S}^{d-1})$ | $= \{\mathbf{v} \in \mathbf{W}^{1,2}(\omega,\mathbb{R}^d) : \lvert\mathbf{v}\rvert = 1$ almost everywhere in $\omega\}$ |
| $\mathbf{W}^{m,p}(\omega,\mathcal{S}^{d-1})$ | $= \mathbf{W}^{1,2}(\omega,\mathcal{S}^{d-1}) \cap \mathbf{W}^{m,p}(\omega,\mathbb{R}^d)$ for $m \geq 1$, $p \geq 2$ |
| $C^\infty_{\mathrm{per}}(\omega)$ | space of periodic functions on $\omega = (0,2D)^2$ |
| $W^{m,p}_{\mathrm{per}}(\omega)$ | $= \overline{C^\infty_{\mathrm{per}}(\omega)}^{W^{m,p}(\omega)}$ , for $\omega = (0,2D)^2$ |
| $\mathbf{W}^{-m,p}_{\mathrm{per}}(\omega,\mathbb{R}^d)$ | $= [\mathbf{W}^{m,p}_{\mathrm{per}}(\omega,\mathbb{R}^d)]^*$ , for $\omega = (0,2D)^2$ |
| $\dot{\mathbf{W}}^{m,p}_{\mathrm{per}}(\omega,\mathbb{R}^d)$ | $= \overline{\{\boldsymbol{\varphi} \in [C^\infty_{\mathrm{per}}(\omega)]^d : \int_\omega \boldsymbol{\varphi}\,\mathrm{d}\mathbf{x} = 0\}}^{\mathbf{W}^{m,p}(\omega,\mathbb{R}^d)}$ , for $\omega = (0,2D)^2$ |
| $\mathbf{W}^{-m,p}_{\mathrm{per}}(\omega,\mathbb{R}^d)$ | $= [\mathbf{W}^{m,p}_{\mathrm{per}}(\omega,\mathbb{R}^d)]^*$ , for $\omega = (0,2D)^2$ |
| $\mathbf{J}_0$ | $= \{\mathbf{v} \in \dot{\mathbf{W}}^{0,2}_{\mathrm{per}}(\omega,\mathbb{R}^2) : \operatorname{div}\mathbf{v} = 0\}$ , for $\omega = (0,2D)^2$ |
| $\tilde{\mathbf{J}}_0$ | $= \{\mathbf{v} \in \mathbf{L}^2(\omega,\mathbb{R}^d) : \operatorname{div}\mathbf{v} = 0,\ \mathbf{v}\big\rvert_{\partial\omega}\cdot\mathbf{n} = 0$, weakly$\}$ |
| $\mathbf{J}_1$ | $= \{\mathbf{v} \in \dot{\mathbf{W}}^{1,2}_{\mathrm{per}}(\omega,\mathbb{R}^2) : \operatorname{div}\mathbf{v} = 0\}$ for $\omega = (0,2D)^2$ |
| $\tilde{\mathbf{J}}_1$ | $= \{\mathbf{v} \in \mathbf{W}^{1,2}_0(\omega,\mathbb{R}^d) : \operatorname{div}\mathbf{v} = 0\}$ |
| $\mathcal{L}^{p,\lambda}$ | Campanato space, with $[u]_{p,\lambda;\Omega} = \left\{\sup_{\mathbf{x}_0\in\Omega, 0<\rho<\operatorname{diam}\Omega} \rho^{-\lambda} \int_{B(\mathbf{x}_0,\rho)} \lvert u(\mathbf{x}) - u_{\mathbf{x}_0,\rho}\rvert^p \,d\mathbf{x}\right\}$ |
| $BMO(\omega)$ | functions of bounded mean oscillation, (see [54, 55]) |
| $L^p(I;X)$ | Bochner space, with $\lVert v \rVert_{L^p(I;X)} = \left(\int_I \lVert v \rVert_X^p \,\mathrm{d}s\right)^{\frac{1}{p}}$, $1 \leq p <$ |
| $L^\infty(I;X)$ | Bochner space, with $\lVert v \rVert_{L^\infty(I;X)} = \operatorname{ess\,sup}_I \lVert v \rVert$ , |
| $\ell^p(I_k;X)$ | discrete Bochner space, with $\lVert \{v^j\} \rVert_{\ell^p(I_k;X)} = \big(k\sum_j \lVert v^j \rVert_X^p$ $1 \leq p < \infty$ |
| $\ell^\infty(I_k;X)$ | discrete Bochner space, with $\lVert \{v^j\} \rVert_{\ell^\infty(I_k;X)} = \max_{I_k} \lVert v^j \rVert$ |

## Time-discretization and Finite elements

| | |
|---|---|
| $\mathbf{P}_{\mathbf{J}_0}$ | Helmholtz projection of $\mathbf{L}^2(\omega, \mathbb{R}^d)$ onto $\mathbf{J}_0$ |
| $\mathbf{A}$ | Stokes operator, $\mathbf{A} : \mathbf{J}_1 \to \mathbf{J}_1^*$ |
| $P_\ell$ | space of all polynomials in $x_1, ..., x_d$ of degree $\leq \ell$ |
| $h, h_{\mathcal{T}}$ | parametrization of regular spatial mesh $\mathcal{T}$ |
| $\mathcal{T}, \mathcal{T}_h$ | regular quasi-uniform triangulation of $\omega$, allowing inverse estimates and $\mathbf{W}^{1,2}$-stability of $\mathbf{P}_{h,\mathbf{L}^2}$ (see [34, 12, 19]) |
| $\mathcal{T}^c$ | cartesian triangulation, aligned with easy axis |
| $\mathcal{T}_{h_0}^{\text{grad}}$ | graduate mesh with refined boundary layer structure |
| $\mathcal{H}_\ell$ | $= \mathcal{H}_\ell(\boldsymbol{\mu}_{h_0}^\ell)$ , mesh refinement strategy |
| $\mathcal{T}\big\|_\omega$ | $= \{K \in \mathcal{T} : K \subset \omega\}$, for $\omega \subset \Omega$ |
| $\mathcal{H}_\ell$ | $= \mathcal{H}_\ell(\boldsymbol{\mu}_{h_0}^\ell)$, mesh refinement strategy |
| $\boldsymbol{\mathcal{T}}_{\mathbf{h}}$ | $= \mathcal{T}_{h_1}^1 \times \mathcal{T}_{h_2}^2$ , for $\mathbf{h} = (h_1, h_2)$ |
| $P_{\mathbf{h}}$ | $= P_{(h_1,h_2)} = P_{h_1}^1 P_{h_2}^2$ |
| $\Xi$ | set of all edges of $\mathcal{T}$ |
| $\Xi^*$ | $= \Xi\|_{\overline{\omega}} \setminus (\Xi\|_{\overline{\omega}} \cap \partial\omega)$ |
| $[\cdot]$ | $= [\cdot]_{\boldsymbol{\xi}}$ jump across inter-element face $\boldsymbol{\xi} \subset \Xi$ |
| $\mathcal{N}$ | set of all nodes in $\Omega$ |
| $\mathcal{K}$ | $:= \mathcal{N} \cap \omega$, set of free nodes, for $\omega \subset \Omega$ |
| $\mathcal{M}$ | set of all midpoints of edges $\boldsymbol{\xi} \in \Xi$ |
| $f_{\mathcal{T}}$ | $= \frac{1}{\|K\|} \int_K f \, \mathrm{d}\mathbf{x} \in \mathcal{L}^0(\omega)$ |
| $(u, v)_h$ | $= \sum_{K \in \mathcal{T}} \int_K u \, v \, \mathrm{d}\mathbf{x}$ |
| $\mathbf{V}_h, \mathbf{W}_h, L^h$ | finite element spaces |
| $\nabla_{\mathcal{T}}, \mathrm{div}_{\mathcal{T}}$ | restricted operators to interior of elements $K \in \mathcal{T}$ |
| $\mathbf{P}_{h,\mathbf{L}^2}\mathbf{v}$ | $(\mathbf{v} - \mathbf{P}_{h,\mathbf{L}^2}\mathbf{v}, \mathbf{w}_h) = 0$, for all $\mathbf{w}_h \in \mathbf{W}_h$ |

| | |
|---|---|
| $P_{h,W^{1,2}}$ | $(v - P_{h,W^{1,2}}v, w_h) = 0$, for all $w_h \in W_h$ |
| $\mathbf{P}_{h,\mathbf{W}^{1,2}}\mathbf{v}$ | $(\mathbf{v} - \mathbf{P}_{h,\mathbf{W}^{1,2}}\mathbf{v}, \mathbf{w}_h)_{\mathbf{W}^{1,2}} = 0$, for all $\mathbf{w}_h \in \mathbf{W}_h$, |
| $\mathcal{L}^0(\omega)$ | $= \{v_h \in L^\infty(\omega) : \forall K \in \mathcal{T}\big\|_\omega, v_h\big\|_K \text{ constant}\}$, for $\omega \subset \Omega$ and all $\mathbf{w}_h \in \mathbf{W}_h$ |
| $\mathcal{S}^1(\omega)$ | $= \{v_h \in C^0(\omega) : v_h\big\|_K \in P_1(K),\, K \in \mathcal{T}\} \subset C^0(\omega)$ |
| $\mathcal{S}^{1,NC}(\omega)$ | $= \{v_h \in L^2(\omega) : v_h\big\|_K \in P_1(K),\, K \in \mathcal{T}\}$ |
| $\mathcal{S}_0^{1,NC}(\omega)$ | $= \{v_h \in \mathcal{S}^{1,NC}(\overline{\omega}) : \forall z \in \mathcal{M} \cap \partial\omega,\, v_h(z) = 0\}$ |
| $\boldsymbol{\mathcal{S}}^1(\omega)$ | $= [\mathcal{S}^1(\omega)]^d$ |
| $\boldsymbol{\mathcal{A}}_h$ | $= \{\boldsymbol{\mu}_h \in \boldsymbol{\mathcal{A}} : \boldsymbol{\mu}_h\big\|_K \in [P_0(K)]^d,\, K \in \mathcal{T}\}$ |
| $(\cdot,\cdot)_h$ | $= \sum_{K\in\mathcal{T}}(\cdot\|_K, \cdot\|_K)$ |
| $k$ | time-step size of equi-distant time-grid |
| $k_j$ | $= t_{j+1} - t_j$ (local) width of time-grid |
| $\tau_j$ | $= \min\{1, t_j\}$ |
| $I_k$ | $= \{t_j\}_{j=1}^J$ (equi-distant) time-grid |
| $d_t v^{j+1}$ | $= k^{-1}\{v^{j+1} - v^j\}$ |
| $\overline{v}^{j+1/2}$ | $= \frac{1}{2}\{v^{j+1} + v^j\}$ |
| $\ell^p(I_k; X)$ | space, with norm $\\| v \\|_{\ell^p(I_k;X)} = \big(k\sum_{j=0}^J \\| v^{j+1} \\|_X^p\big)^{1/p}$ |

# Part I

# Numerical Stationary Micromagnetism

Ferromagnetic materials are used in a variety of modern devices, like permanent magnets, magnetic sensor elements or magnetic recording devices. There are different classes of magnets, depending on the number and orientation of the easy axes of the used material. For instance, multiaxial or, more specialized, planar materials exhibit several easy axes that allow for a relatively strong susceptibility to change a current magnetization according to exterior magnetic fields. A special class of practical interest are cubic materials like iron. Uniaxial materials like samarium-cobalt are strong magnetic materials which only have one easy axis and are therefore preferred in applications that make use of hysteresis effects, such as in permanent magnets and in magnetic recording devices.

In order to explain magnetization patterns in materials or to improve on properties of devices that make use of them, a physical theory on micromagnetism has been developed by Landau and Lifshitz [84], see also [17, 69]. Its formulation is based on a variational principle, where magnetization distributions inside the ferromagnetic body $\omega \subset \mathbb{R}^d$, $d = 2, 3$, with smallest total energy are sought. The energy functional $\mathcal{E}(\cdot)$ comprises four terms, which are known as exchange energy, anisotropic energy, exterior field energy and stray-field (or magnetostatic) energy, respectively, for a given magnetization $\mathbf{m} : \omega \to \mathbb{R}^d$,

$$\begin{aligned} \mathcal{E}(\mathbf{m}) &= \alpha \int_\omega |\nabla \mathbf{m}|^2 \, d\mathbf{x} + \int_\omega \phi(\mathbf{m}) \, d\mathbf{x} \\ &\quad - \int_\omega \langle \mathbf{f}, \mathbf{m} \rangle_{\mathbb{R}^d} \, d\mathbf{x} + \frac{1}{2} \int_{\mathbb{R}^d} |\nabla u|^2 \, d\mathbf{x} . \end{aligned} \tag{I.1}$$

Here, $\alpha > 0$ is the exchange parameter. The non-convex anisotropy density $0 \le \phi(\cdot)$ is zero for $\mathbf{m} \parallel \mathbf{e}_i$, for $i \in I$, with some index set $I$, and positive else. The vectors $\mathbf{e}_i \in \mathbb{R}^d$ are referred to as easy axes. Further, $\mathbf{f} : \omega \to \mathbb{R}^d$ is an applied exterior magnetic field and $u : \mathbb{R}^d \to \mathbb{R}$ stands for the magnetic potential. Both, $\mathbf{m}$ and $u$ are linked through Maxwell's equations that imply the relation

$$\Delta u = \operatorname{div} (\chi_\omega \mathbf{m}) \qquad \text{in } \mathbb{R}^d , \tag{I.2}$$

where $\chi_\omega : \mathbf{x} \mapsto \chi_\omega(\mathbf{x})$ denotes the characteristic function of $\omega$, i.e.,

$$\chi_\omega(\mathbf{x}) = \begin{cases} 1 & \text{if } \mathbf{x} \in \omega, \\ 0 & \text{if } \mathbf{x} \notin \omega. \end{cases}$$

We assume that the material is magnetically saturated, that is, $|\mathbf{m}| = 1$, a.e. in $\omega$, which is normalized for simplicity. We define the set of admissible functions

$$\boldsymbol{\mathcal{A}} = \left\{ \mathbf{m} \in \mathbf{L}^{\infty}(\omega, \mathbb{R}^d) : \ |\mathbf{m}| = 1 \,, \text{ a.e. in } \omega \right\}. \tag{I.3}$$

Then, the **Minimization Problem (P)** in micromagnetism can be stated as follows,

$$\inf_{\boldsymbol{\mu} \in \boldsymbol{\mathcal{A}}} \mathcal{E}(\boldsymbol{\mu})\,. \tag{I.4}$$

As is known from [71, 72], this problem is well-posed for $\alpha > 0$, but no solution needs to exist for $\alpha = 0$ in the case of uniaxial ferromagnets; hence, we have to study minimizing sequences which exhibit microstructure instead (we stick to 'the case $\alpha = 0$' in the following). The limiting structure can be captured in terms of Young measure solutions.

Each of the terms in (I.1) that ensemble the total energy favorize different properties of considered magnetizations. The first term favorizes large domains of equal magnetization. According to the second term, the magnetization should be aligned with the easy axes of a crystal. The stray-field energy is caused by magnetic dipole interactions of the considered magnetization. It is a non-local term, according to (I.2).

As a consequence of different anisotropy functions that correspond to different materials, differently shaped and organized magnetic microstructures are observed in applications: the analysis and understanding of these domains is necessary to explain complex phenomena like hysteresis, where Weissian domains change their shape, or motion of Bloch walls in $\omega$. Model (I.1) includes minimizers which show these wall structures of magnitude $\mathcal{O}(\sqrt{\alpha})$, where the magnetization rotates inside, and it is important to study their motion together with the shape and thickness of related domains, [26, 27].

Domain branching is another phenomenon observed in experiments close to the surface $\partial\omega$ of a ferromagnet: a refined domain pattern is enforced to minimize the overall energy, whereas a wide pattern in the bulk is favored to reduce the wall energy (i.e., exchange energy term). This branching process connects wide and narrow domains in a way that depends on crystal symmetries and in particular on the number of available easy axes; the specific structure is again dependent on the material (i.e., multi-axial, planar or uniaxial). For instance, multi-phase branching in the case of cubic materials (giving rise to closed flux echelon patterns) gives completely different structures and scalings if compared to the two-phase branching where domain refinement is achieved with two magnetization phases only, see also [69, 71].

Domain branching was first introduced to domain theory by Lifshitz [86] in 1944, where patterns are given which were meant to apply to iron (a cubic material) rather than for uniaxial materials where only one easy axis is given. Finally, his domain branching structures were observed in uniaxial materials, while different domain structures were found afterwards in cubic materials that make use of increased symmetry properties with their multiple easy axes. To give some ideas about physical scales, domain branching for samarium-cobalt is typically visible on scales which range from 10 $\mu m$ to 100 $\mu m$ and Bloch walls are of magnitude 1 $nm$ up to 10 $nm$, [69].

In the beginning, the progressive domain refinement towards the surface of the magnetic material by iterated generations of smaller domains was meant to proceed infinitely towards a zero surface domain width. However, it was found that this is not a realistic assumption since finally the gain in creation of energy will be smaller than the additional expense in domain wall energy. Therefore, a theory of two-phase branching has to balance three energy contributions:

1. surface energy contributions caused by fine branching-type structures close to $\partial\omega$.
2. energy contributions add which only depend on the domain width of the last branching generation at the boundary $\partial\omega$.
3. the energy that is connected to internal stray-fields.

Finally, we have to estimate the internal field energy that is connected to the branching process.

In all our subsequent considerations which focus on numerical modeling, we delete the exchange energy contribution from (I.1). This is justified for many macroscopic ferromagnetic samples, where the exchange energy typically ranges between $10^{-12}\frac{J}{m^3}$ and $2 \cdot 10^{-11}\frac{J}{m^3}$, whereas anisotropy and stray-field energy are usually more dominant, giving values of $10^2\frac{J}{m^3}$ up to $2 \cdot 10^7\frac{J}{m^3}$, and zero to $3 \cdot 10^6\frac{J}{m^3}$, respectively [69]. Another motivation to study the limiting case ($\alpha = 0$) is that this is the most difficult one from a mathematical and numerical viewpoint.

In the following, we hence consider the energy functional

$$\mathcal{E}_0(\mathbf{m}) = \int_\omega \phi(\mathbf{m})\,\mathrm{d}\mathbf{x} + \frac{1}{2}\int_{\mathbb{R}^d} |\nabla u|^2\,\mathrm{d}\mathbf{x}\,, \tag{I.5}$$

instead of (I.1); one main result in the subsequent analysis will be that the numerical scaling of computed minimizers may be regarded as an effect which can be interpreted to be caused by an additional, 'artificial' surface energy term. This pollution effect of meshes has to be accounted for if (I.1) is considered directly, taking $\alpha > 0$.

As we already know from the discussion before, domain theory is an important tool if it comes to analyzing and qualifying magnetic materials: for most specimens $\omega \subset \mathbb{R}^3$, there is no way to directly observe interior domains, and physical observations are limited to domains on the surface $\partial\omega$. This makes numerical simulations an indispensable tool in studying ferromagnets.

A mathematical foundation of numerical schemes in stationary micromagnetism started with [95], where a numerical discretization of (I.4) is presented for the uniaxial case which uses piecewise constant functions for the magnetization field. For a polygonal domain $\omega \subset \mathbb{R}^2$, Luskin & Ma then analyzed

$$\min_{\boldsymbol{\mu}_h \in \boldsymbol{A}_h} \mathcal{E}(\boldsymbol{\mu}_h)\,, \tag{I.6}$$

where $\boldsymbol{A}_h \subset \boldsymbol{A}$ is chosen to be

$$\boldsymbol{A}_h = \Big\{\boldsymbol{\mu}_h \in \boldsymbol{A}:\ \boldsymbol{\mu}_h\big|_K \in \big[P_0(K)\big]^2,\ \forall\, K \in \mathcal{T}\Big\}, \tag{I.7}$$

for a Cartesian triangulation $\mathcal{T} = \mathcal{T}^c$ of $\omega$ that *is aligned with the easy axis*, and piecewise constant magnetizations $\mathbf{m}_h\big|_{K\in\mathcal{T}} \in \big[P_0(K)\big]^2$.

The main restriction of their analysis (for uniaxial materials) is the requirement to align the (Cartesian) triangulation $\mathcal{T}^c$ with the easy axis $\mathbf{e} \in \mathbb{R}^d$; the verification of their result is accomplished by constructing a (laminated) magnetization $\tilde{\mathbf{M}}_h \in \boldsymbol{A}_h$ that creates a small amount of energy in terms of the mesh parameter $h = h_{\mathcal{T}}$.

In order to increase applicability of the numerical analysis we consider arbitrary, quasiuniform meshes $\mathcal{T}$. The new idea to verify these error estimates is to construct discrete magnetizations $\tilde{\mathbf{M}}_h \in \boldsymbol{A}_h$ that show multi-scale features, bridging relatively coarse scale solution structures in the interior of the ferromagnet to fine scale structures close to the boundary $\partial\omega$. According to our physical motivation before, the numerical scales can be given some physical meaning: surface energy is created by (triangulation) structures of scale $\mathcal{O}(h)$, whereas closure domains exhibit a scaling $\mathcal{O}(h^\beta)$. Laminated domains in the interior of $\omega$ show a width of magnitude $\mathcal{O}(h^\alpha)$. Finally, magnitudes of branching domains close to the boundary $\partial\Omega$ range from $\mathcal{O}(h)$ across $\mathcal{O}(h^\beta)$ up to $\mathcal{O}(h^\alpha)$; see Table I.1 for arising multiple scales in computed magnetizations for uniaxial vs. cubic magnetizations (for theoretical justification, see Sections 1.1 and 1.2). These observations evidence that a finite element minimizer of (I.5) on general triangulations can be associated with a minimizer of the perturbed energy functional

$$\mathcal{E}_{0,p}(\mathbf{m}) = \int_\omega \tilde{\alpha}(\mathbf{x})\left|\nabla\mathbf{m}\right|^2 \mathrm{d}\mathbf{x} + \int_\omega \phi(\mathbf{m})\,\mathrm{d}\mathbf{x} + \frac{1}{2}\int_{\mathbb{R}^d}\left|\nabla u\right|^2 \mathrm{d}\mathbf{x}\,, \qquad \text{(I.8)}$$

with $0 \leq \tilde{\alpha}(\mathbf{x}) \leq \mathcal{O}(h^2)$, where this function reflects geometrical properties of the underlying mesh; we also refer to [26, 27] on scaling properties of minimizers to the original problem (I.1). Thus, physical exchange energy contributions in the extended model (I.1) may get polluted in case that $\alpha \leq \mathcal{O}(h^2)$.

Symmetry properties are increased in the case of cubic materials, which may be seen as a reason for existence of solutions to the problem [72]

$$\inf_{\boldsymbol{\mu}\in\boldsymbol{\mathcal{A}}} \mathcal{E}(\boldsymbol{\mu})\,.$$

If it comes to numerical analysis of its discrete counterpart (I.6), error statements (see Theorem 1.4) are superior to those of the uniaxial case (see Theorem 1.1); the proof is again by explicit construction of a multi-scale type magnetization $\tilde{\mathbf{M}}_h \in \boldsymbol{\mathcal{A}}_h$ that satisfies the given upper energy bound. However, interior magnetization patterns are different from the uniaxial case.

In another step which is based on the high degree of symmetricity for cubic materials, we can formulate an adaptive algorithm in order to further improve on quality of computed magnetization structures (see Theorem 1.5). As will be explained in Section 1.2, initially used graduated meshes (of refined mesh-size close to the boundary) are refined according to some admissibility criterion, depending on local jumps of previously computed magnetizations.

| magnet | bdy. layer | laminates | cls. domains | wall energy |
|---|---|---|---|---|
| uniaxial | $\mathcal{O}(1)$ | $\mathcal{O}(h^{1/3})$ | $\mathcal{O}(h^{2/3})$ | $\mathcal{O}(h)$ |
| cubic (u.) | $\mathcal{O}(1)$ | $\mathcal{O}(1)$ | $\mathcal{O}(h)$ | $\mathcal{O}(h)$ |
| cubic (a.) | $[0, \mathcal{O}(h_0)]$ | | $\mathcal{O}(h_0^2)$ | $\mathcal{O}(h_0^{3/2})$ |
| | $[\mathcal{O}(h_0), \mathcal{O}(1)]$ | $\mathcal{O}(1)$ | | $\mathcal{O}(h_0)$ |

Table I.1: Comparison of the thickness of (numerical) structural patterns like boundary layers, laminated microstructure, closure domains, walls for the uniaxial and cubic model; results for the adaptive scheme ((a.)) in the cubic case are also included (see Sections 1.1 and 1.2).

From a numerical viewpoint, computing these complicated microstructures in both cases requires a huge effort, and building efficient minimization algorithms that lead to (local) minima is a nontrivial task [95, 96]. Therefore, an alternative approach is to switch to a micromagnetic model that suppresses oscillatory magnetization patterns (like Weissian domains, etc.) by focusing on averaged (macroscopic) magnetization properties: it is presented in Chapter 2 and is considered to be very practical for large, e.g., micron-sized ferromagnets that appear in data storage industries. To obtain those informations it is sufficient to consider a modified minimization problem. The relaxation of the present example is analyzed in [39] and essentially means a convexification of $\phi$ and of the side-constraint $|\mathbf{m}| = 1$ almost everywhere in $\omega$. Let $\varphi(\mathbf{m})$ denote the lower convex envelope $\phi^{**}(\mathbf{m})$ of $\phi$ if $|\mathbf{m}| \leq 1$ and $\varphi(\mathbf{m}) = \infty$ else. The lower convex envelope $\phi^{**}$ is the largest convex function below $\phi$. Then, the **Convexified Problem (CP)** reads: *Seek a minimizer* $\mathbf{m} \in \mathbf{L}^2(\omega, \mathbb{R}^d)$ *of the convexified energy*

$$\tilde{\mathcal{E}}(\mathbf{m}) = \int_\omega \varphi(\mathbf{m})\,\mathrm{d}\mathbf{x} - (\mathbf{f}, \mathbf{m}) + \frac{1}{2}\int_\Omega |\nabla u|^2\,\mathrm{d}\mathbf{x}\,, \tag{I.9}$$

*subject to* (I.2). Owing to convexification, the relaxed energy functional $\tilde{\mathcal{E}}$ is sequentially weakly lower semicontinuous and hence there exist solutions of (CP), [39]. Each solution of (CP) solves the Kuhn-Tucker equations if we involve a further Lagrange multiplier with respect to the constraint $|\mathbf{m}| \leq 1$; we refer to this reformulation of (I.9) as **Problem ($\tilde{\mathbf{P}}$)**: *Seek* $u \in W^{1,2}(\mathbb{R}^d)$,

$\mathbf{m} \in \mathbf{L}^2(\omega, \mathbb{R}^d)$, *and* $\lambda \in L^2(\omega)$ *satisfying*

$$(\nabla u, \nabla w)_{\mathbb{R}^d} = (\mathbf{m}, \nabla w) \quad \forall\, w \in W^{1,2}(\mathbb{R}^d)\,, \tag{I.10}$$

$$\nabla u + D\phi^{**}(\mathbf{m}) + \lambda \mathbf{m} = \mathbf{f} \quad \text{a.e. in } \omega\,, \tag{I.11}$$

$$0 \le \lambda, \quad |\mathbf{m}| \le 1, \quad \text{and} \quad \lambda(1 - |\mathbf{m}|)_+ = 0 \quad \text{a.e. in } \omega\,. \tag{I.12}$$

Here, $(s)_+ := \max\{s, 0\}$ denotes the non-negative part and the last condition in (2.5) states that $\lambda \neq 0$ is possible only for $|\mathbf{m}| = 1$ as a consequence of $\lambda \mathbf{m} \in \partial\psi(\mathbf{m})$ for the convex characteristic functional $\psi : \mathbb{R}^2 \to [0, \infty]$ defined by $\psi(\mathbf{m}) = 0$, if $|\mathbf{m}| \le 1$ and $\psi(\mathbf{m}) = \infty$ if not.

Its stable numerical implementation requires balanced finite element spaces of magnetization and potential field; as will be demonstrated in Section 2.1, the canonical choice of piecewise constants for the magnetization field and conforming affine functions for the magnetostatic potential fails with respect to stability. A rigorous analysis in Section 2.1 will show stable performance of Crouzeix-Raviart elements for the magnetic potential. The key to theoretically verify computationally observed stability of this mixed method is a discrete Helmholtz decomposition principle which is not valid in a numerical scenario where continuous affine functions are used. We complement our studies of a stable mixed method with an adaptive concept to make the numerical method more effective; see Subsections 2.1.3 and 2.1.4.

Conforming affine functions for the magnetic potential create non-physical oscillations, see Figures 2.4, 2.14. An alternative towards non-conforming discretizations which guarantee stability are stabilization techniques for conforming discretizations to keep their (practical) advantages.

The stabilized finite element formulation of (I.10)-(I.12) then reads as follows, for some $\beta > 0$: *Seek* $(u_h, \mathbf{m}_h, \lambda_h) \in \mathcal{S}_0^1(\mathcal{T}) \times \mathcal{L}^0(\mathcal{T}|_\omega)^d \times \mathcal{L}^0(\mathcal{T}|_\omega)$ *that satisfies for every* $\{v_h, \boldsymbol{\mu}_h\} \in \mathcal{S}_0^1(\mathcal{T}) \times \mathcal{L}^0(\mathcal{T}|_\omega)^d$,

$$(\nabla u_h, \nabla w_h)_\Omega - (\mathbf{m}_h, \nabla w_h) = 0\,, \tag{I.13}$$

$$(\nabla u_h, \boldsymbol{\mu}_h) + (D\phi^{**}(\mathbf{m}_h), \boldsymbol{\mu}_h) \tag{I.14}$$
$$+ (\lambda_h \mathbf{m}_h, \boldsymbol{\mu}_h) + t_\ell([\mathbf{m}_h], [\boldsymbol{\mu}_h]) = (\mathbf{f}_\mathcal{T}, \boldsymbol{\mu}_h)\,,$$

$$\lambda_h = \varepsilon^{-1}(|\mathbf{m}_h| - 1)_+ / |\mathbf{m}_h| \quad \text{a.e. in } \omega\,. \tag{I.15}$$

Here, $[\cdot]$ denotes the jump across inter-element interior faces $\boldsymbol{\xi} \subset \Xi^*$, according to a given but arbitrary orientation of triangles $K \in \mathcal{T}$. Alternatively, we deal with the following stabilization strategies, for $\alpha > 0$,[1] and given $\beta_{\boldsymbol{\xi}} > 0$,

[1] In fact, we find $\alpha = 1$ from the convergence analysis.

for all $\boldsymbol{\xi} \in \Xi^*$,

$$t_{\mathrm{A}}([\mathbf{m}_h],[\boldsymbol{\mu}_h]) := \sum_{\xi\in\Xi^*} \beta_\xi h_\xi^\alpha \int_\xi [\langle \mathbf{m}_h, \mathbf{n}\rangle_{\mathbb{R}^d}][\langle \boldsymbol{\mu}_h, \mathbf{n}\rangle_{\mathbb{R}^d}]\, d\mathbf{x}\,, \tag{I.16}$$

$$t_{\mathrm{B}}([\mathbf{m}_h],[\boldsymbol{\mu}_h]) := \sum_{\xi\in\Xi^*} \beta_\xi h_\xi^\alpha \int_\xi \langle [\mathbf{m}_h],[\boldsymbol{\mu}_h]\rangle_{\mathbb{R}^d}\, d\mathbf{x}\,. \tag{I.17}$$

Finally, other numerical schemes fit into this framework that employ element-wise affine, globally continuous magnetizations, $\mathbf{m}_h, \boldsymbol{\mu}_h \in \mathcal{S}^1(\mathcal{T}|_\omega)^d$ in (I.13)-(I.15), and replace $t_\ell([\mathbf{m}_h],[\boldsymbol{\mu}_h])$ in (I.14) by $d_\ell(\mathbf{m}_h,\boldsymbol{\mu}_h)$, for some $\gamma > 0$, and positive $\beta_K > 0$, for any $K \in \mathcal{T}|_\omega$,

$$d_{\mathrm{A}}(\mathbf{m}_h,\boldsymbol{\mu}_h) = \sum_{K\in\mathcal{T}|_\omega} \beta_K h_K^\gamma (\operatorname{div}\mathbf{m}, \operatorname{div}\boldsymbol{\mu}_h)_K\,,$$

$$d_{\mathrm{B}}(\mathbf{m}_h,\boldsymbol{\mu}_h) = \sum_{K\in\mathcal{T}|_\omega} \beta_K h_K^\gamma (\nabla\mathbf{m}_h, \nabla\boldsymbol{\mu}_h)_K\,.$$

These stabilizations are advantageous for several reasons: next to 'cheaper' realization, their theory is much easier (even in 3D) to show the same convergence rates as for the stable scenario using non-conforming elements. Computational results even evidence higher accuracy of these schemes in a sense that is made precise in Subsection 2.2.2.

So far, stable/stabilized mixed methods have been proposed to compute averaged (macroscopic) magnetizations for cases where the convexification of the anisotropy density $\phi^{**}$ is explicitly known; the following a priori error estimates will be proven for both numerical schemes (see Theorem 2.5 and Theorem 2.9, respectively),

$$\|\nabla(u-u_h)\|_{\mathbf{L}^2(\Omega,\mathbb{R}^2)} + \|\langle \mathbf{m}-\mathbf{m}_h, \mathbf{e}_\perp\rangle_{\mathbb{R}^2}\|_{L^2(\omega)} \le C\,h\,, \tag{I.18}$$

provided that $\mathbf{m} \in \mathbf{W}^{2,2}(\omega,\mathbb{R}^2)$. On the other hand, convergence of $\mathbf{m}_h$ in $\mathbf{L}^2(\omega,\mathbb{R}^2)$ (for $h \to 0$) is only weak, in general, because of the degenerate character of the convex hull $\phi^{**}$; see also Figure 2.5 for computational observations. This deficiency of not being provided with strongly converging sequences of finite element functions $\{\mathbf{m}_h\}_{h>0}$ obtained by the above methods on general quasiuniform meshes prevents us to find reliable error statements for the approximation of the magnetic microstructure, where the

Young measure is known from [39],

$$\nu_{\mathbf{x}} = \Gamma(\mathbf{m}(\mathbf{x}))\delta_{\mathbf{m}^+(\mathbf{m}(\mathbf{x}))} + (1 - \Gamma(\mathbf{m}(\mathbf{x})))\delta_{\mathbf{m}^-(\mathbf{m}(\mathbf{x}))}\,, \tag{I.19}$$

$$\mathbf{m}^{\pm}(\mathbf{m}(\mathbf{x})) = \pm(1 - \langle \mathbf{m}(\mathbf{x}), \mathbf{e}_\perp\rangle_{\mathbb{R}^d}^2)^{1/2}\mathbf{e} + \langle \mathbf{m}(\mathbf{x}), \mathbf{e}_\perp\rangle_{\mathbb{R}^d}\mathbf{e}_\perp\,, \tag{I.20}$$

$$\Gamma(\mathbf{m}(\mathbf{x})) = \frac{1}{2} + \frac{\langle \mathbf{m}(\mathbf{x}), \mathbf{e}\rangle_{\mathbb{R}^d}}{2(1 - \langle \mathbf{m}(\mathbf{x}), \mathbf{e}_\perp\rangle_{\mathbb{R}^d}^2)^{1/2}}\,. \tag{I.21}$$

To reach the goal of having schemes at hand which computes magnetic microstructures reliably from the convexified problem (I.10)-(I.12) in situations where $\phi^{**}$ is known, we propose the following ones in Section 2.3: Consider scheme $(\mathrm{P})_h^{ij}$, for $i, j \in \{c, nc\}$, where '$c$' resp. '$nc$' denote conforming resp. non-conforming discretization for the spaces introduced before: Given $\varepsilon_i > 0$, $i = 0, 1, 2$, find $\{u_h, \mathbf{m}_h\} \in \mathcal{X}_0^i(\Omega) \times \boldsymbol{\mathcal{Y}}^j(\omega)$, for $\mathcal{X}_0^i \in \{\mathcal{S}_0^1, \mathcal{S}_0^{1,NC}\}$, $\boldsymbol{\mathcal{Y}}^j \in \{\boldsymbol{\mathcal{S}}^1, \boldsymbol{\mathcal{S}}^{1,NC}\}$ such that for all $\{w_h, \boldsymbol{\chi}_h\} \in \mathcal{X}_0^i(\Omega) \times \boldsymbol{\mathcal{Y}}^j(\omega)$,

$$(\nabla_{\mathcal{T}} u_h, \nabla_{\mathcal{T}} w_h)_\Omega = (\mathbf{m}_h, \nabla_{\mathcal{T}} w_h)\,, \tag{I.22}$$

$$\varepsilon_0 \sum_{\xi \in \Xi^*} \int_\xi \langle [\partial_{\mathbf{n}} \mathbf{m}_h], [\partial_{\mathbf{n}} \boldsymbol{\chi}_h]\rangle_{\mathbb{R}^4}\, d\mathbf{x} + \varepsilon_1 (\nabla \mathbf{m}_h, \nabla \boldsymbol{\chi}_h) \tag{I.23}$$

$$+(\nabla u_h, \boldsymbol{\chi}_h) + \big(\langle \mathbf{m}_h, \mathbf{e}_\perp\rangle_{\mathbb{R}^d}, \langle \boldsymbol{\chi}_h, \mathbf{e}_\perp\rangle_{\mathbb{R}^d}\big) + \big(\lambda_h \mathbf{m}_h, \boldsymbol{\chi}_h\big) = (\mathbf{f}, \boldsymbol{\chi}_h)\,,$$

$$\lambda_h = \frac{1}{\varepsilon_2}\left(|\,\mathbf{m}_h\,| - 1\right)_+ / |\,\mathbf{m}_h\,|\,, \quad \partial_{\mathbf{n}} \mathbf{m}_h\big|_{\partial\omega} = 0\,. \tag{I.24}$$

The key result is then Theorem 2.10 which shows strong convergence at a certain rate for computed sequences $\{\mathbf{m}_h\}_{h>0}$ in cases where $\mathbf{m} \in \mathbf{W}^{2,2}(\omega, \mathbb{R}^2)$, which is the key to show convergence of the computed microstructure; see Corollary 2.1, and Remark 2.9, item 7., on page 92.

According to relaxation theory, convexification leads to magnetizations $\mathbf{m} \in \mathbf{L}^2(\omega, \mathbb{R}^d)$ that are weak limits of minimizing sequences for (P). The key tool here is the explicit knowledge of the convex hull of $\phi$. Then, in a second step, we can recover microstructure in terms of Young measures from algebraic manipulations (I.19)-(I.21) of $\mathbf{m}_h$.

A third approach towards computing magnetic microstructure is by means of relaxation using Young measures: instead of manipulating $\phi$, we extend the set of admissible functions to Young measures [112]. This relaxation of the original problem does not require to compute $\phi^{**}$.

Relaxation by means of Young measures provides averaged macroscopical quantities such as magnetization vectors (the first momenta of the Young measure) as well as information about a minimizing sequence of $\mathcal{E}$. Namely, the support of $\nu_{\mathbf{x}}$ on the unit sphere provides information about what magnetization vectors from $\boldsymbol{\mathcal{A}}$ must be combined in a weakly converging sequence in order to find the observed macroscopical magnetization.

Pedregal [104] (see also [39, 123]) considers

$$\overline{\boldsymbol{\mathcal{A}}} := \Big\{ \nu = \{\nu_{\mathbf{x}}\}_{\mathbf{x}\in\omega} \text{ weakly measurable ; supp } \nu_{\mathbf{x}} \subset \mathcal{S}^{d-1} \text{ f.a.a. } \mathbf{x} \in \omega \Big\}$$

and defines for any $\nu \in \overline{\boldsymbol{\mathcal{A}}}$ and $\mathbf{m}(\mathbf{x}) = \int_{\mathcal{S}^{d-1}} A\nu_{\mathbf{x}}(\mathrm{d}A)$, for almost all $\mathbf{x} \in \omega$, a functional $\bar{\mathcal{E}} : \bar{\mathcal{A}} \to \mathbb{R}$ by

$$\bar{\mathcal{E}}(\nu) = \int_\omega \int_{\mathcal{S}^{d-1}} \phi(A)\nu_{\mathbf{x}}(\mathrm{d}A)\,\mathrm{d}\mathbf{x} - (\mathbf{f}, \mathbf{m}) + \frac{1}{2}\int_{\mathbb{R}^d} |\nabla u|^2\,\mathrm{d}\mathbf{x}\,.$$

Then, $\big\{\bar{\mathcal{E}}, \overline{\boldsymbol{\mathcal{A}}}, \text{ weak* in } L^1(\Omega; C(\mathcal{S}^{d-1}))\big\}$ is a relaxation of $\mathcal{E}$ in the sense stated above. Thus we define the **Relaxed Problem (RP)**

$$\min_{\nu\in\overline{\boldsymbol{\mathcal{A}}}} \bar{\mathcal{E}}(\nu)\,, \tag{I.25}$$

subject to

$$\mathbf{m}(\mathbf{x}) = \int_{\mathcal{S}^{n-1}} A\,\nu_{\mathbf{x}}(\mathrm{d}A)\,, \quad \text{f.a.a.} \quad \mathbf{x} \in \omega\,, \tag{I.26}$$

$$\operatorname{div}\big(-\nabla u + \chi_\omega \mathbf{m}\big) = 0\,, \quad \text{in } W^{-1,2}(\mathbb{R}^d)\,. \tag{I.27}$$

The existence of a solution to (RP) follows from the sequential weak* lower semicontinuity of $\bar{\mathcal{E}}$ and from the sequential weak* compactness of $\bar{\mathcal{A}}$, [104].

The numerical discretization of (I.25)-(I.27) is recommendable for at least two reasons: Firstly, (CP) modifies the anisotropy density $\phi$, and microstructure is only obtained in a post-processing step, whereas the Young measure is computed directly from (RP). Secondly, (CP) is limited to situations where the convex envelope $\phi^{**}$ is known explicitly which may be difficult to compute in many cases of practical interest. For example, the convex envelope is not known yet for $\phi(\mathbf{m}) = c_1 m_1^2 + c_2 m_1^4$, $c_1, c_2 > 0$, see [39].

Also, we mentioned difficulties with (low order) finite element methods to reliably recover the Young measure. This justifies a more general approach using (RP). The main drawback is its high complexity if it comes to solving the minimization problem, because Young measures carry more information compared to Lebesgue measurable functions; therefore, from a numerical point of view it is challenging to devise algorithms that significantly reduce computational effort.

The first numerical discretization of the relaxed problem (RP) was done by Kružík in [80] by means of three-atomic Young measures that are constant for every element of a regular triangulation of $\omega \subset \mathbb{R}^2$.

Here, we propose and analyze a conforming discretization of (RP) by means of element-wise constant multi-atomic Young measures of prescribed support according to a triangulation of $\mathcal{S}^{d-1}$, see Section 3.2. This leads to a quadratic-linear optimization problem which is rather large. In Corollary 3.1, we state convergence of the method with respect to the discretization parameter $\mathbf{h} = (h_1, h_2)$, with $h_1, h_2 > 0$ the mesh-sizes of the triangulations $\mathcal{T}_{h_1}$ of $\omega$ and $\mathcal{T}_{h_2}$ of $\mathcal{S}^{d-1}$, respectively. This result requires mesh-sizes $h_2 = o(h_1^{d/2})$ which makes computations rather costly. To avoid this drawback and reduce the number of unknowns, we make use of an adaptive strategy that (locally) singles out active atoms. Those are found by recently derived optimality conditions for (RP) in [83] to predict the support of a Young measure solution to (RP). The corresponding result concerning convergence of the method is given in Corollary 3.2. We refer to Section 3.3 for details of the algorithm as well as its analysis. This idea of using optimality conditions for numerical purposes was first realized in [21], where a scalar or one-dimensional variational problem is considered.

In Section 3.4, numerical experiments for $d = 2$ are reported which illustrate effectivity of the active set algorithm. Computational experiments are reported, for a density $\phi$ where $\phi^{**}$ is not known explicitly. In all experiments, the active set per element $K \in \mathcal{T}_{h_1}$ consists of no more than five atoms (rather than $\mathcal{O}(h_2^{1-d})$ per element $K$) which is in good agreement with theoretical investigations and shows the efficiency of this adaptive strategy.

# Chapter 1

# Direct Minimization

## 1.1 Error analysis for uniaxial ferromagnets

Direct minimization by finite elements to resolve micromagnetic patterns deals with the non-convex variational problem (I.4). As will turn out, simulations can by rather costly, due to diverse patterns and scales of minimizing magnetizations. Moreover, they can blurr physical information in the case of existing exchange energy contributions; this effect will be illustrated here for uniaxial materials, for the case of absent exterior fields $\mathbf{f} : \omega \to \mathbb{R}^2$, and $\alpha = 0$. All results will be presented for $\omega \subset \mathbb{R}^2$, but can be generalized to $\omega \subset \mathbb{R}^3$ as well.

In this setting, the first contributions towards a numerical analysis of finite element discretizations are [95, 96], where Cartesian triangulations $\mathcal{T}^c$ are employed which are aligned with the easy axis, using piecewise constant magnetizations $\mathbf{m}_h \in \left[P_0(\omega)\right]^2$. Note that solutions of the discrete problem (recall (I.5) for notation),

$$\min_{\boldsymbol{\mu} \in \boldsymbol{\mathcal{A}}_h} \mathcal{E}_0(\boldsymbol{\mu}) \,, \tag{1.1}$$

for

$$\boldsymbol{\mathcal{A}}_h = \left\{ \boldsymbol{\mu}_h \in \boldsymbol{\mathcal{A}} : \; \boldsymbol{\mu}_h\big|_K \in \left[P_0(K)\right]^2, \forall K \in \mathcal{T} \right\}, \tag{1.2}$$

are attained since $\boldsymbol{\mathcal{A}}_h$ is a compact subset of $\boldsymbol{\mathcal{A}}$ and the restriction of $\mathcal{E}_0$ to $\boldsymbol{\mathcal{A}}_h$ is continuous. For the following analysis, we fix $\mathbf{e} \in \mathbb{R}^2$ and assume quadratic growth of the (nonnegative) anisotropy density,

$$\phi(\mathbf{m}) \geq c_0 \min\left\{ |\mathbf{m} - \mathbf{e}|^2, |\mathbf{m} + \mathbf{e}|^2 \right\}, \tag{1.3}$$

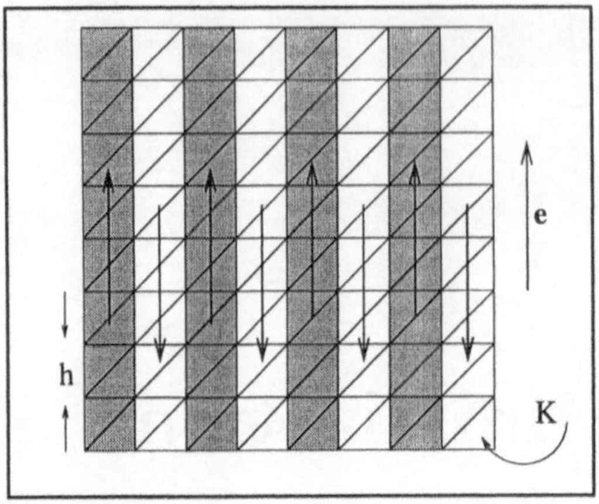


Figure 1.1: Construction of $\tilde{\mathbf{M}}_h \in \boldsymbol{\mathcal{A}}_h$ on a mesh $\mathcal{T}^c$ that is aligned with $\mathbf{e} \in \mathbb{R}^2$.

for $c_0$ a positive constant. — The following result is proved in [95] for uniaxial materials.

**Theorem 1.1** *(from [95]) Suppose we use a Cartesian triangulation $\mathcal{T}^c$ of the polyhedral domain $\omega \subset \mathbb{R}^2$ that is aligned with the easy axis $\mathbf{e} \in \mathbb{R}^2$. Then, there exists a constant $C$ that is independent of the mesh-width $h$, such that holds*

$$\min_{\boldsymbol{\mu}_h \in \boldsymbol{\mathcal{A}}_h} \mathcal{E}_0(\boldsymbol{\mu}_h) \leq C\, h\,.$$

The verification of this theorem in [95] is accomplished by constructing a (laminated) magnetization $\tilde{\mathbf{M}}_h \in \boldsymbol{\mathcal{A}}_h$ that satisfies this upper energy bound. The construction of it as well as the arguments in the proof crucially base on alignment of the triangulation with the easy axis, see Figure 1.1.

It is, however, that magnetization patterns for general quasiuniform triangulations $\mathcal{T}$ will be more complex, and we consider the case of aligned Cartesian meshes as exceptional: these triangulations energetically prefer magnetizations which only exhibit simple laminated structures oscillating on a numerical scale $\mathcal{O}(h)$. In particular, not only laminated structures develop in a computed magnetization $\tilde{\mathbf{M}}_h$ on general meshes $\mathcal{T}$ — which would create an energy contribution of magnitude $\mathcal{O}(\sqrt{h})$ —, but also branching structures close to $\partial\omega$; proving this assertion is the goal of this section.

We continue with the main result in this section.

**Theorem 1.2** *Let $\omega \subset \mathbb{R}^2$ be an uniaxial ferromagnet, and $\mathcal{T}$ its quasiuniform triangulation. Consider the minimization problem*

$$\min_{\boldsymbol{\mu}_h \in \boldsymbol{\mathcal{A}}_h} \mathcal{E}_0(\boldsymbol{\mu}_h) .$$

*There exists a constant $C$ that is independent of $h$, such that*

$$\min_{\boldsymbol{\mu}_h \in \boldsymbol{\mathcal{A}}_h} \mathcal{E}_0(\boldsymbol{\mu}_h) \leq C\Big(1 + \log_3 \frac{1}{h}\Big)\, h^{2/3}.$$

**Remark 1.1** *1. The a priori analysis presented below reflects a 'worst-case' scenario. From this viewpoint, it is justified to expect orders of convergence which range between $\frac{2}{3}$ and 1; this statement combines the a priori results of Theorem 1.1 (valid for aligned meshes) and Theorem 1.2 (valid for general meshes).*

*2. Theorem 1.2 is verified by presenting a candidate $\tilde{\mathbf{M}}_h \in \boldsymbol{\mathcal{A}}_h$ that satisfies the given upper bound for the energy. Surprisingly, its construction is motivated and based on observations in physical experiments, see [69], p. 330, Figure 3.127, and p. 403, Figure 5.6. We emphasize their more complex structure compared to the case of aligned meshes; in particular, its multi-scale branching structure is to bridge relatively coarse scale solution structures in the interior of the ferromagnet and fine scale structures along its boundary $\partial\omega$. This is a consequence of the competing mechanisms during minimization of the energy terms in (I.5).*

*3. A corresponding analysis for cubic ferromagnets will be given in Section 1.2, where (almost) first order of convergence for the energy on arbitrary quasi-uniform meshes is shown. This improved convergence statement reflects enlarged symmetry properties, allowing for different magnetization patterns which create lower energy.*

In order to construct an appropriate magnetization $\tilde{\mathbf{M}}_h \in \boldsymbol{\mathcal{A}}_h$ that satisfies the energy bound from Theorem 1.2, we need to balance both contributions in (I.5) in a way that a small amount of energy is produced from both sources mentioned. For this purpose, the boundary layer is structured according to Figure 1.4, with further modification as indicated in Figure 1.3. The values of the used parameters as well as the verification of the upper bound in Theorem 1.2 are clarified in Subsection 1.1.2.

We conclude this section with a useful lemma. For this purpose, we need some preparations: $\mathcal{L}^{p,\lambda}(\omega)$, for $p \geq 1$ and $\lambda \geq 0$, denotes the space of

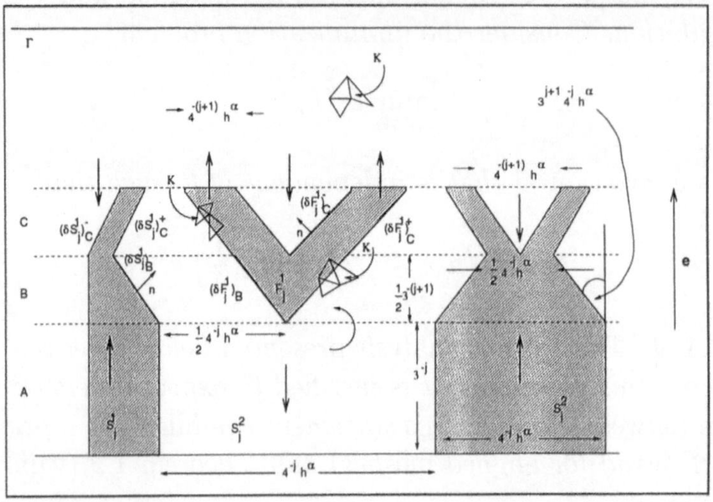


Figure 1.2: Refined microstructure pattern of of $\tilde{\mathbf{M}}$ in $\omega_{BL}$. The sketch shows one refinement indexed by $j$, for $1 \leq j \leq N^\star$, in this area along the direction given by $\mathbf{e}$. The value for $\alpha$ is determined in the proof.

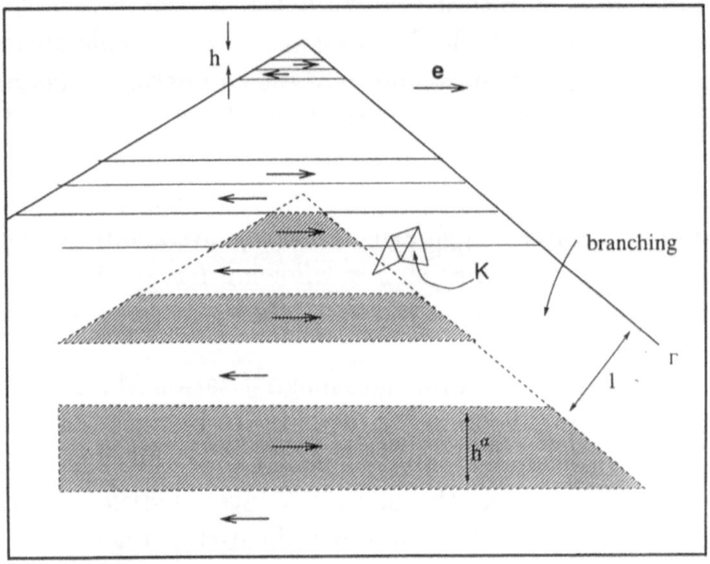


Figure 1.3: Reduced thickness of laminated microstructure constituting $\tilde{\mathbf{M}}_h$ in $\omega_{BL}$.

functions $u \in L^p(\omega)$, such that

$$[u]_{p,\lambda;\omega} = \Big\{ \sup_{\mathbf{x}_0 \in \omega, 0<\rho<\operatorname{diam}\omega} \rho^{-\lambda} \int_{B(\mathbf{x}_0,\rho)} |u(\mathbf{x}) - u_{\mathbf{x}_0,\rho}|^p \,\mathrm{d}\mathbf{x} \Big\}^{1/p} < \infty .$$

These spaces are known as Campanato spaces [55]. Here, $\omega \subset \mathbb{R}^d$ denotes a bounded connected open set. We set $\omega(\mathbf{x},\rho) = \omega \cap B(\mathbf{x},\rho)$, $\operatorname{diam}\omega = \sup\{|\mathbf{x}-\mathbf{y}| : \mathbf{x},\mathbf{y} \in \omega\}$, and

$$u_{\omega(\mathbf{x}_0,\rho)} = \frac{1}{|\omega(\mathbf{x}_0,\rho)|} \int_{\omega(\mathbf{x}_0,\rho)} u(\mathbf{x}) \,\mathrm{d}\mathbf{x} .$$

$\mathcal{L}^{p,\lambda}(\omega)$ is a Banach space with the norm $\|u\|_{\mathcal{L}^{p,\lambda}(\omega)} \equiv \|u\|_{L^p(\omega)} + [u]_{p,\lambda;\omega}$. In particular, there holds $L^\infty(\omega) \subsetneq \mathcal{L}^{p,d}(\omega)$, and $\mathcal{L}^{p,\lambda}(\omega)$ is isomorphic to $C^{0,\alpha}(\omega)$, $\alpha = \frac{\lambda-d}{p}$, for $n < \lambda \leq d+p$, provided $\omega$ is a Lipschitz domain. For cubic domains $\omega \in \mathbb{R}^d$, $\mathcal{L}^{p,d}(\omega)$ is also called $BMO(\omega)$ (i.e., the space of functions of bounded mean oscillation), which is isomorphic to the so-called John-Nirenberg space $\mathcal{E}_0(\omega)$, for all $p > 1$. For a precise definition of it and further details on these spaces, we refer to Chapter 3 of [55].

The following result is due to John and Nirenberg (see [55], Theorem 4.4).

**Lemma 1.1** *Suppose that $\omega_0 \subset \mathbb{R}^d$ is a cube. Then, there exist constants $c_1 > 0$ and $c_2 > 0$ depending only on $d = 2,3$, such that the following holds: If $u \in BMO(\omega_0)$, then for all cubes $\omega \subset \omega_0$, and numbers $t > 0$,*

$$\operatorname{meas}\Big\{\mathbf{x} \in \omega_0 : \ |u - u_\omega|(\mathbf{x}) > t\Big\} \leq c_1 \exp\Big(\frac{-c_2}{|u|_{BMO(\omega_0)}} t\Big) |\omega| .$$

This lemma can now be employed to verify the following statement.

**Lemma 1.2** *Let $\omega_0 \subset \mathbb{R}^d$ be a cube, $d = 2,3$, and $\omega \Subset \omega_0$ a Lipschitz domain such that* $\operatorname{diam}(\omega_0) \leq 2\operatorname{diam}(\omega)$. *For $\boldsymbol{\delta} \in \mathbf{L}^\infty(\omega,\mathbb{R}^d)$ such that $\|\boldsymbol{\delta}\|_{\mathbf{L}^1(\omega,\mathbb{R}^d)} \leq Ch^\gamma$, with $\gamma > 0$, consider the integral*

$$\int_{\omega_0} \langle \nabla u, \chi_\omega \boldsymbol{\delta} \rangle_{\mathbb{R}^d} \,\mathrm{d}\mathbf{x} ,$$

*where*

$$\Delta u = \operatorname{div}(\chi_\omega \mathbf{m}) \quad \text{in } \mathbb{R}^d , \tag{1.4}$$

*for* $\mathbf{m} \in \mathbf{L}^\infty(\omega_0, \mathbb{R}^d)$. *Then, there exists a positive number* $\alpha = \alpha(\omega, \gamma)$, *such that the following bound is valid,*

$$\left| \int_{\omega_0} \langle \nabla u, \chi_\omega \boldsymbol{\delta} \rangle_{\mathbb{R}^d} \, \mathrm{d}\mathbf{x} \right| \leq C h^\gamma \left(1 + \log_3 \frac{1}{h^\alpha}\right).$$

**Proof:**
Setting $T(\chi_\omega \mathbf{m}) = \nabla u$, $T$ is a continuous linear operator

$$T : \mathbf{L}^\infty(\omega_0, \mathbb{R}^d) \to \left[\mathcal{L}^{2,d}(\omega_0)\right]^d \cong \left[BMO(\omega_0)\right]^d,$$

see Theorem 3.3 in [55], as well as Section 4.3 in [55]. Then,

$$\int_{\omega_0} \langle \nabla u, \chi_\omega \boldsymbol{\delta} \rangle_{\mathbb{R}^d} \, \mathrm{d}\mathbf{x} = \left(\nabla u - (\nabla u)_{\omega_0}, \boldsymbol{\delta}\right)_{\omega_0} + \left((\nabla u)_{\omega_0}, \int_\omega \boldsymbol{\delta} \, \mathrm{d}\mathbf{x}\right)_{\omega_0}. \tag{1.5}$$

Thanks to Lemma 1.1, we can choose $\omega = \omega_0$, and $\alpha = \alpha(c_1, c_2, \gamma)$, with $t_0 = \log_3 \frac{1}{h^\alpha}$, such that holds

$$\begin{aligned} \mathrm{meas}\left\{\mathbf{x} \in \omega_0 : \left|\nabla u - (\nabla u)_{\omega_0}\right| > t_0\right\} &\leq c_1 \exp\left(\frac{-c_2 t_0}{\|\nabla u\|_{[\mathcal{L}^{2,d}(\omega_0)]^d}}\right) |\omega_0| \\ &\leq c_1 \exp\left(\frac{-c_2 t_0}{\|\mathbf{m}\|_{\mathbf{L}^\infty(\omega, \mathbb{R}^d)}}\right) |\omega_0| \leq C h^{2\gamma}. \end{aligned}$$

Set $\tilde{\omega} = \left\{\mathbf{x} \in \omega : \ \mathrm{meas}\left\{\mathbf{x} \in \omega_0 : \left|\nabla u - (\nabla u)_{\omega_0}\right| > \log_3 \frac{1}{h^\alpha}\right\}\right\}$. Then we can conclude

$$\begin{aligned} &\left|\left(\nabla u - (\nabla u)_{\omega_0}, \boldsymbol{\delta}\right)\right| \\ &\leq \left| \int_{\omega \setminus (\tilde{\omega} \cap \omega)} \left\langle \nabla u - (\nabla u)_{\omega_0}, \boldsymbol{\delta} \right\rangle_{\mathbb{R}^d} \mathrm{d}\mathbf{x} \right| + \left| \int_{\tilde{\omega} \cap \omega} \left\langle \nabla u - (\nabla u)_{\omega_0}, \boldsymbol{\delta} \right\rangle_{\mathbb{R}^d} \mathrm{d}\mathbf{x} \right| \\ &\leq \left| \int_{\omega \setminus \tilde{\omega}} \left\langle \nabla u - (\nabla u)_{\omega_0}, \boldsymbol{\delta} \right\rangle_{\mathbb{R}^d} \mathrm{d}\mathbf{x} \right| + C \, \|\boldsymbol{\delta}\|_{\mathbf{L}^\infty(\omega, \mathbb{R}^d)} \|\nabla u\|^{1/2}_{\mathbf{L}^2(\omega_0, \mathbb{R}^2)} \sqrt{|\tilde{\omega}|} \\ &\leq C \, \|\nabla u - (\nabla u)_{\omega_0}\|_{\mathbf{L}^\infty(\omega \setminus \tilde{\omega}, \mathbb{R}^d)} \, \|\boldsymbol{\delta}\|_{\mathbf{L}^1(\omega \setminus \tilde{\omega}, \mathbb{R}^d)} + C h^\gamma \qquad (1.6) \\ &\leq C h^\gamma \left(1 + \log_3 \frac{1}{h^\alpha}\right). \end{aligned}$$

This gives sharp bounds for the first term on the right-hand side of (1.5). The estimation of the remaining term by $Ch^\gamma$ is immediate. □

### 1.1.1 A low energy magnetization $\tilde{\mathbf{M}}_h \in \boldsymbol{\mathcal{A}}_h$

We discuss magnetizations for uniaxial materials that exhibit a particular microstructure near the boundary $\partial\omega$; cf. Figure 1.2. In this construction, laminates in the interior of the domain $\omega$ are of thickness $\mathcal{O}(h^\alpha)$, for $0 \leq \alpha \leq 1$, and there is a branching process of them in a $\mathcal{O}(1)$-vicinity of the boundary $\partial\omega$. This two-phase branching scenario is motivated from Figure 3.127, p. 330, in [69] and is depicted in Figure 1.2 for the level $j$, $1 \leq j \leq N_\star$, where $N_\star$ is given below. As we will see from the subsequent studies, the following crucial facts have to be taken into account, if we finally turn to an interpolated version $\tilde{\mathbf{M}}_h \in \boldsymbol{\mathcal{A}}_h$ of $\tilde{\mathbf{M}} \in \boldsymbol{\mathcal{A}}$ for the underlying mesh:

1. Transition between different domains for arbitrary meshes. This can be handled using a perturbation argument and this eventually determines the thickness of laminated microstructure in the interior.

2. Branching in a boundary layer $\omega_{BL}$ of width $\mathcal{O}(1)$. The declination of nonaligned interfaces between places of opposite magnetization as well as thickness of refined domains has to be adjusted to gain a proper candidate $\tilde{\mathbf{M}}_h \in \boldsymbol{\mathcal{A}}_h$.

3. The branching process ends after $N_\star$ steps. Then, boundary integrals from $\partial\omega$ enter our calculation of the total energy for $\tilde{\mathbf{M}}_h$, with the amount depending on the number $N_\star$ (closure domains).

Evidently, contributions due to non-alignment of the underlying mesh also have to be accounted for in 2. and 3.. This summarizes the essential energy contributions to our problem. Subsequently, we verify a first energy bound for the piecewise constant interpolate $\tilde{\mathbf{M}}_h \in \boldsymbol{\mathcal{A}}_h$ of $\tilde{\mathbf{M}} \in \boldsymbol{\mathcal{A}}$.

**Lemma 1.3** *Let $\omega \subset \mathbb{R}^2$ be a uniaxial ferromagnet, and $\mathcal{T}$ its quasiuniform triangulation. There holds*

$$\min_{\boldsymbol{\mu}_h \in \boldsymbol{\mathcal{A}}_h} \mathcal{E}_0(\boldsymbol{\mu}_h) \leq \mathcal{E}_0(\tilde{\mathbf{M}}_h) \leq C\, h^{1/2}\,, \tag{1.7}$$

*for $\omega \subset \mathbb{R}^2$ a polyhedral domain, with $\partial\omega = \bigcup_{i=1}^4 \Gamma_i$, for $\Gamma_i$ an affine face, and $\langle \mathbf{n}_{\Gamma_2}, \mathbf{e}\rangle_{\mathbb{R}^2} = \langle \mathbf{n}_{\Gamma_4}, \mathbf{e}\rangle_{\mathbb{R}^2} = 0$.*

**Proof:**
We study the magnetization $\tilde{\mathbf{M}} \in \boldsymbol{\mathcal{A}}$,

$$\tilde{\mathbf{M}}(\mathbf{x}) = \begin{cases} \mathbf{e} - 2\chi\left(\frac{\langle \mathbf{x}, \mathbf{e}_\perp \rangle_{\mathbb{R}^2}}{2h^\alpha}\right)\mathbf{e}, & \text{for } \operatorname{dist}(\mathbf{x}, \partial\omega) > d = \mathcal{O}(1) , \\ \hat{\mathbf{M}}(\mathbf{x}), & \text{for } \mathbf{x} \in \omega_{BL} , \end{cases} \tag{1.8}$$

using the characteristic function,

$$\chi(s) = \begin{cases} 1 & \text{for all } 0 \leq s \leq \frac{1}{2} , \\ 0 & \text{for all } \frac{1}{2} < s < 1 . \end{cases} \tag{1.9}$$

The construction of $\hat{\mathbf{M}}$ in Figure 1.2 is fixed once we specify $\alpha$ and $N_\star$. Then $\phi(\tilde{\mathbf{M}}) = 0$, almost everywhere. Next, we compute

$$\mathcal{E}_0(\tilde{\mathbf{M}}) = \frac{1}{2} \int_{\mathbb{R}^d} |\nabla \tilde{u}|^2 \, \mathrm{d}\mathbf{x} = \frac{1}{2} (\nabla \tilde{u}, \tilde{\mathbf{M}}) , \tag{1.10}$$

where the tuple $\{\tilde{u}, \tilde{\mathbf{M}}\} \in W^{1,2}(\mathbb{R}^2) \times \boldsymbol{\mathcal{A}}$ solves Maxwell's equation (I.2).

We apply Gauss' theorem; summation over each Weissian domain then gives[1]

$$\begin{aligned} \mathcal{E}_0(\tilde{\mathbf{M}}) &= \frac{1}{2} (\nabla \tilde{u}, \tilde{\mathbf{M}}) \qquad (1.11) \\ &= \frac{1}{2} \sum_{i=1}^{\mathcal{O}(h^{-\alpha})} \sum_{j=1}^{N_\star - 1} \int_{\partial \mathcal{S}_j^i \cup \partial \mathcal{F}_j^i} \langle \tilde{\mathbf{M}}, \mathbf{n} \rangle_{\mathbb{R}^2} \tilde{u} \, \mathrm{d}\mathbf{x} \\ &\quad + \frac{1}{2} \sum_{i=1}^{\mathcal{O}(h^{-\alpha})} \left[ \int_{(\partial \mathcal{S}_{N_\star}^i \cup \partial \mathcal{F}_{N_\star}^i) \cap \partial\omega \neq \emptyset} + \int_{(\partial \mathcal{S}_{N_\star}^i \cup \partial \mathcal{F}_{N_\star}^i) \cap \partial\omega = \emptyset} \right] \langle \tilde{\mathbf{M}}, \mathbf{n} \rangle_{\mathbb{R}^2} \tilde{u} \, \mathrm{d}\mathbf{x} \\ &:= I + II + III , \end{aligned}$$

where any level $L(j)$, $1 \leq j \leq N_\star$, with $\bigcup_{j=1}^{N_\star} \overline{L(j)} = \omega_{BL}$, is assembled from structures $\mathcal{S}_j^i$ and $\mathcal{F}_j^i$, for $1 \leq i \leq \mathcal{O}(h^{-\alpha})$, through

$$L(j) = \bigcup_{i=1}^{\mathcal{O}(h^{-\alpha})} \mathcal{S}_j^i \cup \bigcup_{i=1}^{\mathcal{O}(h^{-\alpha})} \mathcal{F}_j^i ,$$

[1] See also Figure 1.2 for illustration of the used notation.

cf. Figure 1.2. Note that contributions to the overall energy merely come from the magnetization in $\omega_{BL}$, since boundary terms from laminated domains in the interior of $\omega$ vanish.

(i) We deal with $I$ for each $L(j)$, $1 \leq j \leq N_\star - 1$, independently: No contributions to $I$ come from $A$, but there are contributions in $B$ and $C$. In the sequel, we fix vectors $\mathbf{n}_A, \mathbf{n}_B, \mathbf{n}_C$; boundary integrals are defined according to a fixed orientation.

*Contributions from B and C:* Let us come back to the structure of Figure 1.2. Then, contributions from $(\partial\mathcal{S}_j^1)_B \cup (\partial\mathcal{F}_j^1)_B$ are

$$\begin{aligned}
&\int_{(\partial\mathcal{S}_j^1)_B} \langle \mathbf{e}, \mathbf{n}_B\rangle_{\mathbb{R}^2}\tilde{u}\,\mathbf{dx} + \int_{(\partial\mathcal{S}_j^1)_B} \langle -\mathbf{e}, -\mathbf{n}_B\rangle_{\mathbb{R}^2}\tilde{u}\,\mathbf{dx} \\
&\qquad + \int_{(\partial\mathcal{F}_j^1)_B} \langle -\mathbf{e}, \mathbf{n}_B\rangle_{\mathbb{R}^2}\tilde{u}\,\mathbf{dx} + \int_{(\partial\mathcal{F}_j^1)_B} \langle \mathbf{e}, -\mathbf{n}_B\rangle_{\mathbb{R}^2}\tilde{u}\,\mathbf{dx}\,,
\end{aligned} \tag{1.12}$$

and, for $C$ and each single structure $\mathcal{S}_j^i$ or $\mathcal{F}_j^i$, e.g.,

$$\begin{aligned}
&\int_{(\partial\mathcal{F}_j^1)_C^-} \langle -\mathbf{e}, -\mathbf{n}_C\rangle_{\mathbb{R}^2}\tilde{u}\,\mathbf{dx} + \int_{(\partial\mathcal{F}_j^1)_C^-} \langle \mathbf{e}, \mathbf{n}_C\rangle_{\mathbb{R}^2}\tilde{u}\,\mathbf{dx} \\
&\int_{(\partial\mathcal{F}_j^1)_C^+} \langle \mathbf{e}, -\mathbf{n}_C\rangle_{\mathbb{R}^2} + \int_{(\partial\mathcal{F}_j^1)_C^+} \langle -\mathbf{e}, \mathbf{n}_C\rangle_{\mathbb{R}^2}\tilde{u}\,\mathbf{dx}\,.
\end{aligned} \tag{1.13}$$

Next, we exemplify how to control any pairs in $B$ and $C$,

$$\begin{aligned}
&\int_{(\partial\mathcal{S}_j^1)_B} \langle \mathbf{e}, \mathbf{n}_B\rangle_{\mathbb{R}^2}\tilde{u}\,\mathbf{dx} - \int_{(\partial\mathcal{F}_j^1)_B} \langle \mathbf{e}, \mathbf{n}_B\rangle_{\mathbb{R}^2}\tilde{u}\,\mathbf{dx} \\
&\qquad = \langle \mathbf{e}, \mathbf{n}_B\rangle_{\mathbb{R}^2}\Big\{\int_{(\partial\mathcal{S}_j^1)_B} \tilde{u}\,\mathbf{dx} - \int_{(\partial\mathcal{F}_j^1)_B} \tilde{u}\,\mathbf{dx}\Big\}\,.
\end{aligned} \tag{1.14}$$

According to Lemma 1.2, there exists a subset $\tilde{\omega} \subset \omega$, and a positive number $\delta = \delta(\omega,\gamma)$, such that

$$\begin{aligned}
&\tilde{\omega} = \Big\{\mathbf{x} \in \omega :\ \big|\nabla\tilde{u} - (\nabla\tilde{u})_{\omega_0}\big| > \log_3 \frac{1}{h^\delta}\Big\}\,, \\
&\operatorname{meas}\tilde{\omega} \leq Ch^\gamma\,,
\end{aligned}$$

for any chosen value $\gamma > 0$ and any bounded $\omega_0 \Supset \overline{\omega}$. Here, we use the shorthand notation $(\nabla\tilde{u})_{\omega_0} = \frac{1}{\operatorname{meas}(\omega_0)}\int_{\omega_0} \nabla\tilde{u}\,\mathbf{dx}$. Therefore, we can find the following upper bound for (1.14).

$$\leq C\left|\langle \mathbf{e}, \mathbf{n}_B\rangle_{\mathbb{R}^2}\right|\Big(\|\nabla\tilde{u}\|_{L^\infty(\omega\setminus\tilde{\omega})}+1\Big)\mu_1\big((\partial\mathcal{S}_j^1)_B\big)\,\mathrm{dist}\big\{(\partial\mathcal{S}_j^1)_B,(\partial\mathcal{F}_j^1)_B\big\}+Ch^{\gamma/2}$$

thanks to Cauchy's inequality. Correspondingly, we find

$$\begin{aligned}
&\Big|\int_{(\partial\mathcal{F}_j^{i1})_C^-}\langle -\mathbf{e},-\mathbf{n}_C\rangle_{\mathbb{R}^2}\tilde{u}\,\mathbf{dx}-\int_{(\partial\mathcal{F}_j^{i})_C^+}\langle \mathbf{e},-\mathbf{n}_C\rangle_{\mathbb{R}^2}\tilde{u}\,\mathbf{dx}\Big|\\
&\quad+\Big|\int_{(\partial\mathcal{S}_j^{i1})_C^-}\langle -\mathbf{e},-\mathbf{n}_C\rangle_{\mathbb{R}^2}\tilde{u}\,\mathbf{dx}-\int_{(\partial\mathcal{S}_j^{i})_C^+}\langle \mathbf{e},-\mathbf{n}_C\rangle_{\mathbb{R}^2}\tilde{u}\,\mathbf{dx}\Big|\\
&\leq C\left|\langle \mathbf{e},\mathbf{n}_C\rangle_{\mathbb{R}^2}\right|\Big(\|\nabla\tilde{u}\|_{L^\infty(\omega\setminus\tilde{\omega})}+1\Big)\Big\{\mu_1\big((\partial\mathcal{F}_j^i)_C^\pm\big)+\mu_1\big((\partial\mathcal{S}_j^i)_C^\pm\big)\Big\}\\
&\quad\times\Big\{\mathrm{dist}\big((\partial\mathcal{F}_j^i)_C^-,(\partial\mathcal{F}_j^i)_C^+\big)+\mathrm{dist}\big((\partial\mathcal{S}_j^1)_C^-,(\partial\mathcal{S}_j^1)_C^+\big)\Big\}+Ch^{\gamma/2}.
\end{aligned}$$

Correspondingly, subsets of $\partial\mathcal{S}_j^1$ are denoted by $(\partial\mathcal{S}_j^1)_C^+$ and $(\partial\mathcal{S}_j^1)_C^-$. Then, in each segment $\mathcal{Z}_i$ of $\omega_{BL}\equiv\{\mathcal{Z}_i\}_{i=1}^{\mathcal{O}(h^{-\alpha})}$, we have $N_\star=\log_4\frac{1}{h}$ levels of branching. The number of phase transitions on the $j$-th level can be bounded by $C4^j$, while the number of transitions between domains of equal magnetization is bounded by $C2^j$. Finally, we have $\left|\langle\mathbf{e},\mathbf{n}\rangle_{\mathbb{R}^2}\right|=\frac{1}{2}3^{j+1}4^{-j}h^\alpha$, for each connected component and every $1\leq j\leq N_\star$. This then implies

$$\begin{aligned}
&\sum_{j=1}^{N_\star-1}\int_{\partial\mathcal{S}_j^i\cup\partial\mathcal{F}_j^i}\langle\tilde{\mathbf{M}},\mathbf{n}\rangle_{\mathbb{R}^2}u\,\mathbf{dx} &(1.15)\\
&\leq C\sum_{j=1}^{N_\star-1}\frac{1}{2}3^{j+1}4^{-j}h^\alpha\big(1+\log_3\frac{1}{h^\alpha}\big)\frac{1}{2}3^{-(j+1)}4^{-(j+1)}h^\alpha\{2^j+4^j\}+Ch^{\gamma/2}\\
&\leq C\big(1+\log_3\frac{1}{h^\delta}\big)h^{2\alpha}+Ch^{\gamma/2}.
\end{aligned}$$

(ii) The contributions along the boundary $\partial\omega$ can be controlled by Mean Value Theorem, exploiting the scale $\mathcal{O}(h)$ of structures,

$$\begin{aligned}
\int_{\partial\omega}\langle\tilde{\mathbf{M}},\mathbf{n}\rangle_{\mathbb{R}^2}\tilde{u}\,\mathbf{dx} &= \int_{\partial\omega}(-1)^{\chi(\frac{\langle\mathbf{x},\mathbf{e}_\perp\rangle_{\mathbb{R}^2}}{2h})+1}\langle\mathbf{e},\mathbf{n}\rangle_{\mathbb{R}^2}\tilde{u}\,\mathbf{dx} &(1.16)\\
&\leq Ch\,\|\nabla\tilde{u}\|_{L^\infty_{\mu_1}(\partial(\omega\setminus\tilde{\omega}))}+Ch^{\gamma/2}.
\end{aligned}$$

We can now combine this result together with (1.15) and the observation that non-aligned meshes give energy contributions of $\mathcal{O}(h)$ for each transition, at worst. Let $\tilde{\mathbf{M}}_h \in \mathcal{A}_h$ denote an interpolate of $\tilde{\mathbf{M}} \in \mathcal{A}$. Then, we can employ an easy perturbation argument, using (I.2), further set $\gamma = 2$, and we further set $\gamma = 2$. Then, we finally end up having

$$\mathcal{E}_0(\tilde{\mathbf{M}}_h) \leq C\Big(1 + \log_3 \frac{1}{h^\delta}\Big)\Big\{h + h^\alpha + h^{1-\alpha}\Big\}, \tag{1.17}$$

which gives an optimal bound for the choice $\alpha = \frac{1}{2}$. □

**Remark 1.2** *1. This result can already be verified if we use a simple laminated structure. But, as we will see in the refined analysis presented in Subsection 1.1.2, the microstructure introduced here finally leads to convergence results superior to (1.7) in case it is modified in a proper way by using a different scaling of the employed substructures.*

*2. Refinement strategies are given in the proof for a polygonal domain $\omega \subset \mathbb{R}^2$, where $\partial\omega = \bigcup_{\ell=1}^4 \Gamma_\ell$, $\Gamma_\ell$ an affine face, and $\langle \mathbf{n}_{\Gamma_2}, \mathbf{e}\rangle_{\mathbb{R}^2} = \langle \mathbf{n}_{\Gamma_4}, \mathbf{e}\rangle_{\mathbb{R}^2} = 0$. This result can be generalized to arbitrary polygonal domains, where $\omega$ is bounded by $L$ affine manifolds, i.e., $\partial\omega = \bigcup_{l=1}^L \Gamma_l$ through scaling the thickness of the interior laminates in a proper way when approaching a general boundary or corner; cf. Figure 1.3. It only has to be observed in this procedure to balance refinement and wall energy creation due to non-aligned meshes — which is no severe restriction for the construction process.*

*One coarsening strategy to manage a transition of the thickness of laminates inside a polygonal domain $\omega$ from $\mathcal{O}(h)$ (boundary $\partial\omega$) to $\mathcal{O}(h^\alpha)$ (interior) is as follows: For each corner $k$, $1 \leq k \leq N_k$, we count $1 \leq \ell_k \leq \mathcal{O}(h^{-\beta})$ laminates. Starting from the corner point, laminates of thickness $h_{\ell_k} = h\ell_k^{\frac{1}{\alpha}-1}$ may be constructed, that bridge the boundary layer of width $\mathcal{O}(1)$. This requires $\mathcal{M}_k = \mathcal{O}(h^{-\alpha})$ steps, which is due to the following calculation,*

$$\sum_{\ell_k=1}^{\mathcal{M}_k} h_{\ell_k} = h \sum_{\ell_k=1}^{\mathcal{M}_k} \ell_k^{\frac{1}{\alpha}-1} \leq Ch\mathcal{M}_k^{\frac{1}{\alpha}} \overset{!}{\leq} 1.$$

*This strategy does not imply higher energy values which are due to increased wall energy and satisfies our needs of bridging different scales in the interior and the boundary.*

### 1.1.2 Verification of Theorem 1.2

As we see from the previous subsection, the magnetization $\tilde{\mathbf{M}} \in \mathcal{A}$ depicted in the Figures 1.2, 1.3 is not sufficient to prove Theorem 1.2. For this purpose, a *multiple scaling of primal and complementary structures* is necessary; see below for a definition of these structures. This allows for a modified branching, especially with respect to declination of used structures near the boundary. Based on this additional mechanism, we construct a (modified) magnetization $\tilde{\mathbf{M}} \in \mathcal{A}$ that is sufficient for our needs to verify Theorem 1.2. The scaling of $\tilde{\mathbf{M}} \in \mathcal{A}$ close to the boundary $\partial\omega$ is depicted in Figure 1.4 for the subsequent asymptotic analysis. A few remarks are in order to mention certain relevant properties of this magnetization.

1. The interior of $\omega$ is covered with laminated microstructure showing oscillations of width $\mathcal{O}(h^\alpha)$ which is aligned with the easy axis $\mathbf{e}$. Inside a boundary layer $\omega_{BL}$ of width $\mathcal{O}(1)$ the microstructure is refined through branching domains. We distinguish $1 \leq i \leq N_\star$ different levels that are each of thickness $3^{-i}$.

2. Branching Domains: We distinguish between *primal microstructures* ('K'- or 'Y'-shaped 'devices') and the *complementary microstructure* ('V'-shaped 'devices'). The basic primal microstructure is bisected on a scale $h^\alpha$, while the complementary microstructure is always given in terms of $h^\beta$, $0 \leq \alpha \leq \beta \leq 1$; see Figure 1.4.

3. In 'K'-shaped devices, the lower right branch shows a reduced declination (see below), whereas the upper right branch shows a declination of magnitude $3^i h^\beta$. There is no declination on the left-hand side. This device is also used in reflected form. The 'V'-shaped device shows the same declinations on the same level. Its thickness is of magnitude $2^{-i} h^\beta$.

4. The 'Y'-shaped device is designed to avoid energy contributions in the lower half of the device. Note that the branches need not be adjusted in a symmetric way.

5. An interface is called 'active', if it is not aligned with $\mathbf{e} \in \mathbb{R}^2$ and belongs to a primal microstructure. The width of every level is adjusted in a way that the total length of 'active' interfaces on the subsequent

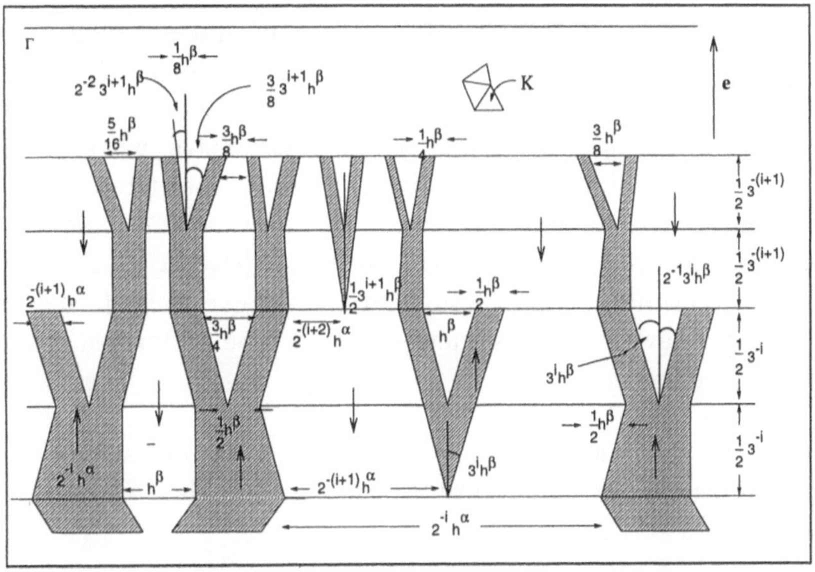


Figure 1.4: Magnetization pattern in $\omega_{BL}$, giving $\tilde{\mathbf{M}}$.

level is bounded through the length of the active interfaces on the present level.

6. Three facts have to be taken into account in the refinement process while approaching the boundary $\partial\omega$:

   (a) At each level, primal structures are build together with complementary structures. This bisects the scale of $2^{-i}h^{\alpha}$ to $2^{-(i+1)}h^{\alpha}$ on original primal microstructures and halves on a scale $h^{\beta}$ in case the primal microstructure originally stems from a complementary one.

   (b) According to the construction, each complementary microstructure on the level indexed by $i$ exhibits a thickness of branches that are of order $2^{-i}h^{\beta}$.

   (c) The primal 'Y'-shaped device is used to manage the energy-free branching inside the primal microstructure, whereas the 'K'-shaped device is used in combination with the complementatory microstructure.

A branching of the primal microstructure of $\tilde{\mathbf{M}}$, starting at level zero, can be accomplished as follows, see also Figure 1.4: The branches of one phase are of width $\frac{1}{2}h^\alpha$ and are separated by the second phase which is of width $h^\beta$. Then, 'K'-type devices may be used for another branching on the first level. Here, those parts of each two 'K'-type devices that lay opposite show no declination with respect to $\mathbf{e} \in \mathbb{R}^2$ in the lower halve, whereas the upper two branches bend at a slope $\frac{3}{2}h^\beta$. This is to avoid interference of branches on subsequent levels and to give place to further branching. The upper right branch with normal $\mathbf{n} \in \mathbb{R}^2$ is adjusted such that $|\sphericalangle(\mathbf{e}, \mathbf{n})| \leq 3h^\beta$.

"Inside" the primal microstructure, branching can then be accomplished using 'Y'-shaped devices: a construction can then be performed that satisfies the following needs:

1. On each level indexed by $i \geq 1$, the thickness of each branch in an original primal microstructure is of magnitude $\mathcal{O}(2^{-i}h^\alpha)$, for a magnetization of a certain phase.

2. New domains which support the opposite phase of width $\mathcal{O}(2^{-i}h^\beta)$ are created on each level $1 \leq i \leq N_\star$.

As far as the refinement of a complementary microstructure is concerned, we proceed correspondingly. Note that each complementary microstructure on level $i$ will turn to a primary one on levels $j \geq i+1$. The only difference is that the branches of it show a thickness that is scaled in $h^\beta$. — Then, for any primary and complementary microstructure we may distinguish between refinement 'inside' (using 'Y'-type devices) and refinement that occurs on the 'boundary' of the structure (using 'K'-type devices). The latter is to allow for complementary structures in order to balance finer structures all along the boundary $\partial\omega$.

Then, the energy of $\tilde{\mathbf{M}} \in \boldsymbol{\mathcal{A}}$ depicted in Figure 1.4 inside $\omega_{BL}$ can be determined as follows. Let $N_1$ denote the number of refinement steps per single complementatory microstructure from level 1, i.e., $2^{N_1}h = h^\beta$. On the other hand, $N_2$ denotes the number of steps to bridge the scale for structures between $\mathcal{O}(h^\alpha)$ and $\mathcal{O}(h^\beta)$ via bisection, i.e., $2^{N_2}h^\beta = h^\alpha$. For $N_\star = N_1 \overset{!}{=} N_2$, this implies

$$2\beta = 1 + \alpha. \tag{1.18}$$

The overall energy contributions for the interpolated version $\tilde{\mathbf{M}}_h \in \boldsymbol{\mathcal{A}}_h$ of $\tilde{\mathbf{M}} \in \boldsymbol{\mathcal{A}}$ is then again assembled from

1. contributions of magnetizations along the boundary $\partial\omega$ of magnitude $\mathcal{O}(h^\beta)$ by using the fundamental theorem of calculus ('closure domains'). Owing to the boundedness $|\nabla u(\mathbf{x})| \le C(1 + \log_3 \frac{1}{h^\delta})$, for almost all $\mathbf{x} \in \omega \setminus \overline{\omega}$, with $\text{meas}(\overline{\omega}) \le Ch^\gamma$, for $\gamma > 0$ arbitrarily large.

2. contributions which are due to switching magnetization directions ('wall energy'). This quantity is less than $Ch^{1-\alpha}(1 + \log\frac{1}{h})$. The logarithmic part is due to the number $N_\star = \log_2 h^{\alpha-\beta}$ of levels that $\omega_{BL}$ is split into. Note that the overall length of boundaries of 'new' microstructures on level $i+1$ (i.e., 'active' and 'non-active' ones) is also less than the length of interfaces on level $i$.

3. The parameters $\beta$ and $\alpha$ are connected through (1.18).

We can now do the summation of the energy contributions like in (1.15). Then, we end up with the upper bound

$$\mathcal{E}(\tilde{\mathbf{M}}_h) \le C\Big\{h^\beta + h^{1-\alpha}\big(1 + \log_3\frac{1}{h}\big)\Big\}\Big|_{2\beta=1+\alpha} = C\Big\{h^\beta + h^{2(1-\beta)}\big(1 + \log_3\frac{1}{h}\big)\Big\},$$

which gives an optimal bound for $\beta = \frac{2}{3}$ and implies $\alpha = \frac{1}{3}$.

## 1.2 Error analysis for cubic ferromagnets

We consider direct minimization for cubic ferromagnets. Those materials possess increased symmetricity properties if compared to uniaxial materials: the corresponding anisotropy density vanishes in cases where the magnetization points in direction of one of the two easy axes $\mathbf{e}_i \in \mathbb{R}^2$, $i = 1, 2$; we again assume quadratic growth away from $\pm\mathbf{e}_i$ like in (1.3). The presented numerical analysis will not require a certain mesh orientation with respect to the easy axes of the problem.

The following two theorems are the first main result in this section.

**Theorem 1.3** *For cubic ferromagnets that cover a Lipschitz domain $\omega \subset \mathbb{R}^2$, and $\mathcal{T}$ a quasiuniform triangulation thereof, we consider the problem*

$$\min_{\boldsymbol{\mu}_h \in \boldsymbol{\mathcal{A}}_h} \mathcal{E}_0(\boldsymbol{\mu}_h)\,.$$

*Let $\mathbf{m}_h \in \boldsymbol{\mathcal{A}}_h$ be a solution to the corresponding finite element variational problem. There exists a constant $c_1 = c_1(\omega)$, independent of $h$, such that holds*

1. *for domains $\omega \in \mathbb{R}^2$, where $\partial\omega = \bigcup_{j=1}^{N} \Gamma_j$, with affine curves $\Gamma_j$, $1 \leq j \leq N$, and $\mathbf{n}_{\Gamma_j}$ normal vectors associated to $\Gamma_j$ that satisfy $\min_i \left|\langle \mathbf{n}_{\Gamma_j}, \mathbf{e}_i\rangle_{\mathbb{R}^2}\right| = 0$,*

$$\mathcal{E}_0(\mathbf{m}_h) = \min_{\boldsymbol{\mu}_h \in \boldsymbol{\mathcal{A}}_h} \mathcal{E}_0(\boldsymbol{\mu}_h) \leq c_1 h \,.$$

2. *for arbitrary Lipschitz domains, there exists a positive number $\alpha = \alpha(\omega) \geq 1$, such that*

$$\mathcal{E}_0(\mathbf{m}_h) = \min_{\boldsymbol{\mu}_h \in \boldsymbol{\mathcal{A}}_h} \mathcal{E}_0(\boldsymbol{\mu}_h) \leq c_1 h^{1/2}\big(1 + \log_3 \frac{1}{h^\alpha}\big) \,.$$

This result suggests decreasing accuracy of the finite element method for general Lipschitz domains $\omega$. Essentially, what happens in more complicated Lipschitz domains is that magnetization patterns exhibiting small-scale structures come up close to the boundary; see Figure 1.5 (b) — opposed to simpler magnetization patterns that are depicted in Figure 1.5 (a). Both Figures are taken from [72], see also [33]. If we take this observation into account in the construction process in our proof, then the following improved convergence statement can be verified for general Lipschitz domains $\Omega$.

**Theorem 1.4** *Suppose that the assumptions of Theorem 1.3 are valid. Then, there exists a positive number $\alpha = \alpha(\omega) \geq 1$, such that the following improved result holds for arbitrary Lipschitz domains and general quasi-uniform triangulations $\mathcal{T}$,*

$$\mathcal{E}_0(\mathbf{m}_h) = \min_{\boldsymbol{\mu}_h \in \boldsymbol{\mathcal{A}}_h} \mathcal{E}_0(\boldsymbol{\mu}_h) \leq c_1 h\big(1 + \log_3 \frac{1}{h^\alpha}\big) \,.$$

Theorems 1.3 and 1.4 are verified in Subsection 1.2.1. The subsequent analysis to show Theorem 1.4 needs to balance two main effects:

1. The general shape of the Lipschitz domain leads to significant energy contributions of discrete magnetizations close to the boundary $\partial\omega$ of the ferromagnetic body.

2. Arbitrary quasi-uniform meshes might create a high amount of energy for magnetizations at Weissian domain interfaces.

Both these sources have to be handled efficiently in a numerical model that aims at improving the convergence rate from Theorem 1.4. For this purpose, we employ two numerical tools:

1. Graduate meshes in a boundary layer bridge a numerical scaling of $\mathcal{O}(h_0^2)$ (which is chosen close to the boundary), and $\mathcal{O}(h_0)$ (which prevails in the interior of the domain).

2. An adaptive grid refinement strategy that is oriented at steep interface jumps of computed magnetizations between two Weissian domains reduces energy contributions that are due to non-aligned meshes.

Since graduate meshes increase the number of finite elements which are decreasing in size close to the boundary, a referencing discretization parameter $h_0$ is needed in order to describe the (initial) mesh $\mathcal{T}_{h_0}^{\text{grad}}$ that consists of $\mathcal{O}(h^{-2})$ elements. Maximum resolution is required close to the boundary and the triangulation there is ensembled from elements $K \in \mathcal{T}_{h_0}^{\text{grad}}$ of diameter $\text{diam}(K) = \mathcal{O}(h_0^2)$. Then, in order to keep the asymptotic overall effort at $\mathcal{O}(h_0^{-2}\log_2\frac{1}{h_0})$, we choose triangulation structures $\mathcal{T}_{h_0}^{\text{grad}}$, and $d_0 = \mathcal{O}(1)$, such that

$$h(\mathbf{x}) = \begin{cases} h_0 & \text{for } \text{dist}(\mathbf{x}, \partial\omega) \geq d_0 \ , \\ h_0^2(m+1) & \text{for } \text{dist}(\mathbf{x}, \partial\omega) = \mathcal{O}\big(\frac{m^2}{2}h_0^2\big) \ , \end{cases} \tag{1.19}$$

for $1 \leq m \leq \mathcal{O}\big(h_0^{-1}\big)$. From this point of view, it is again justified to qualify the resulting numerical scheme by error bounds in terms of the discretization parameter $h_0$.

Secondly, in order to control the energy contributions at interfaces of laminates, an adaptive strategy will be proposed that is based on values of inter-element jumps of magnetizations. The adaptive algorithm then consists of detecting elements where large inter-element jumps occur, and subsequent refinement, which generates a sequence of meshes $\big\{\mathcal{T}_\ell^{\text{grad}}\big\}_{\ell=0}^{\ell_\star}$, where $\ell_\star = \log_2\frac{1}{h_0}$. For the purpose of refining a given triangulation properly, we refer to strategies as they are discussed in [8].

The adaptive algorithm can then be stated in the following way, for $0 \leq \ell \leq \ell_\star$, and with the initial triangulation $\mathcal{T}_0^{\text{grad}} = \mathcal{T}_{h_0}^{\text{grad}}$ given in (1.19):

1. Given $\mu_{h_0}^\ell \in \mathcal{A}_{h_0}^\ell \equiv \prod_{K_\ell \in \mathcal{T}_\ell^{\text{grad}}} \big[P_0(K_\ell)\big]^2$, for $0 \leq \ell \leq \ell_\star$, apply the refinement strategy

$$\mathcal{H}_\ell \equiv \mathcal{H}_\ell(\mu_{h_0}^\ell) : \mathcal{T}_\ell^{\text{grad}} \to \mathcal{T}_{\ell+1}^{\text{grad}} \ ,$$

according to the following criterion: Refine by bisection adjacent elements $K_\ell \in \mathcal{T}_\ell^{\text{grad}}$ that share an edge $\boldsymbol{\xi}_\ell \subset \partial K_\ell$ where the jump of $\boldsymbol{\mu}_{h_0}^\ell$ across $\boldsymbol{\xi}_\ell$ satisfies for a given number $\kappa = \mathcal{O}(1)$,

$$\max_{\mathbf{x}\in\mathcal{F}_\ell} \big| [\boldsymbol{\mu}_{h_0}^\ell](\mathbf{x}) \big| \geq \kappa h_0 \,,$$

where $[\boldsymbol{\mu}_{h_0}^\ell] := \left(\boldsymbol{\mu}_{h_0}^\ell\right)^+ - \left(\boldsymbol{\mu}_{h_0}^\ell\right)^-$ denotes the jump across the inter-element face $\boldsymbol{\xi}_\ell$, for a given orientation of the mesh.

2. Set $\ell := \ell + 1$ and go back to item 1., provided $\ell \leq \ell_\star$. For $\ell = \ell_\star$, go to the next step.

3. For $\ell = \ell_\star$, compute a minimizer $\mathbf{m}_{h_0}^{\ell_\star} \in \boldsymbol{\mathcal{A}}_{h_0}^{\ell_\star}$ of

$$\min_{\boldsymbol{\mu}_{h_0}^{\ell_\star} \in \boldsymbol{\mathcal{A}}_{h_0}^{\ell_\star}} \mathcal{E}_0(\boldsymbol{\mu}_{h_0}^{\ell_\star}). \tag{1.20}$$

In order to ensure a monotonous behavior of the algorithm and thus an optimal performance of the method, the refinement process $\mathcal{H}_\ell$, for $1 \leq \ell \leq \ell_\star - 1$, has to be properly chosen in the following sense:

**Definition 1.1** *Given a quasi-regular triangulation $\mathcal{T}_0^{\text{grad}} \equiv \mathcal{T}_{h_0}^{\text{grad}}$, we call a refinement strategy $\mathcal{H} := \left\{\mathcal{H}_\ell\right\}_{\ell=0}^{\ell_\star - 1}$, with $\mathcal{H}_\ell := \mathcal{T}_\ell^{\text{grad}} \to \mathcal{T}_{\ell+1}^{\text{grad}}$, admissible if there exists a sequence $\left\{\mathcal{T}_\ell^{\text{grad}}, \boldsymbol{\mu}_{h_0}^\ell\right\}_{\ell=0}^{\ell_\star}$, for $\boldsymbol{\mu}_{h_0}^\ell \in \mathcal{A}_{h_0}^\ell$, and a constant $\tilde{C}$ does not dependent on the number of $\ell$, such that*

$$\mathcal{E}_0(\boldsymbol{\mu}_{h_0}^\ell) \leq \frac{\tilde{C}}{2^\ell} h \,, \qquad \text{for} \qquad 0 \leq \ell \leq \ell_\star \,.$$

**Remark 1.3** *A similar adaptive method has been proposed in [109] for crystalline materials that exhibit laminated microstructure. In [109], the energy functional itself was updated at each iteration step to 'tighten the continuity constraints' at interface transitions — something which is not necessary here.*

We are now in a position to state our main result for a numerical method that employs graduate meshes and an adaptive strategy to compute minimizing magnetizations for cubic materials. In the sequel, let $\{\tilde{K}_\ell^\tau\}$ denote the set of all elements on refinement level $\ell \geq 0$, such that $\bigcup \tilde{K}_\ell^\tau = K^\tau \in \mathcal{T}_{h_0}^{\text{grad}}$.

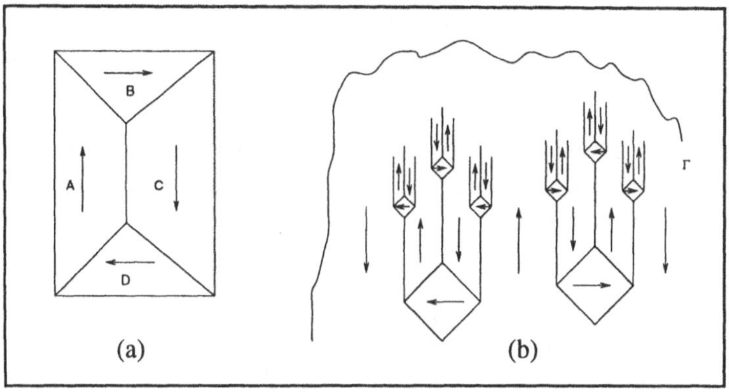


Figure 1.5: Minimizing domain structures. The microstructure depends on the domain shape ((a) simply shaped domain, (b) domain with curved boundary). These constructions are taken from [72], see also [33].

**Theorem 1.5** *Suppose that* $\omega \subset \mathbb{R}^2$ *is a Lipschitz domain, and* $\mathcal{T}_{h_0}^{\mathrm{grad}}$ *is a graduate mesh that satisfies (1.19). There exists an admissible refinement strategy* $\mathcal{H}$ *of* $\mathcal{T}_0^{\mathrm{grad}} = \mathcal{T}_{h_0}^{\mathrm{grad}}$*, such that*

$$\mathcal{T}_{\ell_\star}^{\mathrm{grad}} = \bigcup_{r=1}^{\mathcal{O}(h_0^{-2}\log_2\frac{1}{h_0})} \tilde{K}_\ell^r, \qquad \text{for } \ell_\star = \log_2\frac{1}{h_0},$$

*and there holds*

$$\min_{\mu_{h_0}^{\ell_\star}\in\mathcal{A}_{h_0}^{\ell_\star}} \mathcal{E}_0(\mu_{h_0}^{\ell_\star}) \leq C h_0^2 \left(1+\log_3\frac{1}{h_0^\alpha}\right),$$

*for a positive number* $\alpha = \alpha(\omega) \geq 1$.

This theorem is verified in Subsection 1.2.2.

### 1.2.1 Proofs of Theorems 1.3 and 1.4

Proofs are split into several steps where different error contributions are considered.

The verification of Theorem 1.3 starts with the analysis of a simplified model situation. The general case is then treated by means of perturbation arguments in the steps 2 and 3.

*Proof of Theorem 1.3. 1st step:* We consider a cubic domain $\omega \subset \mathbb{R}^2$, with its boundary aligned to the axes $\mathbf{e}_i$, for $i = 1, 2$, see Figure 1.5 (a). Suppose that $\operatorname{diam}(\omega) = \mathcal{O}(1)$. Further suppose that we are given a triangulation $\mathcal{T}$ such that the lines drawn in Figure 1.5 (a) can be resolved; the diagonal lines in there are supposed to be aligned to $\mathbf{e}_1 \pm \mathbf{e}_2$. According to Figure 1.5 (a), we define the magnetization

$$\wp^1(\mathbf{x}) = \begin{cases} \mathbf{e}_2 & \text{if } \mathbf{x} \in A\ , \\ \mathbf{e}_1 & \text{if } \mathbf{x} \in B\ , \\ -\mathbf{e}_2 & \text{if } \mathbf{x} \in C\ , \\ -\mathbf{e}_1 & \text{if } \mathbf{x} \in D\ , \end{cases} \tag{1.21}$$

for $\omega = A \cup B \cup C \cup D$. Then, the corresponding energy is given by

$$\mathcal{E}_0(\wp^1) = \frac{1}{2} \int_{\mathbb{R}^2} |\nabla u^1|^2 \, \mathbf{dx}\,, \tag{1.22}$$

for $u^1 = \Delta^{-1}\big(\operatorname{div}(\chi_\omega \wp^1)\big) \in W^{1,2}(\mathbb{R}^2)$, and (I.2) implies

$$\int_{\mathbb{R}^2} |\nabla u^1|^2 \, \mathbf{dx} = (\nabla u^1, \wp^1)\,. \tag{1.23}$$

Then, we have by the divergence theorem,

$$\begin{aligned} (\nabla u^1, \wp^1) \;&=\; \int_{\partial A} u^1 n_2 \, \mathbf{dx} + \int_{\partial B} u^1 n_1 \, \mathbf{dx} \\ &\quad - \int_{\partial C} u^1 n_2 \, \mathbf{dx} - \int_{\partial D} u^1 n_1 \, \mathbf{dx}\,, \end{aligned} \tag{1.24}$$

where $\mathbf{n} = (n_1, n_2)^T$ is the unit exterior normal to $\Gamma \in \{\partial A, \partial B, \partial C, \partial D\}$ at $\mathbf{x} \in \Gamma$. Then, the right-hand side of (1.24) is zero, according to our construction.

*2nd step:* We consider the same domain $\omega \subset \mathbb{R}^2$, as in the first step, but the given (quasiuniform) triangulation is now supposed to be arbitrary. If we follow the lines of step 1, the boundary integrals in (1.24) do not cancel out any more and the construction of a magnetization analogous to $\wp^1$ as in (1.21) is not that immediate. In the following, we use the notation of step 1. — In a first approach, fix $u^1 \in W^{1,2}(\mathbb{R}^2)$ that is the unique solution of

$$\Delta u^1 = \operatorname{div}\big(\chi_\omega \wp^1\big)\,, \qquad u^1(\mathbf{x}) \to 0\,, \qquad \text{as } |\mathbf{x}| \to \infty\,, \tag{1.25}$$

Then, $\wp^1$ is modified according to

$$\wp_h^2(\mathbf{x}) = \begin{cases} \wp^1(\mathbf{x}) & \text{for all } K \in \mathcal{T}, \text{ with } K \cap \Gamma = \emptyset \ , \\ \tilde{\wp}^1(\mathbf{x}) & \text{for all } K \in \mathcal{T}, \text{ with } K \cap \Gamma \neq \emptyset \ , \end{cases} \tag{1.26}$$

where $\tilde{\wp}^1\big|_K$ denotes an extension of any of the involved $\wp^1\big|_{K\cap D}$ that are locally in question, for $D \in \{A, B, C, D\}$. Note that $\wp_h^2 \in \mathcal{A}_h$. According to Lemma 1.2, there exists a positive number $\alpha = \alpha(\omega) \geq 1$, such that the following inequality holds true,

$$\big|(\nabla u^1, \wp_h^2 - \wp^1)\big| \leq Ch\big(1 + \log_3 \frac{1}{h^\alpha}\big) \, . \tag{1.27}$$

Then, for $u^2 = \Delta^{-1}\big(\mathrm{div}\,(\wp_h^2 \chi_\omega)\big) \in W^{1,2}(\mathbb{R}^2)$, we obtain from (1.27) and step 1,

$$\begin{aligned} (\nabla u^2, \wp_h^2) &= (\nabla u^1, \wp_h^2) + \big(\nabla(u^2 - u^1), \wp_h^2\big) \\ &\leq Ch\big(1 + \log_3 \frac{1}{h^\alpha}\big) + \Big[\int_A + \int_B + \int_C + \int_D\Big] \langle \nabla u^1, \wp^1 \rangle_{\mathbb{R}^2} \, \mathrm{d}\mathbf{x} \\ &\quad + \big(\nabla(u^2 - u^1), \wp_h^2\big) \\ &= Ch\big(1 + \log_3 \frac{1}{h^\alpha}\big) + \big(\nabla(u^2 - u^1), \wp_h^2\big) \, . \end{aligned} \tag{1.28}$$

Thanks to the reformulation

$$\begin{aligned} \big(\nabla(u^2 - u^1), \wp_h^2\big) &= \int_{\mathbb{R}^2} \langle \nabla(u^2 - u^1), \nabla u^2 \rangle_{\mathbb{R}^2} \, \mathrm{d}\mathbf{x} \\ &= \frac{1}{2} \int_{\mathbb{R}^2} \big\{ |\nabla u^2|^2 - |\nabla u^1|^2 + |\nabla\big(u^2 - u^1\big)|^2 \big\} \, \mathrm{d}\mathbf{x} \, , \end{aligned} \tag{1.29}$$

and a well-known stability result for $\Delta(u^2 - u^1) = \mathrm{div}\big(\chi_\omega(\wp_h^2 - \wp^1)\big)$,

$$\|\nabla(u^2 - u^1)\|_{\mathbf{L}^2(\mathbb{R}^2, \mathbb{R}^2)} \leq C \, \|\wp_h^2 - \wp^1\|_{\mathbf{L}^2(\omega, \mathbb{R}^2)} \, , \tag{1.30}$$

we can continue with estimating the terms on the right-hand side of (1.28). After absorbing terms on the left-hand side to deal with (1.29), we end up with

$$\int_{\mathbb{R}^2} |\nabla u^2|^2 \, \mathrm{d}\mathbf{x} = |(\nabla u^2, \wp_h^2)| \leq Ch\big(1 + \log_3 \frac{1}{h^\alpha}\big) \, . \tag{1.31}$$

*3rd step:* Here, we consider a general Lipschitz domain $\omega \subset \mathbb{R}^2$. We begin with a perturbation argument: Replace $\omega$ by $\hat{\omega} \supset \omega$, according to Figure 1.6 (a), and such that holds $\text{meas}(\hat{\omega} \setminus \omega) \le Ch^\beta$, for $0 \le \beta \le 1$. Then, we employ a magnetization exhibiting laminated structure, $\wp^3 : \hat{\omega} \to \mathbb{R}^2$, as is given in Figure 1.6 (a). Then, we can conclude, for $u^3 = (\Delta)^{-1}\big(\text{div}\,(\wp^3 \chi_\omega)\big) \in W^{1,2}(\mathbb{R}^2)$ and $\hat{u}^3 = \Delta^{-1}\big(\text{div}\,(\wp^3 \chi_{\hat{\omega}})\big) \in W^{1,2}(\mathbb{R}^2)$, respectively,

$$\|\nabla(u^3 - \hat{u}^3)\|^2_{\mathbb{R}^2} \le C\,\|\chi_{\hat{\omega}} - \chi_\omega\|^2_{\hat{\omega}} \le Ch^\beta\,. \tag{1.32}$$

In the following, we refer to laminates $\{\mathcal{L}_i\}_{i=1}^{\mathcal{O}(h^{-\beta})}$ of thickness $\mathcal{O}(h^\beta)$, and tips of length $\mathcal{O}(h^\beta)$. For $\hat{\omega}$, we define a triangulation $\hat{\mathcal{T}}_h \supseteq \mathcal{T}_h$. Then, $\hat{\boldsymbol{A}}_h \ni \hat{\wp}_h^4 : \hat{\omega} \to \mathbb{R}^2$ is defined as follows,

$$\hat{\wp}_h^4(\mathbf{x}) = \begin{cases} \hat{\wp}^3(\mathbf{x}) & \text{for all } K \in \hat{\mathcal{T}}, \text{ s.t. } K \in \partial\mathcal{L}_i \cap \partial\mathcal{L}_j, \\ & i \ne j,\ 1 \le i \le \mathcal{O}(h^{-\beta})\ , \\ \tilde{\wp}^3(\mathbf{x}) & \text{for all } K \in \hat{\mathcal{T}}, \text{ s.t. } K \notin \partial\mathcal{L}_i \cap \partial\mathcal{L}_j, \\ & i \ne j,\ 1 \le i \le \mathcal{O}(h^{-\beta})\ , \end{cases} \tag{1.33}$$

where $\tilde{\wp}^3\big|_K$ is arbitrarily chosen to be $\wp^3\big|_{K\cap\mathcal{L}_i}$ or $\wp^3\big|_{K\cap\mathcal{L}_j}$, $i \ne j$. According to Lemma 1.2, the magnetization $\hat{\wp}_h^4 \in \hat{\boldsymbol{A}}_h$ then satisfies

$$\Big|\int_{\hat{\omega}} \langle \nabla u^3, \hat{\wp}_h^4 - \wp^3 \rangle_{\mathbb{R}^2}\, \mathrm{d}\mathbf{x}\Big| \le Ch^{1-\beta}\big(1 + \log_3 \frac{1}{h^\alpha}\big)\,, \tag{1.34}$$

for a positive constant $\alpha \ge 1$. The upper bound in (1.34) is due to $\mathcal{O}(h^{-\beta})$ laminates. For the following purposes, set $u^4 = \Delta\big(\text{div}\,(\chi_\omega \wp_h^4)\big)$ and $\hat{u}^4 = \Delta\big(\text{div}\,(\chi_{\hat{\omega}} \hat{\wp}_h^4)\big)$, taking $\wp_h^4 := \chi_\omega \hat{\wp}_h^4$. Then, (1.32) and (1.34), together with the argumentation of step 2 imply

$$\begin{aligned} (\nabla u^4, \wp_h^4) &= Ch^\beta + \int_{\hat{\omega}} \langle \nabla \hat{u}^4, \hat{\wp}_h^4 \rangle_{\mathbb{R}^2}\, \mathrm{d}\mathbf{x} \qquad (1.35) \\ &\le Ch^\beta + Ch^{1-\beta}\big(1 + \log_3 \frac{1}{h^\alpha}\big) + \int_{\hat{\omega}} \langle \nabla(\hat{u}^4 - \hat{u}^3), \wp_h^4 \rangle_{\mathbb{R}^2}\, \mathrm{d}\mathbf{x} \\ &\le Ch^\beta + Ch^{1-\beta}\big(1 + \log_3 \frac{1}{h^\alpha}\big)\,, \end{aligned}$$

thanks to the divergence theorem and the cancelation of boundary integrals along the oriented laminates $\bigcup_{i=1}^{\mathcal{O}(h^{-\beta})} \mathcal{L}_i$. Choosing $\beta = \frac{1}{2}$ concludes the verification of Theorem 1.3. □

As we see from the last step in the verification of Theorem 1.3, magnetization structures which are assembled from laminated structures as in (1.33) give an energy contribution that is not bigger than $\mathcal{O}\big(h^{1/2}(1+\log_3\frac{1}{h^\alpha})\big)$ — which is a worse upper bound compared to the case considered in the second step of the proof of Theorem 1.3. To avoid this source of energy, we use magnetizations which possess a modified microstructure close to the boundary; this strategy is motivated by observations in physical experiments, see Figure 1.5 (b). This construction will lead to the improved energy bounds of Theorem 1.4.

*Proof of Theorem 1.4.* In the following, we consider a magnetization $\wp^5$ that shows Weissian domains (i.e., laminates) of thickness $\mathcal{O}(1)$ in the interior of the domain of alternating sign. In a $\mathcal{O}(1)$-neighborhood of the boundary $\partial\omega$, we refine according to Figure 1.6 (b): Fix a segment $\mathcal{Z}_\beta$, with $\big\{\mathcal{Z}_\beta\big\}_{\beta=1}^{\mathcal{O}(1)}$ a set of disjoint domains that covers a strip along the boundary $\partial\omega$ (i.e., a $\mathcal{O}(1)$-neighborhood of $\partial\omega$). Then, each level $1\le i\le i_\star$ consists of rhombic domains $\big\{\mathcal{W}_{ij}^\beta\big\}_{j=1}^{3^i}$ of side-length $3^{-i}$, and laminated domains $\big\{\mathcal{R}_{ij}^\beta\big\}_{j=1}^{2\cdot 3^i}$. One of the edges of each $\mathcal{R}_{ij}^\beta$ is of order $\mathcal{O}(h)$, for all $1\le j\le 2\cdot 3^i$ and $1\le i\le i_\star$, whereas the other one is of length $\frac{1}{2}3^{-i}$. Then, we need $i_\star=\log_3\frac{1}{h}$ steps to bridge the scale for structures of order $\mathcal{O}(1)$ in the interior to $\mathcal{O}(h)$ close to the boundary of the domain. Note that each side of the laminated domains $\mathcal{R}_{i_\star j}^\beta$, for $1\le j\le \mathcal{O}(h^{-1})$ is of length $\mathcal{O}(h)$.

**Remark 1.4** *1. The construction in this proof is given for one prototype segment $\mathcal{Z}_\beta$, see Figure 1.6 (b). The actual length of the 'pistons' depends on the shape of the domain. Since the analysis presented here is based on scaling arguments and the domain is assumed to be of Lipschitz type, this only involves further local constants, leaving the investigations untouched from essential new phenomena.*

*2. Additional integral terms arise in the case of a curved boundary. Therefore, one again has to construct a domain $\omega'\supseteq\omega$, such that each piecewise part $\Gamma_j\subset\partial\omega'$, for $1\le j\le N$, satisfies $\min_i\big|\langle \mathbf{n}_{\Gamma_j},\mathbf{m}_i\rangle_{\mathbb{R}^d}\big|=0$. As we see from below (see formula (1.39)), boundary structures can be resolved up to a scale $\mathcal{O}(h)$, i.e., we have $\mathrm{meas}(\omega'\setminus\omega)\le Ch$ and $\mathrm{meas}(\Gamma_i)=\mathcal{O}(h)$. In order to omit further technicalities on this and since we can employ a perturbation argument corresponding to the one given in the 3rd step of the proof of Theorem 1.4 (for a different scaling) which does not involve further difficulties,* *we confine here to the case* $\omega = \omega'$.

Subsequently, we employ the following notation. The domains $\mathcal{D}_{i_j j} \subset \omega$ of equal magnetization $\pi \in \{\pm\mathbf{e}_1, \pm\mathbf{e}_2\}$ in the interior of $\omega$ satisfy $\operatorname{diam}(\mathcal{D}_{i_j j}) = \mathcal{O}(1)$, for $j = 1, .., 4$. Then,

$$\omega = \bigcup_{j=1}^{4} \bigcup_{i_j=1}^{\mathcal{O}(1)} \mathcal{D}_{i_j j} \cup \bigcup_{\beta=1}^{\mathcal{O}(1)} \mathcal{Z}_\beta, \qquad \text{and} \qquad \mathcal{Z}_\beta = \bigcup_{i=1}^{\log_3 \frac{1}{h}} \Big( \bigcup_{j=1}^{3^i} \mathcal{W}_{ij}^\beta \cup \bigcup_{j=1}^{2\cdot 3^i} \mathcal{R}_{ij}^\beta \Big). \tag{1.36}$$

Then, we construct $\wp_h^6 \in \boldsymbol{A}_h$ from $\wp^5$ according to previous considerations that have been elaborated in the verification of Theorem 1.3: define

$$\mathcal{G} := \bigcup_{j=1}^{4} \bigcup_{i_j=1}^{\mathcal{O}(1)} \partial\mathcal{D}_{i_j j} \cup \bigcup_{\beta=1}^{\mathcal{O}(1)} \bigcup_{i=1}^{\log_3 \frac{1}{h}} \Big( \bigcup_{j=1}^{3^i} \partial\mathcal{W}_{ij}^\beta \cup \bigcup_{j=1}^{2\cdot 3^i} \partial\mathcal{R}_{ij}^\beta \Big). \tag{1.37}$$

Note that $\operatorname{length}(\mathcal{G}) = \mathcal{O}(1 + \log_3 \frac{1}{h})$, according to the construction: the length of the interfaces for each segment $\mathcal{Z}_\beta$ is given by $6 \log_3 \frac{1}{h} + h \sum_{i=1}^{\log_3 \frac{1}{h}} 3^i$. Then,

$$\wp_h^6(\mathbf{x}) = \begin{cases} \wp^5(\mathbf{x}) & \text{for all } K \in \mathcal{T}, \text{ s.t. } K \cap \mathcal{G} = \emptyset \,, \\ \tilde{\wp}^5(\mathbf{x}) & \text{for all } K \in \mathcal{T}, \text{ s.t. } K \cap \mathcal{G} \neq \emptyset \,, \end{cases} \tag{1.38}$$

where $\tilde{\wp}^5 \in \boldsymbol{A}_h$ is chosen from $\wp^5$ through arbitrary constant extension on elements $K \in \mathcal{T}$ along $\mathcal{G}$ from any adjacent Weissian domain (corresponding to prior considerations), and

$$\big|(\nabla u^5, \wp_h^6 - \wp^5)\big| \le Ch\big(1 + \log_3 \frac{1}{h^\alpha}\big)\,, \tag{1.39}$$

for a positive number $\alpha \ge 1$. This construction is possible, thanks to Lemma 1.2, since $\operatorname{length}(\mathcal{G}) = \mathcal{O}(1 + \log_3 \frac{1}{h})$, and thus $\|\wp_h^6 - \wp^5\|_{\mathbf{L}^1(\omega,\mathbb{R}^2)} \le Ch\big(1 + \log_3 \frac{1}{h}\big)$. Here and in the following, we employ the potentials $u^5 = \Delta^{-1}\operatorname{div}\big(\wp^5 \chi_\omega\big) \in W^{1,2}(\mathbb{R}^2)$, and $u^6 = \Delta^{-1}\operatorname{div}\big(\wp_h^6 \chi_\omega\big) \in W^{1,2}(\mathbb{R}^d)$. Then, we can conclude as follows,

$$\begin{aligned} (\nabla u^6, \wp_h^6) &= (\nabla u^5, \wp_h^6) + (\nabla\big(u^6 - u^5\big), \wp_h^6) \qquad (1.40) \\ &\le Ch\big(1 + \log_3 \frac{1}{h^\alpha}\big) + (\nabla u^5, \wp^5) + \int_{\mathbb{R}^2} \langle \nabla\big(u^6 - u^5\big), \nabla u^6 \rangle_{\mathbb{R}^2} \, d\mathbf{x}\,. \end{aligned}$$

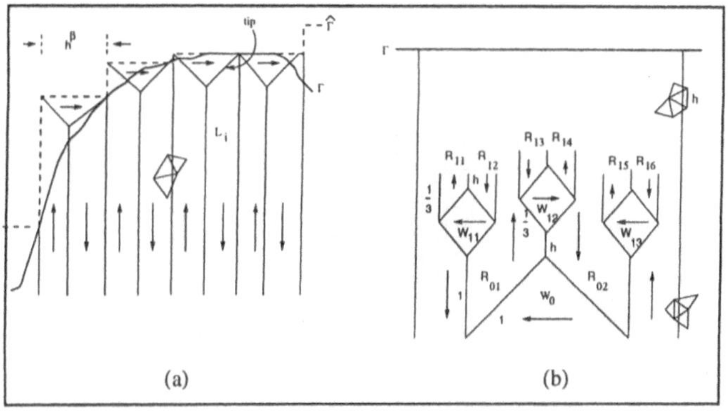


Figure 1.6: (a) Laminated magnetization in an arbitrary Lipschitz domain $\omega$ for a quasi-uniform triangulation. The domain $\omega$ is replaced by $\hat{\omega}$, with $\partial\hat{\omega} = \hat{\Gamma}$. Each of the laminates $\mathcal{L}_i$ are of thickness $\mathcal{O}(h^\beta)$ and the tips are of equal length. (b) Structure of minimizing magnetization on arbitrarily shaped Lipschitz domains.

The second term on the right-hand side is zero, according to the analytic construction of $\wp^5$. The third term can be reformulated in a way that is analogous to (1.29), (1.30), and we thus find

$$\begin{aligned}\left|\int_{\mathbb{R}^2} \langle \nabla (u^6 - u^5), \nabla u^6 \rangle_{\mathbb{R}^2} \, \mathrm{d}\mathbf{x}\right| &\leq C \|\wp_h^6 - \wp^5\|^2_{\mathbf{L}^2(\omega,\mathbb{R}^2)} + \frac{1}{2}\left(\nabla u^6, \wp_h^6\right) \\ &\leq Ch\big(1 + \log_3 \frac{1}{h^\alpha}\big) + \frac{1}{2}\left(\nabla u^6, \wp_h^6\right). \quad (1.41)\end{aligned}$$

Inserting (1.41) into (1.40) furnishes the proof of Theorem 1.4. □

## 1.2.2 Proof of Theorem 1.5

The verification of Theorem 1.5 is established by means of asymptotic scaling argument.

**Proof:**

We start with the construction of a magnetization $\wp^7$ that is appropriate for our needs of graduated meshes $\mathcal{T}_{h_0}^{\mathrm{grad}}$. Therefore, we refer to Figure 1.7 that shows one segment along the boundary being of size $\mathcal{O}(1) \times \mathcal{O}(h_0)$. We consider a stretched pattern for the magnetization $\wp_A^7 \in \mathbf{L}^\infty(\omega, \mathbb{R}^2)$ in this segment, as it is given in Figure 1.7. In fact, we would like to have a

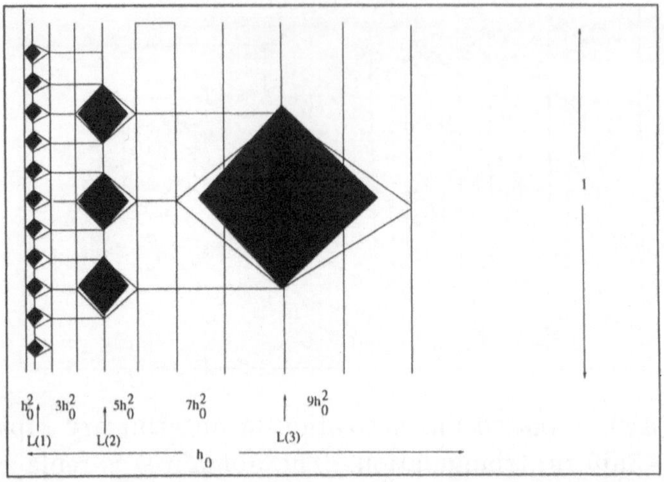


Figure 1.7: Magnetization showing microstructure on a graduated mesh $\mathcal{T}_{h_0}^{\text{grad}}$ in a $\mathcal{O}(h_0)$-neighborhood of the boundary $\partial\omega$. $L(i)$ denotes the level of refinement, for $1 \leq i \leq \log_3 \frac{1}{h_0}$.

magnetization $\wp^7 \in \mathbf{L}^\infty(\omega, \mathbb{R}^2)$ that enjoys the following side constraints to serve our purposes:

1. At each level $L(i) = N_\star - i$, with $1 \leq i \leq N_\star = \log_3 \frac{1}{h_0}$, there exist $\frac{3^{1-i}}{h_0^2}$ rhombic domains.

2. The faces of each of the rhombic domains at level $L(i)$ are of length $3^{i-1} h_0^2$ and aligned to $\pm \mathbf{e}_1 \pm \mathbf{e}_2$.

We confine to constructing a magnetization $\wp^7 \in \mathbf{L}^\infty(\omega, \mathbb{R}^2)$ that satisfies these needs in a $\mathcal{O}(h_0)$ neighborhood of the boundary $\partial\omega$; we then combine this magnetization structure with the one that is presented in the verification of Theorem 1.4 in the interior of the ferromagnet, $\wp^5$; see also Figure 1.6 (b). There holds $h(\mathbf{x}) = \mathcal{O}(h_0^{3/2})$, for $\text{dist}(\mathbf{x}, \partial\omega) = \mathcal{O}(h_0)$.

Note, that the length of the interfaces that are now constructed for $\wp^7$ is of order $\mathcal{O}(1 + \log_3 \frac{1}{h_0})$, due to the construction. Using the shorthand notation $\omega_d = \{\mathbf{x} \in \omega : \text{ dist}(\mathbf{x}, \partial\omega) \geq d\}$, we can proceed as in the proof of Theorem 1.4, by extending the decomposition of $\omega$ into domains of macroscopic structures of magnitude $\mathcal{O}(1)$ in the interior $\omega_1 \Subset \omega$, microstructures on a scale ranging from $\mathcal{O}(1)$ to $\mathcal{O}(h_0)$ on $\omega_{h_0} \setminus \omega_1$, whereas the more refined structures defined above appear on scales ranging in $[h_0^2, h_0]$ occur on

$\omega \setminus \omega_{h_0}$. We omit the elaboration of the arguments that can be given according to those presented in the proof of Theorem 1.4, such that finally holds, for $\wp_h^8 \in \mathcal{A}_{h_0}$ constructed from $\wp^7$ and corresponding potential functions $\{u^7, u^8\} \in W^{1,2}(\mathbb{R}^2)$,

$$\begin{aligned}(\nabla u^8, \wp_{h_0}^8) &= (\nabla u^7, \wp_{h_0}^8) + (\nabla(u^8 - u^7), \wp_{h_0}^8) \\ &\leq C\big(h_0^{3/2} + h_0\big)\big(1 + \log_3 \frac{1}{h^\alpha}\big) + (\nabla u^7, \wp^7) \\ &\quad + \int_{\mathbb{R}^2} \langle \nabla(u^8 - u^7), \nabla u^7 \rangle_{\mathbb{R}^2} \, \mathrm{d}\mathbf{x} \\ &\leq C\big(h_0^{3/2} + h_0\big)\big(1 + \log_3 \frac{1}{h^\alpha}\big) + C\|\wp_{h_0}^8 - \wp^7\|^2_{\mathbf{L}^2(\omega,\mathbb{R}^2)} ,\end{aligned} \tag{1.42}$$

for a positive number $\alpha = \alpha(\omega) \geq 1$, that results from application of Lemma 1.2. In order to reduce energy contributions due to non-aligned meshes, discrete magnetizations have to be constructed, that reduce this error source. This is accomplished in the following admissible refinement process $\mathcal{H}$, where those elements $K \in \mathcal{T}_{h_0}^{\text{grad}}$ in $\omega_1$, $\omega_{h_0}$ as well as $\omega \setminus \omega_{h_0}$ are subject to refinement, where a transition in $\wp^7 \in \mathbf{L}^\infty(\omega, \mathbb{R}^2)$ occurs. Note that the length of the related interface is of size $\mathcal{O}(1 + \log_3 \frac{1}{h_0})$. We construct a sequence $\{\mathcal{T}_\ell^{\text{grad}}, (\wp_{h_0}^8)^\ell\}_{\ell=0}^{\ell_\star}$, with $(\wp_{h_0}^8)^0 \equiv \wp_{h_0}^8$, where each triangulation $\mathcal{T}_\ell^{\text{grad}}$ gives a refined mesh width at the interfaces, and where $(\wp_{h_0}^8)^\ell \in \mathcal{A}_{h_0}^\ell$ is then constructed, corresponding to prior restrictions. This gives an admissible refinement, and we find a constant $C \neq C(\ell)$, such that holds

$$\|(\wp_{h_0}^8)^\ell - \wp^7\|^2_{\mathbf{L}^2(\omega,\mathbb{R}^2)} \leq C \,\text{length}(\mathcal{G}_\ell)\, \frac{h_0}{2^\ell}\,, \qquad \text{for } 1 \leq \ell \leq \ell_\star . \tag{1.43}$$

Here, $\mathcal{G}_\ell$ stands for interfaces separating Weissian domains of different magnetization and where transition occurs, for $1 \leq \ell \leq \ell_\star$, see also (1.37) for a corresponding definition. Again, there holds $\text{length}(\mathcal{G}_\ell) \leq C\big(1 + \log_3 \frac{1}{h_0}\big)$, for $1 \leq \ell \leq \ell_\star$. Taking $\ell_\star = \log_2 \frac{1}{h_0}$, (1.43) and (1.42) — with the latter now holding with the leading term $Ch_0^2$ thanks to Lemma 1.2 — confirms the statement of Theorem 1.5. □

# Chapter 2

# Convexified Micromagnetism

## 2.1 A stable non-conforming discretization[1]

Direct minimization of the non-convex energy in the model of Weiss, Landau et Lifshitz (I.4) has no solution (uniaxial case), in general: minimizing sequences sequences exist, are bounded, contain weakly converging subsequences, but show higher and higher oscillations which prohibit strong convergence; the corresponding numerical analysis for a finite element discretization for both, uniaxial and cubic ferromagnetic materials is given in Chapter 1.

For certain applications, main interest is on simulating averaged (macroscopic) magnetizations. To obtain those informations, it is sufficient to consider a modified minimization problem (CP). This approach uses the lower convex envelope $\phi^{**}(\mathbf{m})$ of $\phi$ in (I.1) (with $\alpha = 0$) and replaces the constraint $|\,\mathbf{m}\,| = 1$ by $|\,\mathbf{m}\,| \leq 1$ in the minimization problem (I.4); it has been analyzed in [39].

More precisely, let $\varphi(\mathbf{m})$ denote the lower convex envelope $\phi^{**}(\mathbf{m})$ of $\phi$ if $|\mathbf{m}| \leq 1$ and $\varphi(\mathbf{m}) = \infty$ if not. The lower convex envelope $\phi^{**}$ is the largest convex function below $\phi$. Then, the convexified problem (CP) reads: Seek a minimizer $\mathbf{m} \in \mathbf{L}^2(\omega, \mathbb{R}^d)$ of the convexified energy

$$\mathcal{E}(\mathbf{m}) = \int_\omega \varphi(\mathbf{m})\,\mathrm{d}\mathbf{x} - \int_\omega \langle \mathbf{f}, \mathbf{m}\rangle_{\mathbb{R}^d}\,\mathrm{d}\mathbf{x} + \frac{1}{2}\int_{\mathbb{R}^d} |\nabla u|^2\,\mathrm{d}\mathbf{x} \qquad (2.1)$$

[1]The material presented in this section this taken from [20].

subject to

$$\Delta u = \operatorname{div}(\chi_\omega \mathbf{m}) \quad \text{in } \mathbb{R}^d . \tag{2.2}$$

Since minimizing sequences of (CP) have bounded magnetization and so a bounded potential, there exist weakly convergent subsequences, and its minimizer solves the following problem, for $\phi \in C^1(\mathbb{R}^d)$: Seek $u \in W_0^{1,2}(\Omega)$, $\mathbf{m} \in \mathbf{L}^2(\omega, \mathbb{R}^2)$, and $\lambda \in L^2(\omega)$ satisfying[2]

$$(\nabla u, \nabla w)_\Omega = (\mathbf{m}, \nabla w) \quad \forall w \in W_0^{1,2}(\Omega) , \tag{2.3}$$

$$\nabla u + D\phi^{**}(\mathbf{m}) + \lambda \mathbf{m} = \mathbf{f} \quad \text{a.e. in } \omega , \tag{2.4}$$

$$0 \le \lambda, \quad |\mathbf{m}| \le 1, \quad \text{and} \quad \lambda(1 - |\mathbf{m}|)_+ = 0 \quad \text{a.e. in } \omega . \tag{2.5}$$

Here, $(s)_+ := \max\{s, 0\}$ denotes the non-negative part and the last condition in (2.5) states that $\lambda \neq 0$ is possible only for $|\mathbf{m}| = 1$ as a consequence of $\lambda \mathbf{m} \in \partial\psi(\mathbf{m})$ for the convex characteristic functional $\psi : \mathbb{R}^2 \to [0, \infty]$ defined by $\psi(\mathbf{m}) = 0$ if $|\mathbf{m}| \le 1$, and $\psi(\mathbf{m}) = \infty$ if not.

**Remark 2.1** *According to (2.3), the stray-field $\nabla u = \mathcal{L}(\chi_\omega \mathbf{m})$, where $\mathcal{L}(\mathbf{m}) = \nabla \Delta_D^{-1} \operatorname{div}(\chi_\omega \mathbf{m})$ and $\Delta_D^{-1} : W^{-1,2}(\Omega) \to W_0^{1,2}(\Omega)$ denotes the solution operator for the Laplace problem with homogeneous Dirichlet boundary data. Then, the Euler-Lagrange equation of (2.1) reads $\mathcal{L}^*\mathcal{L}\mathbf{m} + D\phi^{**}(\mathbf{m}) + \lambda\mathbf{m} = \mathbf{f}$ in $L^2(\omega)^2$. One can prove that this indeed implies (2.4).*

In the sequel, we prove uniqueness of solutions for uniaxial ferromagnets; it is, in general, that the number of solutions may be infinite.

**Theorem 2.1** *In the uniaxial case, $\phi^{**}(\mathbf{m}) = \frac{1}{2}\langle \mathbf{m}, \mathbf{e}_\perp \rangle_{\mathbb{R}^2}^2$ for some unit vector $\mathbf{e} \in \mathbb{R}^2$ and its normal $\mathbf{e}_\perp$, there exists only one solution in (P) and in (CP). In general, there are infinitely many minimizers in the cubic case, where $\phi(\mathbf{m}) = m_1^2 m_2^2$, for $\mathbf{m} = (m_1, m_2)^\top \in \mathbb{R}^2$.*

**Proof:**
Suppose $\{\mathbf{m}_j, \lambda_j, u_j\}$ solves (CP) for $j = 1, 2$ and denote $e := u_2 - u_1$ and $\boldsymbol{\delta} = \mathbf{m}_2 - \mathbf{m}_1$. According to (2.3), we have

$$(\nabla e, \boldsymbol{\delta}) = \| \nabla e \|^2_{\mathbf{L}^2(\Omega, \mathbb{R}^2)} . \tag{2.6}$$

[2]Throughout the remainder of this section, we limit far-field effects of the magnetic potential $u$ to a bounded domain $\Omega \subset \mathbb{R}^2$ which contains $\omega \Subset \Omega$. This is mainly for practical purposes to finally enable standard finite element realizations.

From the monotonicity of the subgradients we infer from $\lambda_j \mathbf{m}_j \in \partial\psi(\mathbf{m}_j)$ that $\langle \lambda_j \mathbf{m}_j, \mathbf{m}_{j+1} - \mathbf{m}_j \rangle_{\mathbb{R}^2} \leq 0$ for $j = 1, 2$ and $\mathbf{m}_3 := \mathbf{m}_1$. This shows $0 \leq \langle \lambda_2 \mathbf{m}_2 - \lambda_1 \mathbf{m}_1, \boldsymbol{\delta} \rangle_{\mathbb{R}^2}$ almost everywhere in $\omega$. Similar arguments show $0 \leq \langle D\phi^{**}(\mathbf{m}_2) - D\phi^{**}(\mathbf{m}_1), \boldsymbol{\delta} \rangle_{\mathbb{R}^2}$ and we deduce with (2.4) and (2.6) that all three terms in

$$\| \nabla e \|^2_{\mathbf{L}^2(\Omega,\mathbb{R}^2)} + \int_\omega \langle D\phi^{**}(\mathbf{m}_2) - D\phi^{**}(\mathbf{m}_1), \boldsymbol{\delta} \rangle_{\mathbb{R}^2} \, d\mathbf{x}$$
$$+ \int_\omega \langle \lambda_2 \mathbf{m}_2 - \lambda_1 \mathbf{m}_1, \boldsymbol{\delta} \rangle_{\mathbb{R}^2} \, d\mathbf{x} = 0 \tag{2.7}$$

are non-negative and hence vanish: $\mathbf{e} = 0$, $\langle \lambda_2 \mathbf{m}_2 - \lambda_1 \mathbf{m}_1, \boldsymbol{\delta} \rangle_{\mathbb{R}^2} = 0$, and $\langle D\phi^{**}(\mathbf{m}_2) - D\phi^{**}(\mathbf{m}_1), \boldsymbol{\delta} \rangle_{\mathbb{R}^2} = 0$ almost everywhere.

In the uniaxial case, we have $D\phi^{**}(\mathbf{m}) = \langle \mathbf{m}, \mathbf{e}_\perp \rangle_{\mathbb{R}^2} \mathbf{e}_\perp$, and so infer

$$0 = \langle D\phi^{**}(\mathbf{m}_2) - D\phi^{**}(\mathbf{m}_1), \boldsymbol{\delta} \rangle_{\mathbb{R}^2} = \langle \boldsymbol{\delta}, \mathbf{e}_\perp \rangle^2_{\mathbb{R}^2} \, . \tag{2.8}$$

On the other hand, $\mathbf{e} = 0$ and (2.6) imply that $\boldsymbol{\delta}$ is divergence-free in the sense of distributions. This means $\operatorname{div} \boldsymbol{\delta} = 0$ almost everywhere in $\omega$ but also $\langle \boldsymbol{\delta}, \mathbf{n} \rangle_{\mathbb{R}^2} = 0$ on the boundary $\partial\omega$, with $\mathbf{n}$ the unit normal vector (in a weak sense according to the formula of integration by parts). Let us extend $\boldsymbol{\delta}$ by zero outside $\omega$. Since normal components are continuous on the boundary $\partial\omega$, the extended function $\boldsymbol{\delta}$ belongs to $\mathbf{H}(\mathrm{div}; \mathbb{R}^2)$ and is divergence-free. Hence, $\boldsymbol{\delta} = \mathbf{curl}\, \eta := (\eta_{,2}, -\eta_{,1})^\top$ for some $\eta \in W^{1,2}(\omega)$, cf. e.g., [58, Theorem 3.1 on page 37]. Outside a ball that includes $\omega$, $\boldsymbol{\delta} = 0$ and so $\eta$ is constant there. Without loss of generality, $\eta = 0$ on the connectivity component $\gamma_0$ of the boundary $\partial\omega$ that includes $\omega$.

Because of (2.8), $\nabla\eta$ is parallel to $\mathbf{e}$, i.e., the directional derivative of $\eta$ in the direction $\mathbf{e}_\perp$ vanishes almost everywhere in $\omega$. Since $\eta = 0$ on $\gamma$ and is constant along almost all lines in parallel to $\mathbf{e}$ inside $\omega$ (and according to the extension of $\boldsymbol{\delta}$, $\eta$ is constant in bounded components of $\mathbb{R}^2 \setminus \omega$) we deduce $\eta = 0$. This shows $\boldsymbol{\delta} = 0$, and even $\lambda_2 = \lambda_1$ by (2.4).

In the cubic case, the convexification $\phi^{**}$ of $\phi$ is zero for $\mathbf{m}$ equal to $(\pm 1, 0)^\top$ and $(0, \pm 1)^\top$. Consequently, $\phi^{**}$ vanishes on their convex hull and so, in particular, on the ball $B_{1/2}(0)$. Given $\mathbf{f} = 0$, one solution with minimal zero energy is $\mathbf{m} = 0$, $u = 0$, $\lambda = 0$. However, for any smooth $\eta$ with compact support in $\omega$ which is small (by scaling with a small factor), i.e., $|\nabla\eta| \leq 1/2$ almost everywhere in $\omega$ the functions $\mathbf{m} = \mathbf{curl}\, \eta$, $u = 0$, and $\lambda = 0$ solve (P). This shows that there are infinitely many solutions in the cubic case. □

The remainder of this section is organized as follows: we discuss the lack of stability of a conforming discretization of (CP) in the form (2.3)-(2.5) in Subsection 2.1.1 and present a stable non-conforming finite element discretization. The side-constraint $|\mathbf{m}| \leq 1$, a.e. in $\omega$, is enforced by penalization and the role of the penalization parameter $\varepsilon = \varepsilon(h)$ is clarified. In Subsection 2.1.3, we present a posteriori error estimators for both, the (non-stable) conforming and (stable) non-conforming discretization scheme, and study their efficiency vs. reliability. Computational experiments that support the theory as well as scientific computations are presented in Subsection 2.1.5.

### 2.1.1 Conforming and non-conforming discretization and penalization

In the first part of this subsection, we consider the lowest order conforming finite element method and show that it is *not* feasible. This favors the use of non-conforming finite element schemes for which we prove stability and optimal a priori error estimates.

The number of degrees of freedom $N = \dim(\mathcal{S}) + 2\dim(\mathcal{L}^0(\mathcal{T}|_\omega)$ serves as a reference to the spatial discretization $\mathcal{T}$, where $\mathcal{S} = \mathcal{S}_0^1(\mathcal{T})$ or $\mathcal{S}_0^{1,NC}(\mathcal{T})$.

The **Discrete Problem** $(\mathbf{P}_N)$:=$(\mathbf{P}|\mathcal{S}_0^1(\mathcal{T}) \times \mathcal{L}^0(\mathcal{T}|_\omega)^2 \times \mathcal{L}^0(\mathcal{T}|_\omega))$ for the conforming finite element method reads as follows: *Seek* $\{u_h, \mathbf{m}_h, \lambda_h\}$ *in* $\mathcal{S}_0^1(\mathcal{T})\times \mathcal{L}^0(\mathcal{T}|_\omega)^2\times \mathcal{L}^0(\mathcal{T}|_\omega)$ *satisfying*

$$(\nabla u_h, \nabla w_h)_\Omega = (\mathbf{m}_h, \nabla w_h) \quad \forall\, w_h \in \mathcal{S}_0^1(\mathcal{T})\,, \tag{2.9}$$

$$\nabla u_h + D\phi^{**}(\mathbf{m}_h) + \lambda_h \mathbf{m}_h = \mathbf{f}_\mathcal{T} \quad \text{a.e. in } \omega\,, \tag{2.10}$$

$$0 \leq \lambda_h\,,\ |\mathbf{m}_h| \leq 1\,,\ \text{and } \lambda_h(1-|\mathbf{m}_h|)_+ = 0 \ \text{ a.e. in } \omega\,. \tag{2.11}$$

To describe the non-conforming finite element method, we define the $\mathcal{T}$-piecewise gradient $\nabla_\mathcal{T}$ by $\nabla_\mathcal{T} U(\mathbf{x}) := \nabla U|_K(\mathbf{x})$, for $\mathbf{x} \in K \in \mathcal{T}$. Therefore, the discrete energy space is $W^{k,2}(\mathcal{T})$,

$$W^{k,2}(\mathcal{T}) := W^{k,2}(\bigcup_{K\in\mathcal{T}} \operatorname{int}(K)) := \Big\{V \in L^2(\Omega) : \forall\, K \in \mathcal{T},\ V|_K \in W^{k,2}(K)\Big\}\,.$$

The **Discrete Problem** $(\mathbf{P}_N^{NC})$:=$(\mathbf{P}|\mathcal{S}_0^{1,NC}(\mathcal{T})\times\mathcal{L}^0(\mathcal{T}|_\omega)^2\times\mathcal{L}^0(\mathcal{T}|_\omega))$ for the non-conforming finite element method reads as follows: *Seek* $\{u_h, \mathbf{m}_h, \lambda_h\}$

*in* $\mathcal{S}_0^{1,NC}(\mathcal{T})\times\,\mathcal{L}^0(\mathcal{T}|_\omega)^2\times\,\mathcal{L}^0(\mathcal{T}|_\omega)$ *satisfying*

$$(\nabla_\mathcal{T} u_h, \nabla_\mathcal{T} w_h)_\Omega = (\mathbf{m}_h, \nabla_\mathcal{T} w_h) \quad \forall\, w_h \in \mathcal{S}_0^{1,NC}(\mathcal{T})\,, \tag{2.12}$$

$$\nabla_\mathcal{T} u_h + D\phi^{**}(\mathbf{m}_h) + \lambda_h \mathbf{m}_h = \mathbf{f}_\mathcal{T} \quad \text{a.e. in } \omega\,, \tag{2.13}$$

$$0 \le \lambda_h\,,\ |\mathbf{m}_h| \le 1\,,\ \text{and } \lambda_h(1-|\mathbf{m}_h|)_+ = 0 \quad \text{a.e. in } \omega\,. \tag{2.14}$$

For a positive $\varepsilon$ and the conforming resp. non-conforming discrete space $\mathcal{S}$, the **Discrete Penalized Problem** $(\mathbf{P}_\varepsilon|\mathcal{S}\times\mathcal{L}^0(\mathcal{T}|_\omega)^2\times\mathcal{L}^0(\mathcal{T}|_\omega))$, shortly $(\mathbf{P}_{\varepsilon,N})$ for $\mathcal{S}=\mathcal{S}_0^1(\mathcal{T})$ resp. $(\mathbf{P}_{\varepsilon,N}^{NC})$ for $\mathcal{S}=\mathcal{S}_0^{1,NC}(\mathcal{T})$ reads as follows: *Seek* $\{u_h, \mathbf{m}_h, \lambda_h\}$ *in* $\mathcal{S}\times\,\mathcal{L}^0(\mathcal{T}|_\omega)^2\times\mathcal{L}^0(\mathcal{T}|_\omega)$ *satisfying*

$$(\nabla_\mathcal{T} u_h, \nabla_\mathcal{T} w_h)_\Omega = (\mathbf{m}_h, \nabla_\mathcal{T} w_h) \quad \forall\, w_h \in \mathcal{S}\,, \tag{2.15}$$

$$\nabla_\mathcal{T} u_h + D\phi^{**}(\mathbf{m}_h) + \lambda_h \mathbf{m}_h = \mathbf{f}_\mathcal{T} \quad \text{a.e. in } \omega\,, \tag{2.16}$$

$$\lambda_h = \varepsilon^{-1}(|\mathbf{m}_h|-1)_+/|\mathbf{m}_h| \quad \text{a.e. in } \omega\,. \tag{2.17}$$

Here, $(1-|\mathbf{m}_h|)_+/|\mathbf{m}_h|$ is understood to vanish if $\mathbf{m}_h = 0$ (and $\nabla_\mathcal{T}$ could be replaced by $\nabla$ in the conforming case $\mathcal{S}=\mathcal{S}_0^1(\mathcal{T})$).

The existence of discrete solutions follows as in the continuous case from the variational problem. The following example illustrates that $(\mathrm{P}_N)$ does allow for multiple solutions for the uniaxial case.

**Example 2.1** Let $\mathbf{m}_h|_{K_j} = (-1)^j(1,1)/\sqrt{2}$, $\omega =\ \mathrm{int}(K_1\cup K_2) = (1/3,2/3)^2 \subset \Omega = (0,1)^2$ and notice by direct calculation that $\int_\omega \langle \mathbf{m}_h, \nabla\varphi_\mathbf{z}\rangle_{\mathbb{R}^2}\,d\mathbf{x} = 0$ for each hat function $\varphi_\mathbf{z}$ (the nodal basis function for conforming $\mathcal{T}$-piecewise affine finite elements, with $\varphi_\mathbf{z}(\mathbf{z}) = 1$). In the uniaxial case with $\mathbf{e} = (1,1)^\top/\sqrt{2}$ we have $\phi^{**}(\mathbf{m}_h) = 0$. Hence, $t\,\mathbf{m}_h$ and $u_h = 0$ satisfy the discrete version of (2.9)-(2.11) for $\mathbf{f}=0$ and any $t\in\mathbb{R}$ with $|t|\le 1$. A discrete Helmholtz decomposition (cf. [6]) shows why: $\mathbf{m}_h$ is the piecewise curl of a non-conforming hat function $\psi_\mathbf{z}$ ($\psi_\mathbf{z} = 1$ on $K_1\cap K_2$ and zero at midpoints of other edges of $\partial\omega$) and is parallel to $\mathbf{e}$. One remedy in this particular situation is to change the mesh by taking the other diagonals in Figure 2.1 or, equivalently, transform the situation to $\mathbf{e} = (1,-1)^\top/\sqrt{2}$.

In contrast, solutions to $(\mathrm{P}_N^{NC})$ are unique in the uniaxial case.

**Theorem 2.2** *In the uniaxial case, where* $\phi^{**}(\mathbf{m}) = \frac{1}{2}\langle \mathbf{m}, \mathbf{e}_\perp\rangle_{\mathbb{R}^2}^2$ *for some unit vector* $\mathbf{e}\in\mathbb{R}^2$ *and its normal* $\mathbf{e}_\perp$, $(P_N^{NC})$ *and* $(P_{\varepsilon,N}^{NC})$ *have unique solutions.*

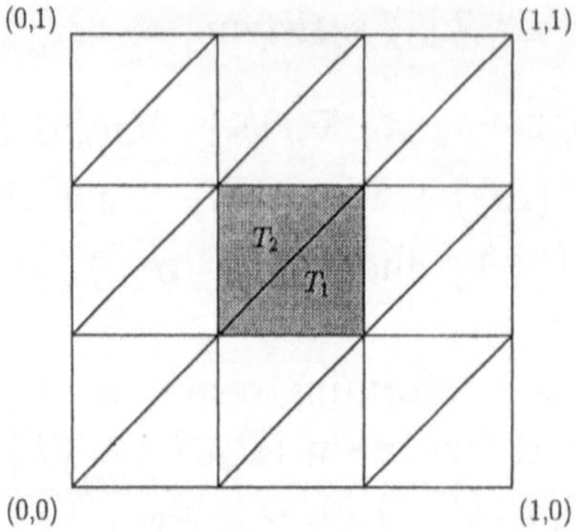


Figure 2.1: A coarse grid where a non-zero magnetization on $K_1 \cup K_2$ provides a solution to the homogeneous discrete problem.

**Proof:**
Suppose $\{u_j, \mathbf{m}_j, \lambda_j\}$ solves $(\mathrm{P}_N^{NC})$ for $j = 1,2$ and denote $e := u_2 - u_1 \in \mathcal{S}_0^{1,NC}(\mathcal{T})$ and $\boldsymbol{\delta} := \mathbf{m}_2 - \mathbf{m}_1 \in \mathcal{L}^0(\mathcal{T}|_\omega)^2$ (we suppress the lower index $h$ in this proof for simplicity). According to (2.15), we have

$$(\nabla_\mathcal{T} e, \boldsymbol{\delta}) = ||\nabla_\mathcal{T} e||^2_{\mathbf{L}^2(\Omega,\mathbb{R}^2)} \,. \tag{2.18}$$

Subtracting the two equations in (2.16) for $j = 1,2$ and multiplying the results with $\boldsymbol{\delta}$ we infer with (2.18) that

$$\begin{aligned} \| \nabla_\mathcal{T} e \|^2_{\mathbf{L}^2(\Omega,\mathbb{R}^2)} + \big(D\phi^{**}(\mathbf{m}_2) - D\phi^{**}(\mathbf{m}_1), \boldsymbol{\delta}\big) \\ + (\lambda_2\, \mathbf{m}_2 - \lambda_1\, \mathbf{m}_1, \boldsymbol{\delta}) = 0\,. \end{aligned} \tag{2.19}$$

All the terms in (2.19) are non-negative and so will vanish separately. This follows for the second term in (2.19) from the convexity of $\phi^{**}$ (i.e., the monotonicity of $D\phi^{**}$) as we have $0 \le \langle D\phi^{**}(\mathbf{b}) - D\phi^{**}(\mathbf{a}), \mathbf{b} - \mathbf{a}\rangle_{\mathbb{R}^2}$, for all $\mathbf{a}, \mathbf{b} \in \mathbb{R}^2$. This is also true for the last term according to (2.17) and the elementary inequality

$$0 \le \left\langle (|\,\mathbf{b}\,| - 1)_+ \frac{\mathbf{b}}{|\,\mathbf{b}\,|} - (|\,\mathbf{a}\,| - 1)_+ \frac{\mathbf{a}}{|\,\mathbf{a}\,|}, \mathbf{b} - \mathbf{a} \right\rangle_{\mathbb{R}^2}, \tag{2.20}$$

for all $\mathbf{a}, \mathbf{b} \in \mathbb{R}^2$. (To prove (2.20), we may assume in the first case $|a| \le 1 < |b|$, where (2.20) is obvious, and in the remaining second case $1 < |a| \le |b|$, where (2.20) follows from a straightforward calculation that shows that

$$(|\mathbf{b}| - 1)(|\mathbf{b}| - |\mathbf{a}|) - (|\mathbf{a}| - 1)(|\mathbf{b}| - |\mathbf{a}|) \geq 0$$

is a lower bound of the right-hand side of (2.20).) Hence, (2.19) implies $\mathbf{e} = 0$ and $\langle \boldsymbol{\delta}, \mathbf{e}_\perp \rangle_{\mathbb{R}^2} = 0$. A discrete Helmholtz decomposition of $\boldsymbol{\delta} \in \mathbf{L}^2(\Omega, \mathbb{R}^2)$ (extended by zero outside of $\omega$) yields

$$\boldsymbol{\delta} = \nabla_{\mathcal{T}} \alpha_h + \mathbf{curl}\, \beta_h \qquad \text{a.e. in } \Omega \tag{2.21}$$

for some $\alpha_h \in \mathcal{S}_0^{1,NC}(\mathcal{T})$ and $\beta_h \in \mathcal{S}^1(\mathcal{T})/\mathbb{R}$ [6]. From (2.15) and $\mathbf{e} = 0$ (with $w_h = \alpha_h$) we deduce $\alpha_h = 0$ according to the $L^2$-orthogonality of $\nabla_{\mathcal{T}} \alpha_h$ and $\mathbf{curl}\, \beta_h$. With $\boldsymbol{\delta} = \mathbf{curl}\, \beta_h$ parallel to $\mathbf{e}$ we conclude that $\partial \beta_h / \partial \mathbf{e}_\perp = 0$ almost everywhere in $\Omega$. Note that $\boldsymbol{\delta} = 0$ and so $\nabla \beta_h = 0$ on $\Omega \setminus \omega$. Hence, $\beta_h$ is constant on the connected open set $\Omega \setminus \overline{\omega}$, without loss of generality, $\beta_h = 0$ on $\Omega \setminus \overline{\omega}$. Integrating along lines parallel to $\mathbf{e}_\perp$ we deduce that $\beta_h = 0$ almost everywhere on $\Omega$ and so $\boldsymbol{\delta} = 0$. The proof is finished for $(\mathrm{P}_{\varepsilon,N}^{NC})$.

The proof of uniqueness for solutions of $(\mathrm{P}_N^{NC})$ is analogous except that the non-negativeness of the last term in (2.19) is verified with the monotone relation in (2.14). An elementary analysis with $\lambda_1 \leq \lambda_2$ reveals $0 \leq \langle \lambda_2\, \mathbf{m}_2 - \lambda_1\, \mathbf{m}_1, \mathbf{m}_2 - \mathbf{m}_1 \rangle_{\mathbb{R}^2}$ pointwise almost everywhere. The remaining details are omitted. □

This subsection is concluded with an example which shows that the magnetization is *not* necessarily convergent in $\mathbf{L}^2(\omega, \mathbb{R}^2)$, in general. As a consequence, we must not expect to prove error estimates for $\mathbf{m} - \mathbf{m}_h$ in the $\mathbf{L}^2$-norm and have to analyze which quantities can be estimated in the following sections. In Section 2.3, we introduce another scheme which has this property.

**Example 2.2** *Suppose $\mathcal{T}|_\omega = \{K_1, K_2, \ldots, K_{2J}\}$ is a structured triangulation of $\omega$ which consists of halved squares $\omega_j = K_{2j-1} \cup K_{2j}$, $j = 1, \ldots, J$, as in Example 2.1 with a diagonal parallel to the easy axis $\mathbf{e} = (1,1)^\top/\sqrt{2}$ in the uniaxial case (cf. Figure 2.1). Suppose $\{u, \mathbf{m}, \lambda\}$ solves (2.3)-(2.5) while $\{u_h, \mathbf{m}_h, \lambda_h\}$ is a solution to $(P_N)$ or $(P_{\varepsilon,N})$. Then,*

$$\sum_{j=1}^{J} \min\Big\{ \|\, 1 - |\mathbf{m}|\, \|_{L^2(K_{2j-1})}^2, \|\, 1 - |\mathbf{m}|\, \|_{L^2(K_{2j})}^2 \Big\} \tag{2.22}$$

$$\leq 4 \max_{\{u_h, \tilde{\mathbf{m}}_h, \lambda_h\}} \|\, \mathbf{m} - \tilde{\mathbf{m}}_h\, \|_{\mathbf{L}^2(\omega, \mathbb{R}^2)}^2,$$

*where* $\{u_h, \tilde{\mathbf{m}}_h, \lambda_h\}$ *denotes an arbitrary solution to* $(P_N)$ *or* $(P_{\varepsilon,N})$. *Note that the left-hand side of* (2.22) *is uniformly positive (for a mesh-size tending to zero) if, for instance, the set* $\{\mathbf{x} \in \omega : |\mathbf{m}(\mathbf{x})| < 1 - \delta\}$ *has no interior point for all* $0 < \delta < 1$.

*Proof of statement* (2.22). *Let* $\mu_j := \max\{|\mathbf{m}_h|_{K_{2j-1}}, |\mathbf{m}_h|_{K_{2j}}\}$ *and suppose without loss of generality that* $\mu_j = |\mathbf{m}_h|_{K_{2j}}$ *for all* $j = 1, \ldots, J$. *For* $\mathbf{m}_h \in \mathcal{L}^0(\mathcal{T})$ *and* $\mathbf{M}_h|_{K_{2j-2+k}} := (1-\mu_j)_+ (-1)^k (1,1)^\top/\sqrt{2}$ *for* $j = 1, \ldots, J$, $k = 1, 2$. *Arguing as in Example 2.1, we observe that* $\{u_h, \mathbf{m}_h \pm \mathbf{M}_h, \lambda_h\}$ *is also a discrete solution. A triangle inequality shows*

$$\begin{aligned} \| \mathbf{M}_h \|_{\mathbf{L}^2(\omega,\mathbb{R}^2)} &\le \frac{1}{2} \| \mathbf{m} - \mathbf{m}_h + \mathbf{M}_h \|_{\mathbf{L}^2(\omega,\mathbb{R}^2)} \\ &+ \frac{1}{2} \| \mathbf{m} - \mathbf{m}_h - \mathbf{M}_h \|_{\mathbf{L}^2(\omega,\mathbb{R}^2)} \le M, \end{aligned} \tag{2.23}$$

*where* $M$ *denotes the right-hand side of* (2.22). *Young's inequality is applied to* $0 \le 1 - |\mathbf{m}| \le 1 - |\mathbf{m}_h| + |\mathbf{m} - \mathbf{m}_h|$ *and shows*

$$\begin{aligned} \frac{1}{2} \sum_{j=1}^{J} \| 1 - |\mathbf{m}| \|^2_{L^2(K_{2j})} &\le \sum_{j=1}^{J} \| 1 - |\mathbf{m}_h| \|^2_{L^2(K_{2j})} + \sum_{j=1}^{J} \| \mathbf{m} - \mathbf{m}_h \|^2_{L^2(K_{2j})} \\ &\le M^2 + \sum_{j=1}^{J} \| \mathbf{M}_h \|^2_{L^2(K_{2j})} \le 2 M^2, \end{aligned} \tag{2.24}$$

*because of the definition of* $\mathbf{M}_h$ *and* (2.23).

**Remark 2.2** *This understanding of instable and stable numerical schemes for problem* $(M_0)$ *might lead to some rigorous insight in the resonance of instable numerical schemes with oscillating microstructures for problem* $(M_\alpha)$.

## 2.1.2 A priori error estimates

To describe the a priori error estimates in the conforming discrete model $(\mathrm{P}_{\varepsilon,N})$, let $(\cdot)_\mathcal{T}$ denote the $\mathcal{T}$-piecewise integral means such as $\mathbf{m}_\mathcal{T} \in \mathcal{L}^0(\mathcal{T}|_\omega)^2$ given by

$$\mathbf{m}_\mathcal{T}|_K = \frac{1}{|K|} \int_K \mathbf{m}\, d\mathbf{x} \qquad \forall K \in \mathcal{T}|_\omega \tag{2.25}$$

and let $P_{h,W^{1,2}(\Omega)} : W_0^{1,2}(\Omega) \to \mathcal{S}_0^1(\mathcal{T})$ denote the Galerkin projector.

**Theorem 2.3** *Suppose $\{u, \mathbf{m}, \lambda\}$ solves (P) and $\{u_h, \mathbf{m}_h, \lambda_h\}$ solves $(P_{\varepsilon,N})$. Then we have*

$$\begin{aligned}
&\frac{1}{2}\| \nabla(u-u_h) \|^2_{\mathbf{L}^2(\Omega,\mathbb{R}^2)} + \frac{1}{2}\int_\omega \varepsilon\lambda_h^2|\mathbf{m}_h|^2\,d\mathbf{x} \\
&\quad + \big(D\phi^{**}(\mathbf{m}) - D\phi^{**}(\mathbf{m}_h), \mathbf{m}-\mathbf{m}_h\big) \\
&\le \frac{1}{2}\int_\omega \varepsilon\lambda^2\,d\mathbf{x} + \frac{1}{2}\| \mathbf{f}-\mathbf{f}_{\mathcal{T}} \|^2_{\mathbf{L}^2(\omega,\mathbb{R}^2)} + \frac{1}{2}\| \mathbf{m}-\mathbf{m}_{\mathcal{T}} \|^2_{\mathbf{L}^2(\omega,\mathbb{R}^2)} \\
&\quad + \| \nabla(u - P_{h,W^{1,2}(\Omega)}u) \|^2_{\mathbf{L}^2(\Omega,\mathbb{R}^2)} \\
&\quad + \| \mathbf{m}-\mathbf{m}_h \|_{\mathbf{L}^2(\omega,\mathbb{R}^2)} \| \nabla(u - P_{h,W^{1,2}(\Omega)}u) \|_{\mathbf{L}^2(\Omega,\mathbb{R}^2)}.
\end{aligned}$$

To establish convergence of the discrete function $\lambda_h\mathbf{m}_h$ towards $\lambda\mathbf{m}$, we suppose that $\phi^{**}$ satisfies for all $\mathbf{m}_1, \mathbf{m}_2 \in \mathbb{R}^2$,

$$\begin{aligned}
&\tilde{c}\,|D\phi^{**}(\mathbf{m}_2) - D\phi^{**}(\mathbf{m}_1)|^2 \\
&\qquad \le \Big\langle D\phi^{**}(\mathbf{m}_2) - D\phi^{**}(\mathbf{m}_1), \mathbf{m}_2-\mathbf{m}_1\Big\rangle_{\mathbb{R}^2},
\end{aligned} \tag{2.26}$$

for some positive constant $\tilde{c}$. Note that this covers the uniaxial case with $\tilde{c}=1$. Let $(\lambda\mathbf{m})_{\mathcal{T}} \in \mathcal{L}^0(\mathcal{T})$ denote the $\mathcal{T}$-piecewise integral mean of $\lambda\mathbf{m}$.

**Theorem 2.4** *Suppose that $\phi^{**}$ satisfies (2.26) and that $\{u, \mathbf{m}, \lambda\}$ solves (P) and $\{u_h, \mathbf{m}_h, \lambda_h\}$ solves $(P_{\varepsilon,N})$. There exists an $h_{\mathcal{T}}$-independent constant $c>0$ such that*

$$\begin{aligned}
&\| \nabla(u-u_h) \|^2_{\mathbf{L}^2(\Omega,\mathbb{R}^2)} + \| D\phi^{**}(\mathbf{m}) - D\phi^{**}(\mathbf{m}_h) \|^2_{\mathbf{L}^2(\omega,\mathbb{R}^2)} \\
&\quad + \| \lambda\mathbf{m} - \lambda_h\mathbf{m}_h \|^2_{\mathbf{L}^2(\omega,\mathbb{R}^2)} \\
&\le c\Big\{ \| \varepsilon\lambda \|^2_{L^2(\omega)} + \| \mathbf{f}-\mathbf{f}_{\mathcal{T}} \|^2_{\mathbf{L}^2(\omega,\mathbb{R}^2)} + \| \mathbf{m}-\mathbf{m}_{\mathcal{T}} \|^2_{\mathbf{L}^2(\omega,\mathbb{R}^2)} \\
&\quad + \| \lambda\mathbf{m} - (\lambda\mathbf{m})_{\mathcal{T}} \|^2_{\mathbf{L}^2(\omega,\mathbb{R}^2)} + \| \nabla(u - P_{h,W^{1,2}(\Omega)}u) \|^2_{\mathbf{L}^2(\Omega,\mathbb{R}^2)} \\
&\quad + \| \mathbf{m}-\mathbf{m}_h \|_{\mathbf{L}^2(\omega,\mathbb{R}^2)} \| \nabla(u - P_{h;W^{1,2}(\Omega)}u) \|_{\mathbf{L}^2(\Omega,\mathbb{R}^2)} \Big\}.
\end{aligned}$$

**Remark 2.3** *1. Since $\phi^{**}$ is convex, $0 \le \langle D\phi^{**}(\mathbf{m}) - D\phi^{**}(\mathbf{M}), \mathbf{m}-\mathbf{M}\rangle_{\mathbb{R}^2}$ for all $\mathbf{m}, \mathbf{M} \in \mathbb{R}^2$, and so $\langle D\phi^{**}(\mathbf{m}) - D\phi^{**}(\mathbf{m}_h), \mathbf{m}-\mathbf{m}_h\rangle_{\mathbb{R}^2}$ is non-negative at almost every $\mathbf{x} \in \omega$. Hence, the lower bound in (2.30) consists of non-negative summands. For the uniaxial case where $\phi^{**}(\mathbf{m}) = \frac{1}{2}\langle \mathbf{m}, \mathbf{e}_\perp\rangle^2_{\mathbb{R}^2}$,*

$$\Big(D\phi^{**}(\mathbf{m}) - D\phi^{**}(\mathbf{m}_h), \mathbf{m}-\mathbf{m}_h\Big) = \| \langle \mathbf{m}-\mathbf{m}_h, \mathbf{e}_\perp\rangle_{\mathbb{R}^2} \|^2_{L^2(\omega)} \tag{2.27}$$

*all the theorems in this subsection yield estimates for the $L^2(\omega)$-norm of the $\mathbf{e}_\perp$-component of the error in the magnetization.*

*2. In case of multiple exact or discrete solutions, any selection of a pair of them is considered in the theorems.*

*3. Under sufficient regularity of the exact solution, the above theorems provides the estimate*

$$\| \nabla(u - u_h) \|_{\mathbf{L}^2(\Omega,\mathbb{R}^2)} + \| \langle \mathbf{m} - \mathbf{m}_h, \mathbf{e}_\perp \rangle_{\mathbb{R}^2} \|_{L^2(\omega)} \leq C \{\varepsilon + \sqrt{h}\} \tag{2.28}$$

*in the uniaxial case and the conforming scheme.*

*4. The last term in the inequality in Theorem 2.3 is the limiting quantity in the upper bound and causes the result to be suboptimal, in general. In fact, the term $\| \mathbf{m} - \mathbf{m}_h \|_{\mathbf{L}^2(\omega,\mathbb{R}^2)}$ can only be controlled by the Lebesgue measure of the domain $\omega$, plus higher order terms (cf. Example 2.2).*

*5. The generic convergence order in (2.28) suggests the choice $\varepsilon = \mathcal{O}(\sqrt{h})$. There is numerical evidence in the examples of Section 2.1.5 for the choice $\varepsilon = \mathcal{O}(h)$.*

To describe our result for the non-conforming scheme, let $I_\mathcal{T} u \in \mathcal{S}_0^{1,NC}(\mathcal{T})$ denote the interpolation for the Crouzeix-Raviart finite element defined by

$$I_\mathcal{T} u(\mathbf{z}) = \frac{1}{h_\xi} \int_\xi u \,\mathrm{ds} \qquad \forall \mathbf{z} \in \boldsymbol{\xi} \cap \mathcal{M}\,,\ \boldsymbol{\xi} \in \Xi\,, \tag{2.29}$$

for any $u \in W_0^{1,2}(\Omega)$. The weight $h_\mathcal{T} \in \mathcal{L}^0(\mathcal{T})$ represents the local mesh-size, $h_\mathcal{T}|_K = h_K := \mathrm{diam}(K)$ for $K \in \mathcal{T}$.

**Theorem 2.5** *Suppose $\{u, \mathbf{m}, \lambda\}$ solves (P) and $\{u_h, \mathbf{m}_h, \lambda_h\}$ solves $(P_{\varepsilon,N}^{NC})$. There holds*

$$\begin{aligned} &\frac{1}{2} \| \nabla_\mathcal{T}(u - u_h) \|^2_{\mathbf{L}^2(\Omega,\mathbb{R}^2)} + \frac{1}{2} \int_\omega \varepsilon \lambda_h^2 |\mathbf{m}_h|^2 \,\mathrm{d}\mathbf{x} \\ &\quad + \big(D\phi^{**}(\mathbf{m}) - D\phi^{**}(\mathbf{m}_h), \mathbf{m} - \mathbf{m}_h\big) \\ &\leq \frac{1}{2} \int_\omega \varepsilon \lambda^2 \,\mathrm{d}\mathbf{x} + \frac{1}{2} \| \mathbf{f} - \mathbf{f}_\mathcal{T} \|^2_{\mathbf{L}^2(\omega,\mathbb{R}^2)} + \| \mathbf{m} - \mathbf{m}_\mathcal{T} \|^2_{\mathbf{L}^2(\omega,\mathbb{R}^2)} \\ &\quad + \frac{3}{2} \| \nabla_\mathcal{T}(u - I_\mathcal{T} u) \|^2_{\mathbf{L}^2(\Omega,\mathbb{R}^2)} + c \| h_\mathcal{T} \nabla_\mathcal{T}(\mathbf{m} - \nabla u) \|^2_{\mathbf{L}^2(\Omega,\mathbb{R}^4)} \,. \end{aligned} \tag{2.30}$$

*The constant $c > 0$ depends only on the shape on the elements but neither on their sizes nor on $\mathbf{f}, \mathbf{m}, u, \lambda, \Omega, \omega, \mathbf{m}_h, u_h,$ or $\lambda_h$.*

**Theorem 2.6** *Suppose that $\phi^{**}$ satisfies (2.26) and that $(u, \mathbf{m}, \lambda)$ solves (P) and $(u_h, \mathbf{m}_h, \lambda_h)$ solves $(P^{NC}_{\varepsilon,N})$. There holds*

$$\begin{aligned}
&\| \nabla_{\mathcal{T}}(u - u_h) \|^2_{\mathbf{L}^2(\Omega,\mathbb{R}^2)} + \| D\phi^{**}(\mathbf{m}) - D\phi^{**}(\mathbf{m}_h) \|^2_{\mathbf{L}^2(\omega,\mathbb{R}^2)} \\
&\quad + \| \lambda\mathbf{m} - \lambda_h\mathbf{m}_h \|^2_{\mathbf{L}^2(\omega,\mathbb{R}^2)} \\
&\le c \Big\{ \int_\omega \varepsilon^2\lambda^2 \,d\mathbf{x} + \| \mathbf{f} - \mathbf{f}_{\mathcal{T}} \|^2_{\mathbf{L}^2(\omega,\mathbb{R}^2)} \\
&\quad + \| \mathbf{m} - \mathbf{m}_{\mathcal{T}} \|^2_{\mathbf{L}^2(\omega,\mathbb{R}^2)} + \| \nabla_{\mathcal{T}}(u - I_{\mathcal{T}}u) \|^2_{\mathbf{L}^2(\Omega,\mathbb{R}^2)} \\
&\quad + \| h_{\mathcal{T}}\nabla_{\mathcal{T}}(\mathbf{m} - \nabla_{\mathcal{T}}u) \|^2_{\mathbf{L}^2(\Omega,\mathbb{R}^4)} + \| \lambda\mathbf{m} - (\lambda\mathbf{m})_{\mathcal{T}} \|^2_{\mathbf{L}^2(\omega,\mathbb{R}^2)} \Big\} .
\end{aligned}$$

*The constant $c > 0$ depends on the shape of the elements but neither on their sizes $h_{\mathcal{T}}$ nor on $\mathbf{f}, \mathbf{m}, u, \lambda, \Omega, \omega, \mathbf{m}_h, u_h$, or $\lambda_h$.*

**Remark 2.4** *1. Under sufficient regularity of the exact solution, the above two theorems provide the estimate*

$$\| \nabla_{\mathcal{T}}(u - u_h) \|_{\mathbf{L}^2(\Omega,\mathbb{R}^2)} + \| \langle \mathbf{m} - \mathbf{m}_h, \mathbf{e}_\perp \rangle_{\mathbb{R}^2} \|_{L^2(\omega)} \le C \{\varepsilon + h\} \qquad (2.31)$$

*in the uniaxial case and the non-conforming scheme.*

*2. The generic convergence order $\mathcal{O}(\varepsilon + h)$ suggests the choice $\varepsilon = \mathcal{O}(h)$ in agreement with numerical experience reported in Subsection 2.1.5 below.*

*3. The proof of Theorem 2.5 reveals that $\int_\omega \varepsilon\lambda^2 \,d\mathbf{x}$ on the right-hand side of (2.30) could be replaced by the smaller contribution $\int_{\omega_h} \varepsilon\lambda^2 \,d\mathbf{x}$ on the smaller domain $\omega_h := \{\mathbf{x} \in \omega : \lambda_h(\mathbf{x}) > 0\}$. This observation will affect the a posteriori error estimates of Subsection 2.1.3.*

The remaining part of this subsection is devoted to the proofs the four theorems of this section. We start with the verification of Theorem 2.5.

**Proof:**

In the first part of the proof, the convexity of the indicator function $\psi$ and the penalization in (2.17) yield the estimate

$$\frac{\varepsilon}{2}|\mathbf{m}_h|^2\lambda_h^2 \le \langle \lambda\mathbf{m} - \lambda_h\mathbf{m}_h, \mathbf{m} - \mathbf{m}_h \rangle_{\mathbb{R}^2} + \frac{\varepsilon}{2}\lambda^2 . \qquad (2.32)$$

Indeed, direct calculations which merely involve $|\mathbf{m}| \le 1$ and Cauchy's inequality reveal

$$\begin{aligned}
-\lambda\langle \mathbf{m}, \mathbf{m} - \mathbf{m}_h \rangle_{\mathbb{R}^2} &\le \lambda(|\mathbf{m}_h| - 1)_+ = \varepsilon\lambda\lambda_h|\mathbf{m}_h| , && (2.33)\\
\varepsilon\lambda_h^2|\mathbf{m}_h|^2 &\le \lambda_h\langle \mathbf{m}_h, \mathbf{m}_h - \mathbf{m} \rangle_{\mathbb{R}^2} . && (2.34)
\end{aligned}$$

Adding (2.33) and (2.34) we obtain (2.32) with Young's inequality, $\lambda\lambda_h\varepsilon|\mathbf{m}_h| \leq \frac{\varepsilon}{2}\lambda^2 + \frac{\varepsilon}{2}|\mathbf{m}_h|^2\lambda_h^2$.

In the second step we provide the identity

$$\big(\mathbf{m}_h - \mathbf{m}_{\mathcal{T}}, \nabla_{\mathcal{T}}(u - I_{\mathcal{T}}u)\big) = 0\,, \tag{2.35}$$

which follows element-wise from (2.29): An integration by parts gives

$$\begin{aligned} \int_K \nabla_{\mathcal{T}}(u - I_{\mathcal{T}}u)\,\mathrm{d}\mathbf{x} &= \int_{\partial K}(u - I_{\mathcal{T}}u)\mathbf{n}\,\mathrm{d}\mathbf{x} \\ &= \sum_{\xi\in\Xi}\mathbf{n}_\xi \int_\xi (u - I_{\mathcal{T}}u)\,\mathrm{d}\mathbf{x} = 0\,, \end{aligned} \tag{2.36}$$

since the normal vector $\mathbf{n}_\xi$ on the edge $\xi \in \Xi$ is constant. Because $\mathbf{m}_h - \mathbf{m}_{\mathcal{T}} \in \mathcal{L}^0(\mathcal{T}|_\omega)^2$, (2.36) proves (2.35).

For the remaining part of the proof, we set $e := u - u_h$, $\boldsymbol{\delta} := \mathbf{m} - \mathbf{m}_h$ and let $w_h := -u_h + I_{\mathcal{T}}u \in \mathcal{S}_0^{1,NC}(\mathcal{T})$. In step three we perform a standard calculation [13] for non-conforming finite elements. The Galerkin-orthogonality for (2.3) and (2.15) leads to

$$(\nabla_{\mathcal{T}}e, \nabla_{\mathcal{T}}w_h)_\Omega - (\boldsymbol{\delta}, \nabla_{\mathcal{T}}w_h) = \int_\Xi \langle\nabla u - \mathbf{m}, \mathbf{n}_\Xi\rangle_{\mathbb{R}^2}\,[w_h]\,\mathrm{d}\mathbf{x}\,, \tag{2.37}$$

where $[w_h]$ denotes the jump of $w_h$ along the edge $\xi \in \Xi$ with normal vector $\mathbf{n}_\xi$, and $\mathbf{m} = 0$ outside of $\omega$.

In step four, we consider the difference of (2.4) and (2.16) and multiply with $\boldsymbol{\delta}$ to obtain finally that

$$\begin{aligned} (\boldsymbol{\delta}, \nabla_{\mathcal{T}}e) &= (\mathbf{m} - \mathbf{m}_{\mathcal{T}}, \mathbf{f} - \mathbf{f}_{\mathcal{T}}) \\ &\quad -(D\phi^{**}(\mathbf{m}) - D\phi^{**}(\mathbf{m}_h), \boldsymbol{\delta}) - (\lambda\mathbf{m} - \lambda_h\mathbf{m}_h, \boldsymbol{\delta})\,. \end{aligned} \tag{2.38}$$

Here, we exploit the result $(\mathbf{m}_h - \mathbf{m}_{\mathcal{T}}, \mathbf{f} - \mathbf{f}_{\mathcal{T}}) = 0$.

In the final step we collect the preceding estimates. Rewriting (2.38) with adding $\|\nabla_{\mathcal{T}}e\|^2_{\mathbf{L}^2(\Omega,\mathbb{R}^2)}$ and then employing (2.37) we obtain with Cauchy's and Young's inequalities

$$\begin{aligned}
&\big(D\phi^{**}(\mathbf{m}) - D\phi^{**}(\mathbf{m}_h), \boldsymbol{\delta}\big) + (\lambda\mathbf{m} - \lambda_h\mathbf{m}_h, \boldsymbol{\delta}) + \| \nabla_{\mathcal{T}} e \|^2_{\mathbf{L}^2(\Omega,\mathbb{R}^2)} \\
&\quad -\frac{1}{2} \| \mathbf{m} - \mathbf{m}_{\mathcal{T}} \|^2_{\mathbf{L}^2(\omega,\mathbb{R}^2)} - \frac{1}{2} \| \mathbf{f} - \mathbf{f}_{\mathcal{T}} \|^2_{\mathbf{L}^2(\omega,\mathbb{R}^2)} \\
&\le (\nabla_{\mathcal{T}} e, \nabla_{\mathcal{T}} e - \boldsymbol{\delta}) \\
&= \int_{\Xi} \langle \nabla u - \mathbf{m}, \mathbf{n}_{\Xi} \rangle_{\mathbb{R}^2} \, [I_{\mathcal{T}} u - u_h] \, \mathrm{d}\mathbf{x} \\
&\quad + \int_{\Omega} \big\langle \nabla_{\mathcal{T}}(u - I_{\mathcal{T}} u), \nabla_{\mathcal{T}} e - \boldsymbol{\delta} \big\rangle_{\mathbb{R}^2} \mathrm{d}\mathbf{x} .
\end{aligned} \tag{2.39}$$

According to (2.35), the last contribution of the right-hand side in (2.39) equals

$$\begin{aligned}
&\big(\nabla_{\mathcal{T}}(u - I_{\mathcal{T}} u), \nabla_{\mathcal{T}} e\big)_{\Omega} - \big(\nabla_{\mathcal{T}}(u - I_{\mathcal{T}} u), \mathbf{m} - \mathbf{m}_{\mathcal{T}}\big) \\
&\le \| \nabla_{\mathcal{T}} e \|_{\mathbf{L}^2(\Omega,\mathbb{R}^2)} \| u - I_{\mathcal{T}} u \|_{\mathbf{L}^2(\Omega)} \\
&\quad + \| u - I_{\mathcal{T}} u \|_{L^2(\omega)} \| \mathbf{m} - \mathbf{m}_{\mathcal{T}} \|_{\mathbf{L}^2(\omega,\mathbb{R}^2)} .
\end{aligned} \tag{2.40}$$

A standard argument for the jumps $[I_{\mathcal{T}} u - u_h]$ with $\Xi$-piecewise integral mean zero shows

$$\begin{aligned}
&\int_{\Xi} \langle \nabla u - \mathbf{m}), \mathbf{n}_{\xi} \rangle_{\mathbb{R}^2} \, [I_{\mathcal{T}} u - u_h] \, \mathrm{d}\mathbf{x} \\
&\qquad \le c \, \| h_{\mathcal{T}} \nabla_{\mathcal{T}} (\nabla u - \mathbf{m}) \|_{\mathbf{L}^2(\Omega,\mathbb{R}^4)} \| \nabla_{\mathcal{T}} e \|_{\mathbf{L}^2(\Omega,\mathbb{R}^2)} ,
\end{aligned} \tag{2.41}$$

with an $h_{\mathcal{T}}$-independent constant $c > 0$ (which only depends on the shapes of the elements) [13]. Using this in (2.39) and owing to (2.32), we finally deduce (2.30) by absorbing the error terms on the right-hand side. □

We continue with the proof of Theorem 2.6.

**Proof:**
Let $(\lambda\mathbf{m})_{\mathcal{T}} \in \mathcal{L}^0(\mathcal{T}|_{\omega})^2$ be defined by $(\lambda\mathbf{m})_{\mathcal{T}}|_K = \frac{1}{|K|} \int_K \lambda\mathbf{m} \, \mathrm{d}\mathbf{x}$ for $K \in \mathcal{T}$. Set

$$\begin{aligned}
A \;&:=\; \| \nabla_{\mathcal{T}}(u - u_h) \|_{\mathbf{L}^2(\Omega,\mathbb{R}^2)} + \| D\phi^{**}(\mathbf{m}) - D\phi^{**}(\mathbf{m}_h) \|_{\mathbf{L}^2(\omega,\mathbb{R}^2)} \\
&\quad + \| \lambda\mathbf{m} - (\lambda\mathbf{m})_{\mathcal{T}} \|_{\mathbf{L}^2(\omega,\mathbb{R}^2)} , \\
B \;&:=\; \| \mathbf{f} - \mathbf{f}_{\mathcal{T}} \|_{\mathbf{L}^2(\omega,\mathbb{R}^2)} + \| \mathbf{m} - \mathbf{m}_{\mathcal{T}} \|_{\mathbf{L}^2(\omega,\mathbb{R}^2)} + \| \lambda\mathbf{m} - (\lambda\mathbf{m})_{\mathcal{T}} \|_{\mathbf{L}^2(\omega,\mathbb{R}^2)} \\
&\quad + \| \nabla_{\mathcal{T}}(u - I_{\mathcal{T}} u) \|_{\mathbf{L}^2(\Omega,\mathbb{R}^2)} + \| h_{\mathcal{T}} \nabla_{\mathcal{T}}(\boldsymbol{\delta} - \nabla_{\mathcal{T}} u) \|_{\mathbf{L}^2(\Omega,\mathbb{R}^2)} .
\end{aligned}$$

Subtract (2.13) from (2.4) and test with the admissible function $(\lambda\mathbf{m})_{\mathcal{T}} - \lambda_h\mathbf{m}_h$ to infer

$$\begin{aligned} &\| \lambda\mathbf{m} - \lambda_h\mathbf{m}_h \|_{\mathbf{L}^2(\omega,\mathbb{R}^2)} \\ &\leq A + \| \mathbf{f} - \mathbf{f}_{\mathcal{T}} \|_{\mathbf{L}^2(\omega,\mathbb{R}^2)} + \| \lambda\mathbf{m} - (\lambda\mathbf{m})_{\mathcal{T}} \|_{\mathbf{L}^2(\omega,\mathbb{R}^2)} \\ &\leq c\Big(B + \Big(\int_\omega \varepsilon\lambda^2\,\mathrm{d}\mathbf{x}\Big)^{1/2}\Big). \end{aligned} \tag{2.42}$$

Because of

$$\begin{aligned} \lambda_h^2|\mathbf{m}_h|^2 - \lambda^2|\mathbf{m}|^2 &= \big(\lambda_h|\mathbf{m}_h| + \lambda|\mathbf{m}|\big)\big(\lambda_h|\mathbf{m}_h| - \lambda|\mathbf{m}|\big) \\ &\leq \big(\lambda_h|\mathbf{m}_h| + \lambda|\mathbf{m}|\big)\big|\lambda_h\mathbf{m}_h - \lambda\mathbf{m}\big| \end{aligned} \tag{2.43}$$

and Cauchy's inequality, we can conclude from (2.42) that

$$\begin{aligned} &\Big|\int_\omega (\lambda_h^2|\mathbf{m}_h|^2 - \lambda^2|\mathbf{m}|^2)\,\mathrm{d}\mathbf{x}\Big| \\ &\leq \Big(2\int_\omega (\lambda_h^2|\mathbf{m}_h|^2 + \lambda^2|\mathbf{m}|^2)\,\mathrm{d}\mathbf{x}\Big)^{1/2}\Big(\int_\omega \big|\lambda_h\mathbf{m}_h - \lambda\mathbf{m}\big|^2\,\mathrm{d}\mathbf{x}\Big)^{1/2} \\ &\leq c\,A\Big(\int_\omega \lambda^2\mathrm{d}\mathbf{x} + B^2\Big)^{1/2}. \end{aligned} \tag{2.44}$$

Multiply (2.44) with $\varepsilon/2$ and recast it into

$$\frac{1}{2}\int_\omega \varepsilon\lambda^2|\mathbf{m}|^2\,\mathrm{d}\mathbf{x} \leq \frac{1}{2}\int_\omega \varepsilon\lambda_h^2|\mathbf{m}_h|^2\,\mathrm{d}\mathbf{x} + cA\Big(\int_\omega \varepsilon^2\lambda^2|\mathbf{m}|^2\,\mathrm{d}\mathbf{x} + \varepsilon^2 B^2\Big)^{1/2}. \tag{2.45}$$

Adding this to (2.30) and employing (2.26) we conclude the proof of Theorem 2.6 by absorbing the first and second contribution in $A$. □

The next proof is to show Theorem 2.3.

**Proof:**

The arguments in the proof of Theorem 2.6 apply to the conforming situation as well and jumps disappear (e.g., in (2.37), (2.39)). From (2.12) we obtain

$$(\nabla e, \nabla e - \boldsymbol{\delta})_\Omega = (\nabla(u - P_1 u), \nabla e - \boldsymbol{\delta})_\Omega\,, \tag{2.46}$$

using the Ritz-Galerkin projection $P_{h,W^{1,2}(\Omega)}$. Note that (2.35) is no longer available and so we end up with

$$(\nabla e, \nabla e - \boldsymbol{\delta})_\Omega \leq \| \nabla(u - P_{h,W^{1,2}(\Omega)}u) \|_{\mathbf{L}^2(\Omega,\mathbb{R}^2)} \times \left( \| \boldsymbol{\delta} \|_{\mathbf{L}^2(\omega,\mathbb{R}^2)} + \| \nabla(u - P_{h,W^{1,2}(\Omega)}u) \|_{\mathbf{L}^2(\Omega,\mathbb{R}^2)} \right). \tag{2.47}$$

□

Finally, we verify Theorem 2.4.

**Proof:**
Following the proof of Theorem 2.3 with the modification

$$\begin{aligned} B := & \| \mathbf{f} - \mathbf{f}_\mathcal{T} \|_{\mathbf{L}^2(\omega,\mathbb{R}^2)} + \| \mathbf{m} - \mathbf{m}_\mathcal{T} \|_{\mathbf{L}^2(\omega,\mathbb{R}^2)} + \| \lambda\mathbf{m} - (\lambda\mathbf{m})_\mathcal{T} \|_{\mathbf{L}^2(\omega,\mathbb{R}^2)} \\ & + \| \nabla(u - P_{h,W^{1,2}(\Omega)}u) \|_{\mathbf{L}^2(\Omega,\mathbb{R}^2)} \| \mathbf{m} - \mathbf{m}_\mathcal{T} \|^{1/2}_{\mathbf{L}^2(\omega,\mathbb{R}^2)} \\ & \times \| \nabla(u - P_{h,W^{1,2}(\Omega)}u) \|^{1/2}_{\mathbf{L}^2(\Omega,\mathbb{R}^2)}, \end{aligned} \tag{2.48}$$

we verify Theorem 2.4. □

### 2.1.3 Reliable or efficient a posteriori error estimates

The a posteriori error estimates differ essentially for conforming and non-conforming schemes. First, our results are stated, then discussed and proved at the end of this subsection. Numerical tests on adaptive algorithms for automatic mesh-refining will be reported on in the subsequent subsection.

The discrete function $\mathbf{m}_h - \nabla u_h$ is $\mathcal{T}$-piecewise constant and its jump across an interior edge $\boldsymbol{\xi} \in \Xi$ with a chosen unit normal vector $\mathbf{n}_\xi$ and length $h_\xi$ is written $\langle [\mathbf{m}_h - \nabla u_h], \mathbf{n}_\xi \rangle_{\mathbb{R}^2}$. We abbreviate the $\Xi$-piecewise constant edge-size and the chosen normal $\mathbf{n}_\xi$ on the skeleton $\Xi$ by $h_\Xi \in L^\infty(\Xi)$ and $\mathbf{n}_\Xi \in \mathbf{L}^\infty(\Xi, \mathbb{R}^2)$ defined by $(h_\Xi)|_\xi := h_\xi := \operatorname{diam}(\boldsymbol{\xi})$ and $(\mathbf{n}_\Xi)|_\xi := \mathbf{n}_\xi$ for $\boldsymbol{\xi} \in \Xi$ in $\Omega$. On the outer boundary we formally set $(\mathbf{n}_\Xi)|_{\partial\Omega} = 0$ such that all terms disappear there. A crucial role plays the sub-domain $\omega_h$ of $\omega$ where $\lambda_h$ is positive,

$$\omega_h = \left\{ \mathbf{x} \in \omega : 0 < \lambda_h(\mathbf{x}) \right\}. \tag{2.49}$$

**Theorem 2.7** *Suppose that $\phi^{**}$ satisfies (2.26), $0 < \varepsilon < \min\{1, c\}$, that $\{u, \mathbf{m}, \lambda\}$ solves (P), and let $\{u_h, \mathbf{m}_h, \lambda_h\}$ solve $(P_{\varepsilon,N})$. Then there exists an $(\varepsilon, h_{\mathcal{T}}, h_{\Xi})$-independent constant $c > 0$ with*

$$\begin{aligned}\| \nabla(u - u_h) \|^2_{\mathbf{L}^2(\Omega,\mathbb{R}^2)} + c \| D\phi^{**}(\mathbf{m}) - D\phi^{**}(\mathbf{m}_h) \|^2_{\mathbf{L}^2(\omega,\mathbb{R}^2)} \\ \leq c \| \varepsilon\lambda_h \mathbf{m}_h \|^2_{\mathbf{L}^2(\omega_h,\mathbb{R}^2)} + (\mathbf{m} - \mathbf{m}_{\mathcal{T}}, \mathbf{f} - \mathbf{f}_{\mathcal{T}}) \\ + \| \mathbf{f} - \mathbf{f}_{\mathcal{T}} \|^2_{\mathbf{L}^2(\omega_h,\mathbb{R}^2)} + c \| h_{\Xi}^{1/2} \langle [\mathbf{m}_h - \nabla u_h], \mathbf{n}_{\Xi} \rangle_{\mathbb{R}^2} \|^2_{L^2(\Xi)} .\end{aligned}$$

*The positive constant c depends on the shape of the elements in $\mathcal{T}$ but neither on their sizes nor on the data $\mathbf{f}$ or solutions $u, \mathbf{m}, \lambda, u_h, \mathbf{m}_h, \lambda_h$.*

The situation is more involved for non-conforming schemes. The non-conformity is controlled by an edge term $[\partial u_h/\partial s]$, where $\partial/\partial s$ denotes the derivative with respect to the arc-length along $\boldsymbol{\xi} \in \Xi$ and $[\partial u_h/\partial s]$ is the jump across $\boldsymbol{\xi}$ of $\partial u_h/\partial s$ from either sides. We regard $[\partial u_h/\partial s]$ as a function on $\Xi$ which is $-\partial u_h/\partial s$ on $\partial\Omega$.

**Theorem 2.8** *Suppose that $\phi^{**}$ satisfies (2.26), $0 < \varepsilon < \min\{1, c\}$, for c positive, and that $\Omega$ is simply connected. Let $\{u, \mathbf{m}, \lambda\}$ solve (P) and let $\{u_h, \mathbf{m}_h, \lambda_h\}$ solve $(P^{NC}_{\varepsilon,N})$. Then there exists an $(\varepsilon, h_{\mathcal{T}}, h_{\Xi})$-independent constant $c > 0$ with*

$$\begin{aligned}\| \nabla_{\mathcal{T}}(u - u_h) \|^2_{\mathbf{L}^2(\Omega,\mathbb{R}^2)} + c \| D\phi^{**}(\mathbf{m}) - D\phi^{**}(\mathbf{m}_h) \|^2_{\mathbf{L}^2(\omega,\mathbb{R}^2)} \\ \leq c \| \varepsilon\lambda_h \mathbf{m}_h \|^2_{\mathbf{L}^2(\omega_h,\mathbb{R}^2)} + \| \mathbf{f} - \mathbf{f}_{\mathcal{T}} \|^2_{\mathbf{L}^2(\omega_h,\mathbb{R}^2)} \\ + 2(\mathbf{m} - \mathbf{m}_{\mathcal{T}}, \mathbf{f} - \mathbf{f}_{\mathcal{T}}) + c \| h_{\Xi}^{1/2} [\partial u_h/\partial s] \|_{L^2(\Xi)} .\end{aligned}$$

*The constant $c > 0$ depends on the shape of the elements in $\mathcal{T}$ but neither on their sizes nor on the data $\mathbf{f}$ or solutions $u, \mathbf{m}, \lambda, u_h, \mathbf{m}_h, \lambda_h$.*

*In case that the exact solution is smooth, i.e., $(\mathbf{m} - \nabla u)|_\omega \in \mathbf{W}^{1,\infty}(\omega, \mathbb{R}^2)$, we have*

$$\begin{aligned}\| \nabla_{\mathcal{T}}(u - u_h) \|^2_{\mathbf{L}^2(\Omega,\mathbb{R}^2)} + c \| D\phi^{**}(\mathbf{m}) - D\phi^{**}(\mathbf{m}_h) \|^2_{\mathbf{L}^2(\omega,\mathbb{R}^2)} \\ \leq c \| \varepsilon\lambda_h \mathbf{m}_h \|^2_{\mathbf{L}^2(\omega_h,\mathbb{R}^2)} + \| \mathbf{f} - \mathbf{f}_{\mathcal{T}} \|^2_{\mathbf{L}^2(\omega_h,\mathbb{R}^2)} \\ + 2\,(\mathbf{m} - \mathbf{m}_{\mathcal{T}}, \mathbf{f} - \mathbf{f}_{\mathcal{T}}) + c \| h_{\Xi}^2 [\partial u_h/\partial s] \|_{L^1(\Xi)}\end{aligned}$$

*for an $(\varepsilon, h_{\mathcal{T}}, h_{\Xi})$-independent constant $c > 0$ which depends on $\| \nabla(\mathbf{m} - \nabla u|_\omega) \|_{\mathbf{W}^{1,\infty}(\omega,\mathbb{R}^2)}$.*

**Remark 2.5** *1. Note that $|\mathbf{m}| \leq 1$ pointwise almost everywhere in $\omega$ implies $\| \mathbf{m} - \mathbf{m}_\mathcal{T} \|_{\mathbf{L}^\infty(\omega,\mathbb{R}^2)} \leq 2$ and so, the first term on the right-hand side of* (2.53) *is estimated by*

$$(\mathbf{m} - \mathbf{m}_\mathcal{T}, \mathbf{f} - \mathbf{f}_\mathcal{T}) \leq 2 \| \mathbf{f} - \mathbf{f}_\mathcal{T} \|_{\mathbf{L}^1(\omega,\mathbb{R}^2)} . \tag{2.50}$$

*In case that $\mathbf{m} \in \mathbf{W}^{1,\infty}(\omega, \mathbb{R}^2)$, a Poincaré-type estimate shows*

$$(\mathbf{m} - \mathbf{m}_\mathcal{T}, \mathbf{f} - \mathbf{f}_\mathcal{T}) \leq \| \nabla \mathbf{m} \|_{\mathbf{L}^\infty(\omega,\mathbb{R}^2)} \| h_\mathcal{T}(\mathbf{f} - \mathbf{f}_\mathcal{T}) \|_{\mathbf{L}^1(\omega,\mathbb{R}^2)} . \tag{2.51}$$

*Note that $\| h_\mathcal{T}(\mathbf{f} - \mathbf{f}_\mathcal{T}) \|_{\mathbf{L}^1(\Omega,\mathbb{R}^2)} \leq C \| h_\mathcal{T} \|_\infty^2$ is of optimal order. Finally, in the uniaxial case $\phi^{**}(\mathbf{m}) = \langle \mathbf{m}, \mathbf{e}_\perp \rangle^2/2$ and for an easy axis $\mathbf{e}$ perpendicular to the exterior magnetic field $\mathbf{f}$ pointwise almost everywhere, we have*

$$\begin{aligned}(\mathbf{m} - \mathbf{m}_\mathcal{T}, \mathbf{f} - \mathbf{f}_\mathcal{T}) &\leq \| \langle \mathbf{m} - \mathbf{m}_\mathcal{T}, \mathbf{e}_\perp \rangle_{\mathbb{R}^2} \|_{L^2(\omega)} \| \mathbf{f} - \mathbf{f}_\mathcal{T} \|_{\mathbf{L}^2(\omega,\mathbb{R}^2)} \\ &= \| D\phi^{**}(\mathbf{m}) - D\phi^{**}(\mathbf{m}_h) \|_{\mathbf{L}^2(\omega,\mathbb{R}^2)} \| \mathbf{f} - \mathbf{f}_\mathcal{T} \|_{\mathbf{L}^2(\omega,\mathbb{R}^2)}\end{aligned} \tag{2.52}$$

*and $\| D\phi^{**}(\mathbf{m}) - D\phi^{**}(\mathbf{m}_h) \|_{\mathbf{L}^2(\omega,\mathbb{R}^2)}$ can be absorbed. This merely results in an additional term $\| \mathbf{f} - \mathbf{f}_\mathcal{T} \|^2_{\mathbf{L}^2(\omega,\mathbb{R}^2)}$ on the right-hand side.*

*2. The estimate of Theorem 2.7 plus* (2.50) *is reliable (i.e., the error is bounded from above by a constant times the computable bound).*

*3. The estimate of Theorem 2.8 plus* (2.50) *is reliable in the sense that the used constant c does not depend on the regularity of $\mathbf{m}$ or $\nabla u$. In case of Theorem 2.7 plus* (2.50)*, the used constant c does depend on the smoothness of the exact solution; the author is unaware of any regularity results on $\mathbf{m}$. Consequently, we have to regard the result of Theorem 2.7 as non-reliable.*

*4. The estimate of Theorem 2.8 is not efficient since the power of the jump contributions is one. This is different for the second estimate in Theorem 2.8, where all the terms on the right-hand side are of optimal order.*

*5. The complementary properties of the error estimates suggest to employ the result from Theorem 2.7 in an adaptive mesh-refining strategy but use the first result of Theorem 2.8 for reliable (but possibly expensive) error estimation.*

*6. Note that $|\langle [\mathbf{m}_h - \nabla_\mathcal{T} u_h], \mathbf{n}_\Xi \rangle_{\mathbb{R}^2}|$ does not appear in the first inequality of Theorem 2.8 while $\|[\partial u_h/\partial s]\|$ is typical in a posteriori error estimates for non-conforming finite element schemes.*

We start with the verification of Theorem 2.8.

**Proof:**
Throughout this proof, we abbreviate $e := u - u_h$, $\boldsymbol{\delta} := \mathbf{m} - \mathbf{m}_h$ and extend $\mathbf{m}$, $\mathbf{m}_h$, and $\boldsymbol{\delta}$ by zero outside of $\omega$.

As in the first step of the proof of the a priori error estimates, we add (2.33) and (2.34) and substitute the resulting estimate for $\langle \lambda\mathbf{m} - \lambda_h \mathbf{m}_h, \boldsymbol{\delta}\rangle_{\mathbb{R}^2}$ in (2.38) and so infer

$$\begin{aligned} &\| \nabla_{\mathcal{T}} e \|^2_{\mathbf{L}^2(\Omega,\mathbb{R}^2)} + \left(D\phi^{**}(\mathbf{m}) - D\phi^{**}(\mathbf{m}_h), \boldsymbol{\delta}\right) \\ &\le (\mathbf{m} - \mathbf{m}_{\mathcal{T}}, \mathbf{f} - \mathbf{f}_{\mathcal{T}}) + \varepsilon\left(\lambda_h|\mathbf{m}_h|, \lambda|\mathbf{m}| - \lambda_h|\mathbf{m}_h|\right) + (\nabla_{\mathcal{T}} e, \nabla_{\mathcal{T}} e - \boldsymbol{\delta})_\Omega , \end{aligned} \tag{2.53}$$

where we added $\| \nabla_{\mathcal{T}} e \|^2_{\mathbf{L}^2(\Omega,\mathbb{R}^2)}$ on both sides.

To bound the second term on the right-hand side of (2.53), we employ (2.4) resp. (2.16) to obtain expressions for $\lambda|\mathbf{m}|$ resp. $\lambda_h|\mathbf{m}_h|$ and subtract the two resulting formulae. This proves

$$\begin{aligned} &\varepsilon\left(\lambda_h|\mathbf{m}_h|, \lambda|\mathbf{m}| - \lambda_h|\mathbf{m}_h|\right) \\ &\quad \le \varepsilon\left(\lambda_h|\mathbf{m}_h|, |\mathbf{f} - \mathbf{f}_{\mathcal{T}} - \nabla_{\mathcal{T}} e - D\phi^{**}(\mathbf{m}) + D\phi^{**}(\mathbf{m}_h)|\right) \\ &\quad \le c\varepsilon^2 \| \lambda_h|\mathbf{m}_h| \|^2_{L^2(\omega_h)} + \frac{1}{2} \| \mathbf{f} - \mathbf{f}_{\mathcal{T}} \|^2_{\mathbf{L}^2(\omega_h,\mathbb{R}^2)} \\ &\qquad + c \| D\phi^{**}(\mathbf{m}) - D\phi^{**}(\mathbf{m}_h) \|^2_{\mathbf{L}^2(\omega_h,\mathbb{R}^2)} + \frac{1}{2} \| \nabla_{\mathcal{T}} e \|^2_{\mathbf{L}^2(\omega_h,\mathbb{R}^2)} . \end{aligned} \tag{2.54}$$

For the last term on the right-hand side of (2.53) we first observe that $\nabla u - \mathbf{m}$ is divergence-free in the sense of distributions on $\Omega$. Hence, there exists a function $b \in W^{1,2}(\Omega)$ with $\nabla u - \mathbf{m} = \mathbf{curl}\, b := (\partial b/\partial x_2, -\partial b/\partial x_1)^\top$.

Let $b_h$ be the Clement-interpolation to $b$ (no boundary conditions); $b_h$ is continuous and $\mathcal{T}$-piecewise affine and, if $b \in W^{\beta+1,2}(\Omega)$, there holds

$$\begin{aligned} &\| h_{\mathcal{T}}^{-\beta} \nabla(b - b_h) \|_{\mathbf{L}^2(\Omega,\mathbb{R}^2)} + \| h_{\mathcal{T}}^{-(\beta+1)}(b - b_h) \|_{L^2(\Omega)} \\ &\quad + \| h_{\Xi}^{-(\beta+1/2)}(b - b_h) \|_{L^2(\Xi)} \le c \| b \|_{\mathbf{W}^{\beta+1,2}(\Omega,\mathbb{R}^2)} . \end{aligned} \tag{2.55}$$

The constant $c > 0$ depends only on $\omega$ and the aspect ratio of the elements, but does not depend on their sizes (or on $b$ or $B$) [30, 13, 128]. An element-wise integration by parts shows

$$(\mathbf{curl}\, b_h, \nabla_{\mathcal{T}} u_h)_\Omega = \int_\Xi [u_h] \langle \mathbf{curl}\, b_h, \mathbf{n}\rangle_{\mathbb{R}^2}\, \mathrm{d}\mathbf{x} = 0 \tag{2.56}$$

since $\langle \mathbf{curl}\, b_h, \mathbf{n}\rangle_{\mathbb{R}^2} = \partial b_h/\partial s$ is continuous in the sense that there is no difference on both sides of $\boldsymbol{\xi}$ and $\langle \mathbf{curl}\, b_h, \mathbf{n}\rangle_{\mathbb{R}^2}$ is constant there while $[u_h]$

has a vanishing integral mean on $\xi$ by construction of the Crouzeix-Raviart elements.

The discrete counterpart $\nabla_{\mathcal{T}} u_h - \mathbf{m}_h$ is perpendicular to $\nabla_{\mathcal{T}} u_h$ in $\mathbf{L}^2(\Omega, \mathbb{R}^2)$ according to (2.15). Surprisingly, $\nabla_{\mathcal{T}} u_h - \mathbf{m}_h$ is perpendicular to $\nabla u$ as well. Indeed, with the interpolation (2.29) and with (2.15), we deduce with an element-wise integration by parts that

$$\begin{aligned}(\nabla u, \nabla_{\mathcal{T}} u_h - \mathbf{m}_h)_\Omega &= (\nabla_{\mathcal{T}}(u - I_{\mathcal{T}} u), \nabla_{\mathcal{T}} u_h - \mathbf{m}_h) \\ &= \int_\Xi \Big\langle \big[(u - I_{\mathcal{T}} u)\,(\nabla_{\mathcal{T}} u_h - \mathbf{m}_h)\big], \mathbf{n}_\Xi \Big\rangle_{\mathbb{R}^2} \mathrm{dx}\,.\end{aligned} \tag{2.57}$$

In the last step, we used that $\mathbf{m}_h$ is $\mathcal{T}$-piecewise constant and $u_h$ is $\mathcal{T}$-piecewise affine such that $\operatorname{div}_{\mathcal{T}}(\nabla_{\mathcal{T}} u_h - \mathbf{m}_h) = 0$. For each edge $\xi \in \Xi$, $u - I_{\mathcal{T}} u$ has integral mean zero on $\xi$ and $\nabla_{\mathcal{T}} u_h - \mathbf{m}_h$ is constant there. Hence, even if the corresponding quantities are discontinuous on $\xi$, we have

$$(\nabla_{\mathcal{T}}\, e, \nabla_{\mathcal{T}} u_h - \mathbf{m}_h)_\Omega = \int_\Xi \Big\langle \big[(u - I_{\mathcal{T}} u)\,(\nabla_{\mathcal{T}} u_h - \mathbf{m}_h)\big], \mathbf{n}_\Xi \Big\rangle_{\mathbb{R}^2} \mathrm{dx} = 0\,. \tag{2.58}$$

From (2.56), (2.58), and $\nabla u - \mathbf{m} = \mathbf{curl}\, b$, we deduce with an element-wise integration by parts and Cauchy's inequality that

$$\begin{aligned}(\nabla_{\mathcal{T}} e, \nabla_{\mathcal{T}} e - \boldsymbol{\delta})_\Omega &= \big(\nabla_{\mathcal{T}} e, \mathbf{curl}\,(b - b_h)\big) \\ &= -\big(\nabla_{\mathcal{T}} u_h, \mathbf{curl}\,(b - b_h)\big) = -\int_\Xi [\partial u_h/\partial s](b - b_h)\,\mathrm{dx} \\ &\le \|\, h_\Xi^{-1/2}(b - b_h)\,\|_{L^2(\Xi)}\, \|\, h_\Xi^{1/2}[\partial u_h/\partial s]\,\|_{L^2(\Xi)} \\ &\le c\,\|\, \nabla b\,\|_{W^{1,2}(\Omega)}\, \|\, h_\Xi^{1/2}[\partial u_h/\partial s]\,\|_{L^2(\Xi)}\,.\end{aligned} \tag{2.59}$$

Notice that, for higher regularity of $b \in W^{2,\infty}(\Omega)$ and with its nodal interpolant $b_h$ the arguments in (2.59) show

$$(\nabla_{\mathcal{T}} e, \nabla_{\mathcal{T}} e - \boldsymbol{\delta})_\Omega \le c\,\|\, \Delta b\,\|_{L^\infty(\Omega)}\, \|\, h_\Xi^2[\partial u_h/\partial s]\,\|_{L^1(\Xi)}\,. \tag{2.60}$$

In the final step we gather all the estimates on the right-hand side of

(2.53) in (2.50), (2.54), (2.59), and (2.60) and eventually obtain,

$$\begin{aligned}
&\frac{1}{2}\| \nabla_{\mathcal{T}} e \|^2_{\mathbf{L}^2(\Omega,\mathbb{R}^2)} + \left(D\phi^{**}(\mathbf{m}) - D\phi^{**}(\mathbf{m}_h), \boldsymbol{\delta}\right) \qquad (2.61)\\
&\leq c\varepsilon^2 \| \lambda_h |\mathbf{m}_h| \|^2_{L^2(\omega)} + (\mathbf{m}-\mathbf{m}_{\mathcal{T}}, \mathbf{f}-\mathbf{f}_{\mathcal{T}}) + \frac{1}{2}\| \mathbf{f}-\mathbf{f}_{\mathcal{T}} \|^2_{\mathbf{L}^2(\omega,\mathbb{R}^2)}\\
&+\frac{\tilde{c}}{2}\| D\phi^{**}(\mathbf{m}) - D\phi^{**}(\mathbf{m}_h) \|^2_{\mathbf{L}^2(\omega,\mathbb{R}^2)}\\
&+c\| b \|_{W^{1+\alpha,2}(\omega)} \| h_{\Xi}^{1/2}[\partial u_h/\partial s] \|_{L^2(\Xi)}\,.
\end{aligned}$$

Absorbing $|D\phi^{**}(\mathbf{m}) - D\phi^{**}(\mathbf{m}_h)|^2$ with (2.26), we conclude the proof of the theorem. We omit details in the remaining case. □

We now show Theorem 2.7.

**Proof:**

Arguing as above we deduce (2.53) and estimate the first and second term on its right-hand side as in (2.50)-(2.54). The last term in (2.53) reads

$$(\nabla e, \nabla e - \boldsymbol{\delta})_\Omega = \left(\nabla(e-e_h), \mathbf{m}_h - \nabla u_h\right)_\Omega, \qquad (2.62)$$

where $e_h \in \mathcal{S}_0^1(\mathcal{T})$ denotes the Clement-interpolation of $e$ which satisfies estimates as in (2.55) (where $b$ resp. $b_h$ is replaced by $e$ resp. $e_h$). According to $\operatorname{div}_{\mathcal{T}}(\mathbf{m}_h - \nabla u_h) = 0$, an integration by parts on the right-hand side in (2.62) shows

$$\begin{aligned}
(\nabla e, \nabla e - \boldsymbol{\delta})_\Omega &= \int_\Xi (e-e_h)\left\langle [\mathbf{m}_h - \nabla u_h], \mathbf{n}_\Xi \right\rangle_{\mathbb{R}^2} \mathrm{d}\mathbf{x} \qquad (2.63)\\
&\leq c\| \nabla e \|_{\mathbf{L}^2(\Omega,\mathbb{R}^2)} \| h_\Xi^{1/2}\left\langle [\mathbf{m}_h - \nabla u_h], \mathbf{n}_\Xi \right\rangle_{\mathbb{R}^2} \|_{L^2(\Xi)}\,.
\end{aligned}$$

The remaining parts in this proof are analogous to those in the previous and hence omitted. □

### 2.1.4 Numerical Realization

Computational examples are provided for the uniaxial case (with the easy axis $\mathbf{e} \in \mathbb{R}^2$) to compare the conforming method and the non-conforming method with respect to stability as well as convergence properties. We consider the minor generalization $(\tilde{\mathrm{P}})$ of (P) on the right-hand side of (2.64).

This is a small modification of Problem (P) stated in (2.3)-(2.5).

**Problem** $(\tilde{P})$: Given $\{g, \mathbf{f}\} \in L^2(\Omega) \times \mathbf{L}^2(\omega, \mathbb{R}^2)$, seek $\{u, \mathbf{m}, \lambda\} \in W_0^{1,2}(\Omega) \times \mathbf{L}^2(\omega, \mathbb{R}^2) \times L^2(\omega)$ that satisfies, for all $w \in W_0^{1,2}(\Omega)$ and $\boldsymbol{\mu} \in \mathbf{L}^2(\omega, \mathbb{R}^2)$,

$$(\nabla u, \nabla w)_\Omega - (\mathbf{m}, \nabla w) = (g, w)_\Omega \,, \tag{2.64}$$
$$(\nabla u, \boldsymbol{\mu}) + \left(\langle \mathbf{m}, \mathbf{e}_\perp \rangle_{\mathbb{R}^2}, \langle \boldsymbol{\mu}, \mathbf{e}_\perp \rangle_{\mathbb{R}^2}\right) + (\lambda \mathbf{m}, \boldsymbol{\mu}) = (\mathbf{f}, \boldsymbol{\mu}) \,, \tag{2.65}$$
$$0 \le \lambda \,, \ |\mathbf{m}| \le 1 \,, \text{ and } \lambda(1 - |\mathbf{m}|)_+ = 0 \quad \text{a.e. in } \omega \,. \tag{2.66}$$

The side constraint $|\,\mathbf{m}\,| \le 1$ is enforced by a penalization strategy and leads to Problem $(\tilde{\mathrm{P}}_\varepsilon)$ and its conforming resp. non-conforming discretization $(\tilde{P}_{\varepsilon,N})$ resp. $(\tilde{P}^{NC}_{\varepsilon,N})$ solved numerically by a Newton-Raphson scheme.

As in Theorems 2.7 and 2.8, we can prove the following bound for the uniaxial case and the conforming scheme $(P_{h_\mathcal{T},N})$, i.e., $\varepsilon = h_\mathcal{T}$,

$$\|\,\nabla(u - u_h)\,\|_{\mathbf{L}^2(\Omega,\mathbb{R}^2)} + \|\,\langle \mathbf{m} - \mathbf{m}_h, \mathbf{e}_\perp \rangle_{\mathbb{R}^2}\,\|_{L^2(\omega)} \le c\,\min\{\eta_C^{(0)}, \eta_C^{(1)}\}, \tag{2.67}$$

and for the non-conforming method $(P^{NC}_{h_\mathcal{T},N})$,

$$\|\,\nabla_\mathcal{T}(u - u_h)\,\|_{\mathbf{L}^2(\Omega,\mathbb{R}^2)} + \|\,\langle \mathbf{m} - \mathbf{m}_h, \mathbf{e}_\perp \rangle_{\mathbb{R}^2}\,\|_{L^2(\omega)} \le c\,\min\{\eta_{NC}^{(0)}, \eta_{NC}^{(1)}\}, \tag{2.68}$$

where the used constants $c > 0$ do not depend on $h_\mathcal{T}$ and the error estimators are, for $\beta = 0, 1$,

$$\eta_C^{(\beta)} := \Big(\|\,h_\mathcal{T}\lambda_h\mathbf{m}_h\,\|^2_{\mathbf{L}^2(\omega,\mathbb{R}^2)} + \|\,\mathbf{f} - \mathbf{f}_\mathcal{T}\,\|^2_{\mathbf{L}^2(\omega,\mathbb{R}^2)} + \|\,h^\beta_\mathcal{T}(\mathbf{f} - \mathbf{f}_\mathcal{T})\,\|_{\mathbf{L}^1(\omega,\mathbb{R}^2)}$$
$$+\|\,h_\mathcal{T} g\,\|^2_{L^2(\Omega)} + \|\,h_\xi^{1/2}\langle[\mathbf{m}_h - \nabla u_h], \mathbf{n}_\xi\rangle_{\mathbb{R}^2}\,\|^2_{L^2(\Xi)}\Big)^{1/2}, \tag{2.69}$$

$$\eta_{NC}^{(0)} := \Big(\|\,h_\mathcal{T}\lambda_h\mathbf{m}_h\,\|^2_{\mathbf{L}^2(\omega,\mathbb{R}^2)} + \|\,\mathbf{f} - \mathbf{f}_\mathcal{T}\,\|^2_{\mathbf{L}^2(\omega,\mathbb{R}^2)} + \|\,\mathbf{f} - \mathbf{f}_\mathcal{T}\,\|_{\mathbf{L}^1(\omega,\mathbb{R}^2)}$$
$$+\|\,h_\mathcal{T} g\,\|^2_{L^2(\Omega)} + \|\,h_\xi^{1/2}[\partial u_h/\partial s]\,\|_{L^2(\Xi)}\Big)^{1/2}, \tag{2.70}$$

$$\eta_{NC}^{(1)} := \Big(\|\,h_\mathcal{T}\lambda_h\mathbf{m}_h\,\|^2_{\mathbf{L}^2(\omega,\mathbb{R}^2)} + \|\,\mathbf{f} - \mathbf{f}_\mathcal{T}\,\|^2_{\mathbf{L}^2(\omega,\mathbb{R}^2)} + \|\,h_\mathcal{T}(\mathbf{f} - \mathbf{f}_\mathcal{T})\,\|_{\mathbf{L}^1(\omega,\mathbb{R}^2)}$$
$$+\|\,h_\mathcal{T} g\,\|^2_{L^2(\Omega)} + \|\,h_\xi^2[\partial u_h/\partial s]\,\|_{L^1(\Xi)}\Big)^{1/2}.$$

The estimates (2.69) resp. (2.71) motivate error indicators for local adaptive mesh-refinement, namely for $\beta = 0, 1$,

$$\begin{aligned}\eta_{K,C}^{(\beta)} &:= \Big( \| h_T \lambda_h \mathbf{m}_h \|^2_{\mathbf{L}^2(K,\mathbb{R}^2)} + \| \mathbf{f} - \mathbf{f}_T \|^2_{\mathbf{L}^2(K,\mathbb{R}^2)} + \| h_T^\beta (\mathbf{f} - \mathbf{f}_T) \|_{\mathbf{L}^1(K,\mathbb{R}^2)} \\ &\quad + \| h_T g \|^2_{L^2(K)} + \| h_\xi^{1/2} \langle [\mathbf{m}_h - \nabla u_h], \mathbf{n}_\Xi \rangle_{\mathbb{R}^2} \|^2_{L^2(\partial K)} \Big)^{1/2}, \qquad (2.71)\\ \eta_{K,NC}^{(1)} &:= \Big( \| h_K \lambda_h \mathbf{m}_h \|^2_{2,K} + \| \mathbf{f} - \mathbf{f}_T \|^2_{\mathbf{L}^2(K,\mathbb{R}^2)} + \| h_T (\mathbf{f} - \mathbf{f}_T) \|_{\mathbf{L}^1(K,\mathbb{R}^2)} \\ &\quad + \| h_T g \|^2_{L^2(K)} + \| h_\xi^2 [\partial u_h / \partial s] \|_{L^1(\partial K)} \Big)^{1/2}. \qquad (2.72)\end{aligned}$$

**Remark 2.6** *1. Note that the a posteriori error estimates (2.69) resp. (2.71) are reliable for $\beta = 0$ in the sense that $c > 0$ does not depend on the data. The estimates are efficient for $\beta = 1$ in the sense that the upper bounds have optimal convergence order.*

*2. The error estimator (2.70) is not a sum of local contributions. For the remaining estimators we have, for $\beta = 0, 1$,*

$$\eta_C^{(\beta)} = \Big( \sum_{K \in \mathcal{T}} (\eta_{K,C}^{(\beta)})^2 \Big)^{1/2} \quad \text{and} \quad \eta_{NC}^{(1)} = \Big( \sum_{K \in \mathcal{T}} (\eta_{K,NC}^{(1)})^2 \Big)^{1/2}. \qquad (2.73)$$

For any choice of $\eta_K = \eta_{K,C}^{(0)}, \eta_{K,C}^{(1)}$ and $\eta = \eta_C^{(0)}, \eta_C^{(1)}$ resp. $\eta_K = \eta_{K,NC}^{(1)}$ and $\eta = \eta_{NC}^{(0)}, \eta_{NC}^{(1)}$, the subsequent mesh-refining algorithm generates a sequence $\mathcal{T}_0, \mathcal{T}_1, \ldots$ of adapted meshes.

**Algorithm 2.1** *1. Start with coarse mesh $\mathcal{T}_0$.*
*2. Solve the discrete problem with respect to $\mathcal{T}_k$.*
*3. Compute $\eta_K$ for all $K \in \mathcal{T}_k$.*
*4. Compute error bound $\eta$ and terminate or goto 5.*
*5. Mark element $K$ red iff $\eta_K \geq \frac{1}{2} \max_{K \in \mathcal{T}_k} \eta_K$.*
*6. Red-green-blue-refinement (cf., e.g., [128]) to avoid hanging nodes, generate mesh $\mathcal{T}_{k+1}$, set $k = k + 1$ and goto 2.*

### 2.1.5 Computational Experiments

#### Academic example for numerical justification of theoretical results

The first example provides experimental evidence for the optimal choice of the penalty parameter $\varepsilon = h^\beta$, $\beta > 0$, and discusses its influence onto the number of iteration steps in Algorithm 1. Stability properties and mesh-dependencies as well as convergence behavior are studied for $(\tilde{\mathrm{P}}_{\varepsilon,N})$ and $(\tilde{\mathrm{P}}^{NC}_{\varepsilon,N})$.

Let $\omega = (1/4, 3/4)^2 \subset \Omega = (0,1)^2$, $\omega_1 := \{(x,y)^\top \in \omega : 1 \le \sin(2\pi(x - .25))\sin(2\pi(y - .25))\}$, and $\mathbf{e} = (e_1, e_2)^\top$. Let $\{u, \mathbf{m}, \lambda\} \in W^{1,2}_0(\Omega) \times \mathbf{L}^2(\omega, \mathbb{R}^2) \times L^2(\omega)$ be the solution of Problem $(\tilde{P})$

$$u(x,y) = \sin(\pi x)\sin(\pi y) \quad \text{and} \quad \mathbf{m} = (\tilde{m}, \tilde{m})^\top, \tag{2.74}$$

$$\tilde{m}(x,y) = \begin{cases} 1 & \text{if } (x,y) \in \omega_1, \\ 5\sin(2\pi(x-1/4))\sin(2\pi(y-1/4)) & \text{if } (x,y) \in \omega_1, \end{cases} \tag{2.75}$$

$$\lambda(x,y) = \begin{cases} 5\left((x-3/2)^2 + (y-1/2)^2\right) & \text{if } (x,y) \in \omega_1, \\ 0 & \text{if } (x,y) \in \omega \setminus \omega_1. \end{cases} \tag{2.76}$$

In order to study the effect of penalization in $(\tilde{P}^{NC}_{\varepsilon,N})$ and $(\tilde{P}_{\varepsilon,N})$, Figure 2.2 displays errors $||\nabla(u-u_h)||_{\mathbf{L}^2(\Omega,\mathbb{R}^2)} + ||\langle \mathbf{m}-\mathbf{m}_h), \mathbf{e}_\perp\rangle_{\mathbb{R}^2}||_{L^2(\omega)}$ versus the degrees of freedom $N$ for different choices of $\varepsilon = h^\beta_\mathcal{T}$, $\beta = 0.25, .., 1.75$, where $\mathbf{e} = (1,0)^\top$. Triangles are added to the plots to indicate the order of convergence which is twice the negative slope. In both pictures, the convergence improves if $\beta$ increases from .25 to 1.0. The convergence behavior for $\beta = 1.0, 1.25, 1.5$, and 1.75 is similar. On the other hand, the computational effort (counted in number of iterations) increases for higher values of $\beta$, see Figure 2.3, which favors the optimal choice $\beta = 1$. Hence, we choose $\varepsilon = h_\mathcal{T}$ in all subsequent computations.

To study the mesh-dependency of the solutions in $(\tilde{P}_{h_\mathcal{T},N})$, a uniform mesh is used with diagonals parallel to $\mathbf{e} = (1,1)^\top/\sqrt{2}$ (aligned) or perpendicular for $\mathbf{e} = (-1,1)^\top/\sqrt{2}$ (nonaligned). The Figure 2.4 shows the approximate magnetization $\mathbf{m}_h$ obtained by the conforming (left) and the non-conforming (right) scheme, with $\tilde{m}(x,y) = 0.8\sin\big(2\pi(x-1/4)\big)\sin\big(2\pi(y-1/4)\big)$ if $(x,y) \in \omega$, instead of (2.75) such that $\lambda = 0$ in (2.74)-(2.76), and instabilities might be enforced. While the right picture shows a reasonable approximation, the left picture indicates instabilities. To assess the quality of the approximation, Figure 2.5 shows the components of the error $\mathbf{m} - \mathbf{m}_h$ in the direction $\mathbf{e}_\perp$ (for which we proved error estimates) and in the direction $\mathbf{e}$ (for which any control lacks). The result in Figure 2.5 supports that $||\,\mathbf{m} - \mathbf{m}_h\,||_{\mathbf{L}^2(\omega,\mathbb{R}^2)}$ does not converge to zero as discussed in Example 2.2.[3] Note that the components in $\mathbf{e}_\perp$-direction converge at experimental convergence rates close to 1.

[3] We refer to Section 2.3 where numerical schemes are presented for which we can verify $\mathbf{m}_h \to \mathbf{m}$ in $\mathbf{L}^2(\omega, \mathbb{R}^2)$ at some rate.

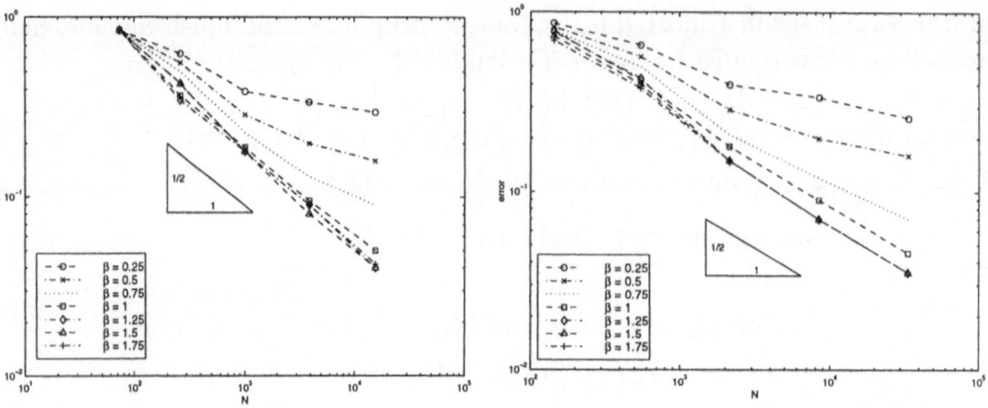


Figure 2.2: Error $= ||\nabla(u - u_h)||_{\mathbf{L}^2(\Omega,\mathbb{R}^2)} + ||\langle \mathbf{m} - \mathbf{m}_h, \mathbf{e}_\perp \rangle_{\mathbb{R}^2}||_{L^2(\omega)}$ versus degrees of freedom $N$ in $(\tilde{P}_{\varepsilon,N})$ (left) and $(\tilde{P}^{NC}_{\varepsilon,N})$ (right) with $\varepsilon = h^\beta$ on a uniform mesh for $\beta = .25, .5, .75, 1, 1.25, 1.5, 1.7$ in the first example.

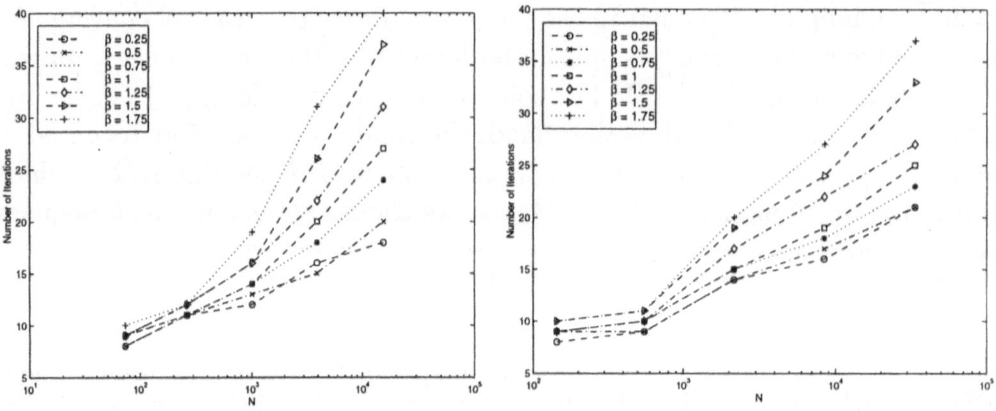


Figure 2.3: Number of iteration steps in Algorithm 1, for solving $(\tilde{P}_{\varepsilon,N})$ (left) and $(\tilde{P}^{NC}_{\varepsilon,N})$ (right) with $\varepsilon = \mathcal{O}(h^\beta)$ for different values of $\beta$ in the first example.

To study the practical performance of the a posteriori error indicators for uniform and adapted meshes generated by Algorithm 2.1, Figure 2.6 resp. 2.7 show the convergence rates for the estimated error contributions (left) and the uncontrolled magnetization error $\| \mathbf{m} - \mathbf{m}_h \|_{\mathbf{L}^2(\omega,\mathbb{R}^2)}$ (right) and some error estimators (2.73) for the conforming (Figure 2.6) and the non-conforming scheme (Figure 2.7) on uniform and adapted meshes. In these figures, a label "$\eta^{(0)}$ ($\eta^{(1)}$-adapted)" indicates that the corresponding symbol displays $\eta^{(0)}$ versus the number of degrees of freedom $N$, for a sequence $\mathcal{T}_0, \mathcal{T}_1, ..,$ generated by Algorithm 2.1 with the error indicator $\eta_K^{(1)}$ in step 5. We observe an experimental convergence rate 1/2 for reliable error estimators with $\beta = 0$ and also for the efficient error estimators with $\beta = 1$. To our surprise, the "efficient" error estimators for $\beta = 1$ do not reflect the experimental linear convergence of the true errors. This is rather pessimistic as the true errors converge linearly. The uncontrolled error $\| \mathbf{m} - \mathbf{m}_h \|_{\mathbf{L}^2(\omega,\mathbb{R}^2)}$ does not seem to converge for the conforming discretization. A linear experimental convergence is deduced for all error components from Figure 2.7 for the non-conforming schemes. The different convergence properties of $\eta_{NC}^{(\beta)}$ are expected at rate 1/2 for $\beta = 0$ and rate 1 for $\beta = 1$. Also, the meshes generated by Algorithm 2.1 seem to be slightly better than a uniform discretization. However, since the exact solution is Lipschitz continuous and at least piecewise smooth, the use of adapted meshes is not important in this example.

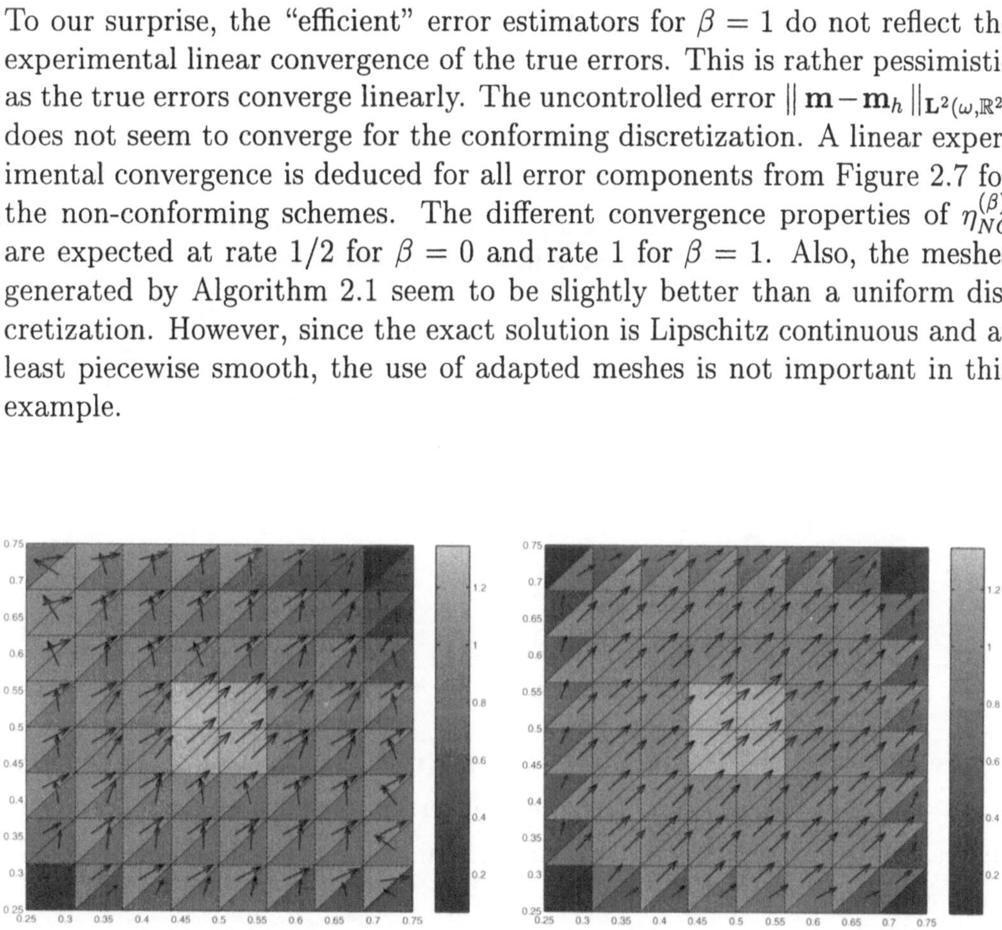


Figure 2.4: Plot of computed magnetization of $(\tilde{P}_{h,233})$ (left) and $(\tilde{P}_{h,520}^{NC})$ (right) for uniformly refined meshes in the first example. The grey-scale shows the modulus of the magnetization.

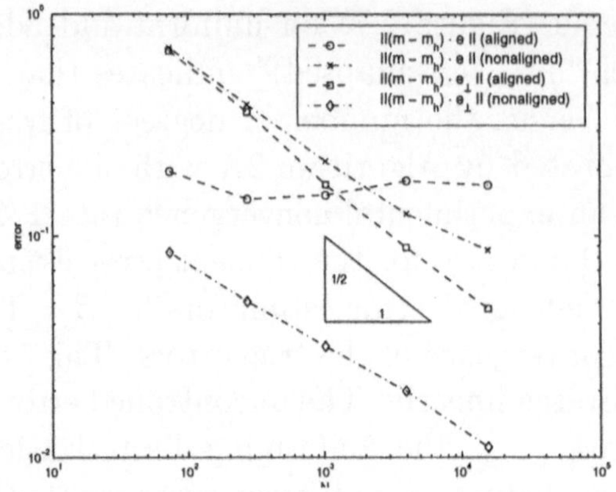


Figure 2.5: Errors $||\mathbf{m} - \mathbf{m}_h||_{\mathbf{L}^2(\omega,\mathbb{R}^2)}$ versus degrees of freedom $N$ in $(P_{h_\mathcal{T},N})$ for aligned and non-aligned meshes in the first example.

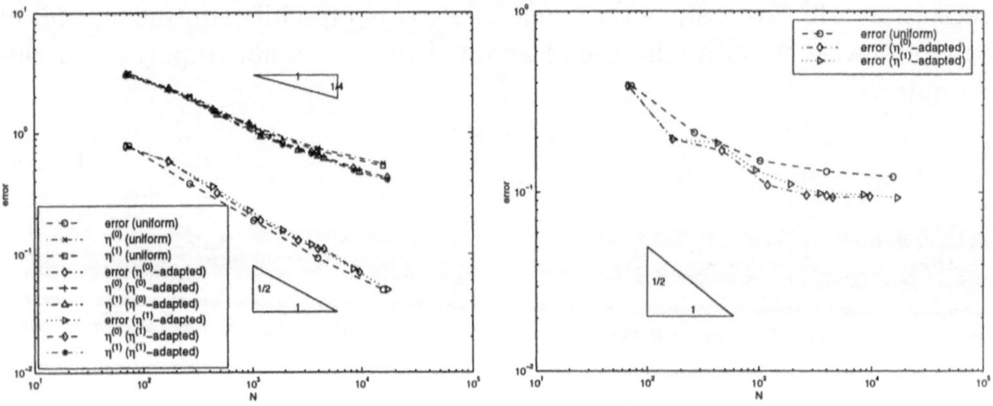


Figure 2.6: Error $= ||\nabla(u - u_h)||_{\mathbf{L}^2(\Omega,\mathbb{R}^2)} + ||\langle \mathbf{m} - \mathbf{m}_h), \mathbf{e}_\perp\rangle_{\mathbb{R}^2}||_{L^2(\omega)}$ (left) and error $= ||\langle \mathbf{m} - \mathbf{m}_h, \mathbf{e}\rangle_{\mathbb{R}^2}||_{L^2(\omega)}$ (right) and error estimators $\eta = \eta_C^{(\beta)}$ versus degrees of freedom $N$ in $(\tilde{P}_{h,N})$ for uniform and $\eta_C^{(\beta)}$-generated meshes in the first example.

### Scientific computing of an uniaxial ferromagnet under a constant magnetization

The uniaxial ferromagnet covers the domain $\omega = (-.5, .5) \times (-2.5, 2.5) \Subset \Omega = (-5.5, 5.5)^2$. It is magnetized by an exterior field $\mathbf{f} = (.6, 0)^\top$, $\mathbf{f} = (.5, .5)^\top$, resp. $\mathbf{f} = (0, .9)^\top$ and $\mathbf{e} = (1, 0)^\top$. The numerical results for $(P^{NC}_{h_\mathcal{T},N})$ on

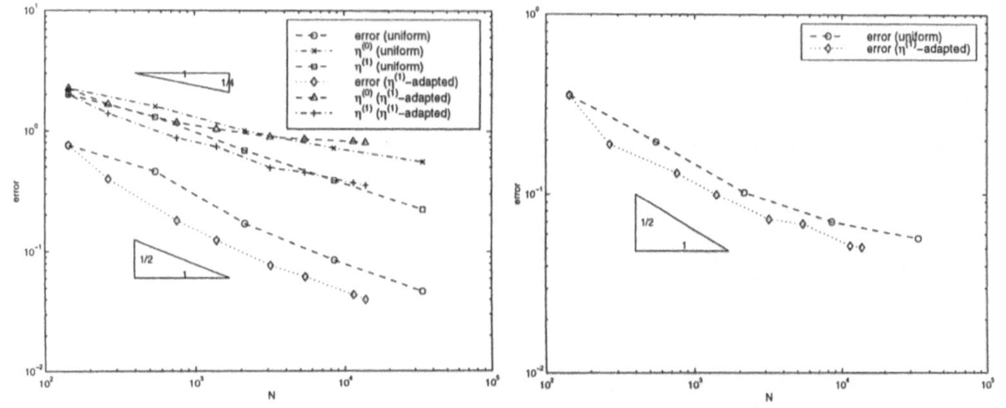


Figure 2.7: Error $= ||\nabla_{\mathcal{T}}(u-u_h)||_{\mathbf{L}^2(\Omega,\mathbb{R}^2)} + ||\langle \mathbf{m}-\mathbf{m}_h, \mathbf{e}_\perp\rangle_{\mathbb{R}^2}||_{L^2(\omega)}$ (left) and error $= ||\langle \mathbf{m}-\mathbf{m}_h, \mathbf{e}\rangle_{\mathbb{R}^2}||_{L^2(\omega)}$ (right) and error estimators $\eta = \eta_{NC}^{(\beta)}$ versus degrees of freedom $N$ in $(\tilde{P}_{h,N}^{NC})$ for uniform and $\eta_{NC}^{(1)}$-generated meshes in the first example.

$\eta_{NC}^{(1)}$-generated meshes are displayed in Figure 2.8, 2.9, resp. 2.10 for three choices of $\mathbf{f}$.

According to the angle between the easy axis vector $\mathbf{e}$ and the constant exterior field $\mathbf{f}$, we arrive at different magnetizations and potential functions. The slightly different choices of $|\,\mathbf{f}\,|$ were made to obtain an intermediate non-fully saturated state with microstructures as indicated in Figure 2.11.

In the first situation, $\mathbf{e} \,||\, \mathbf{f}$ and $\mathbf{m}_h$ is almost uniformly following $\mathbf{f}$ with peak values of the modulus of $\mathbf{m}_h$ forming a cone-like structure at the bottom and the top of the ferromagnet. For $\sphericalangle(\mathbf{e},\mathbf{f}) = \pi/4$ in the second situation, $\mathbf{m}_h$ mimics the direction of $\mathbf{f}$ but is inhomogeneous. The cone-like structure of peak values of $|\,\mathbf{m}_h\,|$ is now distorted. In the final case, a flower-like structure can be observed, with magnetization of large modulus concentrated at the edge points. Note that $\mathbf{f} \perp \mathbf{e}$ so that we could improve the a posteriori error estimates with Remark 2.5 (i).

The Young measure of the original Problem $(M_0)$ is computed with (I.19)-(I.21) where $\mathbf{m}$ is replaced by $(\max\{1, |\,\mathbf{m}_h\,|\})^{-1}\mathbf{m}_h$. Figure 2.11 displays the obtained approximations for the coefficient $\lambda$ in the second examples shown in Figures 2.8-2.10. Note that we described no error estimate for the approximation to $\nu_{\mathbf{x}}$ (which is linked to the lack of control on $\langle \mathbf{m} - \mathbf{m}_h, \mathbf{e}\rangle_{\mathbb{R}^2}$. Nevertheless, there is a weak convergence of the approximations

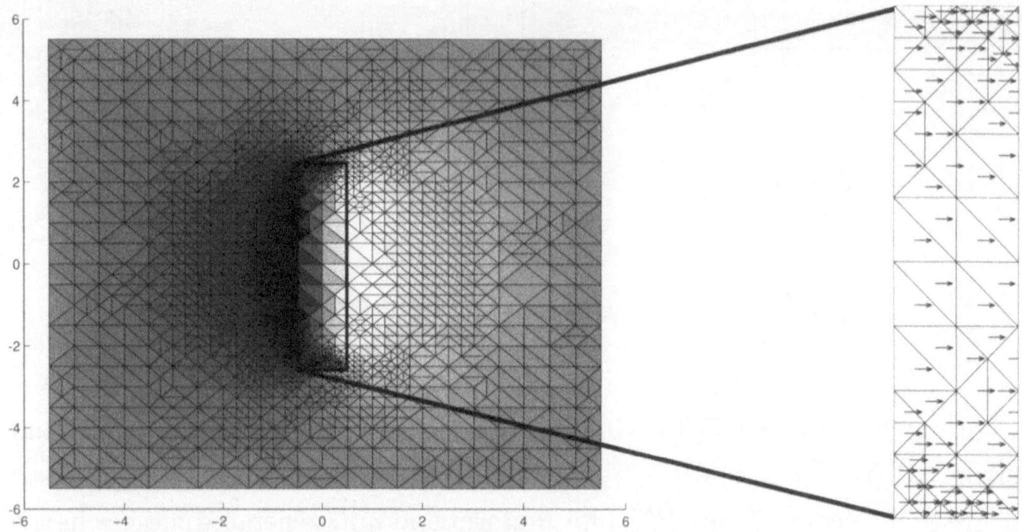


Figure 2.8: Magnetic potential $u_h$ (left) and magnetization $\mathbf{m}_h$ (right) in a ferromagnetic rod, for $(P^{NC}_{h_\mathcal{T},5444})$ on $\eta^{(1)}_{NC}$-generated meshes, for $\mathbf{f} = (.6, 0)^\top \parallel \mathbf{e} = (1, 0)^\top$ in the second example.

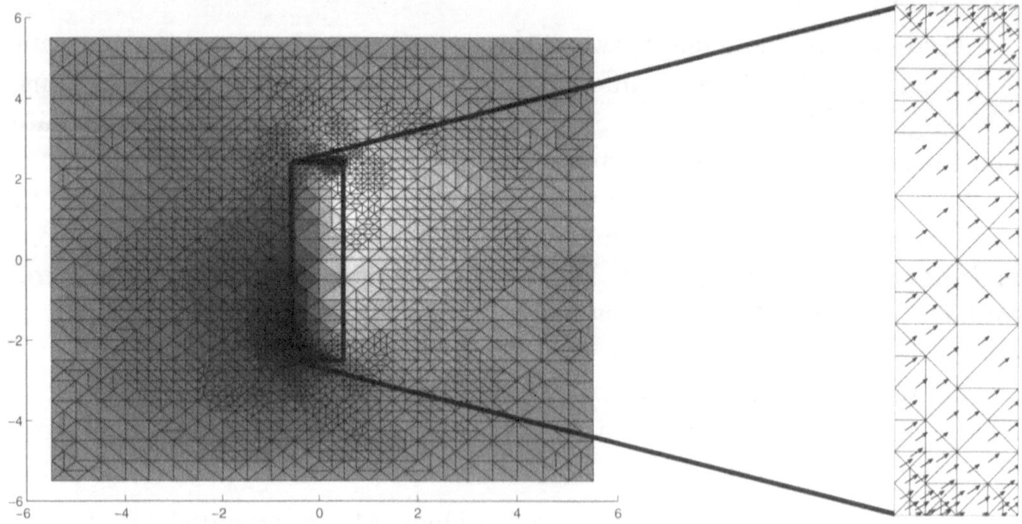


Figure 2.9: Magnetic potential $u_h$ (left) and magnetization $\mathbf{m}_h$ (right) in a ferromagnetic rod, for $(P^{NC}_{h_\mathcal{T},7874})$ on $\eta^{(1)}_{NC}$-generated meshes, for $\mathbf{f} = (.5, .5)^\top$ and $\mathbf{e} = (1, 0)^\top$ in the second example.

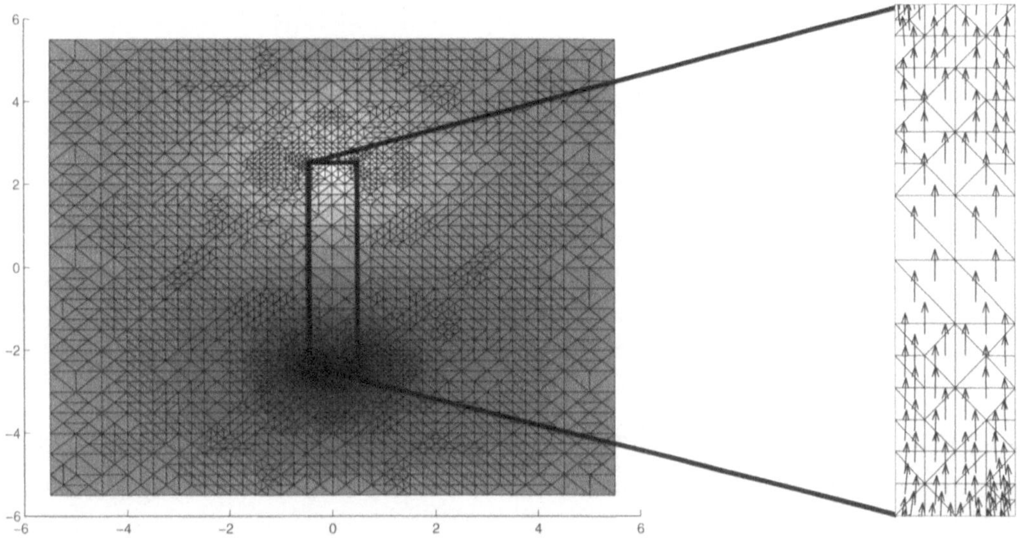


Figure 2.10: Magnetic potential $u_h$ (left) and magnetization $\mathbf{m}_h$ (right) in a ferromagnetic rod, for $(P^{NC}_{h_\mathcal{T},7284})$ on $\eta^{(1)}_{NC}$-generated meshes, for $\mathbf{f} = (0, .9)^\top \perp \mathbf{e} = (1, 0)^\top$ in the second example.

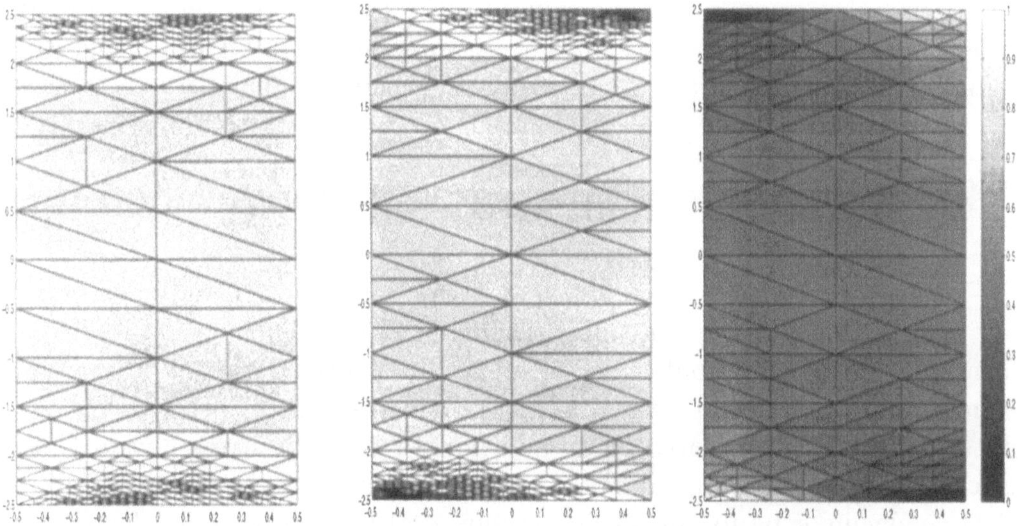

Figure 2.11: Approximate volume fraction $\lambda$ of related Young measures $\nu_\mathbf{x}$ in (I.19) for Problem $(M_0)$ for situations of Figure 2.8 (left), Figure 2.9 (middle), resp. Figure 2.10 (right).

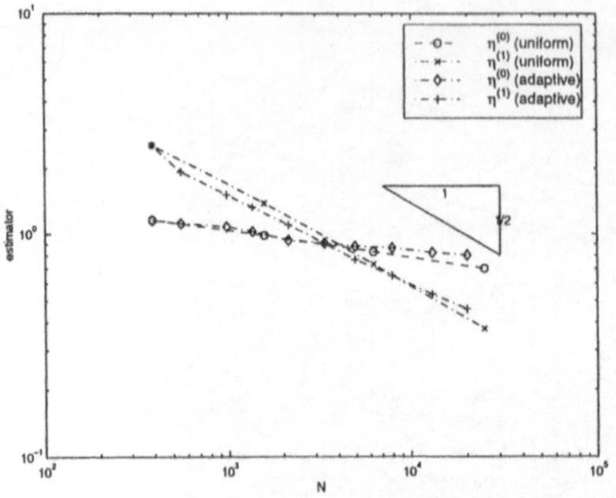


Figure 2.12: Error estimators $\eta_{NC}^{(\beta)}$ versus degrees of freedom $N$ in $(P_{h_{\mathcal{T}},N}^{NC})$ for uniform and $\eta_{NC}^{(1)}$-generated meshes in the second example (cf. Figure 2.9).

and we conjecture that the approximation is accurate on a macroscopic level.[4]

To assess the discretization errors in the approximations of Figure 2.8, 2.9, and 2.10, we computed $\eta_{NC}^{(\beta)}$. Since they show almost identical behavior in the three examples, we only plot the values for the second situation (Figure 2.9) in Figure 2.12. The error estimates show an experimental convergence rate 1 for $\beta = 1$ but a modest convergence behavior for $\beta = 0$.

## 2.2 Stabilization strategies in convexified micromagnetism[5]

Let us come back to the non-stable scheme (2.9)-(2.11) to solve (CP) in the form (2.3)-(2.5), using conforming elements. In a stabilization approach, the equations are modified in a way that non-physical oscillatory behavior of computed solutions are removed by an additional term. We stress physical justification of the following numerical strategy, since it can be motivated from (I.1) (replacing $\phi$ by its convexified version $\phi^{**}$) and positive, $h_{\mathcal{T}}$-dependent exchange energy contribution. The subsequent results are stated for $d = 2, 3$

[4]We refer to Section 2.3 where numerical schemes are presented for which we can verify convergence of computed volume fractions at some rate.

[5]The material presented in this section is taken from [52].

since the analysis is identical in both cases. Moreover, we suppose that $\phi^{**}$ satisfies (2.26) (to be valid for $d = 2,3$) throughout the remainder of this section.

The **Discrete Penalized, Stabilized Problem** $(\mathbf{P}^{\ell}_{\varepsilon}|\mathcal{S}^1_0(\mathcal{T})\times\mathcal{L}^0(\mathcal{T}|_\omega)^d\times\mathcal{L}^0(\mathcal{T}|_\omega))$, or $(\mathbf{P}^{\ell}_{\varepsilon,N})$, and $\ell \in \{\mathrm{A},\mathrm{B}\}$ shortly, reads as follows, for $\beta > 0$: *Seek* $\{u_h, \mathbf{m}_h, \lambda_h\} \in \mathcal{S}^1_0(\mathcal{T})\times \mathcal{L}^0(\mathcal{T}|_\omega)^d\times\mathcal{L}^0(\mathcal{T}|_\omega)$ *that satisfy* for every $\{w_h, \boldsymbol{\mu}_h\} \in \mathcal{S}^1_0(\mathcal{T}) \times \mathcal{L}^0(\mathcal{T}|_\omega)^d$,

$$
\begin{aligned}
&(\nabla u_h, \nabla w_h)_\Omega - (\mathbf{m}_h, \nabla w_h) = 0\,, && (2.77)\\
&(\nabla u_h, \boldsymbol{\mu}_h) + \big(D\phi^{**}(\mathbf{m}_h), \boldsymbol{\mu}_h\big) && (2.78)\\
&\qquad +(\lambda_h \mathbf{m}_h, \boldsymbol{\mu}_h) + t_\ell\big([\mathbf{m}_h], [\boldsymbol{\mu}_h]\big) = (\mathbf{f}_\mathcal{T}, \boldsymbol{\mu}_h)\,, &&\\
&\lambda_h = \varepsilon^{-1}(|\mathbf{m}_h| - 1)_+/|\mathbf{m}_h| \quad \text{a.e. in } \omega\,. && (2.79)
\end{aligned}
$$

Here, $[\cdot]$ denotes the jump across inter-element interior faces $\xi \in \Xi^* := \Xi|_{\overline{\omega}} \setminus \big(\Xi|_{\overline{\omega}} \cap \partial\omega\big)$.

We deal with the following stabilizations, for $\alpha > 0$,

$$
t_\mathrm{A}\big([\mathbf{m}_h], [\boldsymbol{\mu}_h]\big) := \sum_{\xi \subset \Xi^*} \beta_\xi h_\xi^\alpha \int_\xi [\langle \mathbf{m}_h, \mathbf{n}\rangle_{\mathbb{R}^d}][\langle \boldsymbol{\mu}_h, \mathbf{n}\rangle_{\mathbb{R}^d}]\,\mathrm{d}\mathbf{x}\,, \tag{2.80}
$$

$$
t_\mathrm{B}\big([\mathbf{m}_h], [\boldsymbol{\mu}_h]\big) := \sum_{\xi \subset \Xi^*} \beta_\xi h_\xi^\alpha \int_\xi \langle [\mathbf{m}_h], [\boldsymbol{\mu}_h]\rangle_{\mathbb{R}^d}\,\mathrm{d}\mathbf{x}\,, \tag{2.81}
$$

where $0 < \beta_\xi = const$, for all $\xi \in \Xi^*$. Finally, other numerical schemes fit into this framework that employ element-wise affine, globally continuous magnetizations, $\mathbf{m}_h, \boldsymbol{\mu}_h \in \mathcal{S}^1(\mathcal{T}|_\omega)^d$ in (I.13)-(I.15), and where $t_\ell([\mathbf{m}_h], [\boldsymbol{\mu}_h])$ in (2.78) is replaced by $d_\ell(\mathbf{m}_h, \boldsymbol{\mu}_h)$,

$$
\begin{aligned}
d_\mathrm{A}(\mathbf{m}_h, \boldsymbol{\mu}_h) &= \sum_{K \in \mathcal{T}|_\omega} \beta_K h_K^\gamma (\operatorname{div} \mathbf{m}, \operatorname{div} \boldsymbol{\mu}_h)_K\,,\\
d_\mathrm{B}(\mathbf{m}_h, \boldsymbol{\mu}_h) &= \sum_{K \in \mathcal{T}|_\omega} \beta_K h_K^\gamma (\nabla \mathbf{m}_h, \nabla \boldsymbol{\mu}_h)_K\,.
\end{aligned} \tag{2.82}
$$

for some $\gamma > 0$. Subsequently, we confine to investigating the stabilization methods (2.80)-(2.81) which incorporate jumping terms into the scheme which uses element-wise constant magnetizations. Stabilization (2.82) is analyzed in Section 2.3, where piecewise affine magnetizations are used.

**Theorem 2.9** *Let $\mathcal{T}$ be a quasi-uniform triangulation of a polygonal convex domain $\Omega \subset \mathbb{R}^d$, $d = 2, 3$ that satisfies, for $\boldsymbol{\xi}_i \in \Xi^*$, and $\boldsymbol{\sigma} \in \mathcal{N}\big|_{\omega}$,*

$$\exists \boldsymbol{\sigma} = \bigcap_{i=1,..,d} \boldsymbol{\xi}_i \in \omega \quad \text{such that} \int_\omega \langle \nabla \phi_{\boldsymbol{\sigma}}, \mathbf{e} \rangle_{\mathbb{R}^d} \, d\mathbf{x} \neq 0 \,, \tag{2.83}$$

*for any nodal basis function $\phi_{\boldsymbol{\sigma}} \in \mathcal{S}^1_0(\mathcal{T})$ at $\boldsymbol{\sigma}$. Let $\phi^{**}(\mathbf{m}) = \frac{1}{2}\langle \mathbf{m}, \mathbf{e}_\perp \rangle_{\mathbb{R}^d}$. Then, the solution $\{u_h^\ell, \mathbf{m}_h^\ell, \lambda_h^\ell\}$ of scheme $(\mathrm{P}^\ell_{\varepsilon,N})$, $\ell \in \{\mathrm{A}, \mathrm{B}\}$ is unique for $\ell = \mathrm{B}$, but not for $\ell = \mathrm{A}$, in general. Suppose now that $\phi^{**}$ only satisfies (2.26). Provided that the solution of (P) satisfies $\mathbf{m} \in \mathbf{H}(\mathrm{div}, \omega)$ (for $\ell = \mathrm{A}$) or $\mathbf{m} \in \mathbf{W}^{1,2}(\omega, \mathbb{R}^d)$ (for $\ell = \mathrm{B}$), there holds for $\alpha = 1$,*

$$\begin{aligned}
&\| \nabla(u - u_h^\ell) \|_{\mathbf{L}^2(\Omega,\mathbb{R}^d)} + \| D\phi^{**}(\mathbf{m}) - D\phi^{**}(\mathbf{m}_h^\ell) \|_{\mathbf{L}^2(\omega,\mathbb{R}^d)} \\
&\quad + \| \lambda \mathbf{m} - \lambda_h^\ell \mathbf{m}_h^\ell \|_{\mathbf{L}^2(\omega,\mathbb{R}^d)} \\
&\leq c \Big\{ \| \varepsilon\lambda \|^2_{L^2(\omega)} + \| \mathbf{f} - \mathbf{f}_{\mathcal{T}} \|^2_{\mathbf{L}^2(\omega,\mathbb{R}^d)} + \| \lambda\mathbf{m} - (\lambda\mathbf{m})_{\mathcal{T}} \|^2_{\mathbf{L}^2(\omega,\mathbb{R}^d)} \\
&\quad + \| u - P_{h;W^{1,2}(\Omega)} u \|_{L^2(\omega)} + \| \nabla(u - P_{h;W^{1,2}(\Omega)} u) \|^2_{\mathbf{L}^2(\Omega,\mathbb{R}^d)} \\
&\quad + \| \nabla(u - P_{h;W^{1,2}(\Omega)} u) \|^2_{\mathbf{L}^2(\Omega,\mathbb{R}^d)} + \sum_{\boldsymbol{\xi} \in \Xi^*} \frac{1}{h_{\boldsymbol{\xi}}} \int_{\boldsymbol{\xi}} | u - P_{h;W^{1,2}(\Omega)} u |^2 \, d\mathbf{x} \\
&\quad + \sum_{\boldsymbol{\xi} \in \Xi^*} \int_{\boldsymbol{\xi}} | u - \mathrm{P}_{h;W^{1,2}(\Omega)} u | \, d\mathbf{x} + \| \mathbf{m} - \mathbf{P}_{h;\mathbf{L}^2(\omega)} \mathbf{m} \|^2_{\mathbf{L}^2(\omega,\mathbb{R}^d)} \Big\}^{1/2} .
\end{aligned}$$

**Remark 2.7** *1. If we suppose $\{u, \mathbf{m}, \lambda\} \in W_0^{1,2}(\Omega) \cap W^{3,2}(\Omega) \times \mathbf{W}^{2,2}(\omega, \mathbb{R}^d) \times W^{2,2}(\omega)$, the right-hand side is bounded by $C\{\varepsilon + h\}$, which favors to take $\varepsilon = h_{\mathcal{T}}$. If compared to Theorem 2.4 for the non-conforming discretization $(\mathrm{P}^{NC}_{\varepsilon,N})$, we observe the same convergence behavior of $(\mathrm{P}^\ell_{\varepsilon,N})$, $\ell \in \{\mathrm{A}, \mathrm{B}\}$; for the conforming method $(\mathrm{P}_{\varepsilon,N})$, we already know suboptimal convergence behavior (Theorem 2.3).*

*2. Inverse-type inequalities enter the proof of the theorem.*

*3. The convergence analysis for the non-conforming method in Section 2.1 is much more technical: we employed the discrete Helmholtz-decomposition principle, which restricts arguments to the two-dimensional setting and simply-connected domains $\omega \subset \mathbb{R}^2$; it requires further arguments to cover general situations for $d = 3$ for the non-conforming method — in contrast to the stabilized methods $(\mathbf{P}^\ell_{\varepsilon,N})$, $\ell \in \{\mathrm{A}, \mathrm{B}\}$. For example, magnetization states in ring specimens for transforming devices play an important role in applications.*

*4. Condition (2.83) can easily be realized in practice, using e.g. a 'corner element' at $\boldsymbol{\xi} \in \partial\omega$.*

Computational experiments are reported in Section 2.2.2 that illustrate our theoretical results. These studies aim at shedding light at the following controversial subjects:

- comparison of the conforming, non-conforming and both stabilized conforming methods $(\mathrm{P}^{\ell}_{h_{\mathcal{T}},N})$, $\ell \in \{\mathrm{A}, \mathrm{B}\}$,
- mutual dependence of choices for penalization and stabilization parameter,
- performance of the stabilization strategies $(\mathrm{P}^{\ell}_{h_{\mathcal{T}},N})$, $\ell \in \{\mathrm{A}, \mathrm{B}\}$ and some slight modifications in two and three spatial dimensions.

The computational experiments evidence broad flexibility of both stabilization methods $(\mathrm{P}^{\ell}_{h_{\mathcal{T}},N})$, $\ell \in \{\mathrm{A}, \mathrm{B}\}$ with respect to different choices of $\beta_{\boldsymbol{\xi}} = \mathcal{O}(1)$, and optimal convergence behavior supports Theorem 2.9. In our (academic) 2D-examples to test $(\mathrm{P}^{\ell}_{h_{\mathcal{T}},N})$, $\ell \in \{\mathrm{A}, \mathrm{B}\}$, $\mathbf{m}_h$ is a good approximation of $\mathbf{m}$, rather than only $\langle \mathbf{m}_h, \mathbf{e}_\perp \rangle_{\mathbb{R}^2}$. In the case of oscillating solutions having high values of $|\nabla \mathbf{m}|$ but moderate ones for the modulus of $\operatorname{div} \mathbf{m}$, $(\mathrm{P}^{\mathrm{A}}_{h_{\mathcal{T}},N})$ is shown in Section 2.2.2 to give better results than $(\mathrm{P}^{\mathrm{B}}_{h_{\mathcal{T}},N})$. Performance of $(\mathrm{P}^{\mathrm{A}}_{h_{\mathcal{T}},N})$ in 3D-examples deteriorates as far as $\mathbf{m}_h$ is concerned — in contrast to $(\mathrm{P}^{\mathrm{B}}_{h_{\mathcal{T}},N})$ — but $\{\langle \mathbf{m}_h, \mathbf{e}_\perp \rangle_{\mathbb{R}^d}, \nabla u_h\}$ is converging at optimal rate towards their counterparts $\{\langle \mathbf{m}, \mathbf{e}_\perp \rangle_{\mathbb{R}^d}, \nabla u\}$ in $L^2$, as $h \to 0$.

The number of degrees of freedom $N = \dim(\mathcal{S}^1_0) + d \dim(\mathcal{L}^0(\mathcal{T}|_\omega))$ serves as a reference to the spatial discretization $\mathcal{T}$.

### 2.2.1 A priori error analysis for the stabilized schemes $(\mathrm{P}^{\ell}_{\varepsilon,N})$, $\ell \in \{\mathrm{A}, \mathrm{B}\}$

We start this subsection with observations concerning stability of the discrete models $(\mathrm{P}^{\ell}_{\varepsilon,N})$. This can be done at best for uniaxial ferromagnets where $\phi^{**}(\mathbf{m}) = \frac{1}{2} \langle \mathbf{m}, \mathbf{e}_\perp \rangle^2_{\mathbb{R}^d}$, and where uniqueness of the solution of problem (P) is known from Theorem 2.1.

We recall our results from Section 2.1 for the conforming method (i.e., $\beta_{\boldsymbol{\xi}} \equiv 0$, $\forall \boldsymbol{\xi} \in \Xi^*$ in (2.77)-(2.79)) and the non-conforming method (using Crouzeix-Raviart elements for the potential) for $d = 2$. Firstly, the conforming method is shown to give solutions that are non-unique, in general, and statements upon well-posedness depend on the meshing in a crucial way: see Figure 2.13 (a) for the construction of a function $\mathbf{M}_h$, $|\mathbf{M}_h| = 1$, such that

$(0, \gamma \mathbf{M}_h) \in \mathcal{S}_0^1(\mathcal{T}) \times \mathcal{L}^0(\mathcal{T}|_\omega)^d$, for $-1 < \gamma < 1$ solves the discrete problem with $\mathbf{f} \equiv 0$. We refer to Example 2.1 for further details.

On the other hand, the non-conforming method leads to unique solutions for uniaxial materials, a fact, which favorizes this method over the conforming discretization. The key instrument to verify uniqueness is to take benefit from the discrete Helmholtz-decomposition principle [15]. The latter is also the most relevant tool that leads to error estimates for the non-conforming method which are superior to those of the conforming method.

We conclude a deterioration of convergence order for the conforming method through pollution by unphysical oscillatory structures. However, our studies for the stabilization techniques involving $t_\ell(\cdot,\cdot)$, $\ell \in \{\text{A}, \text{B}\}$ show the following 'paradox' scenario: $(\text{P}^{\text{B}}_{\varepsilon,N})$ is well-defined, provided that (2.26) holds, but not $(\text{P}^{\text{A}}_{\varepsilon,N})$. Both methods give an optimal convergence rate $\mathcal{O}(\varepsilon + h)$, provided that $\{ u, \mathbf{m}, \lambda \} \in W^{3,2}(\Omega) \times \mathbf{W}^{2,2}(\omega, \mathbb{R}^d) \times W^{2,2}(\omega)$.

We start with a counterexample for $(\text{P}^{\text{A}}_{\varepsilon,N})$ that excludes uniqueness of $\{ u_h, \mathbf{m}_h, \lambda_h \}$ for $\phi^{**}(\mathbf{m}) = \frac{1}{2}\langle \mathbf{m}, \mathbf{e}_\perp \rangle^2_{\mathbb{R}^d}$, in general.

**Example 2.3** *Let* $\Omega = (0,1)^2 \subset \mathbb{R}^2$, $\omega = \{\mathbb{R}^2 \ni \mathbf{z} := (x,y) : |\mathbf{z} - (.5,.5)^\top|_{l_1} < 1/4\}$, *and* $\mathcal{T}$ *given as depicted in Figure 2.13 (b). It can easily be seen that* $\mathcal{L}^0(\mathcal{T}|_\omega)^2 \ni \mathbf{M}_h = \gamma \text{sign}(x - 0.5)\mathbf{e}$, $0 \leq \gamma < 1$ *solves* $(\text{P}^{\text{A}}_{\varepsilon,N})$ *for* $\phi^{**}(\mathbf{m}) = \frac{1}{2}\langle \mathbf{m}, \mathbf{e}_\perp \rangle^2_{\mathbb{R}^2}$. *Note, in particular, that* $\int_\omega \langle \mathbf{M}_h, \nabla \phi_h \rangle_{\mathbb{R}^2}\, d\mathbf{x} = 0$, *for all* $\phi_h \in \mathcal{S}_0^1(\mathcal{T})$.

This is in contrast to $(\text{P}^{\text{B}}_{\varepsilon,N})$, where uniqueness of solutions can be shown.

**Lemma 2.1** *Let (2.26) be valid, for* $\phi^{**}(\mathbf{m}) = \frac{1}{2}\langle \mathbf{m}, e_\perp \rangle^2_{\mathbb{R}^2}$, *for some unit vector* $\mathbf{e} \in \mathbb{R}^d$, $d = 2, 3$. *Then,* $(\text{P}^{\text{B}}_{\varepsilon,N})$, *for* $\alpha = 1$, *has a unique solution.*

**Proof:**
Let $\{u_h^i, \mathbf{m}_h^i, \lambda_h^i\}$, $i = 1, 2$ be two solutions of $(\text{P}^{\text{B}}_{\varepsilon,N})$, we obtain

$$\| \nabla(u_h^1 - u_h^2) \|^2_{\mathbf{L}^2(\Omega,\mathbb{R}^d)} + \| D\phi^{**}(\mathbf{m}_h^1) - D\phi^{**}(\mathbf{m}_h^2) \|^2_{\mathbf{L}^2(\omega,\mathbb{R}^d)} + \big(\lambda_h^1 \mathbf{m}_h^1 - \lambda_h^2 \mathbf{m}_h^2, \mathbf{m}_h^1 - \mathbf{m}_h^2\big) + t_{\text{B}}\big([\mathbf{m}_h^1 - \mathbf{m}_h^2], [\mathbf{m}_h^1 - \mathbf{m}_h^2]\big) = 0\,.$$

The third term is non-negative due to convexity of the characteristic functional $\psi$, see also the proof of Theorem 2.2. Clearly, this implies $u_h^1 = u_h^2 \in \mathcal{S}_0^1(\mathcal{T})$, and $\langle \mathbf{m}_h^1 - \mathbf{m}_h^2, \mathbf{e} \rangle_{\mathbb{R}^d} = const|_\omega$. Filtering out these modes via (2.83) settles the proof. □

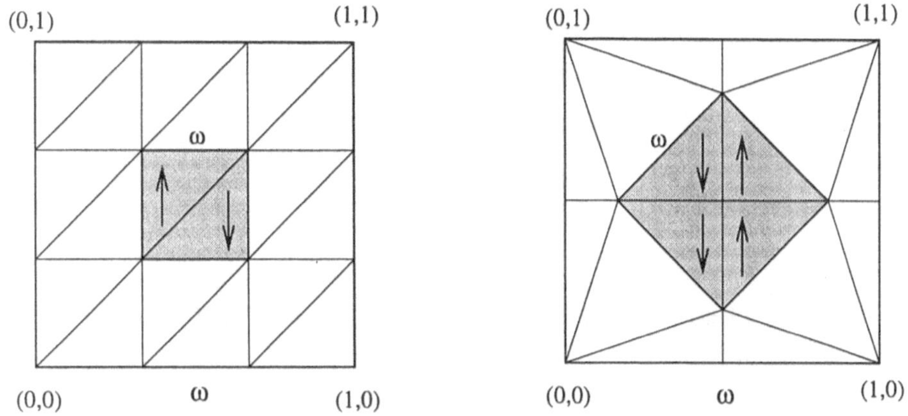


Figure 2.13: Counterexamples giving spurious solutions $\mathbf{M}_h \in \mathcal{L}^0(\mathcal{T}|_\omega)^2$ for the (a) conforming method (aligned mesh) and (b) the stabilized method $(\mathrm{P}^{\mathrm{A}}_{\varepsilon,N})$ (rotated cube) for uniaxial materials.

The remainder of this subsection is devoted to verify Theorem 2.9 for $(\mathrm{P}^{\ell}_{\varepsilon,N})$, $\ell \in \{\mathrm{A},\mathrm{B}\}$, simultaneously. Let $\{u_h, \mathbf{m}_h, \lambda_h\}$ be a solution of the problem (either indexed by $\ell = \mathrm{A}$ or $\mathrm{B}$), and use the shorthand notation $\{e, \boldsymbol{\delta}\} := \{u - u_h, \mathbf{m} - \mathbf{m}_h\} \in W^{1,2}_0(\Omega) \times \mathbf{L}^2(\omega, \mathbb{R}^d)$. Then, $\mathbf{m} \in \mathbf{H}(\mathrm{div}, \omega)$ (for $\ell = \mathrm{A}$) or $\mathbf{m} \in \mathbf{W}^{1,2}(\omega, \mathbb{R}^d)$ (for $\ell = \mathrm{B}$) infer the equations

$$(\nabla e, \nabla w_h)_\Omega - (\boldsymbol{\delta}, \nabla w_h) = 0\,, \tag{2.84}$$

$$(\nabla e, \boldsymbol{\mu}_h) + \big(D\phi^{**}(\mathbf{m}) - D\phi^{**}(\mathbf{m}_h), \boldsymbol{\mu}_h\big) \tag{2.85}$$
$$+(\lambda \mathbf{m} - \lambda_h \mathbf{m}_h, \boldsymbol{\mu}_h) + t_\ell\big([\boldsymbol{\delta}], [\boldsymbol{\mu}_h]\big) = (\mathbf{f} - \mathbf{f}_{\mathcal{T}}, \boldsymbol{\mu}_h)\,,$$

for any $\{w_h, \boldsymbol{\mu}_h\} \in \mathcal{S}^1_0(\mathcal{T}) \times \mathcal{L}^0(\mathcal{T}|_\omega)^d$.

Testing with the admissible functions $\{w_h, \boldsymbol{\mu}_h\} = \{u_h - P_{h,W^{1,2}(\Omega)}u, \mathbf{m}_h - \mathbf{P}_{h,\mathbf{L}^2(\omega,\mathbb{R}^d)}\mathbf{m}\}$, this leads to

$$\begin{aligned}
&\| \nabla e \|^2_{\mathbf{L}^2(\Omega,\mathbb{R}^d)} + \| D\phi^{**}(\mathbf{m}) - D\phi^{**}(\mathbf{m}_h) \|^2_{\mathbf{L}^2(\omega,\mathbb{R}^d)} \\
&\quad +(\lambda \mathbf{m} - \lambda_h \mathbf{m}_h, \boldsymbol{\delta}) + t_\ell\big([\boldsymbol{\delta}], [\boldsymbol{\delta}]\big) \\
&\le |\, (\nabla e, \nabla(u - P_{h,W^{1,2}(\Omega)}u))_\Omega \,| \\
&\quad +|\, \big(D\phi^{**}(\mathbf{m}) - D\phi^{**}(\mathbf{m}_h), \mathbf{m} - \mathbf{P}_{h,\mathbf{L}^2(\omega)}\mathbf{m}\big) \,| \\
&\quad +|\, \big(\boldsymbol{\delta}, \nabla(u - P_{h,W^{1,2}(\Omega)}u)\big) \,| + |\, \big(\nabla e, \mathbf{m} - \mathbf{P}_{h,\mathbf{L}^2(\omega)}\mathbf{m}\big) \,| \\
&\quad +t_\ell\big([\boldsymbol{\delta}], [\mathbf{m} - \mathbf{P}_{h,\mathbf{L}^2(\omega)}\mathbf{m}]\big) + |\, \big(\lambda \mathbf{m} - \lambda_h \mathbf{m}_h, \mathbf{m} - \mathbf{P}_{h,\mathbf{L}^2(\omega)}\mathbf{m}\big)_\omega \,| \\
&\quad +|\, \big(\mathbf{f} - \mathbf{f}_{\mathcal{T}}, \mathbf{m} - \mathbf{m}_{\mathcal{T}}\big) \,|\,,
\end{aligned} \tag{2.86}$$

since $\int_K \langle \mathbf{m}_h - \mathbf{m}_\mathcal{T}, \mathbf{f} - \mathbf{f}_\mathcal{T} \rangle_{\mathbb{R}^d} \, d\mathbf{x} = 0$.

We deal with the terms on the right-hand side of (2.86) independently. Firstly, we recall (2.32) for the present setup,

$$\frac{\varepsilon}{2} \,|\, \mathbf{m}_h \,|^2 \lambda_h^2 \le \langle \lambda \mathbf{m} - \lambda_h \mathbf{m}_h, \mathbf{m} - \mathbf{m}_h \rangle_{\mathbb{R}^d} + \frac{\varepsilon}{2} \lambda^2 . \tag{2.87}$$

Secondly, applying integration by parts on each element $K \in \mathcal{T}|_\omega$ in the third entry on the right-hand side of (2.86) causes

$$\begin{aligned}
&|\,(\boldsymbol{\delta}, \nabla(u - \mathrm{P}_{h,W^{1,2}(\Omega)} u))\,| \le |\,(\operatorname{div} \mathbf{m}, u - \mathrm{P}_{h,W^{1,2}(\Omega)} u)\,| \\
&\quad + \Big| \sum_{\xi \in \Xi^*} \int_\xi \langle \boldsymbol{\delta}, \mathbf{n} \rangle_{\mathbb{R}^2} (u - \mathrm{P}_{h,W^{1,2}(\Omega)} u) \, d\mathbf{x} \Big| \\
&\quad + \Big| \sum_{\xi \cap \partial\omega \neq \emptyset} \int_\xi \langle \boldsymbol{\delta}, \mathbf{n} \rangle_{\mathbb{R}^2} (u - \mathrm{P}_{h,W^{1,2}(\Omega)} u) \, d\mathbf{x} \Big| \\
&\le C \,\|\, u - \mathrm{P}_{h,W^{1,2}(\Omega)} u \,\|_{L^2(\omega)} + \frac{1}{4} t_\ell([\boldsymbol{\delta}], [\boldsymbol{\delta}]) \\
&\quad + \sum_{\xi \in \Xi^*} \frac{1}{h_\xi} \int_\xi |\, u - P_{h,W^{1,2}(\Omega)} u \,|^2 \, d\mathbf{x} + C \sum_{\xi \cap \partial\omega \neq \emptyset} \int_\xi |\, u - P_{h,W^{1,2}(\Omega)} u \,| \, d\mathbf{x} ,
\end{aligned}$$

due to continuity of discrete potential functions. We can now benefit from standard interpolation results (for traces) and a trace inequality to deal with the last term effectively.

Thirdly, in order to bound the last but one term in (2.86), we start testing (2.85) by $\boldsymbol{\mu}_h = (\lambda \mathbf{m})_\mathcal{T} - \lambda_h \mathbf{m}_h$ and employ an inverse-type estimate to conclude

$$\begin{aligned}
\|\, \lambda \mathbf{m} - \lambda_h \mathbf{m}_h \,\|_{\mathbf{L}^2(\omega,\mathbb{R}^d)} &\le c \Big\{ \|\, \lambda \mathbf{m} - (\lambda \mathbf{m})_\mathcal{T} \,\|_{\mathbf{L}^2(\omega,\mathbb{R}^d)} \\
&+ \|\, \nabla e \,\|_{\mathbf{L}^2(\omega,\mathbb{R}^d)} + \|\, D\phi^{**}(\mathbf{m}) - D\phi^{**}(\mathbf{m}_h) \,\|_{\mathbf{L}^2(\omega,\mathbb{R}^d)} \\
&+ t_\ell^{1/2}([\boldsymbol{\delta}], [\boldsymbol{\delta}]) + \|\, \mathbf{f} - \mathbf{f}_\mathcal{T} \,\|_{\mathbf{L}^2(\omega,\mathbb{R}^2)} \Big\} .
\end{aligned} \tag{2.88}$$

Now, using Young's inequality in the last but one term together with (2.88), unknown terms can be absorbed on the left-hand side of (2.86). Hence, we have

$$\begin{aligned}
&\| \nabla e \|^2_{\mathbf{L}^2(\Omega,\mathbb{R}^d)} + \| D\phi^{**}(\mathbf{m}) - D\phi^{**}(\mathbf{m}_h) \|^2_{\mathbf{L}^2(\omega,\mathbb{R}^d)} \\
&\quad + \int_\omega \frac{\varepsilon}{2} | \mathbf{m}_h |^2 \lambda_h^2 \, d\mathbf{x} + t_\ell([\delta],[\delta]) \\
&\le c \Big\{ \| u - P_{h,W^{1,2}(\Omega)} u \|_{L^2(\Omega)} + \sum_{\xi \in \Xi^*} \frac{1}{h_\xi} \int_\xi | u - P_{h,W^{1,2}(\Omega)} u |^2 \, d\mathbf{x} \\
&\quad + \| \nabla (u - P_{h,W^{1,2}(\Omega)} u) \|^2_{\mathbf{L}^2(\Omega,\mathbb{R}^d)} + C \| u - P_{h;W^{1,2}(\Omega)} u \|_{L^2(\partial\omega)} \\
&\quad + \| \mathbf{m} - \mathrm{P}_{h,\mathbf{L}^2(\omega,\mathbb{R}^d)} \mathbf{m} \|^2_{\mathbf{L}^2(\omega,\mathbb{R}^d)} + \| (\lambda\mathbf{m}) - (\lambda\mathbf{m})_{\mathcal{T}} \|^2_{\mathbf{L}^2(\omega,\mathbb{R}^d)} \\
&\quad + \| \mathbf{f} - \mathbf{f}_{\mathcal{T}} \|^2_{\mathbf{L}^2(\omega,\mathbb{R}^2)} \Big\} + \int_\omega \frac{\varepsilon}{2} \lambda^2 \, d\mathbf{x} \, .
\end{aligned} \tag{2.89}$$

A bootstrapping argument helps to improve the right-hand side of (2.89) with respect to the latter argument, using

$$\lambda_h^2 | \mathbf{m}_h |^2 - \lambda^2 | \mathbf{m} |^2 \le \{ \lambda_h | \mathbf{m}_h | + \lambda | \mathbf{m} | \} | \lambda_h \mathbf{m}_h - \lambda \mathbf{m} | ,$$

and hence

$$\Big| \int_\omega \frac{\varepsilon}{2} (\lambda_h^2 | \mathbf{m}_h |^2 - \lambda^2 | \mathbf{m} |^2) \, d\mathbf{x} \Big| \le \int_\omega \frac{\varepsilon^2}{2} \lambda^2 \, d\mathbf{x} + \frac{1}{4} \| \lambda\mathbf{m} - \lambda_h \mathbf{m}_h \|^2_{\mathbf{L}^2(\omega,\mathbb{R}^d)} \, . \tag{2.90}$$

(Part of) The latter term can be absorbed via (2.88) on the left-hand side of (2.89). This furnishes the proof of Theorem 2.9.

### 2.2.2 Computational Experiments

To illustrate our theoretical results for the stabilizations $(\mathrm{P}^\ell_{h_{\mathcal{T}},N})$, $\ell \in \{\mathrm{A}, \mathrm{B}\}$, we consider four numerical examples, including a comparison of methods $(\mathrm{P}^\ell_{h_{\mathcal{T}},N})$, $\ell \in \{\mathrm{A}, \mathrm{B}\}$ to the non-conforming and the non-stabilized conforming method presented in Subsection 2.1.1. For this purpose, we consider the uniaxial case, setting $\phi^{**}(\mathbf{m}) = \frac{1}{2} \langle \mathbf{m}, \mathbf{e}_\perp \rangle^2_{\mathbb{R}^2}$.

We start with some remarks on the numerical implementation of a slightly generalized version $(\tilde{\mathrm{P}})$ of Problem (P) where we take into account a non-vanishing function $s \in L^2(\Omega)$. The **Discrete Penalized, Stabilized, Generalized Problem** $(\tilde{\mathbf{P}}^\ell_{h_{\mathcal{T}},N} | \mathcal{S}^1_0(\mathcal{T}) \times \mathcal{L}^0(\mathcal{T}|_\omega)^d \times \mathcal{L}^0(\mathcal{T}|_\omega))$, shortly $(\tilde{\mathbf{P}}^\ell_{h_{\mathcal{T}},N})$, for $\ell \in \{\text{A}, \text{B}\}$, reads as follows, for $\beta > 0$: *Seek* $\{u_h, \mathbf{m}_h, \lambda_h\} \in \mathcal{S}_0^1(\mathcal{T}) \times \mathcal{L}^0(\mathcal{T}|_\omega)^d \times \mathcal{L}^0(\mathcal{T}|_\omega)$ *satisfying*

$$(\nabla u_h, \nabla w_h)_\Omega - (\mathbf{m}_h, \nabla w_h) = (s, w_h)_\Omega\,, \tag{2.91}$$

$$(\nabla u_h, \boldsymbol{\mu}_h) + \big(\langle \mathbf{m}_h, \mathbf{e}_\perp\rangle_{\mathbb{R}^d}, \langle \boldsymbol{\mu}_h, \mathbf{e}_\perp\rangle_{\mathbb{R}^d}\big) + (\lambda_h \mathbf{m}_h, \boldsymbol{\mu}_h) \tag{2.92}$$
$$+t_\ell\big([\mathbf{m}_h], [\boldsymbol{\mu}_h]\big) = (\mathbf{f}, \boldsymbol{\mu}_h)\,,$$

$$\lambda_h = h_\mathcal{T}^{-1}(|\mathbf{m}_h| - 1)_+/|\mathbf{m}_h| \quad \text{a.e. in } \omega\,, \tag{2.93}$$

for any $\{w_h, \boldsymbol{\mu}_h\} \in \mathcal{S}_0^1(\mathcal{T}) \times \mathcal{L}^0(\mathcal{T}|_\omega)^d$. Correspondingly, we refer to $(\tilde{\text{P}}^C_{h_\mathcal{T},N})$ and $(\tilde{\text{P}}^{NC}_{h_\mathcal{T},N})$ as (non-stabilized) conforming and non-conforming discretizations, respectively.

The discrete, nonlinear problems $(\tilde{\text{P}}^\ell_{h_\mathcal{T},N})$, $(\tilde{\text{P}}^C_{h_\mathcal{T},N})$, and $(\tilde{\text{P}}^{NC}_{h_\mathcal{T},N})$ are solved numerically by Newton-Raphson's method. If we abstract from the specific form of $(\tilde{\text{P}}^\ell_{h_\mathcal{T},N})$ by setting

$$\mathcal{F}([u_h, \mathbf{m}_h]; [w_h, \boldsymbol{\mu}_h]) = 0 \quad \forall \{w_h, \boldsymbol{\mu}_h\} \in \mathcal{S}_0^1(\mathcal{T}) \times \mathcal{L}^0(\mathcal{T}|_\omega)^d\,,$$

this amounts to running the following iteration, for $n \geq 1$,

$$D\mathcal{F}\big([u_h^n, \mathbf{m}_h^n], [w_h, \boldsymbol{\mu}_h]; [u_h^n - u_h^{n+1}, \mathbf{m}_h^n - \mathbf{m}_h^{n+1}]\big) = \mathcal{F}\big([u_h^n, \mathbf{m}_h^n]; [w_h, \boldsymbol{\mu}_h]\big)\,,$$

for all tuples $[w_h, \boldsymbol{\mu}_h] \in \mathcal{S}_0^1(\mathcal{T}) \times \mathcal{L}^0(\mathcal{T}|_\omega)^d$, until a certain tolerance is reached. This calculation is followed at each step by an algebraic update to gain $\{u_h^{n+1}, \mathbf{m}_h^{n+1}\}$.

In the present situation and for given values $\{u_h, \mathbf{m}_h\}$, the bilinear form $D\mathcal{F}([u_h, \mathbf{m}_h], [w_h, \boldsymbol{\mu}_h]; [V_h, \mathbf{M}_h])$ is of the form

$$(\nabla V_h, \boldsymbol{\mu}_h) + \big(\langle \mathbf{M}_h, \mathbf{e}_\perp\rangle_{\mathbb{R}^d}, \langle \boldsymbol{\mu}_h, \mathbf{e}_\perp\rangle_{\mathbb{R}^d}\big) \tag{2.94}$$
$$+\Big(\frac{1}{h_\mathcal{T}} H(|\,\mathbf{m}_h\,| - 1)\langle \text{sgn}(\mathbf{m}_h), \mathbf{M}_h\rangle_{\mathbb{R}^d} \mathbf{m}_h, \boldsymbol{\mu}_h\Big)$$
$$+\big(\frac{1}{h_\mathcal{T}}(\mathbf{m}_h - 1)_+ \mathbf{M}_h, \boldsymbol{\mu}_h\big) + t_\ell\big([\mathbf{m}_h], [\boldsymbol{\mu}_h]\big)\,,$$

for tuples $\{V_h, \mathbf{M}_h\}$, which satisfy $(\nabla V_h, \nabla w_h)_\Omega - (\mathbf{M}_h, \nabla w_h) = 0$, for all $w_h \in \mathcal{S}_0^1(\mathcal{T})$. Here, $H : \mathbb{R} \to \mathbb{R}_0^+$ denotes the Heaviside function.

**Remark 2.8** *According to Example 2.3,* $(\tilde{\text{P}}^\text{A}_{h_\mathcal{T},N})$ *does not necessarily imply uniqueness of solutions. For the purpose of computations, we single out one of them by adding* $\gamma(\mathbf{m}_h, \boldsymbol{\mu}_h)$ *to* (2.94)*, with a small, positive number* $\gamma$ *(i.e.,* $\gamma = 10^{-6}$*).*

In Examples 2.4, 2.6 below the error of the potential $u$ and the magnetization $\mathbf{m}$ are known explicitly. Hence the $L^2(\Omega)$-norm of $u-u_h$, $\nabla(u-u_h)$, $\langle\mathbf{m}-\mathbf{m}_h,\mathbf{e}\rangle_{\mathbb{R}^d}$, and $\langle\mathbf{m}-\mathbf{m}_h,\mathbf{e}_\perp\rangle_{\mathbb{R}^d}$ can be calculated via a 7-point quadrature rule on any triangle to run convergence studies.

**Example 2.4** *Consider* $(\tilde{\mathbf{P}})$ *on* $\Omega=(-1,2)^2$ *and* $\omega=(0,1)^2$ *with exact solution*

$$u(x,y)=(x^2-x-2)(y^2-y-2) \quad \textit{and} \quad \mathbf{m}_k(x,y)=(w_k(x,y),w_k(y,x))^\top, \tag{2.95}$$

*where*

$$w_k(x,y):=\begin{cases}\widetilde{w}_k(x,y) & , \textit{ if } \widetilde{w}_k(x,y)^2+\widetilde{w}_k(y,x)^2\le 1\\ \widetilde{w}_k(x,y)/\sqrt{\widetilde{w}_k(x,y)^2+\widetilde{w}_k(y,x)^2} & , \textit{ else,}\end{cases} \tag{2.96}$$

*for* $\widetilde{w}_k(x,y):=\sin(\pi x)\sin(k\pi y)$. *We choose* $\lambda=0$ *on* $\{\mathbf{x}\in\omega\,|\,|\mathbf{m}_k|<1\}$, *and* $\lambda=1$, *else.*

*The prescribed potential $u$ is smooth and $|\operatorname{div}\mathbf{m}|$ is bounded, but $|\nabla\mathbf{m}|$ is unbounded as $k$ tends to infinity. Given $\mathbf{e}$, the right-hand sides $s$ and $\mathbf{f}$ are computed by (2.91)-(2.92) from $u$ and $\mathbf{m}$ above. For those data, we compute the discrete solution of $(\tilde{\mathrm{P}}^{\ell}_{h_\mathcal{T},N})$, $\ell\in\{\mathrm{A},\mathrm{B}\}$, thus testing both versions of stabilization.*

*Let the direction of the easy axis be $\mathbf{e}=(2,1)^\top/\sqrt{5}$ in $\omega$. As initial mesh we use a partition of $\Omega$ into triangles with $h_\mathcal{T}=\sqrt{2}/15$ which are halved squares. If not stated otherwise, we choose $\beta_{\boldsymbol{\xi}}=1$, $\forall\boldsymbol{\xi}\in\Xi^*$, in all discussed numerical examples below.*

*Figure 2.14 shows the computed magnetization $\mathbf{m}_h$ for a meshing $\mathcal{T}$ aligned to $\mathbf{e}$ and obtained by $(\tilde{\mathrm{P}}^{C}_{h_\mathcal{T},N})$ (left) and $(\tilde{\mathrm{P}}^{\mathrm{A}}_{h_\mathcal{T},N})$ (right), where the first one shows spurious oscillations that are avoided by the stabilized analogon. The approximated magnetization is depicted by vectors given in the center of each triangle and the color-bars show the modulus of the magnetization.*

*Tables 2.1 and 2.2 gather convergence rates experimentally obtained for $u-u_h$ in the $W_0^{1,2}(\Omega)$-seminorm and the $L^2(\Omega)$-norm, for $k=4$ and both types of stabilization $(\tilde{\mathrm{P}}^{\ell}_{h_\mathcal{T},N})$, $\ell\in\{\mathrm{A},\mathrm{B}\}$. The computed convergence rates $\xi_h$ coincide with the theoretical predictions in Theorem 2.9 in both cases. Further, Figures 2.15 and 2.16 provide a direct comparison of all numerical schemes discussed above for $\mathbf{e}=(1,0)^\top$ and $k=4$. These results show a*

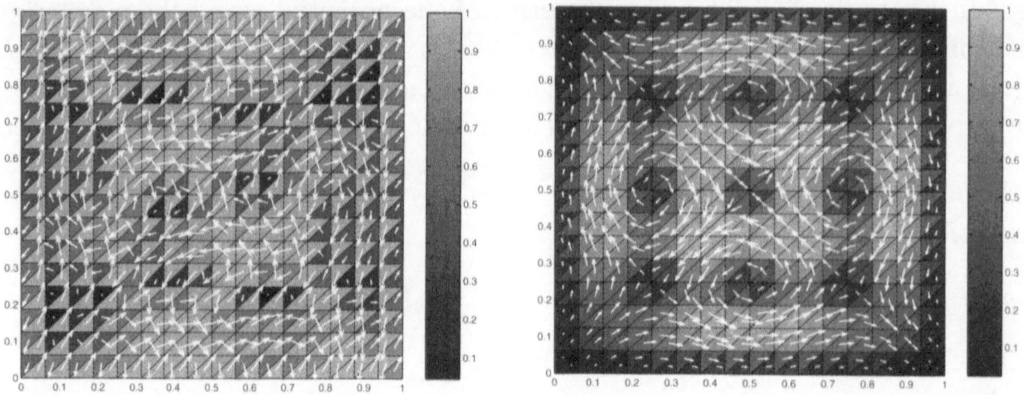

Figure 2.14: Plot of magnetization $\mathbf{m}_h$ computed with no stabilization (left) and stabilization $(\tilde{\mathrm{P}}^{\mathrm{A}}_{h_\mathcal{T},N})$ (right) for $\mathbf{e} = (1,0)^\top$ and $k = 4$.

*convergence behavior of first order for both methods* $(\tilde{\mathrm{P}}^{\ell}_{h_\mathcal{T},N})$, $\ell \in \{\mathrm{A},\mathrm{B}\}$ *for error* $= \| \nabla(u - u_h) \|_{\mathbf{L}^2(\Omega,\mathbb{R}^2)} + \| \langle \mathbf{m} - \mathbf{m}_h, \mathbf{e}_\perp \rangle_{\mathbb{R}^2} \|_{L^2(\omega)}$. *Corresponding results hold for the conforming (non-stabilized) method* $(\tilde{\mathrm{P}}^{C}_{h_\mathcal{T},N})$ *and the non-conforming method* $(\tilde{\mathrm{P}}^{NC}_{h_\mathcal{T},N})$ *as well, although shifting of corresponding curves reflects an increased number of unknowns in the latter case. (A slope* $-1/2$, *resp.* $-1$ *in Figures 2.15 and 2.16 corresponds to an experimental convergence rate 1, resp. 2 owing to* $N \propto h^{-2}$ *in two dimensions.) As to the error* $\| u - u_h \|_{L^2(\Omega)}$, *improved convergence behavior for both methods* $(\tilde{\mathrm{P}}^{\ell}_{h_\mathcal{T},N})$, $\ell \in \{\mathrm{A},\mathrm{B}\}$ *over the remaining two is observed: numerical experiments for both stabilized schemes* $(\tilde{\mathrm{P}}^{\ell}_{h_\mathcal{T},N})$, $\ell \in \{\mathrm{A},\mathrm{B}\}$ *show double the rate of convergence for* $\| u - u_h \|_{L^2(\Omega)}$, *if compared to* $\| \nabla(u - u_h) \|_{\mathbf{L}^2(\Omega,\mathbb{R}^2)}$.

*The present example of a meshing that is aligned with* $\mathbf{e}$ *has been chosen to reflect non-stable behavior of* $(\tilde{\mathrm{P}}^{C}_{h_\mathcal{T},N})$ *as already observed in Subsection 2.1.4. This can be seen from Figure 2.16 (left), which shows poor convergence of this method. However, the stabilizing effect in* $(\tilde{\mathrm{P}}^{\ell}_{h_\mathcal{T},N})$, $\ell \in \{\mathrm{A},\mathrm{B}\}$ *is evident from corresponding curves in this figure. Numerical experiments for stabilization* $(\tilde{\mathrm{P}}^{\mathrm{B}}_{h_\mathcal{T},N})$ *show asymptotically the same convergence behavior as* $(\tilde{\mathrm{P}}^{\mathrm{A}}_{h_\mathcal{T},N})$ *or the non-conforming scheme* $(\tilde{\mathrm{P}}^{NC}_{h_\mathcal{T},N})$, *but the error* $\|\langle \mathbf{m} - \mathbf{m}_h, \mathbf{e}\rangle_{\mathbb{R}^2}\|_{L^2(\omega)}$ *is larger up to a factor 7 for this kind of stabilization. This observation concerning performance of* $(\tilde{\mathrm{P}}^{\ell}_{h_\mathcal{T},N})$, $\ell \in \{\mathrm{A},\mathrm{B}\}$ *has been made in all test studies for 2D-examples, involving different aligned and non-aligned meshes.*

| N | $\| \nabla(u - u_h) \|_{L^2(\Omega)}$ | $\xi_h$ | $\| u - u_h \|_{L^2(\Omega)}$ | $\zeta_h$ |
|---|---|---|---|---|
| 65 | 3.2506 | | 0.7733 | |
| 233 | 1.6377 | 1.07 | 0.1487 | 2.58 |
| 881 | 0.8236 | 1.03 | 0.0340 | 2.22 |
| 3425 | 0.4122 | 1.02 | 0.0071 | 2.31 |
| 13505 | 0.2058 | 1.01 | 0.0016 | 2.20 |
| 53633 | 0.1028 | 1.07 | 0.0005 | 1.76 |

Table 2.1: Error $\| \nabla(u - u_h) \|_{\mathbf{L}^2(\Omega,\mathbb{R}^2)}$, resp. $\| u - u_h \|_{L^2(\Omega)}$ and convergence rate $\xi_h$, resp. $\zeta_h$ in Example 2.4 for $(\tilde{\mathrm{P}}^{\mathrm{A}}_{h_\mathcal{T},N})$ $(k = 4)$.

| N | $\| \nabla(u - u_h) \|_{L^2(\Omega)}$ | $\xi_h$ | $\| u - u_h \|_{L^2(\Omega)}$ | $\zeta_h$ |
|---|---|---|---|---|
| 65 | 3.2516 | | 0.7777 | |
| 233 | 1.6392 | 1.07 | 0.1551 | 2.53 |
| 881 | 0.8272 | 1.03 | 0.0376 | 2.13 |
| 3425 | 0.4194 | 1.00 | 0.0104 | 1.90 |
| 13505 | 0.2118 | 0.99 | 0.0042 | 1.30 |
| 53633 | 0.1056 | 1.02 | 0.0017 | 1.31 |

Table 2.2: Error $\| \nabla(u - u_h) \|_{\mathbf{L}^2(\Omega,\mathbb{R}^2)}$, resp. $\| u - u_h \|_{L^2(\Omega)}$ and convergence rate $\xi_h$, resp. $\zeta_h$ in Example 2.4 for $(\tilde{\mathrm{P}}^{\mathrm{B}}_{h_\mathcal{T},N})$ $(k = 4)$.

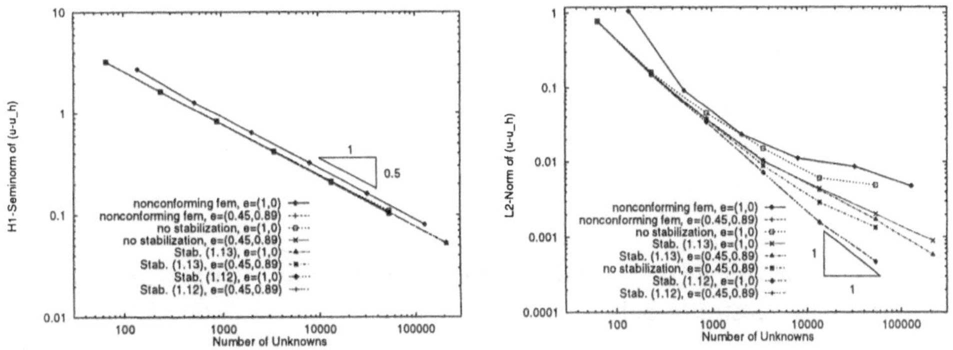


Figure 2.15: Errors $\| \nabla(u - u_h) \|_{\mathbf{L}^2(\Omega,\mathbb{R}^2)}$ and $\| u - u_h \|_{L^2(\Omega)}$ vs. number of unknowns $N$ in Example 2.4 $(k = 4)$.

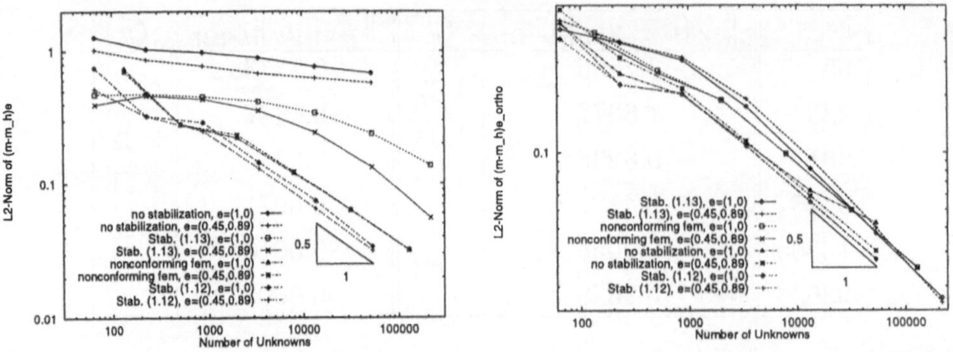


Figure 2.16: Errors $\| \langle \mathbf{m} - \mathbf{m}_h, \mathbf{e} \rangle_{\mathbb{R}^2} \|_{\mathbf{L}^2(\omega,\mathbb{R}^2)}$ and $\| \langle \mathbf{m} - \mathbf{m}_h, \mathbf{e}_\perp \rangle_{\mathbb{R}^2} \|_{\mathbf{L}^2(\omega,\mathbb{R}^2)}$ vs. number of unknowns $N$ in Example 2.4 ($k = 4$).

*In a next step, we study the performance of the stabilized methods* $(\tilde{\mathrm{P}}^{\ell}_{h_\mathcal{T},N})$, $\ell \in \{\mathrm{A}, \mathrm{B}\}$ *in the case of highly oscillatory magnetizations. For a fixed number of unknowns* ($N = 13505$ *and* $N = 53653$), *the dependence of* $\|\langle \mathbf{m} - \mathbf{m}_h, \mathbf{e} \rangle_{\mathbb{R}^2}\|_{L^2(\omega)}$ *and* $\|\langle \mathbf{m} - \mathbf{m}_h, \mathbf{e}_\perp \rangle_{\mathbb{R}^2}\|_{L^2(\omega)}$ *on the parameter* $k$ *is shown in Figure 2.17.*

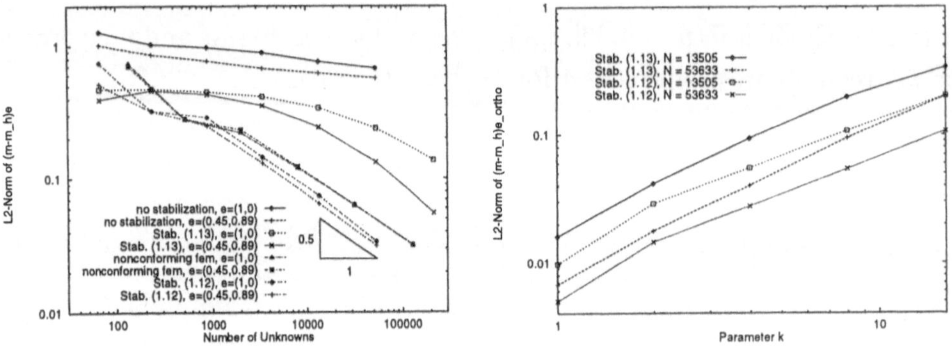


Figure 2.17: Errors $\| \langle \mathbf{m} - \mathbf{m}_h, \mathbf{e} \rangle_{\mathbb{R}^2} \|_{L^2(\omega)}$ and $\| \langle \mathbf{m} - \mathbf{m}_h, \mathbf{e}_\perp \rangle_{\mathbb{R}^2} \|_{L^2(\omega)}$ vs parameter $k$ in Example 2.4.

*As mentioned before,* $|\operatorname{div} \mathbf{m}|$ *is bounded and* $|\nabla \mathbf{m}|$ *becomes unbounded as* $k$ *tends to infinity. This implies reduced convergence behavior for* $\|\langle \mathbf{m} - \mathbf{m}_h, \mathbf{e}_\perp \rangle_{\mathbb{R}^2}\|_{L^2(\omega)}$ *in the case of* $(\tilde{\mathrm{P}}^{\mathrm{B}}_{h_\mathcal{T},N})$ *opposed to* $(\tilde{\mathrm{P}}^{\mathrm{A}}_{h_\mathcal{T},N})$, *as* $k$ *grows larger, illustrating a higher degree of robustness of* $(\tilde{\mathrm{P}}^{\mathrm{A}}_{h_\mathcal{T},N})$ *for non-smooth solutions.*

*The following two Figures 2.18 and 2.19 show numerical results to analyze the dependence of* $\| \nabla(u-u_h) \|_{\mathbf{L}^2(\Omega,\mathbb{R}^2)}$, $\| u-u_h \|_{L^2(\Omega)}$, $\|\langle \mathbf{m}-\mathbf{m}_h, \mathbf{e}\rangle_{\mathbb{R}^2}\|_{L^2(\omega)}$, *and* $\|\langle \mathbf{m}-\mathbf{m}_h, \mathbf{e}_\perp\rangle_{\mathbb{R}^2}\|_{L^2(\omega)}$ *on the stabilization factor* $\beta$, *for* $\mathbf{e} = (2,1)^\top/\sqrt{5}$, *and* $k = 4$.

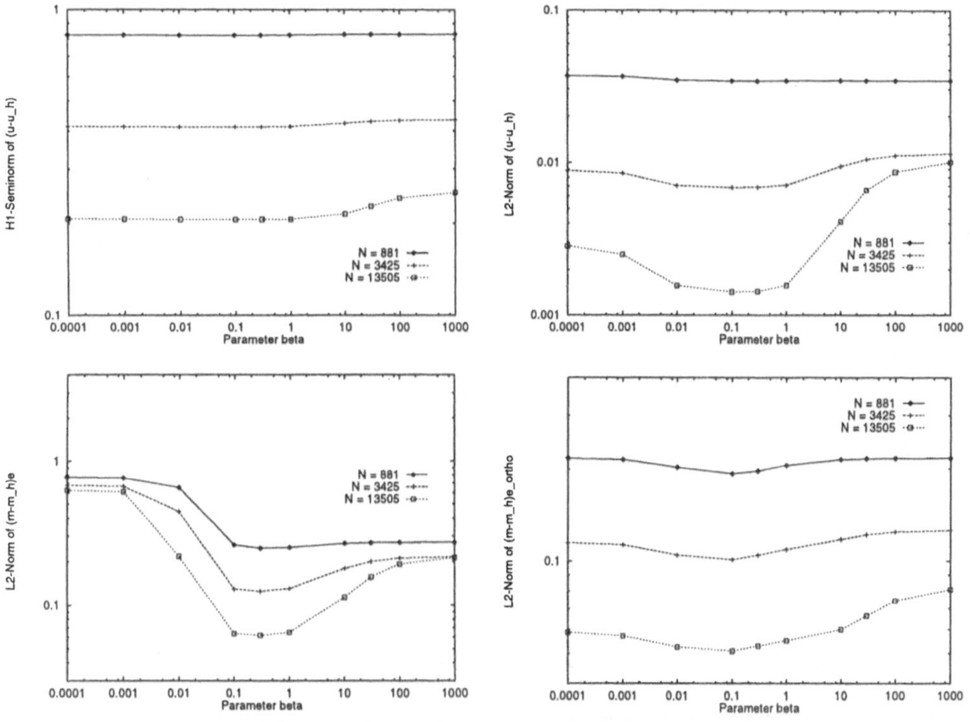


Figure 2.18: Errors vs. stabilization factor $\beta$ in Example 2.4 for $(\tilde{\mathrm{P}}^{\mathrm{A}}_{h_\mathcal{T},N})$.

*As to* $(\tilde{\mathrm{P}}^{\mathrm{A}}_{h_\mathcal{T},N})$, *all considered quantities behave almost insensitive on a broad range of* $\beta$, *and optimal choices are* $\beta = \beta_{\boldsymbol{\xi}} \in (0.1, 1)$. *A slightly more sensitive behavior can be stated for* $(\tilde{\mathrm{P}}^{\mathrm{B}}_{h_\mathcal{T},N})$. *Hence, these observations underline flexibility of the parameter choice for* $\beta$ *with respect to the nonlinear mechanisms inherent to the problem for both stabilized schemes.*

**Example 2.5** *As a more applied example, we consider the following problem, where* $\Omega = (-5.5, 5.5)^2$, $\omega = (-0.5, 0.5) \times (-2.5, 2.5)$, $\mathbf{f} = (0, 1.1)^\top$ *and* $\mathbf{e} = (2,1)^\top$ *in* $(\mathrm{P}^{\mathrm{A}}_{h_\mathcal{T},N})$. *The setting of* $\Omega$, $\omega$, $\mathbf{f}$ *and* $\mathbf{e}$ *is depicted in Figure 2.20. This problem models the potential and the magnetization in a symmetric configuration where the easy axis* $\mathbf{e}$ *is not aligned with the coordinate axis. A plot of the solution potential* $u_h$ *with* $h = \frac{1}{4}\sqrt{2}$ *and the computed magnetization* $\mathbf{m}_h$ *is shown in Figure 2.20.*

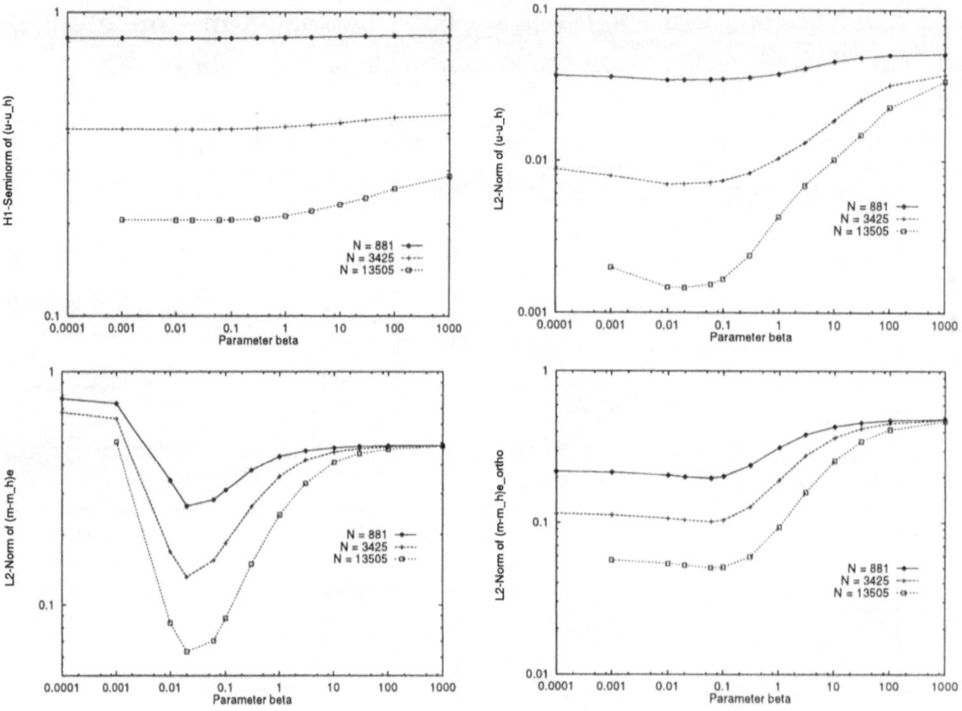


Figure 2.19: Errors vs. stabilization factor $\beta$ in Example 2.4 for $(\tilde{\mathrm{P}}^{\mathrm{B}}_{h_{\mathcal{T}},N})$.

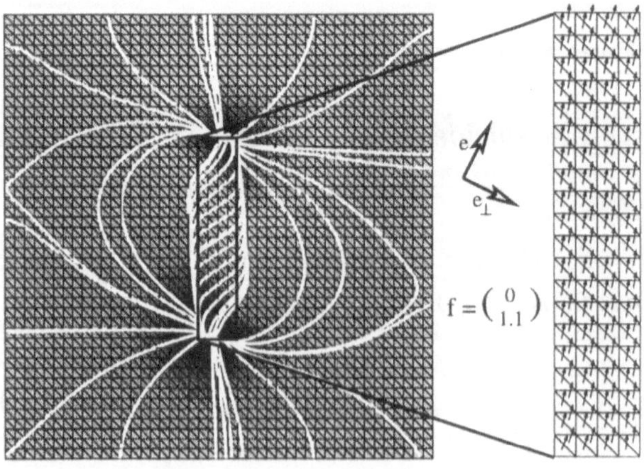


Figure 2.20: Solution of Example 2.5, using $(\mathrm{P}^{\mathrm{A}}_{h_{\mathcal{T}},N})$.

*The streamlines in Figure 2.20 provide some insight into gradients of the magnetic potential. Due to the non-alignment of coordinate axis and easy axis* $\mathbf{e}$*, the streamlines are not symmetric with respect to the coordinate axis.*

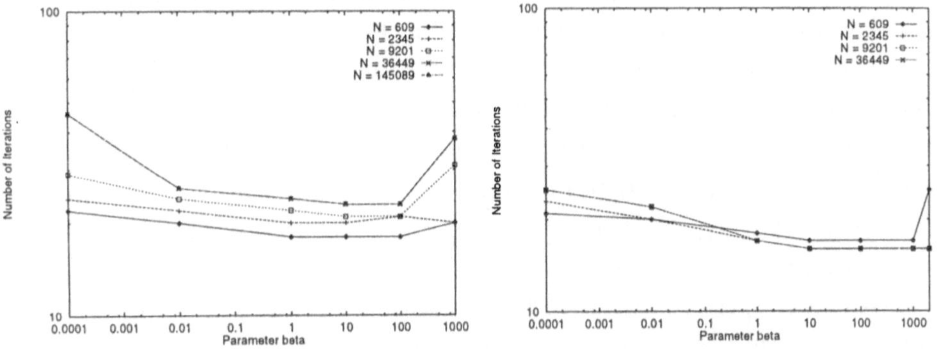


Figure 2.21: Number of Newton-iterations vs. number of unknowns $N$ in Example 2.4 for $(\tilde{\mathrm{P}}^{\mathrm{A}}_{h_\mathcal{T},N})$ (left) and $(\tilde{\mathrm{P}}^{\mathrm{B}}_{h_\mathcal{T},N})$ (right).

**Example 2.6** *We consider a 3D-example for* $(\tilde{\mathrm{P}}^{\ell}_{h_\mathcal{T},N})$*,* $\ell \in \{\mathrm{A}, \mathrm{B}\}$ *and calculate approximations in the case of an aligned mesh* $(\mathbf{e} = (0,0,1)^\top)$ *and an unaligned mesh* $(\mathbf{e} = (1,2,3)^\top/\sqrt{14})$*. Let* $\Omega = (-1,2)^3$ *and* $\omega = (0,1)^3$*. The right-hand sides* $s$ *and* $\mathbf{f}$ *are determined by the exact solution*

$$u(x,y,z) = (x^2 - x - 2)(y^2 - y - 2)(z^2 - z - 2)\,, \tag{2.97}$$

$$\mathbf{m}_k(x,y,z) = \begin{cases} w_k(x,y,z) & , \text{ if } |w_k(x,y,z)| \le 1 \\ w_k(x,y,z)/|w_k(x,y,z)| & , \text{ else} \end{cases} \tag{2.98}$$

*where*

$$w_k(x,y,z) = \begin{pmatrix} \sin(k\pi x)\sin(\pi y)\sin(\pi z) \\ \sin(\pi x)\sin(k\pi y)\sin(\pi z) \\ \sin(\pi x)\sin(\pi y)\sin(\pi z) \end{pmatrix},$$

*and*

$$\lambda(x,y,z) = \begin{cases} 0 & , \text{ if } |\mathbf{m}_k(x,y,z)| < 1\,, \\ 1 & , \text{ else}\,. \end{cases} \tag{2.99}$$

*Similar to Example 2.4, the prescribed magnetic potential $u$ is smooth and $|\operatorname{div}\mathbf{m}|$ is bounded, but $|\nabla\mathbf{m}|$ is unbounded as $k$ tends to infinity.*

*Division of $\Omega$ into equally sized cubes and subdivision of these cubes into five tetrahedra give the uniform triangulation $\mathcal{T}$ used in our computations. To integrate quantities like $\int_\Omega s\eta\,\mathrm{d}\mathbf{x}$, $\int_\omega f\theta\,\mathrm{d}\mathbf{x}$, etc. we use a Gaussian quadrature rule of order 6 [75] where $\eta$, $\theta$ denote basis functions.*

*For any $\mathbf{e}_\perp \in \operatorname{span}\{\mathbf{e}_\perp^1, \mathbf{e}_\perp^2\}$, such that $\mathbb{R}^3 = \operatorname{span}\{\mathbf{e}, \mathbf{e}_\perp^1, \mathbf{e}_\perp^2\}$, we compute $\|\langle\mathbf{m}-\mathbf{m}_h, \mathbf{e}_\perp\rangle_{\mathbb{R}^3}\|^2_{L^2(\omega)} := \|\langle\mathbf{m}-\mathbf{m}_h, \mathbf{e}_\perp^1\rangle_{\mathbb{R}^3}\|^2_{L^2(\omega)} + \|\langle\mathbf{m}-\mathbf{m}_h, \mathbf{e}_\perp^2\rangle_{\mathbb{R}^3}\|^2_{L^2(\omega)}$. For two easy axes $\mathbf{e} = (0,0,1)^\top$ and $\mathbf{e} = (1,2,3)^\top/\sqrt{14}$, we compute approximations $\{u_h, \mathbf{m}_h, \lambda_h\}$ for $(\mathrm{P}^{\mathrm{A}}_{h_\mathcal{T},N})$, $(\mathrm{P}^{\mathrm{B}}_{h_\mathcal{T},N})$, and without stabilization. The corresponding numerical convergence results for $\|\nabla(u-u_h)\|_{\mathbf{L}^2(\Omega,\mathbb{R}^3)}$, $\|u-u_h\|_{L^2(\Omega)}$, $\|\langle\mathbf{m}-\mathbf{m}_h, \mathbf{e}\rangle_{\mathbb{R}^3}\|_{L^2(\omega)}$, and $\|\langle\mathbf{m}-\mathbf{m}_h, \mathbf{e}_\perp\rangle_{\mathbb{R}^3}\|_{L^2(\omega)}$ vs. the number of unknowns $N$ are plotted in Figure 2.22.*

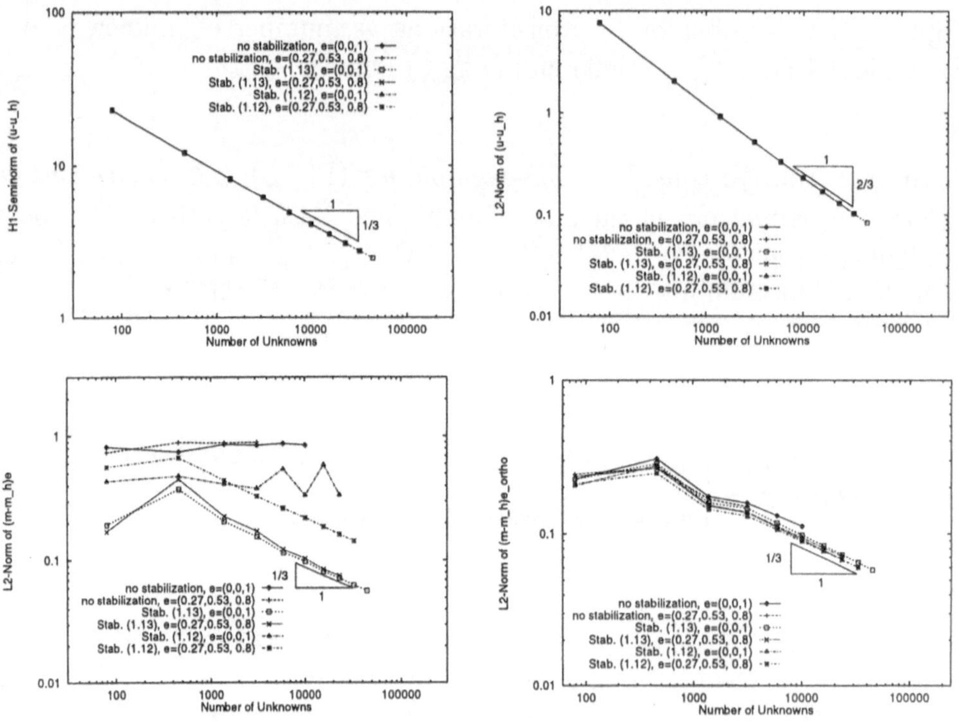


Figure 2.22: Errors $\|\nabla(u-u_h)\|_{\mathbf{L}^2(\Omega,\mathbb{R}^3)}$, $\|u-u_h\|_{L^2(\Omega)}$, $\|\langle\mathbf{m}-\mathbf{m}_h, \mathbf{e}\rangle_{\mathbb{R}^3}\|_{L^2(\omega)}$, and $\|\langle\mathbf{m}-\mathbf{m}_h, \mathbf{e}_\perp\rangle_{\mathbb{R}^3}\|_{L^2(\omega)}$ vs. number of unknowns $N$ in Example 2.6 ($k=3$).

*For a sequence of uniform meshes, we verify experimentally first order of convergence for error* $= \| \nabla(u - u_h) \|_{\mathbf{L}^2(\Omega,\mathbb{R}^3)} + \| \langle \mathbf{m} - \mathbf{m}_h, \mathbf{e}_\perp \rangle_{\mathbb{R}^3} \|_{L^2(\omega)}$ *for both stabilized methods. (A slope* $-1/3$*, resp.* $-2/3$ *in Figure 2.22 corresponds to an experimental convergence rate 1, resp. 2 owing to* $N \propto h^{-3}$ *in three dimensions.) Our numerical experiments show second order convergence rate for* $\| u - u_h \|_{L^2(\Omega)}$ *and all meshes (*$N \leq 44791$*). Without stabilization we observe no convergence of* $\|\langle \mathbf{m} - \mathbf{m}_h, \mathbf{e} \rangle_{\mathbb{R}^3}\|_{L^2(\omega)}$ *for both easy axes* $\mathbf{e}$ *and only poor convergence for* $(\mathbf{P}^{\mathrm{A}}_{h\mathcal{T},N})$ *if* $\mathbf{e}$ *is aligned with the mesh* $\mathcal{T}$*. If* $\mathbf{e}$ *is not aligned with* $\mathcal{T}$*, we observe convergence which is asymptotically first order in* $(\mathrm{P}^{\mathrm{A}}_{h\mathcal{T},N})$*. Only for stabilization method* $(\mathrm{P}^{\mathrm{A}}_{h\mathcal{T},N})$*, we get a first order convergence rate for both easy axes* $\mathbf{e}$*. Again, there is no theoretical justification concerning convergence rates for* $\|\langle \mathbf{m} - \mathbf{m}_h, \mathbf{e} \rangle_{\mathbb{R}^3}\|_{L^2(\omega)}$.

**Example 2.7** *Similar to Example 2.5, the approximate magnetization is computed in a 3D ferromagnetic rod by means of a stabilized method, using the data* $\Omega = (0,3)^3$, $\omega = (1,2) \times (4/3, 5/3)^2$, $\mathbf{f} = (1,0,0)^\top$, $\mathbf{e} = (1,2,3)^\top/\sqrt{14}$*, and* $\beta = 0.1$ *in* $(\mathrm{P}^{\mathrm{B}}_{h\mathcal{T},N})$*. A plot of the magnetic potential* $u_h$ *with* $h = 0.2$ *as well as some magnified details of the computed magnetization* $\mathbf{m}_h$ *are shown in Figure 2.23.*

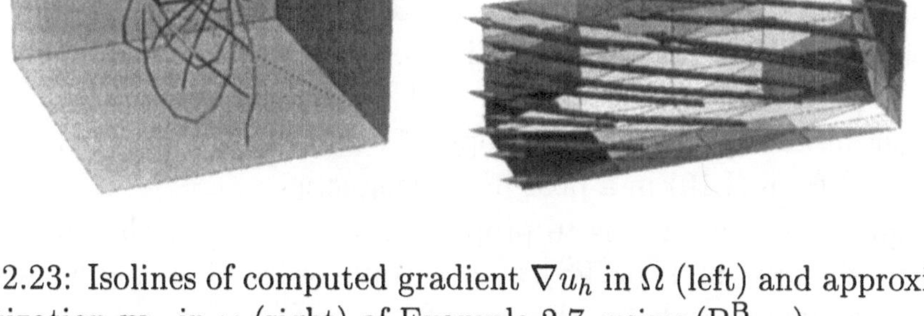

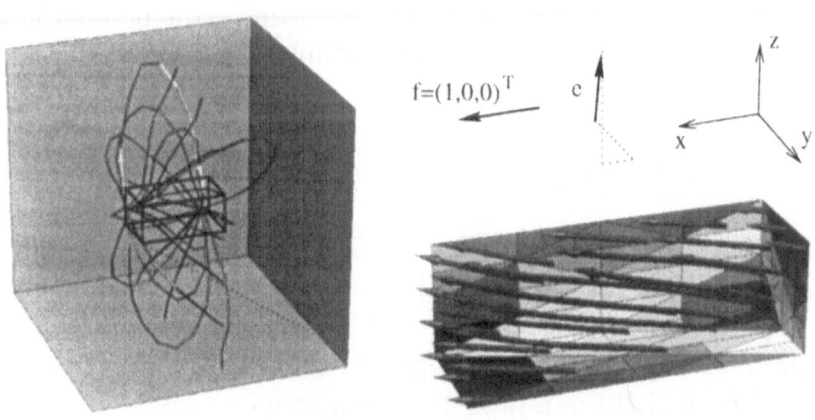

Figure 2.23: Isolines of computed gradient $\nabla u_h$ in $\Omega$ (left) and approximated magnetization $\mathbf{m}_h$ in $\omega$ (right) of Example 2.7, using $(\mathrm{P}^{\mathrm{B}}_{h\mathcal{T},N})$.

## 2.3 Strong convergence in convexified micromagnetics

System (I.10)-(I.12) resolves essential macroscopic properties of rigid (ferro-) magnetic materials by deleting fine-scale oscillations in the classical (non-convex) micromagnetic model of Weiss, Landau, and Lifshitz, see [17]. According to Sections 2.1 and 2.2, there are efficient, stable schemes available to compute its solution numerically.[6]

On the other hand, physicists have considerable interest in the study of magnetic microstructures like Weissian domains inside the ferromagnetic rigid body which can be described in terms of Young measures. The main impedience towards reaching this goal is degeneracy of system (I.10)-(I.12) which prevents us to verify results concerning strong convergence of finite element solutions $\mathbf{m}_h$ towards its solution $\mathbf{m}$ (in $\mathbf{L}^2(\omega, \mathbb{R}^2)$). It is this missing link that prevents us to find reliable statements on approximation of the microstructure for the methods in [20, 52], see the discussion at the end of Subsection 2.1.5 and also below. In the following, we confine to *uniaxial* materials like cobalt with easy axis $\mathbf{e} \in \mathbb{R}^2$. Then, $\phi^{**}(\mathbf{m}) = \frac{1}{2}\langle \mathbf{m}, \mathbf{e}_\perp \rangle_{\mathbb{R}^2}^2$, and uniqueness of solutions to (I.10)-(I.12) is known from Theorem 2.2. The corresponding Young measure is derived from (I.19).

For many interesting uniaxial materials, $\phi^{**}$ is known, which is the basis of numerical analysis for (2.3)-(2.5). The following error bounds have been verified in Sections 2.1 and 2.2 for both, stable non-conforming as well as stabilized continuous finite elements,

$$\| \nabla(u - u_h) \|_{\mathbf{L}^2(\Omega,\mathbb{R}^2)} + \| \langle \mathbf{m} - \mathbf{m}_h, \mathbf{e}_\perp \rangle_{\mathbb{R}^2} \|_{L^2(\omega)} \leq C\, h\,, \tag{2.100}$$

provided that $u \in W^{2,2}(\Omega)$, $\mathbf{m} \in \mathbf{W}^{1,2}(\omega, \mathbb{R}^2)$. In particular, no convergence statement is available for the magnetization $\mathbf{m}_h$, which is conjectured to be weak in $\mathbf{L}^2(\omega, \mathbb{R}^2)$; see Subsection 2.1.5. This lack prevents us from drawing conclusions on the approximation properties of the (discrete) Young measure $\nu_h$ computed from (I.19) in a post-processing step.

The goal of this section is to propose schemes $(\mathrm{P})_h^{i,j}$ such that solutions $\mathbf{m}_h \to \mathbf{m}$ strongly in $\mathbf{L}^2(\omega, \mathbb{R}^2)$ at a certain order of $h$, giving way to a strong convergence result for the mass-fraction $\Gamma(\mathbf{m}_h)$ to reliably compute

---

[6] The analysis in this section is given for the case $d = 2$. A corresponding analysis for $d = 3$ is more involved.

the discrete Young measure $\nu_h$, for $\Gamma_h(\mathbf{m}_h) = \frac{1}{2} + \frac{\langle \mathbf{m}_h, \mathbf{e}\rangle_{\mathbb{R}^2}}{2(|\mathbf{m}_h|_\bullet^2 - \langle \mathbf{m}_h, \mathbf{e}_\perp\rangle_{\mathbb{R}^2}^2)^{1/2}}$, and $(s)_\bullet := \max\{1, s\}$,

$$\nu_h \equiv \Gamma_h(\mathbf{m}_h)\delta_{\mathbf{m}^+(\mathbf{m}_h)} + \big(1 - \Gamma_h(\mathbf{m}_h)\big)\delta_{\mathbf{m}^-(\mathbf{m}_h)} \,.$$

We expect our results presented below to also apply to situations where $\phi^{**}$ is known and satisfies the monotonicity property (2.26).

Consider scheme $(\mathbf{P})_h^{i,j}$, for $i,j \in \{c, nc\}$, where '$c$' resp. '$nc$' denote conforming resp. non-conforming discretization for the spaces introduced before: Given $\varepsilon_0, \varepsilon_1, \varepsilon_2 > 0$, find $\{u_h, \mathbf{m}_h\} \in \mathcal{X}_0^i(\Omega) \times \mathcal{Y}^j(\omega)$, for $\mathcal{X}_0^i \in \{\mathcal{S}_0^1, \mathcal{S}_0^{1,NC}\}$, $\mathcal{Y}^j \in \{\mathcal{S}^1, \mathcal{S}^{1,NC}\}$ such that for all $\{w_h, \chi_h\} \in \mathcal{X}_0^i(\Omega) \times \mathcal{Y}^j(\omega)$,

$$(\nabla_{\mathcal{T}} u_h, \nabla_{\mathcal{T}} w_h)_\Omega = (\mathbf{m}_h, \nabla_{\mathcal{T}} w_h)\,, \tag{2.101}$$

$$\varepsilon_0 \sum_{\xi \in \Xi^*} \int_\xi \langle [\partial_{\mathbf{n}} \mathbf{m}_h], [\partial_{\mathbf{n}} \chi_h]\rangle_{\mathbb{R}^4} \, \mathrm{d}\mathbf{x} + \varepsilon_1 (\nabla \mathbf{m}_h, \nabla \chi_h) \tag{2.102}$$

$$+(\nabla u_h, \chi_h) + \big(\langle \mathbf{m}_h, \mathbf{e}_\perp\rangle_{\mathbb{R}^d}, \langle \chi_h, \mathbf{e}_\perp\rangle_{\mathbb{R}^d}\big) + \big(\lambda_h \mathbf{m}_h, \chi_h\big) = (\mathbf{f}, \chi_h)\,,$$

$$\partial_{\mathbf{n}} \mathbf{m}_h\big|_{\partial\omega} = 0\,, \tag{2.103}$$

where $\lambda_h = \frac{1}{\varepsilon_2} \frac{(|\mathbf{m}_h| - 1)_+}{|\mathbf{m}_h|}$. Note that we require unphysical homogeneous Neumann boundary data for the computed magnetization. The main result of this section is stated in the following theorem.

**Theorem 2.10** *Let* $\phi(\mathbf{m}) = \frac{1}{2}\langle \mathbf{m}, \mathbf{e}_\perp\rangle_{\mathbb{R}^2}^2$. *Suppose that* $\{\mathbf{m}, u, \lambda\}$ *solves (I.10)-(I.12) for* $d = 2$, *and* $\{\mathbf{m}_h, u_h, \lambda_h\}$ *solves* $(P)_h^{c,c}$. *Let* $\varepsilon_0 = \mathcal{O}(h)$, $\varepsilon_1 = \mathcal{O}(h^{3/2})$, $\varepsilon_2 = \mathcal{O}(h^2)$. *There exists a constant* $C$ *independent of* $h$, *such that for* $\{\mathbf{m}, u\} \in \mathbf{W}^{3,2}(\omega, \mathbb{R}^2) \times W^{2,2}(\Omega)$,

$$\text{(a)} \qquad \|\nabla(u - u_h)\|_{\mathbf{L}^2(\Omega,\mathbb{R}^2)} + \|\langle \mathbf{m} - \mathbf{m}_h, \mathbf{e}_\perp\rangle_{\mathbb{R}^2}\|_{L^2(\omega)}$$
$$+\frac{h}{2}\|\lambda_h \mathbf{m}_h\|_{\mathbf{L}^2(\omega,\mathbb{R}^2)} + \frac{\sqrt{h}}{2}\Big(\sum_{\xi\in\Xi^*}\int_\xi |[\mathbf{m} - \mathbf{m}_h]|^2 \, \mathrm{d}\mathbf{x}\Big)^{1/2}$$
$$+h^{1/2}\|\mathbf{m} - \mathbf{m}_h\|_{\mathbf{L}^2(\omega,\mathbb{R}^2)} + h^{3/4}\|\nabla(\mathbf{m} - \mathbf{m}_h)\|_{\mathbf{L}^2(\omega,\mathbb{R}^4)} \le C\,h\,.$$

*Moreover, there holds for* $\omega' \Subset \omega$, *such that* $\mathrm{dist}(\partial\omega, \omega') \ge \mathcal{O}(\sqrt{h})$,

$$\text{(b)} \qquad \|\mathbf{m} - \mathbf{m}_h\|_{\mathbf{L}^2(\omega',\mathbb{R}^2)} \le C\Big\{h^{5/8} + \mathrm{dist}^{-1/2}(\partial\omega, \omega')h^{3/4}\Big\}\,.$$

**Remark 2.9** *1. We find $\varepsilon_2 = \mathcal{O}(h^2)$, in contrast to $\varepsilon_2 = \mathcal{O}(h)$ in Sections 2.1 and 2.2, where piecewise constant magnetizations $\mathbf{m}_h$ are used; this changed scenario comes from different interaction between regularization (scaled by $\varepsilon_0, \varepsilon_1$) and penalization term (scaled by $\varepsilon_2$) for the present finite element scheme.*

*2. Note the reduced order of convergence for $\mathbf{m}_h$ if compared to $\langle \mathbf{m}_h, \mathbf{e}_\perp \rangle_{\mathbb{R}^2}$ which reflects degeneracy of the problem.*

*3. This convergence result implies an $\mathbf{L}^\infty(\omega, \mathbb{R}^2)$-result for the magnetization $\mathbf{m}_h \in C^0(\omega, \mathbb{R}^2)$ on quasi-uniform meshes that allow for inverse estimates [29]: by inverse inequality, we can conclude that the term $\| \nabla_h \mathbf{m}_h \|_{\mathbf{L}^{8/3}(\omega,\mathbb{R}^4)}$ is uniformly bounded for $h \to 0$. Next, $W^{1,q}(\omega) \hookrightarrow L^\infty(\omega)$, for any $q > 2$, and $\omega \subset \mathbb{R}^2$ bounded. Interpolation of $L^q(\omega)$, $2 < q \leq \frac{8}{3}$, between $L^2(\omega)$ and $L^{8/3}(\omega)$ then implies existence of a constant $C_q$ (with $\lim_{q\to 2} C_q = \infty$), such that*

$$\begin{aligned} \| \mathbf{m} - \mathbf{m}_h \|_{\mathbf{L}^\infty(\omega,\mathbb{R}^2)} \leq \; & C_q \| \nabla(\mathbf{m} - \mathbf{m}_h) \|_{\mathbf{L}^2(\omega,\mathbb{R}^2)}^{\frac{8-3q}{q}} \\ & \times \| \nabla_h(\mathbf{m} - \mathbf{m}_h) \|_{\mathbf{L}^{8/3}(\omega,\mathbb{R}^2)}^{\frac{4q-8}{q}} . \end{aligned} \tag{2.104}$$

*Hence, we can asymptotically find convergence order strictly less than $\frac{1}{4}$.*

*4. The results in (a) are also valid in case that the leading term in (2.102) gives only exchange interaction along certain directions, i.e., $\varepsilon_1(\nabla_\mathcal{T} \mathbf{m}_h, \nabla_\mathcal{T} \chi_h)$ is replaced by $\varepsilon_1(\nabla_\mathcal{T} \mathbf{m}_h \cdot \mathbf{e}, \nabla_\mathcal{T} \chi_h \cdot \mathbf{e})$. This limits the stabilization effect to the direction along the easy axis and reduces computational work. By evidence, this implies the control of $\| \nabla_\mathcal{T}(\mathbf{m} - \mathbf{m}_h) \cdot \mathbf{e} \|_{\mathbf{L}^2(\omega,\mathbb{R}^2)} \leq C\, h^{1/4}$, rather than for the whole gradient.*

*5. Choosing discontinuous finite elements in $(P)_h^{i,j}$, $i = nc$ and/or $j = nc$ is possible. In the analysis, this leads to additional terms, stemming e.g. from Sobolev-type estimates, or from the necessity to verify error bounds in $\mathbf{W}^{-1,2}(\omega, \mathbb{R}^2)$.*

*6. The assumption of strong solutions to (I.10)-(I.12) is an open question; the subsequent analysis is provided for solutions $\{\mathbf{m}, u\} \in \mathbf{W}^{3,2}(\omega, \mathbb{R}^2) \times W_0^{1,2}(\Omega) \cap W^{2,2}(\Omega)$.*

*7. We can drop the term in (2.102) that is scaled by $\varepsilon_0 > 0$. This would imply weaker convergence statements in Theorem 2.10 (even under weakened regularity assumptions $\mathbf{m} \in \mathbf{W}^{2,2}(\omega, \mathbb{R}^2)$). In contrast, we cannot verify convergence rates for $\nabla \mathbf{m}$ in $\mathbf{L}^2(\omega, \mathbb{R}^4)$ – which we need to verify (2.104) as a prerequisite for the convergence rates given in the subsequent Corollary 2.1.*

*As a summary, scheme* $(P)_h^{i,j}$ *shares convergence behavior of the schemes in Sections 2.1 and 2.2 for basic regularities (see Theorems 2.4 and 2.6) but is capable to turn increased (local) regularity properties of the solution to* (I.10)-(I.12) *into strong convergence* $\mathbf{m}_h \to \mathbf{m}$ *in* $\mathbf{L}^2(\omega, \mathbb{R}^2)$ *at a certain rate — a property that is missing in the considered schemes in Sections 2.1 and 2.2.*

The mass fraction $\Gamma : B^2 \to \mathbb{R}$ stated in (I.19) is discontinuous at $\mathbf{m} = \langle \mathbf{m}, \mathbf{e}_\perp \rangle_{\mathbb{R}^2} \mathbf{e}_\perp \in \mathbf{L}^2(\omega, B^2)$, for $B^2 \subset \mathbb{R}^2$ the unit ball. This is the reason why we leave out these points in our approximation result for the mass fraction $\Gamma_h(\mathbf{m}_h)$.

**Corollary 2.1** *Suppose that the assumptions of Theorem 2.10 are satisfied. For any* $\delta > 0$ *and any* $\xi < \frac{1}{8}$, *there exists* $h_0 \equiv h_0(\delta)$ *and a constant* $C_\xi$ *independent of* $h$, *such that for all* $h \leq h_0$ *holds*

$$\| \Gamma(\mathbf{m}) - \Gamma_h(\mathbf{m}_h) \|_{L^\infty(\omega)} \leq \frac{C_\xi}{\delta} h^\xi$$

*for all*

$$\frac{\mathbf{m}_h}{|\, \mathbf{m}_h \,|} \in \mathcal{D} := \Big\{ \mathbf{b} \in B^2 : 1 - |\, \langle \mathbf{b}, \mathbf{e}_\perp \rangle_{\mathbb{R}^2} \,|^2 > \delta \Big\}. \tag{2.105}$$

**Remark 2.10** *1. Alternatively,* (2.105) *can be phrased for* $\mathbf{m} \in \mathcal{D}$ *rather than* $\mathbf{m}_h \in \mathcal{D}$.

*2. We expect improved convergence rates in the interior of* $\omega$ *from Theorem 2.10 but skip its validation.*

The proof of Theorem 2.10 is split into four parts: we start with a first result for quantities given in the first line of (a). Its verification is by mainly dealing with the penalization in $\lambda_h$, adopting arguments that have been developed in Subsection 2.1.2. To verify a first convergence result for $\mathbf{m}_h$ in $\mathbf{L}^2(\omega, \mathbb{R}^2)$, we use Helmholtz-decomposition for $\mathbf{L}^2(\omega, \mathbb{R}^2)$-functions, and make use of increased stability of the scheme $(\mathrm{P})_h^{i,j}$, together with estimates available for $\| \mathbf{m} - \mathbf{m}_h \|_{\mathbf{W}^{-1,2}(\omega, \mathbb{R}^2)}$, for $\big[W_0^{1,2}(\omega)\big]^* = W^{-1,2}(\omega)$. Thirdly, we employ a localization strategy inside $\omega$ to verify error bounds in $\mathbf{L}^2(\omega', \mathbb{R}^2)$, for any $\omega' \Subset \omega$, for $\mathbf{m} - \mathbf{m}_h$. Then, we estimate the thickness of the arising boundary layer. Finally, a bootstrapping argument settles the proof.

The verification of Corollary 2.1 then uses Remark 2.9, item 3.

### 2.3.1 Proofs of Theorem 2.10 and Corollary 2.1

We start with the verification of Theorem 2.10.

**Proof:**

*1st step:* The error equations read, for $e := u - u_h \in W_0^{1,2}(\Omega)$, $\boldsymbol{\eta} := \mathbf{m} - \mathbf{m}_h \in \mathbf{L}^2(\omega, \mathbb{R}^2)$, for $\partial_{\mathbf{n}}\boldsymbol{\eta}\big|_{\partial\omega} = \partial_{\mathbf{n}}\mathbf{m}\big|_{\partial\omega}$, and all $\{w_h, \boldsymbol{\chi}_h\} \in \mathcal{S}_0^1(\Omega) \times \boldsymbol{\mathcal{S}}^1(\omega)$,

$$(\nabla e, \nabla w_h)_\Omega = (\boldsymbol{\eta}, \nabla w_h)\,, \tag{2.106}$$

$$\varepsilon_0 \sum_{\xi\in\Xi^*} \int_\xi \langle [\partial_{\mathbf{n}}\boldsymbol{\eta}], [\partial_{\mathbf{n}}\boldsymbol{\chi}_h] \rangle_{\mathbb{R}^4} \,\mathrm{d}\mathbf{x} + \varepsilon_1 (\nabla\boldsymbol{\eta}, \nabla\boldsymbol{\chi}_h) + (\nabla e, \boldsymbol{\chi}_h) \tag{2.107}$$
$$+\big(\langle \boldsymbol{\eta}, \mathbf{e}_\perp\rangle_{\mathbb{R}^2}, \langle \boldsymbol{\chi}_h, \mathbf{e}_\perp\rangle_{\mathbb{R}^2}\big) + (\lambda\mathbf{m} - \lambda_h\mathbf{m}_h, \boldsymbol{\chi}_h) = \varepsilon_1 (\nabla\mathbf{m}, \nabla\boldsymbol{\chi}_h)\,.$$

We choose $\{w_h, \boldsymbol{\chi}_h\} = \{P_{W_0^{1,2}(\Omega)} e, \mathbf{P}_{\mathbf{W}^{1,2}(\omega,\mathbb{R}^2)}\boldsymbol{\eta}\}$, where $P_{W_0^{1,2}(\Omega)} : W_0^{1,2}(\Omega) \to \mathcal{S}_0^1(\Omega)$ and $\mathbf{P}_{\mathbf{W}^{1,2}(\omega,\mathbb{R}^2)} : \left[W^{1,2}(\omega)/\mathbb{R}\right]^2 \to \left[\mathcal{S}^1/\mathbb{R}\right]^2$ are corresponding Ritz-projections. We obtain

$$\begin{aligned}
&\| \nabla e \|^2_{\mathbf{L}^2(\Omega,\mathbb{R}^2)} + \varepsilon_1 \| \nabla\boldsymbol{\eta} \|^2_{\mathbf{L}^2(\omega,\mathbb{R}^4)} + \| \langle \boldsymbol{\eta}, \mathbf{e}_\perp\rangle_{\mathbb{R}^2} \|^2_{L^2(\omega)} \\
&\quad + \frac{\varepsilon_0}{2} \sum_{\xi\in\Xi^*} \int_\xi | [\partial_{\mathbf{n}}\boldsymbol{\eta}] |^2 \,\mathrm{d}\mathbf{x} + (\lambda\mathbf{m} - \lambda_h\mathbf{m}_h, \boldsymbol{\eta}) \\
&= -\big(\boldsymbol{\eta}, \nabla(u - P_{W_0^{1,2}(\Omega)}u)\big) + (\nabla e, \mathbf{m}_h - \mathbf{P}_{\mathbf{W}^{1,2}(\omega,\mathbb{R}^2)}\mathbf{m}) \\
&\quad + \big(\nabla e, \nabla(u - P_{W_0^{1,2}(\Omega)}u)\big) + \varepsilon_1\big(\nabla\boldsymbol{\eta}, \nabla(\mathbf{m} - \mathbf{P}_{\mathbf{W}^{1,2}(\omega,\mathbb{R}^2)}\mathbf{m})\big) \\
&\quad + \big(\langle\boldsymbol{\eta}, \mathbf{e}_\perp\rangle_{\mathbb{R}^2}, \langle \mathbf{m} - \mathbf{P}_{\mathbf{W}^{1,2}(\omega,\mathbb{R}^2)}\mathbf{m}, \mathbf{e}_\perp\rangle_{\mathbb{R}^2}\big) \\
&\quad + \frac{\varepsilon_0}{2} \sum_{\xi\in\Xi^*} \int_\xi | [\partial_{\mathbf{n}}(\mathbf{m} - \mathbf{P}_{\mathbf{W}^{1,2}(\omega,\mathbb{R}^2)}\mathbf{m})] |^2 \,\mathrm{d}\mathbf{x} \\
&\quad + \big(\lambda\mathbf{m} - \lambda_h\mathbf{m}_h, \boldsymbol{\eta} + \mathbf{m} - \mathbf{P}_{\mathbf{W}^{1,2}(\omega,\mathbb{R}^2)}\mathbf{m}\big) \\
&\quad + \varepsilon_1\big(\nabla\mathbf{m}, \nabla(\boldsymbol{\eta} + \mathbf{m} - \mathbf{P}_{\mathbf{W}^{1,2}(\omega,\mathbb{R}^2)}\mathbf{m})\big)\,.
\end{aligned} \tag{2.108}$$

Critical terms are treated separately.

$$\begin{aligned}
&\left|\big(\boldsymbol{\eta}, \nabla(P_{W_0^{1,2}(\Omega)}u - u)\big)\right| \\
&= \left|(\operatorname{div}\boldsymbol{\eta}, P_{W_0^{1,2}(\Omega)}u - u) + \int_{\partial\omega} \langle\boldsymbol{\eta}, \mathbf{n}\rangle_{\mathbb{R}^2}\{P_{W_0^{1,2}(\Omega)}u - u\}\,\mathrm{d}s\right| \\
&\le \| \nabla\boldsymbol{\eta} \|_{\mathbf{L}^2(\omega,\mathbb{R}^4)} \| u - P_{W_0^{1,2}(\Omega)}u \|_{L^2(\Omega)} \\
&\quad + C \| \nabla\boldsymbol{\eta} \|_{\mathbf{L}^2(\omega,\mathbb{R}^4)} \| u - P_{W_0^{1,2}(\Omega)}u \|_{L^2(\partial\omega)}\,.
\end{aligned} \tag{2.109}$$

In the sequel, we employ the result $\| u - P_{W_0^{1,2}(\Omega)} u \|_{L^2(\partial\omega)} \leq C h^{3/2} \| \Delta u \|_{L^2(\Omega)}$. — Note that, for the convex indicator function $\psi : \mathbb{R}^2 \to [0, \infty]$, defined by $\psi(\mathbf{m}) = 0$ if $|\mathbf{m}| \leq 1$, and $\psi(\mathbf{m}) = \infty$ else, there holds $\lambda \mathbf{m} \in \partial\psi(\mathbf{m})$. Together with the definition of $\lambda_h$ next to (2.102), we conclude (see also (2.32) )

$$\frac{\varepsilon_2}{2} | \mathbf{m}_h |^2 \lambda_h^2 \leq \langle \lambda \mathbf{m} - \lambda_h \mathbf{m}_h, \boldsymbol{\eta} \rangle_{\mathbb{R}^2} + \frac{\varepsilon_2}{2} \lambda^2 . \tag{2.110}$$

Then, we are led to the following result, supposing sufficient regularity of $\{\mathbf{m}, u\}$, and any $\gamma > 0$ ,

$$\begin{aligned} &\| \nabla e \|^2_{\mathbf{L}^2(\Omega,\mathbb{R}^2)} + \frac{\varepsilon_1}{2} \| \nabla \boldsymbol{\eta} \|^2_{\mathbf{L}^2(\omega,\mathbb{R}^4)} + \| \langle \boldsymbol{\eta}, \mathbf{e}_\perp \rangle_{\mathbb{R}^2} \|^2_{L^2(\omega)} \\ &\quad + \frac{\varepsilon_2}{2} \| \lambda_h \mathbf{m}_h \|^2_{\mathbf{L}^2(\omega,\mathbb{R}^2)} + \frac{\varepsilon_0}{2} \sum_{\xi \in \Xi^*} \int_\xi | [\partial_{\mathbf{n}} \boldsymbol{\eta}] |^2 \, \mathrm{dx} \\ &\leq \frac{\varepsilon_2}{2} \| \lambda \mathbf{m} \|^2_{\mathbf{L}^2(\omega,\mathbb{R}^2)} + \frac{h^4}{\varepsilon_1} + h^4 + C \varepsilon_1 \| \nabla \mathbf{m} \|^2_{\mathbf{L}^2(\omega,\mathbb{R}^4)} + C \big( \varepsilon_0 + \varepsilon_1 \big) h \\ &\quad + \frac{1}{\varepsilon_1} \| u - P_{W_0^{1,2}(\Omega)} u \|^2_{L^2(\Omega)} + \frac{1}{\varepsilon_1} \| u - P_{W_0^{1,2}(\Omega)} u \|^2_{L^2(\partial\omega)} \\ &\quad + \sqrt{\varepsilon_2} \| \lambda \mathbf{m} - \lambda_h \mathbf{m}_h \|_{\mathbf{L}^2(\omega,\mathbb{R}^2)} \frac{1}{\sqrt{\varepsilon_2}} \| \mathbf{m} - \mathbf{P}_{\mathbf{W}^{1,2}(\omega,\mathbb{R}^2)} \mathbf{m} \|_{\mathbf{L}^2(\omega,\mathbb{R}^2)} . \end{aligned}$$

Then, we find[7]

$$\begin{aligned} &\| \nabla e \|^2_{\mathbf{L}^2(\Omega,\mathbb{R}^2)} + \varepsilon_1 \| \nabla \boldsymbol{\eta} \|^2_{\mathbf{L}^2(\omega,\mathbb{R}^4)} + \frac{\varepsilon_0}{2} \sum_{\xi \in \Xi^*} \int_\xi | [\partial_{\mathbf{n}} \boldsymbol{\eta}] |^2 \, \mathrm{dx} \\ &\quad + \| \langle \boldsymbol{\eta}, \mathbf{e}_\perp \rangle_{\mathbb{R}^2} \|^2_{L^2(\omega)} + \frac{\varepsilon_2}{2} \| \lambda_h \mathbf{m}_h \|^2_{\mathbf{L}^2(\omega,\mathbb{R}^2)} \\ &\leq C \Big\{ \varepsilon_0 h + \varepsilon_1 + \varepsilon_2 + \frac{h^3}{\varepsilon_1} + h^4 \{ \varepsilon_1^{-1} + \varepsilon_2^{-1} \} \Big\} . \end{aligned} \tag{2.111}$$

To see $\| \mathbf{m}_h \|_{\mathbf{L}^2(\omega,\mathbb{R}^2)} \leq C$, we employ the bound $\sqrt{\varepsilon_2} \| \lambda_h \mathbf{m}_h \|_{\mathbf{L}^2(\omega,\mathbb{R}^2)} \leq C$ from (2.111). Let $\omega := \mathcal{B} \cup \mathcal{C}$, where $\mathcal{B} = \{ \mathbf{x} \in \omega \| \mathbf{m}_h(\mathbf{x}) | \geq 2 \}$. Then

$$\| \mathbf{m}_h \|_{\mathbf{L}^2(\mathcal{B},\mathbb{R}^2)} \leq 2 \| \frac{(| \mathbf{m}_h | - 1)_+}{| \mathbf{m}_h |} \mathbf{m}_h \|_{\mathbf{L}^2(\mathcal{B},\mathbb{R}^2)} \leq C \sqrt{\varepsilon_2} , \tag{2.112}$$

[7] At this stage, we cannot follow the argumentation next to (2.42) which was successful in the context of element-wise constant magnetizations to show a convergence rate for $\| \lambda \mathbf{m} - \lambda_h \mathbf{m}_h \|_{\mathbf{L}^2(\omega,\mathbb{R}^2)}$, the reason being $\Delta_h \boldsymbol{\eta} \notin \mathbf{L}^2(\omega, \mathbb{R}^2)$.

which shows the desired result.

*2nd step:* By Helmholtz-decomposition principle, we have $\mathbf{L}^2(\omega,\mathbb{R}^2) \ni \boldsymbol{\eta} = \nabla a + \mathbf{curl}\, b$, where $b \in W_0^{1,2}(\omega)$, $a \in W^{1,2}(\omega)/\mathbb{R}$. Note that

$$\| \nabla a \|_{\mathbf{L}^2(\omega,\mathbb{R}^2)} \leq \| \operatorname{div} \boldsymbol{\eta} \|_{\mathbf{W}^{-1,2}(\omega,\mathbb{R}^2)} \,. \tag{2.113}$$

The second contribution can be bounded as follows,

$$\begin{aligned} \| \mathbf{curl}\, b \|_{\mathbf{W}^{-1,2}(\omega,\mathbb{R}^2)} &\leq C \| b \|_{L^2(\omega)} \leq C \| \langle \nabla b, \mathbf{e} \rangle_{\mathbb{R}^2} \|_{L^2(\omega)} \\ &= C \| \langle \mathbf{e}_\perp, \mathbf{curl}\, b \rangle_{\mathbb{R}^2} \|_{L^2(\omega)} \\ &\leq \| \langle \boldsymbol{\eta}, \mathbf{e}_\perp \rangle_{\mathbb{R}^2} \|_{L^2(\omega)} + \| \langle \nabla a, \mathbf{e}_\perp \rangle_{\mathbb{R}^2} \|_{L^2(\omega)} \\ &\leq C \| \langle \boldsymbol{\eta}, \mathbf{e}_\perp \rangle_{\mathbb{R}^2} \|_{L^2(\omega)} + \| \operatorname{div} \boldsymbol{\eta} \|_{W^{-1,2}(\omega)} \,. \end{aligned} \tag{2.114}$$

In the sequel, we make use of (2.101) and restrict it to $\mathcal{S}_0^1(\omega) \subset W_0^{1,2}(\omega)$. Let $\varphi_h \equiv P_{W_0^{1,2}(\omega)} \varphi$ denote the Ritz projection of $\varphi \in W_0^{1,2}(\omega)$ onto $\mathcal{S}_0^1(\omega)$.

$$\begin{aligned} (\boldsymbol{\eta}, \nabla \varphi) &= (\nabla u, \nabla \varphi) - \big(\mathbf{m}_h, \nabla(\varphi - \varphi_h)\big) - (\nabla u_h, \nabla \varphi_h) \\ &= \big(\nabla(u - u_h), \nabla \varphi\big) - \big(\mathbf{m}_h, \nabla(\varphi - \varphi_h)\big) \\ &\quad - \big(\nabla(u - u_h), \nabla(\varphi - \varphi_h)\big) + \big(\nabla(u - P_{W_0^{1,2}(\Omega)} u), \nabla(\varphi - \varphi_h)\big) \,. \end{aligned} \tag{2.115}$$

Then, by conformity of the scheme, (2.111), and regularity properties of $\{u, \mathbf{m}\}$, we can bound $\| \operatorname{div}(\mathbf{m} - \mathbf{m}_h) \|_{W^{-1,2}(\omega)}$. Together with this result, and (2.113), (2.114), we finally find

$$\| \boldsymbol{\eta} \|_{\mathbf{W}^{-1,2}(\omega,\mathbb{R}^2)} \leq C \Big\{ \sqrt{\varepsilon_0 h} + \sqrt{\varepsilon_1} + \sqrt{\varepsilon_2} + \frac{h^{3/2}}{\sqrt{\varepsilon_1}} + h^2 \{ \varepsilon_1^{-1/2} + \varepsilon_2^{-1/2} \} \Big\} \,. \tag{2.116}$$

*3rd step:* Define a cut-off function $\sigma \in C_0^0(\omega, [0,1])$, for $\omega' \Subset \omega$, such that $\sigma\big|_{\omega'} = 1$, and $\| \nabla \sigma \|_{\mathbf{L}^\infty(\omega \setminus \omega', \mathbb{R}^2)} \leq \delta^{-1} \equiv 2 \operatorname{dist}(\omega', \partial\omega)^{-1}$. Consider

$$\begin{aligned} \| \sqrt{\sigma} \boldsymbol{\eta} \|^2_{\mathbf{L}^2(\omega,\mathbb{R}^2)} &\leq C \| \nabla \{ \sigma \boldsymbol{\eta} \} \|_{\mathbf{L}^2(\omega,\mathbb{R}^2)} \| \boldsymbol{\eta} \|_{\mathbf{W}^{-1,2}(\omega,\mathbb{R}^2)} \\ &\leq C \Big\{ \| \nabla \boldsymbol{\eta} \|_{\mathbf{L}^2(\omega,\mathbb{R}^4)} + \delta^{-1} \| \boldsymbol{\eta} \|_{\mathbf{L}^2(\omega,\mathbb{R}^2)} \Big\} \\ &\quad \times \Big\{ \sqrt{\varepsilon_0 h} + \sqrt{\varepsilon_1} + \sqrt{\varepsilon_2} + \frac{h^{3/2}}{\sqrt{\varepsilon_1}} + h^2 \{ \varepsilon_1^{-1/2} + \varepsilon_2^{-1/2} \} \Big\} \,, \end{aligned} \tag{2.117}$$

thanks to (2.116). Note that for an appropriate constraint $F(\varepsilon_0, \varepsilon_1, \varepsilon_2; h) > 0$, we can make sure $\| \nabla \boldsymbol{\eta} \|_{\mathbf{L}^2(\omega,\mathbb{R}^4)} \leq C$, with $C > 0$ independent of $h$.

On the other hand, we find for $q, q' > 2$, such that $q^{-1} + (q')^{-1} = 2^{-1}$, by interpolation of $L^q$ between $L^2$ and $W^{1,2}$, generalized Young's inequality, and $\mu_2(\omega') \leq C\,\delta$,

$$\begin{aligned} \| (1-\sqrt{\sigma})\boldsymbol{\eta} \|_{\mathbf{L}^2(\omega',\mathbb{R}^2)} &\leq C \| 1-\sqrt{\sigma} \|_{L^{q'}(\omega',\mathbb{R}^2)} \| \boldsymbol{\eta} \|_{\mathbf{L}^q(\omega',\mathbb{R}^2)} \\ &\leq C\,\delta^{\frac{1}{q'}} \| \boldsymbol{\eta} \|^{2/q}_{\mathbf{L}^2(\omega,\mathbb{R}^2)} \| \boldsymbol{\eta} \|^{1-2/q}_{\mathbf{W}^{1,2}(\omega,\mathbb{R}^2)} \\ &\leq C_\alpha \Big( \delta^{\frac{1}{q'}} \| \boldsymbol{\eta} \|^{1-\frac{2}{q}}_{\mathbf{W}^{1,2}(\omega,\mathbb{R}^2)} \Big)^\alpha + \frac{1}{2} \| \boldsymbol{\eta} \|^{\frac{2\alpha'}{q}}_{\mathbf{L}^2(\omega,\mathbb{R}^2)} \quad \text{for } \alpha^{-1} + (\alpha')^{-1} = 1\,. \end{aligned} \tag{2.118}$$

We choose $\alpha' = \frac{q}{2}$ and combine (2.117), (2.118).

$$\begin{aligned} \| \boldsymbol{\eta} \|_{\mathbf{L}^2(\omega,\mathbb{R}^2)} &\leq C\,\delta^{1/2} \| \boldsymbol{\eta} \|_{\mathbf{W}^{1,2}(\omega,\mathbb{R}^2)} \\ &+ C \Big\{ \sqrt[4]{\varepsilon_0 h} + \sqrt[4]{\varepsilon_1} + \sqrt[4]{\varepsilon_2} + \sqrt[4]{\frac{h^3}{\varepsilon_1}} + h\{\varepsilon_1^{-1/4} + \varepsilon_2^{-1/4}\} \Big\} \\ &+ \delta^{-1} \Big\{ \sqrt{\varepsilon_0 h} + \sqrt{\varepsilon_1} + \sqrt{\varepsilon_2} + \sqrt{\frac{h^3}{\varepsilon_1}} + h^2\{\varepsilon_1^{-1/2} + \varepsilon_2^{-1/2}\} \Big\}. \end{aligned} \tag{2.119}$$

We set $\delta^{1/2} = \delta^{-1}\{\sqrt{\varepsilon_0 h} + \sqrt{\varepsilon_1} + \sqrt{\varepsilon_2} + \sqrt{\frac{h^3}{\varepsilon_1}} + h^2\{\varepsilon_1^{-1/2} + \varepsilon_2^{-1/2}\}\}$, and suppose $\varepsilon_1 \gg \varepsilon_2, h^2$ (to be made precise in the sequel). This implies $\varepsilon_1^{1/6}$ as upper bound for the right-hand side. Integrating by parts on every triangle, we obtain, by continuity of finite element functions,

$$\begin{aligned} &\| \nabla \boldsymbol{\eta} \|^2_{\mathbf{L}^2(\omega,\mathbb{R}^4)} \\ &\leq \Big( \sum_{K \in \mathcal{T}} \| \Delta \boldsymbol{\eta} \|^2_{\mathbf{L}^2(\omega,\mathbb{R}^2)} \Big)^{1/2} \| \boldsymbol{\eta} \|_{\mathbf{L}^2(\omega,\mathbb{R}^2)} + C \sum_{\xi \in \Xi^*} \int_\xi \big\langle [\partial_{\mathbf{n}} \mathbf{m}], \boldsymbol{\eta} \big\rangle_{\mathbb{R}^2} \, \mathrm{dx} \\ &\leq C \| \mathbf{m} \|_{\mathbf{W}^{2,2}(\omega,\mathbb{R}^2)} \| \boldsymbol{\eta} \|_{\mathbf{L}^2(\omega,\mathbb{R}^2)} + C \sum_{\xi \in \Xi^*} \int_\xi | [\partial_{\mathbf{n}} \mathbf{m}] |^2 \, \mathrm{dx} + \frac{1}{4} \| \boldsymbol{\eta} \|^2_{\mathbf{W}^{1,2}(\omega,\mathbb{R}^2)}\,. \end{aligned}$$

Hence, accounting for the bound

$$\Big( \sum_{\xi \in \Xi^*} \int_\xi | [\partial_{\mathbf{n}} \mathbf{m}] |^2 \, \mathrm{dx} \Big)^{1/2} \leq C \Big\{ \sqrt{h} + \sqrt{\frac{\varepsilon_1}{\varepsilon_0}} + \sqrt{\frac{h^3}{\varepsilon_0 \varepsilon_1}} + \sqrt{\frac{h^4}{\varepsilon_0 \varepsilon_2}} \Big\} \tag{2.120}$$

from (2.111), we obtain

$$\| \nabla \boldsymbol{\eta} \|_{\mathbf{L}^2(\omega,\mathbb{R}^4)} \le C \Big\{ \| \boldsymbol{\eta} \|^{1/2}_{\mathbf{L}^2(\omega,\mathbb{R}^2)} + \big(\sqrt{h} + \sqrt{\frac{\varepsilon_1}{\varepsilon_0}} + \sqrt{\frac{h^3}{\varepsilon_0\varepsilon_1}} + \sqrt{\frac{h^4}{\varepsilon_0\varepsilon_2}}\big) \Big\}.$$

To simplify our studies, we suppose that the right-hand side of this inequality can be controlled by $C\varepsilon_1^{\frac{1}{12}}$, see above. This amounts to a constraint $F(\varepsilon_0, \varepsilon_1, \varepsilon_2; h) > 0$ that is met for the final choices of $\varepsilon_i$, $i = 0, 1, 2$ (check with the end of step 4). We summarize these considerations to

$$\| \nabla \boldsymbol{\eta} \|_{\mathbf{L}^2(\omega,\mathbb{R}^4)} \le C\, \varepsilon_1^{\frac{1}{12}} . \tag{2.121}$$

*4th step:* We proceed by bootstrapping, sharpening the bound for the crucial term $\varepsilon_1(\nabla\boldsymbol{\eta}, \nabla\mathbf{m}) \le \varepsilon_1 \| \nabla\mathbf{m} \|_{\mathbf{L}^2(\omega,\mathbb{R}^4)} \| \nabla\boldsymbol{\eta} \|_{\mathbf{L}^2(\omega,\mathbb{R}^4)}$ in (2.108) by means of (2.121). In a first step, we find

$$\begin{aligned} &\| \nabla e \|^2_{\mathbf{L}^2(\Omega,\mathbb{R}^2)} + \varepsilon_1 \| \nabla\boldsymbol{\eta} \|^2_{\mathbf{L}^2(\omega,\mathbb{R}^4)} + \| \langle \boldsymbol{\eta}, \mathbf{e}_\perp \rangle_{\mathbb{R}^2} \|^2_{L^2(\omega)} + \frac{\varepsilon_2}{2} \| \lambda_h \mathbf{m}_h \|^2_{\mathbf{L}^2(\omega,\mathbb{R}^2)} \\ &\le C \Big\{ \varepsilon_1^{1+\xi_i} + \varepsilon_0 h + \varepsilon_2 + \frac{h^3}{\varepsilon_1} + h^4 \{ \varepsilon_1^{-1} + \varepsilon_2^{-1} \} \Big\}, \end{aligned} \tag{2.122}$$

with $\xi_i$ set equal to $\frac{1}{6}$. A consequence is the following modification of (2.116), again for $\xi_i$ set equal to $\frac{1}{6}$,

$$\| \boldsymbol{\eta} \|_{\mathbf{W}^{-1,2}(\omega,\mathbb{R}^2)} \le C \Big\{ \varepsilon_1^{\frac{1+\xi_i}{2}} + \sqrt{\varepsilon_0 h} + \sqrt{\varepsilon_2} + \sqrt{\frac{h^3}{\varepsilon_1}} + h^2 \{ \varepsilon_1^{-1/2} + \varepsilon_2^{-1/2} \} \Big\}. \tag{2.123}$$

Accordingly, (2.117) takes the form (for $\xi_i = \frac{1}{6}$)

$$\begin{aligned} \| \sqrt{\sigma}\boldsymbol{\eta} \|^2_{\mathbf{L}^2(\omega,\mathbb{R}^2)} &\le C \| \nabla\{\sigma\boldsymbol{\eta}\} \|_{\mathbf{L}^2(\omega,\mathbb{R}^2)} \| \boldsymbol{\eta} \|_{\mathbf{W}^{-1,2}(\omega,\mathbb{R}^2)} \\ &\le C \Big\{ \varepsilon_1^{\frac{\xi_i}{2}} + \delta_i^{-1} \| \boldsymbol{\eta} \|_{\mathbf{L}^2(\omega,\mathbb{R}^2)} \Big\} \\ &\quad \times \Big\{ \varepsilon_1^{\frac{1+\xi_i}{2}} + \sqrt{\varepsilon_0 h} + + \sqrt{\varepsilon_2} + \sqrt{\frac{h^3}{\varepsilon_1}} + h^2 \{ \varepsilon_1^{-1/2} + \varepsilon_2^{-1/2} \} \Big\}. \end{aligned} \tag{2.124}$$

Finally,

$$\begin{aligned}\| \boldsymbol{\eta} \|_{\mathbf{L}^2(\omega,\mathbb{R}^2)} \leq\ & \delta_i^{-1}\Big\{\varepsilon_1^{\frac{1+\xi_i}{2}} + \sqrt{\varepsilon_0 h} + \sqrt{\varepsilon_2} + \sqrt{\frac{h^3}{\varepsilon_1}} + h^2\{\varepsilon_1^{-1/2} + \varepsilon_2^{-1/2}\}\Big\} \\ & + C_q\, \delta_i^{1/2}\, \varepsilon_1^{\xi_i/2} + C\, \varepsilon_1^{\frac{\xi_i}{4}} \Big\{\varepsilon_1^{\frac{1+\xi_i}{4}} + \sqrt[4]{\varepsilon_2} \\ & + \sqrt[4]{\varepsilon_0 h} + \sqrt[4]{\frac{h^3}{\varepsilon_1}} + h\{\varepsilon_1^{-1/4} + \varepsilon_2^{-1/4}\}\Big\}. \end{aligned} \tag{2.125}$$

We balance $\delta_i^{1/2}\, \varepsilon_1^{\frac{\xi_i}{2}} = \delta_i^{-1}\, \varepsilon_1^{\frac{1+\xi_i}{2}}$, which implies $\delta_i = \varepsilon_1^{1/3}$. The right-hand side of (2.125) is bounded by $\varepsilon_1^{\frac{\xi_i}{2}+\frac{1}{6}}$. This leads to the sequence $\xi_{i+1} = \frac{\xi_i}{2} + \frac{1}{6}$, for $\xi_0 = \frac{1}{6}$, and $\lim_{i\to\infty} \xi_i = \frac{1}{3}$. The first contribution on the right-hand side of (2.125) is balanced for values $2 < q \leq \frac{5}{2}$. This asymptotic study implies

$$\begin{aligned}& \| \nabla e \|^2_{\mathbf{L}^2(\Omega,\mathbb{R}^2)} + \varepsilon_1 \| \nabla \boldsymbol{\eta} \|^2_{\mathbf{L}^2(\omega,\mathbb{R}^4)} + \| \langle \boldsymbol{\eta}, \mathbf{e}_\perp \rangle_{\mathbb{R}^2} \|^2_{L^2(\omega)} + \frac{\varepsilon_2}{2} \| \lambda_h \mathbf{m}_h \|^2_{\mathbf{L}^2(\omega,\mathbb{R}^2)} \\ & \leq C\left\{\varepsilon_1^{4/3} + \varepsilon_0 h + \varepsilon_2 + h^2 + h^4\{\varepsilon_1^{-1} + \varepsilon_2^{-1}\}\right\}. \end{aligned} \tag{2.126}$$

Hence we choose $\varepsilon = \mathcal{O}(h)$, $\varepsilon_1 = \mathcal{O}(h^{3/2})$ and $\varepsilon_2 = \mathcal{O}(h^2)$ for equilibration of the arising terms.

By (2.124), and $i \to \infty$, for $\omega' \Subset \omega$, such that $\operatorname{dist}(\partial\omega, \omega') \geq \mathcal{O}(\sqrt{h})$, we obtain

$$\| \boldsymbol{\eta} \|^2_{\mathbf{L}^2(\omega',\mathbb{R}^2)} \leq C\left\{h^{5/8} + \operatorname{dist}^{-1/2}(\partial\omega, \omega') h^{3/4}\right\}. \tag{2.127}$$

□

Finally, we show Corollary 2.1.
**Proof:**

$$\begin{aligned} 2\{\Gamma(\mathbf{m}) - \Gamma_h(\mathbf{m}_h)\} &= \frac{\langle \mathbf{m}, \mathbf{e}\rangle_{\mathbb{R}^2}}{\sqrt{1 - \langle \mathbf{m}, \mathbf{e}_\perp\rangle^2_{\mathbb{R}^2}}} - \frac{\langle \mathbf{m}_h, \mathbf{e}\rangle_{\mathbb{R}^2}}{\sqrt{| \mathbf{m}_h |^2_\bullet - \langle \mathbf{m}_h, \mathbf{e}_\perp\rangle^2_{\mathbb{R}^2}}} \\ &= \frac{\langle \mathbf{m}, \mathbf{e}\rangle_{\mathbb{R}^2}\sqrt{| \mathbf{m}_h |^2_\bullet - \langle \mathbf{m}_h, \mathbf{e}_\perp\rangle^2_{\mathbb{R}^2}} - \langle \mathbf{m}_h, \mathbf{e}\rangle_{\mathbb{R}^2}\sqrt{1 - \langle \mathbf{m}, \mathbf{e}_\perp\rangle^2_{\mathbb{R}^2}}}{\sqrt{1 - \langle \mathbf{m}, \mathbf{e}_\perp\rangle^2_{\mathbb{R}^2}}\sqrt{| \mathbf{m}_h |^2_\bullet - \langle \mathbf{m}_h, \mathbf{e}_\perp\rangle^2_{\mathbb{R}^2}}} \\ &\leq \frac{\sqrt{| \mathbf{m}_h |^2_\bullet - \langle \mathbf{m}_h, \mathbf{e}_\perp\rangle^2_{\mathbb{R}^2}}\langle \boldsymbol{\eta}, \mathbf{e}\rangle_{\mathbb{R}^2} + | \langle \mathbf{m}_h, \mathbf{e}\rangle_{\mathbb{R}^2} | \sqrt{| \mathbf{m}_h |^2_\bullet - 1 + \langle \boldsymbol{\eta}, \mathbf{e}_\perp\rangle^2_{\mathbb{R}^2}}}{\sqrt{1 - \langle \mathbf{m}, \mathbf{e}_\perp\rangle^2_{\mathbb{R}^2}}\sqrt{| \mathbf{m}_h |^2_\bullet - \langle \mathbf{m}_h, \mathbf{e}_\perp\rangle^2_{\mathbb{R}^2}}}, \end{aligned}$$

where we used the bound $\sqrt{a}-\sqrt{b} \leq \sqrt{a-b}$, for $0 \leq b \leq a$.

For $\max\{\|\boldsymbol{\eta}\|^2_{\mathbf{L}^\infty(\omega,\mathbb{R}^2)}, \|\,|\mathbf{m}_h|^2_\bullet - 1\,\|_{L^\infty(\omega)}\} \leq \frac{\delta}{2}$, we find

$$\|\Gamma(\mathbf{m}) - \Gamma_h(\mathbf{m}_h)\|_{L^\infty(\omega)} \leq C \frac{\|\boldsymbol{\eta}\|_{\mathbf{L}^\infty(\omega,\mathbb{R}^2)} + \|\boldsymbol{\eta}\|^{1/2}_{\mathbf{L}^\infty(\omega,\mathbb{R}^2)}}{\sqrt{\delta - \|\boldsymbol{\eta}\|^2_{\mathbf{L}^\infty(\omega,\mathbb{R}^2)}}\sqrt{\delta + |\mathbf{m}_h|^2_\bullet - 1}}.$$

Then, (2.104) finishes the proof. □

# Chapter 3

# Relaxed Micromagnetism using Young Measures

## 3.1 Relaxed micromagnetism[1]

Equilibrium magnetization states in ferromagnetic materials are characterized by minima of the energy $\mathcal{E}$ in (I.1) (for $\alpha = 0$); as the set $\mathcal{A}$ in (I.3) is non-convex, it is not weakly closed in $\mathbf{L}^2(\omega; \mathbb{R}^d)$, and a solution to (I.4) does not have to exist for uniaxial materials; cf. for instance [71]. An implication is that highly oscillatory minimizing sequences of $\mathcal{E}$ do not have weak limits in $\mathcal{A}$. One way to overcome this problem is to convexify the anisotropy energy, see Chapter 2. In this chapter, we consider another scheme which leaves $\mathcal{E}$ in (I.1) (for $\alpha = 0$); unmodified and extends the set of admissible functions to Young measures instead. This alternative relaxation of the problem is of practical interest for cases where $\phi^{**}$ is not known or where it is not desirable to compute it; moreover, the microstructure is computed directly in terms of Young measure-valued solutions — rather than from certain averages in a post-processing step.

Relaxation means that we seek for a set $\mathcal{B}$ endowed with some (relative) topology $\tau$ and a functional $F : \mathcal{B} \to \mathbb{R}$ such that (i) $\min_{\mathcal{B}} F$ exists, (ii) $\min_{\mathcal{B}} F = \inf_{\mathcal{A}} \mathcal{E}$, and (iii) minimizers of $F$ are $\tau$-limits of minimizing sequences of $\mathcal{E}$; conversely, we require minimizing sequences of $\mathcal{E}$ to tend to minimizers of $F$ in the topology $\tau$. In Chapter 2, we discussed the result of DeSimone [39] (cf. also [35, 44]) who showed that we can set $F := \tilde{\mathcal{E}}$ and

[1] The material presented in this chapter is taken from [82].

$\mathcal{B} := \tilde{\mathcal{A}}$, with $\tau$ the weak topology of $\mathbf{L}^2(\omega;\mathbb{R}^d)$. Then

$$\inf_{\boldsymbol{\mu}\in\mathcal{A}} \mathcal{E}(\boldsymbol{\mu}) = \min_{\boldsymbol{\mu}\in\tilde{\mathcal{A}}} \tilde{\mathcal{E}}(\boldsymbol{\mu}),$$

where

$$\tilde{\mathcal{A}} = \Big\{\boldsymbol{\mu} \in \mathbf{L}^2(\omega;\mathbb{R}^d) : \ |\boldsymbol{\mu}(\mathbf{x})| \le 1, \text{ f.a.a. } \mathbf{x}\in\omega\Big\},$$

together with

$$\tilde{\mathcal{E}}(\boldsymbol{\mu}) = \int_\omega \phi^{**}(\boldsymbol{\mu})\,\mathrm{d}\mathbf{x} - (\mathbf{f},\boldsymbol{\mu}) + \frac{1}{2}\|\nabla u\|^2_{\mathbf{L}^2(\mathbb{R}^d,\mathbb{R}^d)}. \tag{3.1}$$

Here,

$$\phi^{**} : \mathbb{R}^d \to \mathbb{R}, \quad \text{where} \quad \phi^{**} = \sup\{f;\ f \le \hat{\phi},\ \ f \text{ is convex}\}$$

is the convex envelope of $\phi : \mathbb{R}^d \to \mathbb{R}\cup\{+\infty\}$,

$$\hat{\phi}(\boldsymbol{\mu}) = \begin{cases} \phi(\boldsymbol{\mu}) & \text{if } |\boldsymbol{\mu}| = 1, \\ +\infty & \text{otherwise}, \end{cases}$$

and the triple $(\tilde{\mathcal{E}}, \tilde{\mathcal{A}},$ weak in $\mathbf{L}^2(\omega,\mathbb{R}^d))$ satisfies all the requirements that are imposed on $(F,\mathcal{B},\tau)$. Physically, the passage from $\phi$ to $\phi^{**}$ means that we get only macroscopic properties of the solution instead of microscopic ones which are reflected in minimizing sequences of $\mathcal{E}$.

Another way to obtain a relaxation of $\mathcal{E}$ that does not require the explicit knowledge of $\phi^{**}$ is by means of Young measures generated by minimizing sequences of $\mathcal{E}$. There is a result (see [7, 35, 132]) that for an open bounded set $\omega \subset \mathbb{R}^d$ and from any sequence $\{\mathbf{w}_k\}_{k\in\mathbb{N}}$ of measurable functions $\mathbf{w}_k \in \mathbf{L}^\infty(\omega,\mathbb{R}^d)$, $|\mathbf{w}_k| = 1$ almost everywhere, we can extract a subsequence (denoted by the same indices) such that there exists a family of probability measures $\nu = \{\nu_\mathbf{x}\}_{\mathbf{x}\in\omega}$, with supp $\nu_\mathbf{x} \subset \mathcal{S}^{d-1}$ and

$$\lim_{k\to\infty} v \circ \mathbf{w}_k = \nu \bullet v \quad \text{weakly* in } \mathbf{L}^\infty(\omega,\mathbb{R}^d), \tag{3.2}$$

for any continuous function $v : \mathcal{S}^{d-1} \to \mathbb{R}$, and

$$[\nu \bullet v](\mathbf{x}) = \int_S v(A)\nu_\mathbf{x}(\mathrm{d}A), \quad \text{for almost all } \mathbf{x}\in\omega.$$

Conversely, having a family of probability measures supported on $\mathcal{S}^{d-1}$ which is weakly measurable, i.e., $\mathbf{x} \mapsto [\nu \bullet v](\mathbf{x}) = \int_{\mathcal{S}^{d-1}} v(A)\nu_{\mathbf{x}}(\mathrm{d}A)$ is measurable for any $v \in C(\mathcal{S}^{d-1})$, then there is a sequence $\{\mathbf{w}_k\}_{k\in\mathbb{N}}$, with $\mathbf{w}_k \in \mathbf{L}^\infty(\omega, \mathbb{R}^d)$, $|\mathbf{w}_k| = 1$ almost everywhere, such that (3.2) is fulfilled. A family of parameterized probability measures obtained this way is called a Young measure; we refer to [112] for more details on the subject of Young measures.

By the mapping $\mathbf{w} \mapsto \{\delta_{\mathbf{w}(\mathbf{x})}\}_{\mathbf{x}\in\omega}$, the set $\{\mathbf{w} \in \mathbf{L}^\infty(\omega; \mathbb{R}^d);\ |\mathbf{w}| = 1\}$ can be embedded into the set of Young measures supported on the unit sphere. For the sake of simplicity we will identify a Young measure $\nu = \{\nu_{\mathbf{x}}\}_{\mathbf{x}\in\omega}$ such that $\nu_{\mathbf{x}}$ is supported on the unit sphere in $\mathbb{R}^d$ for almost all $\mathbf{x} \in \omega$ with the functional in $L^1(\omega; C(\mathcal{S}^{d-1}))^*$ by the formula

$$\langle \nu, h\rangle = \int_\omega \int_{\mathcal{S}^{d-1}} h(\mathbf{x}, A)\nu_{\mathbf{x}}(\mathrm{d}A)\,\mathrm{d}\mathbf{x} \qquad \forall\, h \in L^1(\omega; C(\mathcal{S}^{d-1}))\,. \tag{3.3}$$

Pedregal [104] (see also [39, 123]) showed that if one sets

$$\bar{\mathcal{A}} := \Big\{\nu = \{\nu_{\mathbf{x}}\}_{\mathbf{x}\in\omega} \text{ weakly measurable ; supp } \nu_{\mathbf{x}} \subset \mathcal{S}^{d-1} \text{ f.a.a. } \mathbf{x} \in \omega\Big\},$$

and defines for any $\nu \in \bar{\mathcal{A}}$ and $\mathbf{m}(\mathbf{x}) = \int_{\mathcal{S}^{d-1}} A\nu_{\mathbf{x}}(\mathrm{d}A)$, for almost all $\mathbf{x} \in \omega$, a functional $\bar{\mathcal{E}} : \bar{\mathcal{A}} \to \mathbb{R}$ by

$$\bar{\mathcal{E}}(\nu) = \int_\omega \int_{\mathcal{S}^{d-1}} \phi(A)\nu_{\mathbf{x}}(\mathrm{d}A)\,\mathrm{d}\mathbf{x} - (\mathbf{f}, \mathbf{m}) + \frac{1}{2}\, \|\, \nabla u\, \|^2_{\mathbf{L}^2(\mathbb{R}^d, \mathbb{R}^d)}\,,$$

then $\big(\bar{\mathcal{E}}, \bar{\mathcal{A}},$ weak* in $L^1(\omega; C(\mathcal{S}^{d-1}))\big)$ is a relaxation of $\mathcal{E}$ in the sense stated above. Thus we define the **Relaxed Problem (RP)**:

$$\min_{\nu\in\bar{\mathcal{A}}} \bar{\mathcal{E}}(\nu)\,, \tag{3.4}$$

subject to

$$\mathbf{m}(\mathbf{x}) = \int_{\mathcal{S}^{d-1}} A\nu_{\mathbf{x}}(\mathrm{d}A) \quad \text{ f.a.a. } \mathbf{x} \in \omega\,, \tag{3.5}$$

$$\operatorname{div}\big(-\nabla u + \chi_\omega \mathbf{m}\big) = 0 \quad \text{ in } \mathbb{R}^d\,. \tag{3.6}$$

The existence of a solution to (RP) follows from the sequential weak* lower semicontinuity of $\bar{\mathcal{E}}$ and from the sequential weak* compactness of $\bar{\mathcal{A}}$.

The first numerical discretization of the relaxed problem (**RP**) has been done by Kružík in [80] by means of three-atomic Young measures that are constant for every element from a regular triangulation of $\omega \subset \mathbb{R}^2$.

We conclude this introduction with a short outline of the main results in this chapter. In a first step, we propose and analyze a conforming discretization of (RP) by means of element-wise constant multi-atomic Young measures of prescribed support according to a triangulation of $\mathcal{S}^{d-1}$, see Section 3.2. This leads to a quadratic-linear optimization problem which is rather large. In Corollary 3.1, we state convergence of the method with respect to the discretization parameter $\mathbf{h} = (h_1, h_2)$, with $h_1, h_2 > 0$ the mesh-sizes of the triangulations $\mathcal{T}_{h_1}$ of $\omega$ and $\mathcal{T}_{h_2}$ of $\mathcal{S}^{d-1}$, respectively. This result amounts to a condition $h_2 = o(h_1^{d/2})$, which makes computations rather costly. To avoid this drawback and decrease the number of unknowns, we make use of an adaptive strategy that singles out the so-called active atoms. Those are found by recently derived optimality conditions for (RP) (in [83]) to predict the support of a Young measure solution to (RP). The corresponding result concerning convergence of the method is given in Corollary 3.2. We refer to Section 3.3 for details of the algorithm as well as its analysis. Let us remark that this idea of using optimality conditions for numerical purposes is due to Carstensen & Roubíček in [21], where a scalar or one-dimensional variational problem is considered.

Section 3.4 presents numerical experiments for $d = 2$ and for a density $\phi$ where $\phi^{**}$ is not explicitly known. In all experiments, the active set per element $K \in \mathcal{T}_{h_1}$ consists of no more but five atoms (rather than $\mathcal{O}(h_2^{1-d})$ per element $K$) which is in good agreement with theoretical investigations and, more importantly, shows the efficiency of the adaptive strategy.

## 3.2 Conforming finite element approximation of (RP)

We write $h = g \otimes f$ if $h(\mathbf{x}, A) = g(\mathbf{x})f(A)$ for almost all $\mathbf{x} \in \omega$ and all $A \in \mathcal{S}^{d-1}$, $g \in L^1(\omega)$ and $f \in C(\mathcal{S}^{d-1})$.

Let $\mathcal{T}_{h_1}^1$ be a uniform triangulation on $\omega$ that consists of elements with diameter not exceeding $h_1 > 0$. We also suppose that for $\tilde{h}_1 < h_1$ we have $\mathcal{T}_{h_1}^1 \subset \mathcal{T}_{\tilde{h}_1}^1$, i.e., $\mathcal{T}_{\tilde{h}_1}^1$ is a refinement of $\mathcal{T}_{h_1}^1$. Similarly, we define a finite element triangulation $\mathcal{T}_{h_2}^2$ on $\mathcal{S}^{d-1}$ which is assembled from elements with a spherical diameter not exceeding $h_2$. Again we suppose that $\mathcal{T}^2_{\tilde{h}_2}$ is a refinement of $\mathcal{T}^2_{h_2}$, once $\tilde{h}_2 < h_2$. Therefore, this gives a triangulation $\mathcal{T}_{\mathbf{h}} = \mathcal{T}^1_{h_1} \times \mathcal{T}^2_{h_2}$ of $\omega \times \mathcal{S}^{d-1}$, with the mesh parameter $\mathbf{h} = (h_1, h_2) \in [0, +\infty)^2$.

Now, let us still define a projector $P^1_{h_1} : L^1(\omega; C(\mathcal{S}^{d-1})) \to L^1(\omega; C(\mathcal{S}^{d-1}))$ by

$$[P^1_{h_1} h](\mathbf{x}, A) = \frac{1}{|K|} \int_K h(\mathbf{x}, A)\, \mathrm{d}\mathbf{x} \qquad \forall \mathbf{x} \in K \in \mathcal{T}^1_{h_1}, \tag{3.7}$$

and $[P^2_{h_2} h](\mathbf{x}, \cdot)$ a projector that assigns to each $h(\mathbf{x}, \cdot)$ its $\mathcal{T}^2_{h_2}$-element-wise affine interpolation, i.e.,

$$[P^2_{h_2} h](\mathbf{x}, A) = \sum_{i=1}^{L_{h_2}} h(\mathbf{x}, A_i) v_i(A),$$

where the basis functions $v_i$ are nonnegative and $\sum_{i=1}^{L_{h_2}} v_i(A) = 1$ for all $A \in \mathcal{S}^{d-1}$, and $L_{h_2} = \mathcal{O}(h_2^{1-d})$.

Then, the composed projector

$$P_{\mathbf{h}} = P_{(h_1, h_2)} = P^1_{h_1} P^2_{h_2} = P^2_{h_2} P^1_{h_1}$$

provides an $\omega$-element-wise constant and $\mathcal{S}^{d-1}$-element-wise affine approximation.

Moreover, the adjoint projector $P^*_{\mathbf{h}} : \bar{\mathcal{A}} \to \bar{\mathcal{A}}$ is then defined by

$$\langle \nu, P_{\mathbf{h}} h \rangle = \langle P^*_{\mathbf{h}} \nu, h \rangle ,$$

and we denote $\bar{\mathcal{A}}_{\mathbf{h}} = P^*_{\mathbf{h}} \bar{\mathcal{A}} \subset \bar{\mathcal{A}}$; cf. [112, Sec. 3.5].

It is easy to see that for any $h \in L^1(\omega; C(\mathcal{S}^{d-1}))$ we have $\lim_{\mathbf{h}\to 0} \|P_{\mathbf{h}} h - h\|_{L^1(\omega; C(\mathcal{S}^{d-1}))} = 0$; cf. [112, Sec. 3.5]. Therefore,

$$\begin{aligned} |\langle \nu - P^*_{\mathbf{h}} \nu, h \rangle| &= |\langle \nu, h - P_{\mathbf{h}} h \rangle| \\ &\le \|\nu\|_{L^1(\omega; C(\mathcal{S}^{d-1}))^*} \|h - P_{\mathbf{h}} h\|_{L^1(\omega; C(\mathcal{S}^{d-1}))} \to 0, \end{aligned}$$

for $\mathbf{h} \to 0$, and thus

$$\text{weak*-}\lim_{\mathbf{h}\to 0} P^*_{\mathbf{h}} \nu = \nu \quad \text{in } L^1(\omega; C(\mathcal{S}^{d-1}))^*.$$

It can be found in [112, Sec. 3.5] that $\nu \in \bar{\mathcal{A}}_{\mathbf{h}}$ if and only if

$$\nu_{\mathbf{x}} = \sum_{i=1}^{L_{h_2}} \lambda_i(\mathbf{x})\delta_{A_i}\,, \quad \mathbf{x} \in \omega\,, \tag{3.8}$$

with $\lambda_i : \omega \to [0,1]$, $\sum_{i=1}^{L_{h_2}} \lambda_i = 1$, and $\lambda_i\big|_K \equiv \lambda_{K,i}$ is $\mathcal{T}^1_{h_1}$-element-wise constant for any $1 \le i \le L_{h_2}$ and $A_i \in \mathcal{S}^{d-1}$ for any $1 \le i \le L_{h_2}$. Hence, we have

$$\bar{\mathcal{A}}_{\mathbf{h}} := \left\{ \nu^{\mathbf{h}} = \{\nu^{\mathbf{h}}_K\}_{K\in\mathcal{T}^1_{h_1}},\ \nu^{\mathbf{h}}_K = \sum_{i=1}^{L_{h_2}} \lambda_{K,i}\delta_{A_i} \quad \forall K \in \mathcal{T}^1_{h_1} \right\}.$$

The **Discrete Relaxed Problem** ($\mathbf{R^hP}$) is now given in the form

$$\min_{\mu^{\mathbf{h}}\in\bar{\mathcal{A}}_{\mathbf{h}}} \left\{ \int_\omega \int_{\mathcal{S}^{d-1}} \phi(A)\mu^{\mathbf{h}}(\mathrm{d}A)\,\mathrm{d}\mathbf{x} - (\mathbf{f},\mathbf{m}_{\mathbf{h}}) + \frac{1}{2} \| \nabla u_{\mathbf{h}} \|^2_{\mathbf{L}^2(\mathbb{R}^d,\mathbb{R}^d)} \right\}, \tag{3.9}$$

subject to

$$\mathbf{m}_{\mathbf{h}}\big|_{K\in\mathcal{T}^1_{h_1}} = \int_{\mathcal{S}^{d-1}} A\mu^{\mathbf{h}}_K(\mathrm{d}A)\,, \tag{3.10}$$

$$\mathrm{div}(-\nabla u_{\mathbf{h}} + \chi_\omega \mathbf{m}_{\mathbf{h}}) = 0 \quad \text{in } \mathbb{R}^d\,. \tag{3.11}$$

($\mathrm{R^hP}$) is a conforming discretization of (RP). In the sequel, we discuss well-posedness and approximation properties of ($\mathrm{R^hP}$).

**Lemma 3.1** *Problem* ($\mathrm{R^hP}$) *has a solution in* $\bar{\mathcal{A}}_{\mathbf{h}}$.

**Proof:**
The sequential weak* compactness of $\bar{\mathcal{A}}$ (see [112, Corollary 3.1.7.]) together with sequential weak* lower semicontinuity of $\bar{\mathcal{E}}$ and weak* continuity of $P^*_{\mathbf{h}}$ ensures the existence of a solution to ($\mathrm{R^hP}$) for any $\mathbf{h}$ with $h_1, h_2 > 0$. □

**Lemma 3.2** *Let* $\phi \in C^0(\mathcal{S}^{d-1})$ *and* $\mathbf{f} \in \mathbf{L}^2(\omega,\mathbb{R}^d)$. *If* $\mathbf{h} = (h_1,h_2) \to 0$ *with* $h_2/h_1^{d/2} \to 0$, *then* $\min(\mathrm{R^hP}) \to \min(\mathrm{RP})$.

**Proof:**
Let $\mathcal{E}(\nu) = \min_{\mu\in\bar{\mathcal{A}}} \mathcal{E}(\mu)$ be a solution to (RP), let $\nu^{\mathbf{h}} = P^*_{\mathbf{h}}\nu \in \bar{\mathcal{A}}_{\mathbf{h}}$ and let $\mu^{\mathbf{h}} \in \bar{\mathcal{A}}_{\mathbf{h}}$ solve ($\mathrm{R}^{\mathbf{h}}$P). We have

$$\begin{aligned}
0 &\le \bar{\mathcal{E}}(\mu^{\mathbf{h}}) - \bar{\mathcal{E}}(\nu)\\
&\le \bar{\mathcal{E}}(\nu^{\mathbf{h}}) - \bar{\mathcal{E}}(\nu) = \langle \nu^{\mathbf{h}} - \nu, 1\otimes\phi\rangle\\
&\quad + \langle \nu - \nu^{\mathbf{h}}, \mathbf{f}\otimes\mathbf{id}\rangle + \frac{1}{2}\left\{\|\nabla u_{\mathbf{h}}\|_{\mathbf{L}^2(\mathbb{R}^d;\mathbb{R}^d)} - \|\nabla u\|^2_{\mathbf{L}^2(\mathbb{R}^d;\mathbb{R}^d)}\right\}\\
&= \langle \nu, P_{\mathbf{h}}(1\otimes\phi) - 1\otimes\phi\rangle + \langle \nu, P_{\mathbf{h}}(\mathbf{f}\otimes\mathbf{id}) - \mathbf{f}\otimes\mathbf{id}\rangle\\
&\quad + \frac{1}{2}\left\{\|\nabla u_{\mathbf{h}}\|^2_{\mathbf{L}^2(\mathbb{R}^d;\mathbb{R}^d)} - \|\nabla u\|^2_{\mathbf{L}^2(\mathbb{R}^d;\mathbb{R}^d)}\right\}\\
&\le \|\nu\|_{L^1(\omega;C(\mathcal{S}^{d-1}))^*}\left\{\|P_{\mathbf{h}}(1\otimes\phi) - 1\otimes\phi\|_{L^1(\omega;C(\mathcal{S}^{d-1}))}\right.\\
&\quad \left. + \|P_{\mathbf{h}}(\mathbf{f}\otimes\mathbf{id}) - \mathbf{f}\otimes\mathbf{id}\|_{L^1(\omega;C(\mathcal{S}^{d-1}))}\right\}\\
&\quad + \|\mathbf{m} - \mathbf{m}_{\mathbf{h}}\|_{L^2(\omega;\mathbb{R}^d)}\,,
\end{aligned}$$

where $\mathbf{m}_{\mathbf{h}} = \nu^{\mathbf{h}} \bullet \mathbf{id}$, $\mathbf{m} = \nu \bullet \mathbf{id}$, $\Delta u_{\mathbf{h}} = \operatorname{div}(\chi_\omega \mathbf{m}_{\mathbf{h}})$ and $\Delta u = \operatorname{div}(\chi_\omega \mathbf{m})$. Clearly, $\lim_{\mathbf{h}\to 0}\|P_{\mathbf{h}}(1\otimes\phi) - 1\otimes\phi\|_{L^1(\omega;C(\mathcal{S}^{d-1}))} = \lim_{\mathbf{h}\to 0}\|P_{\mathbf{h}}(\mathbf{f}\otimes\mathbf{id}) - \mathbf{f}\otimes\mathbf{id}\|_{L^1(\omega;C(\mathcal{S}^{d-1}))} = 0$. Further, we have

$$\begin{aligned}
\|\mathbf{m}_{(h_1,h_2)} - \mathbf{m}\|_{\mathbf{L}^2(\omega;\mathbb{R}^d)} &\le \|\mathbf{m}_{(h_1,h_2)} - \mathbf{m}_{(h_1,0)}\|_{\mathbf{L}^2(\omega;\mathbb{R}^d)}\\
&\quad + \|\mathbf{m}_{(h_1,0)} - \mathbf{m}\|_{\mathbf{L}^2(\omega;\mathbb{R}^d)}\,.
\end{aligned}$$

The second term on the right-hand side converges to zero as $h_1 \to 0$. The first term is less or equal to $\frac{h_2}{h_1^{d/2}}$. To see this, we realize that we look for $\nu^{(h_1,h_2)}$ in the form $\nu^{(h_1,h_2)} = \sum_{i=1}^{L_{h_2}} \lambda_i\delta_{A_i}$ with $\{A_i\}_{i=1}^{L_{h_2}}$ constituting the mesh $\mathcal{T}^2_{h_2}$, and $\min_{(i,j),i\neq j}|A_i - A_j| \le Ch_2$, $C>0$. Thus, $\mathbf{m}_{(h_1,h_2)}\Big|_{K\in\mathcal{T}^1_{h_1}} = \sum_{i=1}^{L_{h_2}}\lambda_i A_i$. Moreover, $\mathbf{m}_{(h_1,0)}\Big|_{K\in\mathcal{T}^1_{h_1}} = \int_{\mathcal{S}^{d-1}} A\nu_K(\mathrm{d}A)$. Then one has, on any element $K\in\mathcal{T}^1_{h_1}$,

$$\|\mathbf{m}_{(h_1,0)} - \mathbf{m}_{(h_1,h_2)}\|^2_{\mathbf{L}^\infty(K,\mathbb{R}^d)} \le C\,h_2^2 \qquad \forall K\in\mathcal{T}^1_{h_1}\,.$$

Since we have $\mathcal{O}(h_1^{-d})$ elements in $\mathcal{T}^1_{h_1}$, we obtain

$$\|\mathbf{m}_{(h_1,h_2)} - \mathbf{m}_{(h_1,0)}\|_{\mathbf{L}^2(\omega;\mathbb{R}^d)} \le C\frac{h_2}{h_1^{d/2}}\,.$$

□

In fact, the following result of convergence is valid for sufficiently smooth data as well as solutions $\mathbf{m}$.

**Corollary 3.1** *Let $\phi$ be Lipschitz, $\mathbf{f} \in \mathbf{W}^{1,2}(\omega;\mathbb{R}^d)$, and $\nu^{\mathbf{h}} = \operatorname{argmin}_{\bar{\mathcal{A}}_{\mathbf{h}}} \bar{\mathcal{E}}$ solves $(\mathrm{R}^{\mathbf{h}}\mathrm{P})$, for $\mathbf{h} = (h_1, h_2)$. Further, let $\nu = \operatorname{argmin}_{\bar{\mathcal{A}}} \bar{\mathcal{E}}$ and suppose $\mathbf{m} \in W^{1,2}(\omega;\mathbb{R}^d)$. Then, there exists a constant $C > 0$ that does not depend on $\mathbf{h} \in \mathbb{R}^2$, such that*

$$| \bar{\mathcal{E}}(\nu) - \bar{\mathcal{E}}(\nu^{\mathbf{h}}) | \leq C\Big\{h_1 + h_2 + \frac{h_2}{h_1^{d/2}}\Big\}.$$

**Proof:**
The proof follows from the estimate in the proof of Lemma 3.2, from the smoothness of $\phi$ and $\mathbf{f}$ and from the standard approximation results for Sobolev mappings in $\mathbf{W}^{1,2}(\omega;\mathbb{R}^d)$; cf. [29]. □

**Remark 3.1** *1. No regularity properties are known to the authors that make the assumption $\mathbf{m} \in \mathbf{W}^{1,2}(\omega;\mathbb{R}^d)$ hold in a general setting.*

*2. We need to adjust both scales $h_1, h_2$ in order to ensure convergence of the method.*

## 3.3 The Active Set based Scheme $(\mathrm{R}^{\mathbf{h}^j}\mathrm{P}_Q)_{j\in\mathbb{N}}$

$(\mathrm{R}^{\mathbf{h}}\mathrm{P})$ is a convex optimization problem over $\{\lambda_{K,i}\}$, which has a linear-quadratic structure. On the other hand, the number of unknowns is of order $\mathcal{O}\big((h_1^d h_2^{d-1})^{-1}\big)$, which makes the problem rather large. Moreover, Corollary 3.1 implies usage of a large number of atoms per element $K \in \mathcal{T}_{h_1}^1$ to achieve reasonable approximation of the minimum energy.

Hence, the remainder of this chapter is devoted to proposing and analyzing an adaptive method to reduce computational effort, and which only uses a small set of active atoms per each element $K \in \mathcal{T}_{h_1}$. It is based on optimality conditions for $(\mathrm{R}^{\mathbf{h}}\mathrm{P})$ which have first been used by Carstensen & Roubíček in [21] for a scalar or one-dimensional variational problem. The main idea is that a discrete Young measure solution to $(\mathrm{R}^{\mathbf{h}}\mathrm{P})$ is typically supported at a very few atoms. Note that Carathéodory's theorem predicts that there is always a solution $\nu \in \bar{\mathcal{A}}$ to (RP) such that for almost every point $\mathbf{x} \in \omega$, the measure $\nu_{\mathbf{x}}$ is supported at $(d+1)$ points, at most. Let us first formulate optimality conditions for $(\mathrm{R}^{\mathbf{h}}\mathrm{P})$. They are stated in a form of the Weierstrass-type maximum principle.

**Lemma 3.3** *Let $\mathbf{f} \in \mathbf{L}^2(\omega;\mathbb{R}^d)$, $\phi : \mathbb{R}^d \to \mathbb{R}$ be continuous, and $(\nu^{\mathbf{h}}, \mathbf{m}_{\mathbf{h}}) \in \bar{\mathcal{A}}_{\mathbf{h}} \times \mathbf{L}^2(\omega;\mathbb{R}^d)$ solves $(\mathrm{R}^{\mathbf{h}}\mathrm{P})$. Then,*

$$\int_{\mathcal{S}^{d-1}} \mathcal{H}^{\mathbf{h}}_{\boldsymbol{\ell}_{\mathbf{h}}}(\mathbf{x}, A)\nu^{\mathbf{h}}_K(\mathrm{d}A) = \max_{A\in\mathcal{S}^{d-1}} \mathcal{H}^{\mathbf{h}}_{\boldsymbol{\ell}_{\mathbf{h}}}(\mathbf{x}, A) \qquad \forall\, \mathbf{x} \in K \in \mathcal{T}^1_{h_1} ,$$

*where the Hamiltonian is defined to be*

$$\mathcal{H}^{\mathbf{h}}_{\boldsymbol{\ell}_{\mathbf{h}}} := P_{\mathbf{h}}(\boldsymbol{\ell}_{\mathbf{h}} \otimes \mathbf{id} - \phi) ,$$

*for $\boldsymbol{\ell}_{\mathbf{h}} = \mathbf{f} - \nabla u_{\mathbf{h}}$, and $u_{\mathbf{h}}$ solves (3.6).*

**Proof:**
The verification of this lemma is the same as of [83, Prop. 4] where one uses Young measures from $\bar{\mathcal{A}}_{\mathbf{h}}$ instead of $\bar{\mathcal{A}}$. □

**Lemma 3.4** *Let $\mathbf{h} = (h_1, h_2)$ be a discretization parameter, let $\nu^{(h_1,0)} \in \bar{\mathcal{A}}$ be a solution to $(\mathrm{R}^{(h_1,0)}\mathrm{P})$ and let $\nu^{\mathbf{h}} \in \bar{\mathcal{A}}_{\mathbf{h}}$ be a solution to $(\mathrm{R}^{\mathbf{h}}\mathrm{P})$. Further, let $\mathbf{m}_{(h_1,0)} = \nu^{(h_1,0)} \bullet \mathbf{id}$ and $\mathbf{m}_{\mathbf{h}} = \nu^{\mathbf{h}} \bullet \mathbf{id}$, with $u_{(h_1,0)}$ and $u_{\mathbf{h}}$ the corresponding solutions to the Poisson equation (3.6). Then, $\lim_{h_2\to 0} \mathbf{m}_{\mathbf{h}} = \mathbf{m}_{(h_1,0)}$ in the norm topology of $\mathbf{L}^2(\omega;\mathbb{R}^d)$ and $\lim_{h_2\to 0} \boldsymbol{\ell}_{(h_1,h_2)} = \boldsymbol{\ell}_{(h_1,0)}$ a.e. in $\omega$, where $\boldsymbol{\ell}_{\mathbf{h}}$ and $\boldsymbol{\ell}_{(h_1,0)}$ are the Lagrange multipliers for $(\mathrm{R}^{\mathbf{h}}\mathrm{P})$ and $(\mathrm{R}^{(h_1,0)})$, respectively.*

**Proof:**
The fact, that $\lim_{h_2\to 0} \mathbf{m}_{\mathbf{h}} = \mathbf{m}_{(h_1,0)}$ follows from the proof of Lemma 3.2 where we proved that $\|\mathbf{m}_{(h_1,h_2)} - \mathbf{m}_{(h_1,0)}\|_{\mathbf{L}^2(\omega;\mathbb{R}^d)} \leq C\frac{h_2}{h_1^{d/2}}$.

The strong convergence $\nabla u_{(h_1,h_2)} \to \nabla u_{(h_1,0)}$ in $\mathbf{L}^2(\mathbb{R}^d;\mathbb{R}^d)$ then follows from the estimate

$$\|\nabla u_{(h_1,0)} - \nabla u_{(h_1,h_2)}\|_{\mathbf{L}^2(\mathbb{R}^d,\mathbb{R}^d)} \leq \|\mathbf{m}_{(h_1,0)} - \mathbf{m}_{(h_1,h_2)}\|_{\mathbf{L}^2(\omega;\mathbb{R}^d)} .$$

Moreover, we can identify $\nabla u_{(h_1,h_2)}$, $h_2 \geq 0$, restricted to $\omega$ with its $\mathcal{T}^1_{h_1}$-element-wise averages because it enters the energy functional only via

$$\big(\nabla u_{(h_1,h_2)}, \mathbf{m}_{(h_1,h_2)}\big) = \|\nabla u_{(h_1,h_2)}\|^2_{\mathbf{L}^2(\mathbb{R}^d;\mathbb{R}^d)},$$

and $\mathbf{m}_{(h_1,h_2)}$ is $\mathcal{T}^1_{h_1}$-element-wise constant. By the same argument we can identify the external field $\mathbf{f}$ restricted to $\omega$ with its $\mathcal{T}^1_{h_1}$-element averages. This altogether shows that $\boldsymbol{\ell}_{\mathbf{h}} = \boldsymbol{\ell}_{(h_1,h_2)} = \mathbf{f} - \nabla u_{(h_1,h_2)}$ converges to $\boldsymbol{\ell}_{(h_1,0)} = \mathbf{f} - \nabla u_{(h_1,0)}$ strongly in $\mathbf{L}^2(\omega;\mathbb{R}^d)$. Since $\boldsymbol{\ell}_{(h_1,h_2)}$, $h_2 \geq 0$, are $\mathcal{T}^1_{h_1}$-element-wise constant, the convergence is even pointwise. □

Following [112], we define the support of a Young measure $\nu \in \bar{A}$ via

$$\operatorname{supp} \nu = \Big\{(\mathbf{x}, A) \in \omega \times \mathcal{S}^{d-1};\ A \in \operatorname{supp} \nu_{\mathbf{x}}\Big\},$$

and state a new problem $(\mathrm{R}^{\mathbf{h}}\mathrm{P}_Q)$ in which we restrict the support of competing Young measures from $\bar{A}_{\mathbf{h}}$ to a set $Q \subset \omega \times \mathcal{S}^{d-1}$. Hence, $Q$ is assembled from selected nodes $\{A_i\}_K$, for each element $K \in \mathcal{T}^1_{h_1}$. Following [21], we refer to the **Discrete Relaxed Problem on the Active Set** $Q$, shortly $(\mathbf{R}^{\mathbf{h}}\mathbf{P}_Q)$, which reads:

$$\min_{\mu^{\mathbf{h}} \in \bar{A}_{\mathbf{h}}} \left\{ \int_\omega \int_{\mathcal{S}^{d-1}} \phi(A)\mu^{\mathbf{h}}(\mathrm{d}A)\,\mathrm{d}\mathbf{x} - (\mathbf{f}, \mathbf{m}_{\mathbf{h}}) + \frac{1}{2} \| \nabla u_{\mathbf{h}} \|^2_{\mathbf{L}^2(\mathbb{R}^d,\mathbb{R}^d)} \right\}, \tag{3.12}$$

subject to

$$\mathbf{m}_{\mathbf{h}}\big|_{K \in \mathcal{T}^1_{h_1}} = \int_{\mathcal{S}^{d-1}} A\mu^{\mathbf{h}}_K(\mathrm{d}A), \quad \text{for} \quad \operatorname{supp} \mu^{\mathbf{h}} \subset Q, \tag{3.13}$$

$$\operatorname{div}\big(-\nabla u_{\mathbf{h}} + \chi_\omega \mathbf{m}_{\mathbf{h}}\big) = 0 \quad \text{in } \mathbb{R}^d. \tag{3.14}$$

In the sequel, the Lemma 3.5 through 3.8 can be seen as versions of Propositions 4, 6, 8 and Corollary 1 from [21], respectively, by adopting them to our specific situation. Set for $\varepsilon > 0$

$$Q = \Big\{(\mathbf{x}, A) \in \omega \times \mathcal{S}^{d-1};\ h(\mathbf{x}, A) \geq \max_{B \in \mathcal{S}^{d-1}} h(\mathbf{x}, B) - \varepsilon\Big\}. \tag{3.15}$$

**Lemma 3.5** *Let $Q \subset \omega \times \mathcal{S}^{d-1}$ satisfy*

$$\Big\{(\mathbf{x}, A) \in \omega \times \mathcal{S}^{d-1};\ A \text{ is a grid point of } \mathcal{T}^2_{h_2},$$
$$\mathcal{H}^{\mathbf{h}}_{\boldsymbol{\ell}}(\mathbf{x}, A) = \max_{\tilde{A} \in \mathcal{S}^{d-1}} \mathcal{H}^{\mathbf{h}}_{\boldsymbol{\ell}}(\mathbf{x}, \tilde{A})\Big\} \subset Q,$$

*and let* $\boldsymbol{\ell}_{\mathbf{h}}$ *be a corresponding Lagrange multiplier to the solution to* $(\mathrm{R}^{\mathbf{h}}\mathrm{P})$. *Then, every solution to* $(\mathrm{R}^{\mathbf{h}}\mathrm{P}_Q)$ *also solves* $(\mathrm{R}^{\mathbf{h}}\mathrm{P})$.

**Proof:**
According to Lemma 3.3, any solution to $(\mathrm{R}^{\mathbf{h}}\mathrm{P})$ is supported on $Q$ and so is feasible for $(\mathrm{R}^{\mathbf{h}}\mathrm{P}_Q)$. Therefore, $\min(\mathrm{R}^{\mathbf{h}}\mathrm{P}) \geq \min(\mathrm{R}^{\mathbf{h}}\mathrm{P}_Q)$. On the other hand, $\min(\mathrm{R}^{\mathbf{h}}\mathrm{P}) \leq \min(\mathrm{R}^{\mathbf{h}}\mathrm{P}_Q)$. If $\nu^{\mathbf{h}} \in \bar{\mathcal{A}}_{\mathbf{h}}$ solves $(\mathrm{R}^{\mathbf{h}}\mathrm{P}_Q)$, then $\bar{\mathcal{E}}(\nu^{\mathbf{h}}) = \min(\mathrm{R}^{\mathbf{h}}\mathrm{P}_Q) = \min(\mathrm{R}^{\mathbf{h}}\mathrm{P})$ and thus $\nu^{\mathbf{h}}$ solves $(\mathrm{R}^{\mathbf{h}}\mathrm{P})$, as well. □

The following results justify the approximation of the Hamiltonian in our active set based scheme given below, since we will not have $\mathcal{H}^{\mathbf{h}}_{\ell}$ at our disposal in each iteration step.

**Lemma 3.6** *Let* $\|h-\mathcal{H}^{\mathbf{h}}_{\boldsymbol{\ell}_{\mathbf{h}}}\|_{L^\infty(\omega;\Sigma)} \leq \varepsilon/2$ *for some solution* $\nu^{\mathbf{h}}$ *to* $(\mathrm{R}^{\mathbf{h}}\mathrm{P})$ *with a Lagrange multiplier* $\boldsymbol{\ell}_{\mathbf{h}}$ *and for some* $\Sigma \subset \mathcal{S}^{d-1}$ *such that for* a.a. $\mathbf{x} \in \omega$

$$\arg\max \mathcal{H}^{\mathbf{h}}_{\boldsymbol{\ell}_{\mathbf{h}}}(\mathbf{x},\cdot) \cup \arg\max h(\mathbf{x},\cdot) \subset \Sigma\,. \tag{3.16}$$

*Let* $Q$ *be chosen as above. Then any solution to* $(\mathrm{R}^{\mathbf{h}}\mathrm{P}_Q)$ *solves* $(\mathrm{R}^{\mathbf{h}}\mathrm{P})$.

**Proof:**
We will verify assumptions of the Lemma 3.5. Let us take $\nu^{\mathbf{h}} \in \bar{\mathcal{A}}_{\mathbf{h}}$ such that $\nu^{\mathbf{h}}_K$ is supported on $\arg \mathcal{H}^{\mathbf{h}}_{\boldsymbol{\ell}_{\mathbf{h}}}(\mathbf{x},\cdot)$ for almost all $K \in \mathcal{T}^1_{h_1}$. Then $\operatorname{supp}(\nu^{\mathbf{h}}) \subset \omega \times \Sigma$. Taking $(\mathbf{x}, A) \in \operatorname{supp}\nu$, we have from (3.16)

$$\begin{aligned} h(\mathbf{x},A) &\geq \mathcal{H}^{\mathbf{h}}_{\boldsymbol{\ell}_{\mathbf{h}}}(\mathbf{x},A) - \frac{\varepsilon}{2} = \max_{\tilde{A}\in\mathcal{S}^{d-1}} \mathcal{H}^{\mathbf{h}}_{\boldsymbol{\ell}_{\mathbf{h}}}(\mathbf{x},\tilde{A}) - \frac{\varepsilon}{2} \\ &\geq \max_{\tilde{A}\in\mathcal{S}^{d-1}} h(\mathbf{x},\tilde{A}) - \varepsilon\,, \end{aligned}$$

which shows that Lemma 3.5 may be applied. □

**Lemma 3.7** *For any* $\varepsilon > 0$ *there exists* $\tilde{h}_2 > 0$ *such that if* $h_2 < \tilde{h}_2$, *then* $\|\mathcal{H}^{(h_1,h_2)}_{\boldsymbol{\ell}_{(h_1,h_2)}} - \mathcal{H}^{(h_1,\tilde{h}_2)}_{\boldsymbol{\ell}_{(h_1,\tilde{h}_2)}}\|_{L^\infty(\omega\times\mathcal{S}^{d-1})} \leq \varepsilon/2$ .

**Proof:**
We know from Lemma 3.4 that $\{\boldsymbol{\ell}_{(h_1,h_2)}\}_{h_2>0}$ is a Cauchy net and so is $\{\boldsymbol{\ell}_{(h_1,h_2)}\otimes\mathbf{id}-\phi\}_{h_2>0}$ and also $\{P_{\mathbf{h}}(\boldsymbol{\ell}_{(h_1,h_2)}\otimes\mathbf{id}-\phi)\}_{h_2>0} = \left\{\mathcal{H}^{(h_1,h_2)}_{\boldsymbol{\ell}_{(h_1,h_2)}}\right\}_{h_2>0}$.
This proves the assertion. □

**Lemma 3.8** *For every* $\varepsilon > 0$ *there exists* $\mathbf{b} = (h_1, \tilde{h}_2)$ *such that for all* $h_2 < \tilde{h}_2$ *any solution to* $(\mathrm{R}^{\mathbf{h}}\mathrm{P}_Q)$, *with* $\mathbf{h} = (h_1, h_2)$, *solves* $(\mathrm{R}^{\mathbf{h}}\mathrm{P})$ *provided that*

$$Q = \left\{ (\mathbf{x}, A) \in \omega \times \mathcal{S}^{d-1};\ \mathcal{H}^{\mathbf{b}}_{\boldsymbol{\ell}_{\mathbf{b}}}(\mathbf{x}, A) \geq \max_{\tilde{A} \in \mathcal{S}^{d-1}} \mathcal{H}^{\mathbf{b}}_{\boldsymbol{\ell}_{\mathbf{b}}}(\mathbf{x}, \tilde{A}) - \varepsilon \right\}.$$

**Proof:**
Take $\tilde{h}_2$ such that Lemma 3.7 holds. Then use Lemma 3.5 with $h = \mathcal{H}^{\mathbf{b}}_{\boldsymbol{\ell}_{\mathbf{b}}}$. □

We may now check the optimality condition stated in Lemma 3.3 in an element-wise manner, for every $K \in \mathcal{T}_{h_1}$, to single out active atoms in the minimizing process. This leads to the following **Active Set based Scheme** $\{(\mathrm{R}^{\mathbf{h}^j}\mathbf{P}_Q)\}_{j\in\mathbb{N}}$: We choose a sequence of $\{\varepsilon_j\}_{j\in\mathbb{N}} \subset (0, +\infty)$, an initial discretization parameter $\mathbf{h}^1 = (h_1^1, h_2^1)$ and a stopping criterion $\mathtt{TOL} > 0$. Furthermore, we denote by $\nu^j \in \bar{\mathcal{A}}_{\mathbf{h}^j}$ a solution to $(\mathrm{R}^{\mathbf{h}^j}\mathrm{P})$, for $\mathbf{h}^j = (h_1^1, 2^{1-j}h_2^1)$, and we abbreviate $\mathcal{H}^j := \mathcal{H}^{\mathbf{h}^j}_{\boldsymbol{\ell}_{\mathbf{h}^j}}$, with $\boldsymbol{\ell}_{\mathbf{h}^j}$ the Lagrange multiplier corresponding to $\nu^j$.

**Algorithm 3.1** *1. Put* $j := 1$, $h := 0$ $\varepsilon := \varepsilon_j$.
*2. Construct* $\mathcal{T}_{\mathbf{h}^j}$ *and compute* $Q \subset \mathcal{T}_{\mathbf{h}^j}$ *from (3.15).*
*3. Compute* $\nu^j := \arg\min_{\bar{\mathcal{A}}_{\mathbf{h}^j}} (\mathrm{R}^{\mathbf{h}^j}\mathbf{P}_Q)$.
*4. Check whether the maximum principle in Lemma 3.3 is satisfied for* $\nu^j$, *then go to 6., else continue.*
*5. Increase the tolerance (i.e., the number of considered atoms) by setting* $\varepsilon := 2\varepsilon$, *and go to (2).*
*6. If* $j = 1$ *then go to 8., else continue.*
*7. If* $\bar{\mathcal{E}}(\nu^{j-1}) - \bar{\mathcal{E}}(\nu^j) < \mathtt{TOL}$ *end, else continue.*
*8. Update* $j := j + 1$, *and set* $h := \mathcal{H}^{j-1}$, $\mathbf{h}^j := (h_1, h_2^{j-1}/2)$, $\varepsilon := \varepsilon_j$, *then go to 2.*

**Remark 3.2** *1. The number* $\varepsilon > 0$ *is a tuning parameter in the iterative scheme that may be suited to the basic discretization parameter* $\mathbf{h}_1$, *see Lemma 3.8. Note that small values of* $\varepsilon$ *restrict enrichment of* $Q$ *in the* $j$*th iteration cycle.*
*2. Checking the Weierstrass maximum principle for a set* $Q$ *that is assembled from elements and nodes of* $\mathcal{T}_{\mathbf{h}^j}$ *amounts to solving a discrete finite optimization problem for each* $K \in \mathcal{T}^1_{h_1}$.

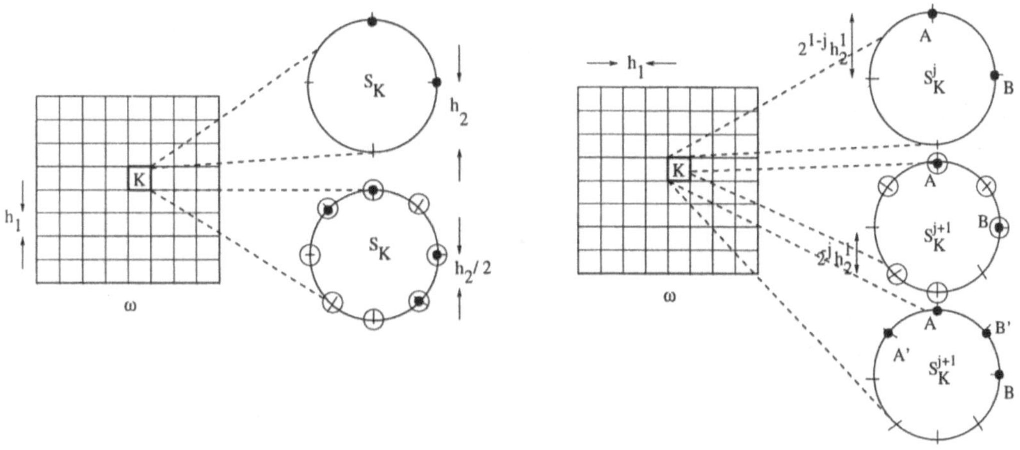


(a) (b)

Figure 3.1: (a) Uniform discretization of $\mathcal{S}^{d-1}$, (b) adaptive active set strategy via Weierstrass maximum principle: the set of active atoms for $K \in \mathcal{T}^1_{h_1}$ is enlarged from $\{A, B\}$ to $\{A, B, A', B'\}$. (The support of $\nu^{\mathbf{h}}_K$, $K \in \mathcal{T}^1_{h_1}$ solving $(\mathrm{R^h P})$ is indicated by '•' and supports of competing Young measures by '○').

At each step $j \in \mathbb{N}$ in $\{(\mathrm{R}^{\mathbf{h}^j}\mathrm{P}_Q)\}_{j\in\mathbb{N}}$, the set of active atoms per each element $K \in \mathcal{T}^1_{h_1}$ is assembled from nodes that are singled out from the triangulation $\mathcal{T}^2_{h^j_2}$. It is step 4. in the Algorithm 3.1 that ensures optimality of the set $Q$ for the present triangulation $\mathcal{T}^j_{\mathbf{h}}$, i.e., $\nu^j := \arg\min_{\bar{\mathcal{A}}_{\mathbf{h}^j}}(\mathrm{R}^{\mathbf{h}^j}\mathrm{P}_Q)$ $= \arg\min_{\bar{\mathcal{A}}_{\mathbf{h}^j}}(\mathrm{R}^{\mathbf{h}^j}\mathrm{P})$. Hence, we may apply Lemma 3.3 to verify the following result.

**Corollary 3.2** *Let the assumptions of Corollary 3.1 hold. Then in the $j$-th cycle of the previous algorithm we have, for $\nu^j := \arg\min_{\bar{\mathcal{A}}_{\mathbf{h}^j}}(\mathrm{R}^{\mathbf{h}^j}\mathrm{P}_Q)$,*

$$|\,\bar{\mathcal{E}}(\nu) - \bar{\mathcal{E}}(\nu^j)\,| \le C\Big\{h^1_1 + 2^{1-j}h^1_2 + \frac{2^{1-j}h^1_2}{(h^1_1)^{d/2}}\Big\}.$$

## 3.4 Computational Experiments

We show the efficiency of the scheme $\{(\mathrm{R}^{\mathbf{h}^j}\mathrm{P}_Q)\}_{j\in\mathbb{N}}$ in 2D computational experiments. Let us take $\omega = (0,1)^2$. Alternatively, we may compute the magnetostatic energy $\frac{1}{2}\|\nabla u\|^2_{\mathbf{L}^2(\mathbb{R}^2;\mathbb{R}^2)}$ through $\frac{1}{2}(\mathbf{m}, \nabla u)$ thanks to (3.6), which uses an integration only over $\omega$ instead of $\mathbb{R}^2$.

We have (see [56])

$$u_\mathbf{h}(\mathbf{x}) = \frac{1}{2\pi}\mathrm{div} \sum_{K\in\mathcal{T}^1_{h_1}} \int_K \log(|\mathbf{x}-\mathbf{y}|)\mathbf{m}_\mathbf{h}(\mathbf{y})\,\mathrm{d}\mathbf{y} \qquad \forall \mathbf{x}\in\omega\,.$$

Since we work only with an element-wise constant magnetization $\mathbf{m}_\mathbf{h}$, it is sufficient to work with element-wise averages of $\nabla u_\mathbf{h}$. We follow the approach by Ma [95, 96] to get

$$\begin{aligned} u_\mathbf{h}(\mathbf{x}) &= \frac{1}{2\pi}\mathrm{div}\int_\omega \log(|\mathbf{x}-\mathbf{y}|)\mathbf{m}_\mathbf{h}(\mathbf{y})\,\mathrm{d}\mathbf{y} \\ &= \frac{1}{2\pi}\int_\omega \frac{\partial \log|\mathbf{x}-\mathbf{y}|}{\partial \mathbf{x}_1}(\mathbf{m}_\mathbf{h})_1(\mathbf{y})\,\mathrm{d}\mathbf{y} + \int_\omega \frac{\partial \log|\mathbf{x}-\mathbf{y}|}{\partial \mathbf{x}_2}(\mathbf{m}_\mathbf{h})_2(\mathbf{y})\,\mathrm{d}\mathbf{y} \\ &= -\frac{1}{2\pi}\sum_{K\in\mathcal{T}^1_{k_1}}\Big\langle \mathbf{m}_\mathbf{h}\big|_K, \int_K \nabla_\mathbf{y}\log|\,\mathbf{x}-\mathbf{y}\,|\,\mathrm{d}\mathbf{y}\Big\rangle_{\mathbb{R}^2} = \sum_{K\in\mathcal{T}^1_{h_1}} \mathbf{m}_\mathbf{h}\big|_K G_K(\mathbf{x})\,, \end{aligned}$$

where

$$G_K(\mathbf{x}) = -\frac{1}{2\pi}\int_K \nabla_\mathbf{y}\log|\mathbf{x}-\mathbf{y}|\,\mathrm{d}\mathbf{y} = -\frac{1}{2\pi}\int_{\partial K}\log(|\mathbf{x}-\mathbf{y}|)\mathbf{n}\,\mathrm{d}\mathbf{x}\,,$$

with $\mathbf{n}\in\mathbb{R}^2$ being the outward unit normal to $\partial K$. We compute

$$\begin{aligned} \nabla u_\mathbf{h}\big|_K &= \frac{1}{|K|}\int_K \nabla u_\mathbf{h}(\mathbf{x})\,\mathrm{d}\mathbf{x} = \frac{1}{|K|}\int_{\partial K} u_\mathbf{h}\mathbf{n}\,\mathrm{d}\mathbf{x} \\ &= \frac{1}{|K|}\int_{\partial K}\Big(\sum_{K\in\mathcal{T}^1_{h_1}} \mathbf{m}_h\big|_K G_K\Big)\mathbf{n}\,\mathrm{d}\mathbf{x} = \frac{1}{|K|}\sum_{K\in\mathcal{T}^1_{h_1}} \mathbf{m}_\mathbf{h}\big|_K\int_K G_K\otimes\mathbf{n}\,\mathrm{d}\mathbf{x}\,, \end{aligned}$$

where $G_K$ can be computed in advance by the symbolic manipulator "Mathematica", for instance.

The problem ($\mathrm{R}^\mathbf{h}\mathrm{P}$) is quadratic-linear, with a linear constraint which can be written in the form

$$\min f(\mathbf{y}) = \frac{1}{2}\langle\mathbf{y}, \mathbf{C}\mathbf{y}\rangle + \langle\mathbf{e},\mathbf{y}\rangle\,, \tag{3.17}$$

subject to $0 \le y_i \le 1$, $1 \le i \le N$, and $\mathbf{Fy} + \mathbf{b} = 0$, where $N$ is a number of variables. Let us denote for simplicity by $M \in \mathbb{N}$ the number of elements in $\mathcal{T}^1_{h_1}$ and by $N_i$ the number of active atoms at $\mathcal{T}^2_{h_2}$ for any $1 \le i \le M$. Thus $N = \sum_{i=1}^{M} N_i$. Now $\mathbf{C} \in \mathbb{R}^{N\times N}$ is a matrix coming from the evaluation of the magnetostatic energy, $\mathbf{e} \in \mathbb{R}^N$ is a vector coming from the evaluation of anisotropy and interaction energies. Further, let $\mathbf{b} \in \mathbb{R}^M$ be $\mathbf{b} = (-1, -1, \dots, -1)^\top$. Finally, $\mathbf{F} \in \mathbb{R}^{M\times N}$, for any $1 \le i \le M$, and $F_{ij} = 1$ if $1 + \sum_{k=1}^{i-1} N_k \le j \le \sum_{k=1}^{i} N_k$ and $F_{ij} = 0$ otherwise. This enforces that $\sum_{k=1+N_{i-1}}^{N_i} y_k = 1$ which ensures that in each element in $\mathcal{T}^1_{h_1}$ we work with a probability measure. The problem (3.17) was solved by means of the QLD routine [114]. The solution also provides us with a corresponding Lagrange multiplier. Then, checking the maximum principle on the $i$-th element in $\mathcal{T}^1_{h_1}$ consists of the evaluation of the Hamiltonian at $N_i$ points.

In the following, we discuss two examples of energy densities that model uniaxial magnetic materials.

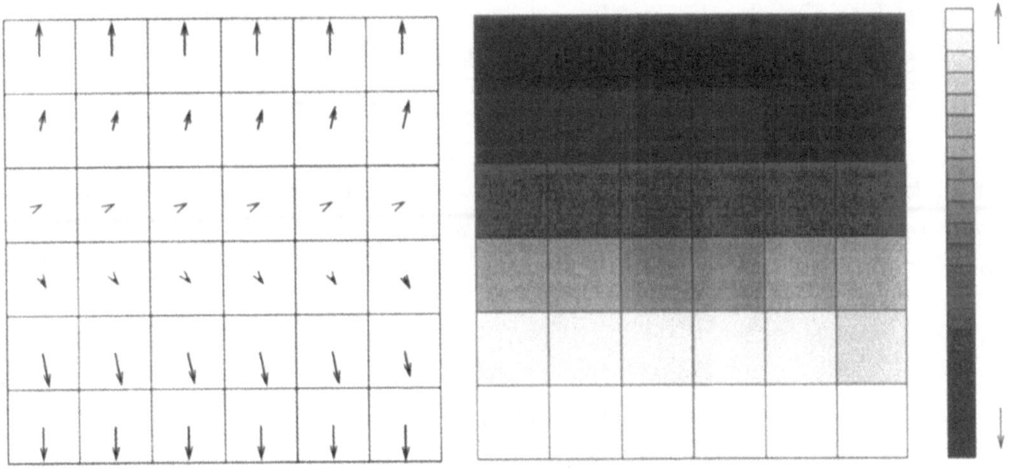

Figure 3.2: A solution of $\{(\mathrm{R}^{\mathbf{h}^j}\mathrm{P}^j_Q)\}_{1\le j\le J}$. The magnetization $\mathbf{m}_{\mathbf{h}^J}$ (left) and portions of phases (right).

**Example 3.1** *Let us take* $\phi(\mathbf{m}) = 10^{-2}\{m_1^2 + (m_2^2 - 1)^2\}$ *where we do not know the corresponding* $\phi^{**}$. *Moreover, we set* $\varepsilon_j = 2^{-j}/80$, TOL=*0.0001*, $\mathbf{f}(\mathbf{x}) = (3.5 \times 10^{-2} \times x_2(x_2 - 1), 2 \times 10^{-3}x_2 - 10^{-3})^\top$, $\mathbf{x} \in \omega$ *and* $\mathbf{h}^1 =$

| j | number of atoms/element | average number of active atoms |
|---|---|---|
| 1. | 8 | 3.1 |
| 2. | 16 | 3.4 |
| 3. | 32 | 3.7 |
| 4. | 64 | 3.5 |
| 5. | 128 | 4.0 |
| 6. | 256 | 4.2 |

| j | number of atoms/element | average number of active atoms |
|---|---|---|
| 1. | 8 | 3.2 |
| 2. | 16 | 3.7 |
| 3. | 32 | 3.9 |
| 4. | 64 | 4.2 |
| 5. | 128 | 4.3 |
| 6. | 256 | 4.3 |

Table 3.1: Comparison of used atoms on the uniform discretization $(\mathrm{R}^{\mathbf{h}^j}\mathrm{P})$, for $1 \le j \le J$, and the active set strategy $\{(\mathrm{R}^{\mathbf{h}^j}\mathrm{P}_Q)\}_{1\le j\le J}$, for $\mathbf{h}^j = (h_1^1, 2^{1-j}h_2^1)$ (top: Example 3.1, bottom: Example 3.2).

*$(1/6, 2\pi/8)$. The density $\phi$ was already used in [95, 96]. Shades of the grey color on the picture on the right-hand side pictures show portions of two 'phases' represented by vectors $(0,-1)^\top$ and $(0,1)^\top$. On a given element $K \in \mathcal{T}_{h_1}^1$, we evaluate*

$$\psi(K) := \sum_k \lambda_k \frac{|A_k - (0,1)^\top|^2}{|A_k - (0,1)^\top|^2 + |A_k - (0,-1)^\top|^2}, \tag{3.18}$$

*where $k = k(K)$ runs from 1 to a number of used atoms on the element $K$. Note that $0 \le \psi(K) \le 1$. Then the shade is the interpolation between white ($\psi(K) = 0$) and black ($\psi(K) = 1$). The left-hand side picture shows the mean magnetization on every element, i.e., $\mathbf{m_h}\Big|_K = \sum_k \lambda_k A_k$.*

*The results in Table 3.1 illustrate the advantage of the active set strategy $\{(\mathrm{R}^{\mathbf{h}^j}\mathrm{P}_Q)\}_{j\in\mathbb{N}}$ over $(\mathrm{R}^{\mathbf{h}^j}P)$, for $\mathbf{h}^j = (h_1^1, 2^{1-j}h_2^1)$ and $1 \le j \le N$: the active set based method only employs approximately four atoms per element $K \in \mathcal{T}_{h_1}^1$ in the minimization process, which takes approximately 90% of the overall computing time. Note that this number of atoms does not dependent* *on the scaling value $h_2$. This is in contrast to a uniform discretization, where admissible measures in the minimization are constituted from $8 \cdot 2^{j-1}$ degrees of freedom per element $K \in \mathcal{T}^1_{h_1}$ to reach a prescribed tolerance* TOL= *0.0001 convergence with respect to the energy.*

**Example 3.2** *We take $\phi(\mathbf{m}) = 10^{-2}(m_1^2 + 2m_1^4)$. This type of $\phi$ was suggested in [39]. Again, we are not aware of any result giving the formula for $\phi^{**}$. Further, we put $\varepsilon_j = 2^{-j}/80$ and* TOL=*0.0001, $\mathbf{f} = (10^{-2}, 0)^\top$ and $\mathbf{h}^1 = (1/6, 2\pi/8)$. On three chosen elements in $\omega$ we also plot (by arrows) points where the Young measure solution is supported. The thickness of stripes is proportional to weights of those points in the Young measure; see also Table 3.1. As the maximum principle predicts that the solution is supported on two atoms we singled out two leading directions which are plotted. This is often referred as the microstructure and it is closely related to some minimizing sequence of the original problem. See [39] for details.*

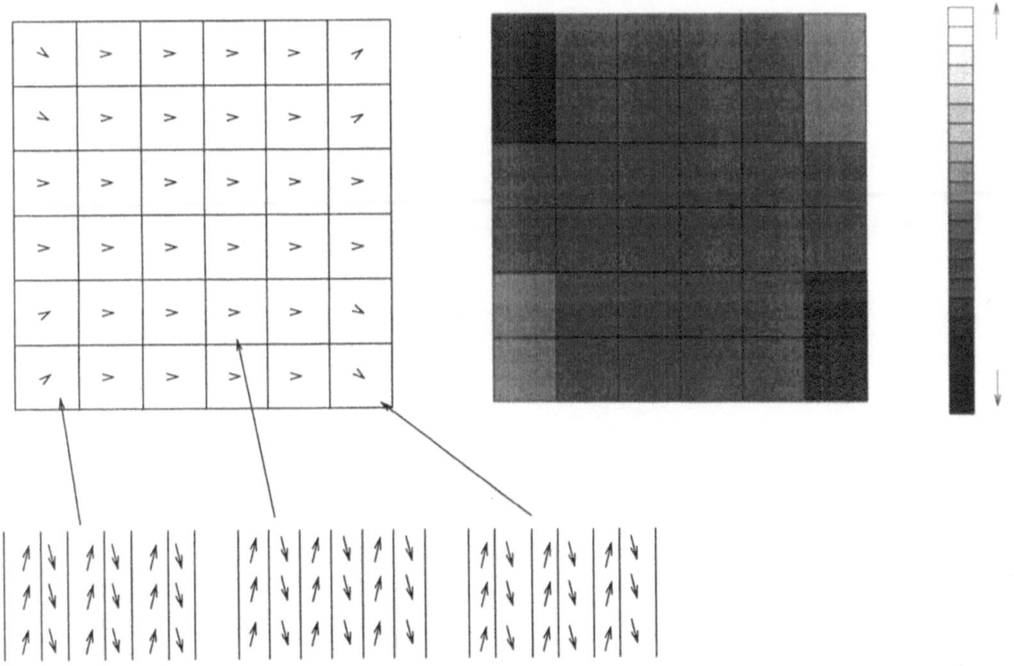

Figure 3.3: A solution of $\{(\mathrm{R}^{\mathbf{h}^j}\mathrm{P}^j_Q)\}_{1\leq j\leq J}$. The magnetization $\mathbf{m}_{\mathbf{h}^J}$ (left) and portions of phases (right).

# Part II

# Numerical Nonstationary Micromagnetism

Modeling small-scale dynamical processes in mechanics and fluid flows is an area of active research and has challenging applications in high technology industries, like improving 'smart materials', data storage devices, or laptop displays; physical phenomena on scales of nanometers up to micrometers are typically different from those that occur on larger scales, and mathematical models have to be able to describe dynamical behavior of microstructures, defects, etc.

Numerical schemes to study the dynamics of magnetic microstructure have to cope with strong nonlinearities of the problem, restricted regularities of solutions, going along with non-convex constraints, which makes their construction a non-trivial endeavor. In this part, we consider three problems:

1. Nonstationary micromagnetics in ferromagnets (Landau-Lifshitz-Gilbert equation).

2. Nonstationary micromagnetics in electrically conducting ferromagnets (Maxwell-Landau-Lifshitz-Gilbert equation).

3. Liquid crystal fluid flow ([modified] Leslie-Ericksen equation).

Despite of their different origin, these problems share common features, like non-convex constraints for the involved order parameter. It is the aim of this work to provide a numerical analysis for these problems, dealing with these constraints efficiently in terms of penalization: more specifically, we (i) construct projection type time (splitting) discretizations, as well as (ii) present a mathematical analysis of these numerical strategies in a 2D setting. In another step, most of the analyzed schemes are implemented to study the necessity of assumptions that are needed in theory and compare theoretical statements with convergence rates that we obtained computationally, as well as to check applicability of derived results to situations that are not covered by theory (like, e.g., related boundary value problems).

Dynamic micromagnetic simulations for ferromagnetic materials play an important role when designing new or improved (high density) devices like hard disks to satisfy nowadays needs of increased data storage. At sub-micron and nano-scale dimensions the properties of magnetic devices are strongly affected by their size and shape, resulting from the interplay between different types of magnetic energy. Arrays of nanomagnets may then be used for ultra-high density storage on hard disks or for fast and dense, non-volatile solid-state memory. Storage applications are possible because hysteresis in the nanomagnets creates two oppositely magnetized states which are stable in zero applied field and can store binary data.

The key recent discovery in magnetic materials was giant magnetoresistance (GMR) in 1988, a phenomenon in which a material undergoes a large change in resistance of typically 5 – 40% when an external field is applied. When there is a large GMR response to a small field very sensitive magnetic field sensors can be produced. GMR sensors are already making inroads into magnetic data storage in the recording heads for hard disks, see Figure II.1. The basic components of a hard disk system are the disk itself and the recording head. At the heart of an advanced recording head is the magnetorestrictive sensor element. This is a small piece of a thin film magnetic material which is a sensitive detector of magnetic field. The data is written by a tiny electromagnet, also built into the recording head, which magnetizes a small part of the disk to create a bit. Data is encoded as 'transitions' which are changes between two bits magnetized in opposite directions. The presence of a transition indicates a binary 1 and the absence of a transition (i.e., two adjacent bits written in the same direction) a binary 0. When data is being read, the vertical magnetic field at the transition changes the orientation of the magnetization in the sensor, causing an increase or decrease in resistance.

To predict and improve the properties of miniaturized magnetic systems we need to understand the nanomagnets which are their active components. The dynamics of magnetization is described by the **Landau-Lifshitz equation in Gilbert form (LLG)**, see [57, 84],

$$\mathbf{m}_t = \gamma' \, \mathbf{m} \times \mathbf{f}_{\text{eff}} - \frac{\alpha\gamma'}{M_s} \mathbf{m} \times \left(\mathbf{m} \times \mathbf{f}_{\text{eff}}\right), \tag{II.1}$$
$$|\,\mathbf{m}\,| = 1 \,, \qquad \partial_{\mathbf{n}} \mathbf{m}\big|_{\partial\omega} = 0 \,,$$

for given $\gamma' > 0$ and the saturation magnetization $M_s$. We recall the Landau-Lifshitz free energy, for $\mu_0 > 0$,

$$\begin{aligned} \mathcal{E}(\mathbf{m}) \;&=\; \int_\omega \big\{ \alpha \,|\, \nabla \mathbf{m} \,|^2 + \phi(\mathbf{m}) \\ &\qquad + \mu_0 \, \langle \nabla u, \mathbf{m} \rangle_{\mathbb{R}^3} - \mu_0 M_s \, \langle \mathbf{f}, \mathbf{m} \rangle_{\mathbb{R}^3} \big\} \, \mathrm{d}\mathbf{x} \,. \end{aligned} \tag{II.2}$$

The effective field is given by

$$\mathbf{f}_{\text{eff}} \equiv -\frac{\delta \mathcal{E}}{\delta \mathbf{m}} = -D\phi(\mathbf{m}) - \mu_0 \nabla u + 2\alpha \Delta \mathbf{m} + \mu_0 M_s \mathbf{f} \tag{II.3}$$

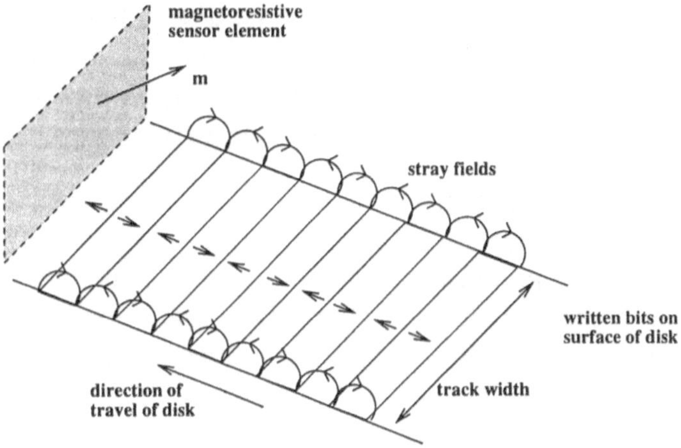


Figure II.1: Magnetoresistive sensor element reading bits on a hard disk. The magnetization $\mathbf{m}$ of a sensor element changes its direction due to the stray field (pointing up or down) at the transitions between bits. An alternating pattern of bits, as shown here, would be read as a sequence of binary ones (adopted from [75]).

and drives the precession in (II.1), attributing to magneto-crystalline anisotropy, nonlocal (de-)magnetization, exchange interaction and applied external fields $\mathbf{f}$. The constants $\mu_0, \alpha > 0$ are given, and $D = D(\cdot)$ denotes the Frechet-derivative of the (possibly non-convex) anisotropic energy. For the magnetostatic case, the magnetic potential $u$ and magnetization $\mathbf{m}$ are related through

$$\Delta u = \operatorname{div}(\chi_\omega \mathbf{m}) \qquad \text{in } \mathbb{R}^2\,, \tag{II.4}$$

where $\omega \subset \mathbb{R}^2$ is the domain covered by the ferromagnet, and $\chi_\omega = 1$ on $\omega$ and 0 elsewhere.[2]

Problem (II.1)-(II.4) is a strongly coupled degenerated quasilinear parabolic system which makes it hard to analyze mathematically. Hence, mathematical works mainly deal with the Landau-Lifshitz equation in one dimen-

[2]The Landau-Lifshitz-Gilbert equation is the most frequently used model to describe nonstationary micromagnetism [49, 41]. However, there are different models to describe a rate-independent dissipation mechanism in ferromagnetism, see e.g. [130, 131, 113]. For a numerical analysis and implementation of the model proposed in [113], using (nonstationary) Young measures in combination with an active set strategy (see Chapter 3), we refer the reader to [113, 81].

sion, and it is only recently that the 2D and 3D setting have been studied successfully [3, 9, 24, 25, 60, 61]. In [60], unique strong (local) solutions to (II.1)-(II.4) for the prototype case $\mathbf{f}_{\text{eff}} = \Delta\mathbf{m}$ have been verified for Riemannian surfaces by using results from the theory of harmonic maps, [120, 122, 121].

In engineering disciplines, computer codes are developed that are based on finite elements [118, 119, 67, 40] or finite differences; we also refer to [1], Chapter 11, and [69], pp. 122-129 for a survey of contributions in this direction. However, most of these schemes make use of explicit time-stepping strategies like the explicit Euler method to reduce computational efford and lack a thorough mathematical unterstanding concerning convergence properties.

Robust numerical strategies to solve the nonstationary problem (II.1)-(II.4) have to take into account the saturation constraint $|\,\mathbf{m}\,| = 1$, apart from efficient treatment of the strong nonlinearities; literature on this subject is rather limited. In [43], first steps were made to construct and analyze stable and more efficient numerical schemes, where the authors discuss projection methods of first and second order. In the cited work, also some error statements were given under the assumption of smooth solutions. Another numerical work is [106], where a penalization strategy for the Landau-Lifshitz-Gilbert equation is used to study blow-up of solutions in 2D computationally.

The analyses given in the cited works crucially rely on the assumption of classical solutions to (LLG) that do not exist in general. More importantly, no evidence is given in existing studies on how penalization strategies (if existing) affect stability of the method; a question in this direction is how to choose penalization parameters with dependence on time and spatial discretization parameters. Hence, the first goal of this work is to construct and analyze new different **penalization strategies** and present a convergence analysis valid in general situations for these numerical strategies for the prototype case $\mathbf{f}_{\text{eff}} = \Delta\mathbf{m}$ in (LLG); we consider the following methods of (II.1) in semidiscretized form, using the identity $\mathbf{m}\times(\mathbf{m}\times\Delta\mathbf{m}) = -|\,\nabla\mathbf{m}\,|^2\mathbf{m} - \Delta\mathbf{m}$, and setting $\gamma' = 1$,

$$d_t\mathbf{m}^{j+1} - \alpha\,\Delta\mathbf{m}^{j+1} + \frac{1}{\varepsilon}\phi(\mathbf{m}^j, \mathbf{m}^{j+1})\mathbf{m}^{j+1} \qquad \text{(II.5)}$$
$$= \alpha\,|\,\nabla\mathbf{m}^j\,|^2\mathbf{m}^{j+1} + \mathbf{m}^{j+1}\times\Delta\mathbf{m}^{j+1}\,,$$

for penalization functions $\phi(\cdot) = \phi_i(\cdot)$, $i = 1, 2, 3$,

$$\begin{aligned} \phi_1(\mathbf{m}^j, \mathbf{m}^{j+1}) &= \left( |\, \mathbf{m}^{j+1} \,|^2 - 1 \right), \\ \phi_2(\mathbf{m}^j, \mathbf{m}^{j+1}) &= \left( 1 - \frac{1}{|\, \mathbf{m}^j \,|^2} \right), \\ \phi_3(\mathbf{m}^j, \mathbf{m}^{j+1}) &= \left( 1 - \frac{1}{|\, \mathbf{m}^j \,|} \right), \end{aligned} \tag{II.6}$$

and, more general,

$$\phi_3^\gamma(\mathbf{m}^j, \mathbf{m}^{j+1}) = \left( 1 - \frac{1}{|\, \mathbf{m}^j \,|^{2-\gamma}} \right), \quad \gamma \in \mathbb{N}_0 \,, \tag{II.7}$$

where $\mathbf{m}^0 = \mathbf{m}_0$. We remark that the penalty ansatz $\phi_1$ ('Ginzburg-Landau approximation') has already been used in [106] and is frequently used in analytical works to show existence of solutions to (LLG) or 'related problems' in the 3D case. — As it will turn out from the analysis, $\phi_3^\gamma$ is the best choice with respect to stability properties of the schemes, whereas $\phi_1$ performs in the poorest way, see Table II.1 for a survey of the results. Here, we describe increased freedom in stable choices of pairs $\{\varepsilon, k\}$ for $\phi_3^\gamma$ by a constant $C_\gamma > 0$. For any such $C_\gamma$, we can find $\gamma = \mathcal{O}(1)$ such that this method is stable.

The sharpness of these results for stable pairs $\{\varepsilon, k\}$ is supported by computational experiments. Finally, optimal performance of $\mathbf{W}^{1,2}(\omega, \mathbb{R}^3)$-conforming, piecewise affine finite elements is stated in Theorem 4.10 under a moderate stability constraint $F(\varepsilon, k, h) > 0$.

An alternative numerical tool compared to penalization is **projection** as proposed by E & Wang [43, 2], where each iteration step is split into two substeps (for $\gamma = 0$). Let $\tilde{\mathbf{m}}^0 = \mathbf{m}^0 = \mathbf{m}_0$.

1. Given $\mathbf{m}^j, \tilde{\mathbf{m}}^j$, compute $\tilde{\mathbf{m}}^{j+1}$ from

$$\frac{1}{k}\{\tilde{\mathbf{m}}^{j+1} - \mathbf{m}^j\} - \alpha\, \Delta \tilde{\mathbf{m}}^{j+1} = \alpha \,|\, \nabla \tilde{\mathbf{m}}^j \,|^2 \tilde{\mathbf{m}}^{j+1} + \tilde{\mathbf{m}}^{j+1} \times \Delta \tilde{\mathbf{m}}^{j+1} \,.$$

2. Compute $\mathbf{m}^{j+1} = \frac{\tilde{\mathbf{m}}^{j+1}}{|\, \tilde{\mathbf{m}}^{j+1} \,|^{2-\gamma}}$, for $\gamma \in \mathbb{N}_0$ .

For the purpose of its analysis, we reformulate this scheme $(\mathbf{P})_B^\gamma$ as a semi-implicit penalization method using $\phi_3$, which leads to optimal convergence behavior stated in Theorem 4.9; see also Theorem 4.11 for fully space-time discretization.

| $\phi_i$ | 0 | $\phi_1$ | $\phi_2$ | $\phi_3$ | $\phi_3^\gamma$ |
|---|---|---|---|---|---|
| $F(\varepsilon, k) > 0$ | | $\varepsilon^{-1} = o(k^{-1})$ | $\varepsilon > 1.9k$ | $\varepsilon \geq k$ | $\varepsilon \geq C_\gamma k$ |
| order of convergence (in $k^\alpha$) for | | | | | |
| $\mathbf{m}^j$ $(\ell^\infty(I_k; \mathbf{L}^2(\omega, \mathbb{R}^3)))$ | 1 | 1 | 1 | 1 | 1 |
| $\mathbf{m}^j$ $(\ell^4(I_k; \mathbf{L}^4(\omega, \mathbb{R}^3)))$ | | $< 0.75$ | 0.75 | 0.75 | 0.75 |
| $\mathbf{m}^j$ $(\ell^2(I_k; \mathbf{W}^{1,2}(\omega, \mathbb{R}^3)))$ | 1 | 1 | 1 | 1 | 1 |
| $1 - \lvert \mathbf{m}^j \rvert$ $(\ell^\infty(I_k; L^2(\omega)))$ | 0.5 | $< 1$ | 1 | 1 | 1 |
| Theorem | 4.4 | 4.5 | 4.6 | 4.7 | 4.8 |
| Corollary | 4.1 | 4.2 | 4.3 | 4.4 | |

Table II.1: Orders of convergence of penalization schemes (II.5)-(II.7) for the stability constraint $F(\varepsilon, k) > 0$, and $\mathbf{m}_0 \in \mathbf{W}^{2,2}(\omega, \mathcal{S}^2)$, $\omega = \mathbb{R}^2/\mathbb{Z}^2$.

It is then from the analysis of the penalization functions $\phi_i$, $i = 1, 2, 3$ in (II.6) that we can come up with a scaled projection method as presented above for values $\gamma > 0$, which possesses increased stability properties; this observation is elaborated in Section 4.2.2.

These studies are carried out for the case $\mathbf{f}_{\text{eff}} = \Delta\mathbf{m}$ and $\omega = \mathbb{R}^2/\mathbb{Z}^2$ the flat torus, as in all cited papers above because questions addressing existence of solutions immediately apply to the more complex scenario (II.3), (II.4); numerically, efficient realization of the case (II.3), (II.4) is an important task which is accessed in Section 4.4; here, we discuss a numerical scheme that combines the ideas which guarantee the saturation constraint via projection and splitting the computation of magnetization and magnetic potential at each iteration step. In a second step, a corresponding finite element analysis shows asymptotic optimal convergence behavior under realistic regularity assumptions for the solution; see Theorem 4.12. We present simulations that are based on this scheme and show the dynamics of (point) defects inherent to an initial magnetization.

Equations (II.1)-(II.4) model physical phenomena like hysteresis or ferromagnetic resonance properly. In electrically conducting ferromagnets, however, eddy currents have to be taken into account, giving rise to crucial changes in hysteresis loop shapes already at low magnetization frequencies (eddy current losses, domain wall bowing), see [10, 59]. A proper mathematical model is **Maxwell's equations together with (LLG)**, shortly **(MLLG)**, to describe electromagnetism for such materials. More practically relevant applications can be found in high-frequency applications. — We use Ohm's law to state (MLLG) as follows, for $\kappa, \sigma, \beta > 0$, and magnetizations that satisfy $|\mathbf{m}| = 1$ ,

$$\mathbf{m}_t = \mathbf{m} \times \big(\Delta\mathbf{m} + \mathbf{H}\big) - \kappa\, \mathbf{m} \times \big(\mathbf{m} \times (\Delta\mathbf{m} + \mathbf{H})\big)\,, \tag{II.8}$$

$$\nabla \times \mathbf{H} = \mathbf{E}_t + \sigma\, \mathbf{E}\,, \tag{II.9}$$

$$\nabla \times \mathbf{E} = -\mathbf{H}_t - \beta\, \mathbf{m}_t\,, \tag{II.10}$$

$$\operatorname{div} \mathbf{H} + \beta \operatorname{div} \mathbf{m} = 0\,, \qquad \operatorname{div} \mathbf{E} = 0\,, \tag{II.11}$$

where $\mathbf{H}$ is the magnetic field.

The numerical realization of this problem builds the second part. First numerical studies of this problem using discretization via time-splitting and finite elements can be found in [73, 98], where the authors analyze stability and convergence of a semidiscretization in space via finite elements; in fact, they consider the case of no exchange energies in the magnetic energy formulation which is more delicate to access, supposing smoothness of solutions to (MLLG) for their analysis. In Chapter 5, we analyze two penalized time discretizations of (MLLG) for $\omega = (0, 2D)^2$. The first scheme $\mathbf{(MLLG)}_{k,\varepsilon}^{E_1}$ reads: Find $\{\mathbf{m}^j, \mathbf{H}^j, \mathbf{E}^j\}_{j=0}^{J} \in \ell^2(I_k; \mathbf{W}^{1,2}_{\mathrm{per}}(\omega)) \times \big[\ell^2(I_k; \mathbf{H}(\mathbf{curl}, \omega))\big]^2$ that solves

$$\begin{aligned} d_t\mathbf{m}^{j+1} - \kappa\, \Delta\mathbf{m}^{j+1} - \mathbf{l}_\varepsilon(\mathbf{m}^{j+1})\mathbf{m}^{j+1} &= \kappa\, |\nabla\mathbf{m}^j|^2\mathbf{m}^{j+1} \\ &\quad -\gamma'\, \mathbf{m}^j \times \big(\Delta\mathbf{m}^{j+1} + \mathbf{H}^{j+1}\big) + \kappa\, \big(\mathbf{H}^{j+1} - \langle \mathbf{m}^j, \mathbf{H}^j\rangle_{\mathbb{R}^3}\mathbf{m}^{j+1}\big)\,, \end{aligned} \tag{II.12}$$

$$\nabla \times \mathbf{H}^{j+1} = d_t\mathbf{E}^{j+1} + \sigma\, \mathbf{E}^{j+1}\,, \tag{II.13}$$

$$\nabla \times \mathbf{E}^{j+1} = -d_t\mathbf{H}^{j+1} - \beta\, \big(d_t\mathbf{m}^{j+1} - \mathbf{l}_\varepsilon(\mathbf{m}^{j+1})\mathbf{m}^{j+1}\big)\,, \tag{II.14}$$

$$\operatorname{div} \mathbf{E}^{j+1} = 0\,, \tag{II.15}$$

where $\mathbf{l}_\varepsilon(\varphi) = \frac{1}{\varepsilon}\big(|\varphi|^2 - 1\big)$. Note that this scenario requires corresponding penalization terms in (LLG) *and* Faraday's law (II.14). However, the idea of penalization partly interferes with the physical principle of nonexistence of magnetic monopoles (i.e., $(\text{II.11})_1$), such that choosing the penalization parameter becomes a subtle question. We refer to Table II.2 for results of convergence verified for this scheme.

Convergence results for $(\mathrm{MLLG})_{k,\varepsilon}^{E_1}$ can only be verified in weak norms, which is due to the limited regularity of the solution to (MLLG) at $t = 0$. To overcome this problem, we employ stretched time-grids $\mathcal{G}_2(k_{j+1})$ from [107] that refine near the origin by preserving the same overall numerical effort as equi-distant time-grids,

$$k : j \mapsto k_{j+1} \equiv \begin{cases} (j+1)k_0^2 \,, & \text{for } 0 \le t_{j+1} \le 1, \\ \gamma k_0 \,, & \text{for } t_{j+1} \ge 1, \end{cases} \qquad . \tag{II.16}$$

with $k_0 > 0$ the basic grid size and $\gamma = \mathcal{O}(1)$. — The second scheme $(\mathbf{MLLG})_{k,\varepsilon}^{E_2}$ that we analyze is then $(\mathrm{MLLG})_{k,\varepsilon}^{E_1}$ for $\mathcal{G}_2(k_{j+1})$, with $\mathbf{l}_\varepsilon(\mathbf{m}^{j+1})\mathbf{m}^{j+1}$ replaced by $-\tilde{\Delta}^{-1}\left[\mathbf{l}_\varepsilon(\mathbf{m}^j)\mathbf{m}^j\right]$. We refer to Table II.2 for a survey of improved convergence results for $(\mathrm{MLLG})_{\varepsilon,k}^{E_2}$.

In a next step, we discuss two (scaled) **projection schemes** $(\mathbf{MLLG})_k^{P_\ell}$, $\ell = 1, 2$ on equi-distant time-grids. The first scheme splits the computation of $\mathbf{m}^{j+1}$ and $\{\mathbf{H}^{j+1}, \mathbf{E}^{j+1}\}$ at each iteration step. — Let $\tilde{\mathbf{m}}^0 = \mathbf{m}^0 = \mathbf{m}_0$, $\mathbf{H}^0 = \mathbf{H}_0$, $\mathbf{E}^0 = \mathbf{E}_0$.

1. Given $\mathbf{m}^j, \tilde{\mathbf{m}}^j, \mathbf{H}^j$, find $\tilde{\mathbf{m}}^{j+1} \in \mathbf{W}^{1,2}_{\mathrm{per}}(\omega, \mathbb{R}^3)$ that solves

$$\begin{aligned} \frac{1}{k}\{\tilde{\mathbf{m}}^{j+1} - \mathbf{m}^j\} - \kappa\,\Delta\tilde{\mathbf{m}}^{j+1} &= \kappa\,|\,\nabla\tilde{\mathbf{m}}^j\,|^2\tilde{\mathbf{m}}^{j+1} \\ &+ \tilde{\mathbf{m}}^j \times \left(\Delta\tilde{\mathbf{m}}^{j+1} + \mathbf{H}^j\right) + \kappa\left(\mathbf{H}^j - \langle\tilde{\mathbf{m}}^j, \mathbf{H}^j\rangle_{\mathbb{R}^3}\tilde{\mathbf{m}}^{j+1}\right). \end{aligned} \tag{II.17}$$

2. Given $\mathbf{m}^j, \tilde{\mathbf{m}}^j, \mathbf{H}^j$, compute $\left\{\mathbf{H}^{j+1}, \mathbf{E}^{j+1}\right\} \in \left[\mathbf{H}(\mathbf{curl}, \omega))\right]^2$ from

$$\nabla \times \mathbf{H}^{j+1} = d_t\mathbf{E}^{j+1} + \sigma\,\mathbf{E}^{j+1}\,, \tag{II.18}$$

$$\nabla \times \mathbf{E}^{j+1} = -d_t\mathbf{H}^{j+1} - \frac{\beta}{k}\{\tilde{\mathbf{m}}^{j+1} - \mathbf{m}^j\}\,. \tag{II.19}$$

3. Compute $\mathbf{m}^{j+1} = \frac{\tilde{\mathbf{m}}^{j+1}}{|\,\tilde{\mathbf{m}}^{j+1}\,|^{2-\gamma}}$, for $\gamma \in \mathbb{N}_0$.

The second scheme decouples the computation of each iterate at one iteration step. For this purpose, we need a stabilization term in Ampere's law (II.9). Then $(\mathrm{MLLG})_k^{P_2}$ reads as follows. Let $\mathbf{m}^0 = \tilde{\mathbf{m}}^0 = \mathbf{m}_0$, $\mathbf{E}^0 = \mathbf{E}_0$, $\mathbf{H}^0 = \mathbf{H}_0$.

1. Given $\mathbf{m}^j, \tilde{\mathbf{m}}^j, \mathbf{H}^j$, find $\tilde{\mathbf{m}}^{j+1} \in \mathbf{W}^{1,2}_{\mathrm{per}}(\omega, \mathbb{R}^3)$ that solves

$$\begin{aligned} \frac{1}{k}\{\tilde{\mathbf{m}}^{j+1} - \mathbf{m}^j\} - \kappa\,\Delta\tilde{\mathbf{m}}^{j+1} &= \kappa\,|\,\nabla\tilde{\mathbf{m}}^j\,|^2\tilde{\mathbf{m}}^{j+1} \\ &- \kappa\,\tilde{\mathbf{m}}^j \times \left(\Delta\tilde{\mathbf{m}}^{j+1} + \mathbf{H}^j\right) + \kappa\left(\mathbf{H}^j - \langle\tilde{\mathbf{m}}^j, \mathbf{H}^j\rangle_{\mathbb{R}^3}\tilde{\mathbf{m}}^{j+1}\right), \end{aligned} \tag{II.20}$$

2. Given $\mathbf{E}^j, \mathbf{H}^j$, determine $\mathbf{E}^{j+1} \in \mathbf{W}^{1,2}_{\text{per}}(\omega, \mathbb{R}^3)$ from

$$d_t\mathbf{E}^{j+1} - \delta k\, \Delta\mathbf{E}^{j+1} + \sigma\, \mathbf{E}^{j+1} = \nabla \times \mathbf{H}^j\,, \qquad 0 < \delta = \mathcal{O}(1)\,. \quad \text{(II.21)}$$

3. Given $\mathbf{H}^j, \mathbf{m}^j, \tilde{\mathbf{m}}^j$, update $\mathbf{H}^{j+1} \in \mathbf{W}^{0,2}_{\text{per}}(\omega, \mathbb{R}^3)$, and $\mathbf{m}^{j+1} \in \mathbf{W}^{1,2}_{\text{per}}(\omega, \mathbb{R}^3)$, for $\gamma \in \mathbb{N}_0$, via

$$d_t\mathbf{H}^{j+1} = -\nabla \times \mathbf{E}^{j+1} - \frac{\beta}{k}\{\tilde{\mathbf{m}}^{j+1} - \mathbf{m}^j\}\,, \qquad \text{(II.22)}$$
$$\mathbf{m}^{j+1} = \frac{\tilde{\mathbf{m}}^{j+1}}{|\,\tilde{\mathbf{m}}^{j+1}\,|^{2-\gamma}}\,.$$

The main results for these four time discretization schemes are collected in Table II.2. Moreover, convergence results on a **stabilized finite element** realization of $(\text{MLLG})_k^{P_2}$ using $\mathbf{W}^{1,2}(\omega, \mathbb{R}^3)$-conforming piecewise affine finite elements for each, magnetization, electric and magnetic field are presented in Theorem 5.7. Computational studies that compare magnetizations obtained from $(\text{P})_B^\gamma$ for (LLG) and $(\text{MLLG})_k^{P_2}$ for electrically conducting ferromagnets conclude Chapter 5.

The last chapter concerns numerical analysis of nematic liquid crystals. Phases of a liquid crystal are intermediate between liquid and solid phases; they are ordered fluids composed of elongated molecules that are either rod- or disklike. Liquid crystals are mainly categorized as thermo- and lyotropic, depending on whether phase transition from isotropic to nematic is induced by temperature or concentration of the mesogenic material, see Figure II.2.

The molecular size of a liquid crystal is typically on the scale of nanometers. By contrast, liquid crystal devices are usually on a micron scale. The Leslie-Ericksen continuum theory in hydrodynamics was introduced to describe equilibrium director configurations in rod-like nematic liquid crystals in the presence of competing forces (like surface, electric fields).

The construction of versatile display technology certainly ranks amongst the main applications of liquid crystals nowadays: a laptop computer screen typically consists of 640×480 pixels. At each of them, a voltage can be applied, and it is due to the electric susceptibility of the nematic to (re-)align the director in this cell, affecting optical properties of it; when the voltage

| method | $(\mathrm{MLLG})^{E_1}_{\varepsilon,k}$ | $(\mathrm{MLLG})^{E_2}_{\varepsilon,k}$ | $(\mathrm{MLLG})^{P_1}_{\varepsilon,k}$ | $(\mathrm{MLLG})^{P_2}_{\varepsilon,k}$ |
|---|---|---|---|---|
| $F(\varepsilon,k)>0$ | $\varepsilon^{-1}=o(k^{-1})$ | $\varepsilon_j^{-1}=o(k_j^{-1})$ | | |
| order of convergence (in $k^\alpha$) for | | | | |
| $\mathbf{m}^j\ (\ell^\infty(I_k;\mathbf{L}^2(\omega,\mathbb{R}^3)))$ | 1 | 1 | 1 | 1 |
| $\mathbf{m}^j\ (\ell^\infty(I_k;\mathbf{W}^{1,2}(\omega,\mathbb{R}^3)))$ | 0.5 | 1 | 0.5 | 0.5 |
| $\mathbf{m}^j\ (\ell^2(I_k;\mathbf{W}^{1,2}(\omega,\mathbb{R}^3)))$ | 1 | 1 | 1 | 1 |
| $\mathbf{m}^j\ (\ell^2(I_k;\mathbf{W}^{2,2}(\omega,\mathbb{R}^3)))$ | 0.5 | 1 | 0.5 | 0.5 |
| $\mathbf{m}^j\ (\ell^4(I_k;\mathbf{L}^4(\omega,\mathbb{R}^3)))$ | $<0.75$ | $<0.75$ | 0.75 | 0.75 |
| $\mathbf{H}^j,\mathbf{E}^j\ (\ell^\infty(I_k;\mathbf{W}^{-1,2}_{\mathrm{per}}(\omega,\mathbb{R}^3)))$ | 1 | 1 | 1 | 1 |
| $\mathbf{H}^j,\mathbf{E}^j\ (\ell^\infty(I_k;\mathbf{L}^2(\omega,\mathbb{R}^3)))$ | 0.5 | 1 | 0.5 | 0.5 |
| $1-\lvert\mathbf{m}^j\rvert\ (\ell^\infty(I_k;L^2(\omega)))$ | 0.5 | 0.5 | 0.5 | 0.5 |
| $\mathrm{div}\,(\mathbf{H}^j+\beta\,\mathbf{m}^j)\ (\ell^\infty(I_k;L^2(\omega)))$ | 0.5 | 0.5 | 0 | 0 |
| $\mathrm{div}\,(\mathbf{H}^j+\beta\,\mathbf{m}^j)\ (\ell^\infty(I_k;W^1(\omega)))$ | 0 | 0.5 | 0 | 0 |
| Theorem | 5.4 | 5.5 | 5.6 | 5.6 |

Table II.2: Orders of convergence of penalization schemes to solve (II.8)-(II.11), for the stability constraint $F(\varepsilon,k)>0$, $\mathbf{m}_0\in\mathbf{W}^{2,2}_{\mathrm{per}}(\omega,\mathcal{S}^2)$, and $\mathbf{H}_0,\mathbf{E}_0\in\mathbf{W}^{1,2}_{\mathrm{per}}(\omega,\mathbb{R}^3)$, for $\omega=(0,2D)^2$. Convergence rates are given for values $\varepsilon=k^{1-\delta}$, $\delta>0$ arbitrarily small.

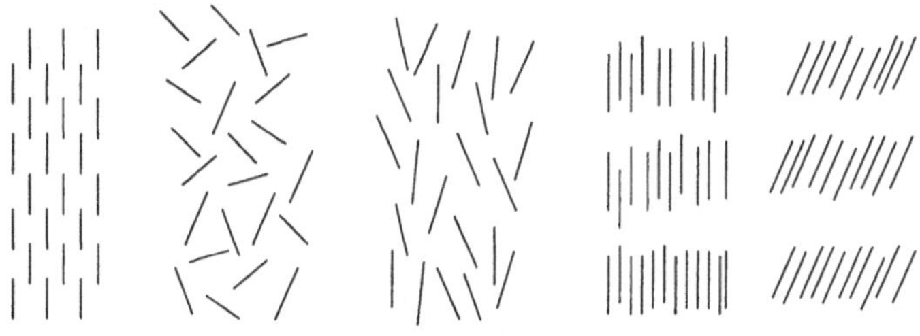


Figure II.2: Diverse phases of a liquid crystal: molecular order in a crystalline (a), isotropic (b), nematic (c), and two smectic phases, characterized by orientational and positional symmetry (adopted from [22]).

is off, the pixel can transmit light, and when it is on, light transmission is blocked. The result is a flat display screen, where typical cells are of the order of $20\mu$m thick.

For more details on liquid crystals, we refer to the monographs [22, 38, 53, 127].

In [89], a **simplified Ericksen-Leslie model** has been proposed that basically models the most significant features of the original model.[3] In effect, these are modified Navier-Stokes equations that take account of liquid crystallinity,

$$\mathbf{u}_t - \nu\,\Delta\mathbf{u} + \mathbf{u}\cdot\nabla\mathbf{u} + \nabla p + \lambda\,\mathrm{div}\big(\nabla\mathbf{d}\odot\nabla\mathbf{d}\big) = \mathbf{f}\,, \tag{II.23}$$

$$\mathbf{d}_t - \varrho\,\Delta\mathbf{d} + \mathbf{u}\cdot\nabla\mathbf{d} = \varrho\,|\,\nabla\mathbf{d}\,|^2\mathbf{d}\,, \tag{II.24}$$

$$\mathrm{div}\,\mathbf{u} = 0\,, \quad |\,\mathbf{d}\,| = 1\,, \tag{II.25}$$

where $\{\mathbf{u}, p, \mathbf{d}\}$ are velocity field, pressure, and director field of the fluid flow, respectively. The analysis of this simplified Ericksen-Leslie model in penalized formulation is accomplished in [89]; however, some of these results do not carry over to (II.23)-(II.25) since relevant a priori bounds in strong norms depend on the penalization parameter $\varepsilon > 0$.

In recent papers [93, 94], C. Liu and N. Walkington presented a first numerical analysis of (II.23)-(II.25), by using the Ginzburg-Landau approximation term $\phi_1$ from (II.6) in (II.24). Convergence analyses for *Hermite*-type discretizations (using $\mathbf{W}^{2,2}(\omega,\mathbb{R}^d)$-conforming finite elements for the director

[3] We also refer to [90] for the analysis of the original Ericksen-Leslie model.

field) in [93] or mixed methods [94] to treat the corresponding Euler scheme are derived for sufficiently smooth solutions, containing a stability parameter in their bounds that shows *exponential* dependence on the penalization parameter $\varepsilon > 0$.

We start Chapter 6 by proving local existence of strong solutions to (II.23)-(II.25) for $\omega = (0, 2D)^2$ by using a fixed point argument, and global existence for small initial data. In this analysis, we are particularly interested in the verification of bounds for $\mathbf{d} \in L^2(I; \mathbf{W}^{2,2}_{\text{per}}(\omega))$. For the **implicit Euler method**

$$
\begin{aligned}
& d_t\mathbf{u}^{j+1} - \nu\,\Delta\mathbf{u}^{j+1} + \mathbf{u}^j \cdot \nabla\mathbf{u}^{j+1} \\
& \qquad + \lambda\,\mathrm{div}\big(\nabla\mathbf{d}^j \odot \nabla\mathbf{d}^{j+1}\big) + \nabla p^{j+1} = \mathbf{f}^{j+1}\,, && \text{(II.26)}\\
& d_t\mathbf{d}^{j+1} - \varrho\,\Delta\mathbf{d}^{j+1} + \mathbf{l}_\varepsilon(\mathbf{d}^{j+1}) + \mathbf{u}^j \cdot \nabla\mathbf{d}^{j+1} = \varrho\,|\,\nabla\mathbf{d}^j\,|^2\mathbf{d}^{j+1}\,, && \text{(II.27)}\\
& \mathrm{div}\mathbf{u}^{j+1} = 0\,, \quad \mathbf{u}^0 = \mathbf{u}_0\,, \quad \mathbf{d}^0 = \mathbf{d}_0\,, && \text{(II.28)}
\end{aligned}
$$

and $\mathbf{l}_\varepsilon(\varphi) = \frac{1}{\varepsilon}\big(|\,\varphi\,|^2 - 1\big)$, we verify different convergence statements that involve constants which do *not* depend on the penalization parameter $\varepsilon$ which is identified to be of magnitude $k$; see Table II.3 for a survey of obtained results. We continue with a convergence analysis for general stable velocity/pressure space pairings that satisfy the Ladyshenskaja-Babuška-Brezzi condition.

In a next step, we propose and analyze a **splitting/projection scheme** that uses the ideas of (scaled) projection[4]

1. in the sense mentioned for (LLG), i.e., to satisfy $|\,\mathbf{d}\,| = 1$,

2. in the sense of Chorin's projection method [28, 123], to ensure $\mathrm{div}\,\mathbf{u} = 0$.

In every step, the method computes iterates $\{\mathbf{u}^{j+1}, \mathbf{d}^{j+1}, p^{j+1}\}$ independently:

1. Given $\{\mathbf{d}^j, \tilde{\mathbf{d}}^j, \mathbf{u}^j\}$, compute $\tilde{\mathbf{d}}^{j+1}$ from

$$
\frac{1}{k}\{\tilde{\mathbf{d}}^{j+1} - \mathbf{d}^j\} + \mathbf{u}^j \cdot \nabla\tilde{\mathbf{d}}^{j+1} - \varrho\,\Delta\tilde{\mathbf{d}}^{j+1} = \varrho\,|\,\nabla\tilde{\mathbf{d}}^j\,|^2\tilde{\mathbf{d}}^{j+1}\,. \qquad \text{(II.29)}
$$

[4]In [2, 32, 31, 37], the authors propose a procedure in which each iteration step to compute discrete minimizers of the Oseen-Frank energy functional is split into two substeps, with the latter being a projection onto the sphere.

2. Given $\{\mathbf{u}^j, \tilde{\mathbf{u}}^j, \tilde{\mathbf{d}}^j\}$, determine $\tilde{\mathbf{u}}^{j+1} \in \dot{\mathbf{W}}^{1,2}_{\text{per}}(\omega, \mathbb{R}^3)$ from

$$\frac{1}{k}\{\tilde{\mathbf{u}}^{j+1} - \mathbf{u}^j\} - \nu\,\Delta\tilde{\mathbf{u}}^{j+1} + \mathbf{u}^j \cdot \nabla\tilde{\mathbf{u}}^{j+1} \qquad \text{(II.30)}$$
$$+\lambda\,\mathrm{div}\big(\nabla\tilde{\mathbf{d}}^{j+1} \odot \nabla\tilde{\mathbf{d}}^{j+1}\big) = \mathbf{f}^{j+1}\,.$$

3. Given $\{\tilde{\mathbf{u}}^j, \tilde{\mathbf{d}}^j\} \in \dot{\mathbf{W}}^{1,2}_{\text{per}}(\omega, \mathbb{R}^3) \times \mathbf{W}^{2,2}_{\text{per}}(\omega, \mathbb{R}^3)$, compute $\{\mathbf{u}^{j+1}, \mathbf{d}^{j+1}\}$ via

$$\frac{1}{k}\{\mathbf{u}^{j+1} - \tilde{\mathbf{u}}^{j+1}\} + \nabla p^{j+1} = 0, \quad \mathrm{div}\,\mathbf{u}^{j+1} = 0\,, \qquad \text{(II.31)}$$
$$\mathbf{d}^{j+1} = \frac{\tilde{\mathbf{d}}^{j+1}}{|\tilde{\mathbf{d}}^{j+1}|^{2-\gamma}}, \quad \gamma \in \mathbb{N}_0\,.$$

Note that $(\text{II}.31)_{1,2}$ can be reformulated as a Poisson problem for the pressure, $\Delta p^{j+1} = \frac{1}{k}\,\mathrm{div}\,\tilde{\mathbf{u}}^{j+1}$ on $\omega$. — For analytical purposes, this scheme can be reinterpreted as a semi-implicit penalization method involving penalization/stabilization terms in the original system (II.23)-(II.24) that attribute to the constraints (II.25). The main results of this analysis are reported in Table II.3. We present computational experiments in Section 6.5 that show sharpness of our results.

| method | penalization | projection |
|---|---|---|
| $F(\varepsilon, k) > 0$ | $\varepsilon^{-1} = o(k^{-1})$ | |
| order of convergence (in $k^\alpha$) for | | |
| $\mathbf{u}^j$ (in $\ell^\infty(I_k; \dot{\mathbf{W}}^{-1,2}_{\text{per}}(\omega, \mathbb{R}^2))$) | 1 | 1 |
| $\mathbf{u}^j$ (in $\ell^2(I_k; \mathbf{L}^2(\omega, \mathbb{R}^2))$) | 1 | 1 |
| $\mathbf{d}^j$ (in $\ell^\infty(I_k; \mathbf{L}^2(\omega, \mathbb{R}^2))$) | 1 | 1 |
| $\mathbf{d}^j$ (in $\ell^2(I_k; \mathbf{W}^{1,2}(\omega, \mathbb{R}^2))$) | 1 | 1 |
| $\mathbf{d}^j$ (in $\ell^4(I_k; \mathbf{L}^4(\omega, \mathbb{R}^2))$) | 0.75 | 0.75 |
| $p^j$ (in $\ell^\infty(I_k; W^{-1,2}_{\text{per}}(\omega))$) | 0.5 | 0.5 |
| $p^j$ (in $\ell^2(I_k; L^2(\omega)/\mathbb{R})$) | 0.5 | 0.5 |
| Theorem | 6.5 | 6.7 |

Table II.3: Orders of convergence for penalization scheme (II.26)-(II.28) and projection scheme (II.29)-(II.31) for the stability constraint $F(\varepsilon, k) > 0$, and $\{\mathbf{u}_0, \mathbf{d}_0\} \in \mathbf{J}_1 \cap \mathbf{W}^{2,2}_{\text{per}} \times \mathbf{W}^{3,2}(\omega, \mathcal{S}^1)$, $\omega = (0, 2D)^2$.

# Chapter 4

# The Landau-Lifshitz-Gilbert Equation

The Landau-Lifshitz-Gilbert equation describes the evolution of spin fields in continuum ferromagnets [84], and reads

$$\mathbf{m}_t = \mathbf{m} \times \mathbf{H}_{\text{eff}} - \alpha\, \mathbf{m} \times (\mathbf{m} \times \mathbf{H}_{\text{eff}})\,, \tag{4.1}$$

where $\alpha > 0$ is the Gilbert damping constant. The effective field $\mathbf{H}_{\text{eff}}$ typically contains contributions from exchange interaction, crystalline anisotropy, magnetostatic selfenergy, and external magnetic fields. $\mathbf{m} : \omega \times [0, \infty] \to \mathbb{R}^3$ stands for the spin vector (also referred to as magnetization), for $\omega \subset \mathbb{R}^2$ a bounded domain. The modulus of the spin is supposed to stay constant, i.e., $|\,\mathbf{m}\,| = 1$.

In the following, we consider a typical form of $\mathbf{H}_{\text{eff}}$ that corresponds to the pure isotropic case without external magnetic fields, $\mathbf{H}_{\text{eff}} = \Delta\mathbf{m}$. We study dynamic behavior of magnetizations on the flat torus $\omega := \mathbb{R}^2/\mathbb{Z}^2$. Then, the Cauchy problem for (4.1) reads, for $|\,\mathbf{m}_0\,| = 1$ almost everywhere,

$$\mathbf{m}_t = -\alpha\, \mathbf{m} \times \big(\mathbf{m} \times \Delta\mathbf{m}\big) + \mathbf{m} \times \Delta\mathbf{m}\,, \quad \mathbf{m}\big|_{t=0} = \mathbf{m}_0\,. \tag{4.2}$$

A crucial observation is that $|\,\mathbf{m}(\mathbf{x}, t)\,| = 1$, for (almost) all $(\mathbf{x}, t) \in \omega \times [0, \infty]$, provided that the solution to (4.2) is sufficiently smooth. This immediately results from scalar multiplication with $\mathbf{m}$. Therefore, we may benefit from the vector cross product formula

$$\mathbf{a} \times \big(\mathbf{b} \times \mathbf{c}\big) = \langle \mathbf{a}, \mathbf{c} \rangle_{\mathbb{R}^3} \mathbf{b} - \langle \mathbf{a}, \mathbf{b} \rangle_{\mathbb{R}^3} \mathbf{c} \tag{4.3}$$

to conclude

$$\mathbf{m} \times (\mathbf{m} \times \Delta\mathbf{m}) = \langle \mathbf{m}, \Delta\mathbf{m} \rangle_{\mathbb{R}^3} \mathbf{m} - |\, \mathbf{m} \,|^2 \Delta\mathbf{m} \,. \tag{4.4}$$

Since $\nabla |\, \mathbf{m} \,|^2 = 0$ almost everywhere, we are led to

$$\mathbf{m} \times (\mathbf{m} \times \Delta\mathbf{m}) = -|\, \nabla\mathbf{m} \,|^2 \mathbf{m} - \Delta\mathbf{m} \,, \tag{4.5}$$

and (4.2) can be rewritten in the form which — if stated as Cauchy problem on $\omega = \mathbb{R}^2/\mathbb{Z}^2$ — will be addressed as (LLG) in the following,

$$\mathbf{m}_t - \alpha \Delta\mathbf{m} = \alpha |\, \nabla\mathbf{m} \,|^2 \mathbf{m} + \mathbf{m} \times \Delta\mathbf{m} \,. \tag{4.6}$$

The main difficulty in (LLG) is the strongly nonlinear character of the problem and we review and extend existing results on this matter in Section 4.1. In particular, we show existence of smooth, global solutions to the Cauchy problem (LLG) on the flat torus $\omega$ in Section 4.1.1, for *small* initial data $\mathbf{m}_0 \in \mathbf{W}^{1,2}(\omega, \mathcal{S}^2)$, following [60]. For general initial data, one may benefit from the theory of harmonic maps briefly outlined in Section 4.1.2. In [120, 122], see also [121], M. Struwe shows existence and regularity of (weak) heat flows for Riemannian surfaces by controlling energy contributions (locally) in time and space. This approach has been applied successfully to (LLG) by Guo and Hong in [60] to show existence and uniqueness of locally strong and globally weak solutions to the 2D problem. We briefly sum up their studies in Section 4.1.3 to illustrate possible blow up of solutions to (LLG), i.e., $\Delta\mathbf{m} \notin L^2(I; \mathbf{L}^2(\omega, \mathbb{R}^3))$ for $T > 0$, in general. It is, however, that solutions to (LLG) are smooth locally in time: we develop this scenario in Section 4.1.3 and derive a priori bounds for solutions to (LLG) with initial values $\mathbf{m}_0 \in \mathbf{W}^{2,2}(\omega, \mathcal{S}^2)$. We remark that corresponding results are not available in a 3D setting, where much less is known regarding regularity of solutions; we refer to [122] for a survey in the context of harmonic mappings of Riemannian manifolds of dimension $d \geq 3$. This is the reason for our numerical studies of (LLG) for $\omega \subset \mathbb{R}^2$.

The analysis of (LLG) builds the first part of this chapter and ends with Theorem 4.3, which is the basis for the numerical part, where different penalized first order time discretization schemes to solve (4.6) in 2D are proposed and analyzed; this program is realized in Section 4.2.

Firstly, we analyze a method that does not take into account the saturation constraint explicitly; for $\mathbf{m}^0 = \mathbf{m}_0$, let

$$d_t\mathbf{m}^{j+1} - \alpha\,\Delta\mathbf{m}^{j+1} = \alpha\,|\,\nabla\mathbf{m}^j\,|^2\mathbf{m}^{j+1} + \mathbf{m}^{j+1}\times\Delta\mathbf{m}^{j+1}\,. \tag{4.7}$$

We refer to Section 4.2.1 for an analysis of this scheme. Although it is first order accurate, it suffers from severe instabilities caused by the non-fulfillment of the side-constraint $|\,\mathbf{m}^j\,| = 1$; this is illustrated by the following two examples.

**Example 4.1** *(Perturbation analysis of harmonic heat flow, from [43]) Consider the heat flow of harmonic maps to $\mathcal{S}^2$,*

$$\mathbf{m}_t - \Delta\mathbf{m} = |\,\nabla\mathbf{m}\,|^2\mathbf{m}\,. \tag{4.8}$$

*Set $e = |\,\mathbf{m}\,|^2 - 1$. Then $e$ satisfies*

$$e_t - \Delta e = 2\,|\,\nabla\mathbf{m}\,|^2 e\,.$$

*This calculation shows that if $e$ is not identically zero, then $e$ grows exponentially fast. Since the solution to (4.7) (deleting the torque $\mathbf{m}\times\Delta\mathbf{m}$ from it) does not preserve the normalization exactly, $|\,\mathbf{m}^j\,|^2 - 1$ can grow exponentially fast. — This argument is supported by computational experiments in [43].*

**Example 4.2** *(Violation of the saturation constraint by the implicit Euler scheme)*
*(a) For $\omega = \mathbb{R}^2/\mathbb{Z}^2$, consider the scheme (fully implicit version of (4.6))*

$$d_t\mathbf{m}^{j+1} - \alpha\,\Delta\mathbf{m}^{j+1} = \alpha\,|\,\nabla\mathbf{m}^{j+1}\,|^2\mathbf{m}^{j+1} + \mathbf{m}^{j+1}\times\Delta\mathbf{m}^{j+1}\,. \tag{4.9}$$

*We proceed by contradiction, and suppose $|\,\mathbf{m}^j\,| = 1$ almost everywhere: multiplication by $\mathbf{m}^{j+1}$ and (4.5) lead to*

$$d_t|\,\mathbf{m}^{j+1}\,|^2 + k\,|\,d_t\mathbf{m}^{j+1}\,|^2 = 0\,, \quad \text{for a.e. } \mathbf{x}\in\omega\,. \tag{4.10}$$

*Note that $|\,\mathbf{m}^j\,| = 1$ almost everywhere in $\omega$ is violated, in general — in contrast to the continuous equation (4.6). We remark that the scheme*

$$d_t\mathbf{m}^{j+1} = -\alpha\,\mathbf{m}^{j+1}\times\big(\mathbf{m}^{j+1}\times\Delta\mathbf{m}^{j+1}\big) + \mathbf{m}^{j+1}\times\Delta\mathbf{m}^{j+1} \tag{4.11}$$

*is not identical to* (4.9), *in general.*

*(b) A discretization of* (4.2) *that satisfies* $|\,\mathbf{m}^j\,| = 1$, *almost everywhere in* $\omega$ *is*

$$d_t\mathbf{m}^{j+1} = -\alpha\,\overline{\mathbf{m}}^{j+1/2} \times (\overline{\mathbf{m}}^{j+1/2} \times \Delta\mathbf{m}^{j+1}) + \overline{\mathbf{m}}^{j+1/2} \times \Delta\mathbf{m}^{j+1}\,, \tag{4.12}$$

*where* $\overline{\mathbf{m}}^{j+1/2} := \frac{1}{2}\{\mathbf{m}^{j+1} + \mathbf{m}^j\}$. *However, we cannot recast* (4.12) *into the form* (4.9).

Based on these observations, we propose different penalization strategies in Section 4.2.2 that enforce the saturation constraint. Let $\mathbf{m}^0 = \mathbf{m}_0$. The considered schemes update $\mathbf{m}^j \in \ell^\infty(I_k; \mathbf{W}^{1,2}(\omega, \mathbb{R}^3))$, defined on a mesh $I_k$ by its points $0 = t_0 < t_1 < ... \leq t_J$ via

$$\begin{aligned} d_t\mathbf{m}^{j+1} - \alpha\,\Delta\mathbf{m}^{j+1} + \frac{1}{\varepsilon}\phi(\mathbf{m}^j, \mathbf{m}^{j+1})\mathbf{m}^{j+1} & \\ = \alpha\,|\,\nabla\mathbf{m}^j\,|^2\mathbf{m}^{j+1} + \mathbf{m}^{j+1} \times \Delta\mathbf{m}^{j+1}\,, & \end{aligned} \tag{4.13}$$

for penalization functions $\phi(\cdot) = \phi_i(\cdot)$, $i = 1, 2, 3$,

$$\begin{aligned} \phi_1(\mathbf{m}^j, \mathbf{m}^{j+1}) &= (|\,\mathbf{m}^{j+1}\,|^2 - 1)\,, \\ \phi_2(\mathbf{m}^j, \mathbf{m}^{j+1}) &= (1 - \frac{1}{|\,\mathbf{m}^j\,|^2})\,, \\ \phi_3(\mathbf{m}^j, \mathbf{m}^{j+1}) &= (1 - \frac{1}{|\,\mathbf{m}^j\,|})\,, \end{aligned} \tag{4.14}$$

and, more general,

$$\phi_3^\gamma(\mathbf{m}^j, \mathbf{m}^{j+1}) = (1 - \frac{1}{|\,\mathbf{m}^j\,|^{2-\gamma}})\,, \quad \gamma \in \mathbb{N}_0\,. \tag{4.15}$$

We recall the importance of the already known first penalty ansatz ('Ginzburg-Landau approximation') to show existence of solutions to (LLG) in the 3D case. Our main results in this direction are:

1. $\varepsilon^{-1} = o(\frac{1}{k})$ is a stable choice in $\phi_1$, giving optimal convergence behavior. In fact, constraints can be sharpened to $\varepsilon > 1.9k$ (for $\phi_2$) and $\varepsilon \geq k$ (for $\phi_3$).

2. From a numerical viewpoint, $\phi_3^\gamma$ is the most favorable choice of the considered penalizations, for $\varepsilon = \mathcal{O}(k)$ and $\gamma \in \mathbb{N}_0$ sufficiently large; this is because of more proper balancing of stability and accuracy properties, i.e., the choice of $\varepsilon$ is more flexible with respect to the time-step size $k$.

3. Optimal convergence of $\mathbf{W}^{1,2}$-conforming finite elements on quasi-uniform meshes requires $\frac{1}{\sqrt{k}} = o(\frac{1}{h})$.

These statements are made precise in Theorems 4.5 through 4.7 as well as Corollaries 4.2 to 4.4 in Section 4.2.2.

The time discretization schemes considered so far encounter a significant computational effort to satisfy the saturation constraint via implicit penalization strategies; its reduction is the motivation of a projection method $(\mathrm{P})_A^\gamma$ proposed by E & Wang [43] (for $\gamma = 0$). The torque $\mathbf{m} \times \Delta\mathbf{m}$ is dealt with in linearized form, whereas the side-constraint is enforced in an explicit manner, and each iteration step is split into two substeps. — Let $\tilde{\mathbf{m}}^0 = \mathbf{m}^0 = \mathbf{m}_0$.

1. Given $\mathbf{m}^j, \tilde{\mathbf{m}}^j$, compute $\tilde{\mathbf{m}}^{j+1} \in \mathbf{W}^{1,2}(\omega, \mathbb{R}^3)$ from

$$\frac{1}{k}\{\tilde{\mathbf{m}}^{j+1} - \mathbf{m}^j\} - \alpha\,\Delta\tilde{\mathbf{m}}^{j+1} = \alpha\,|\,\nabla\tilde{\mathbf{m}}^j\,|^2\tilde{\mathbf{m}}^{j+1} + \tilde{\mathbf{m}}^{j+1} \times \Delta\tilde{\mathbf{m}}^{j+1}\,.$$

2. Compute $\mathbf{W}^{1,2}(\omega, \mathbb{R}^3) \ni \mathbf{m}^{j+1} = \frac{\tilde{\mathbf{m}}^{j+1}}{|\,\tilde{\mathbf{m}}^{j+1}\,|^{2-\gamma}}$, for $\gamma \in \mathbb{N}_0$ .

For $\gamma = 0$, we obtain $\mathbf{m}^{j+1} \in \mathbf{W}^{1,2}(\omega, \mathcal{S}^2)$, where

$$\mathbf{W}^{1,2}(\omega, \mathcal{S}^2) := \big\{\varphi \in \mathbf{W}^{1,2}(\omega, \mathbb{R}^3) :\ |\,\mathbf{m}(\mathbf{x})\,| = 1 \text{ f.a.e. } \mathbf{x} \in \omega\big\}\,, \tag{4.16}$$

and $\mathbf{W}^{k,p}(\omega, \mathcal{S}^2) = \mathbf{W}^{1,2}(\omega, \mathcal{S}^2) \cap \mathbf{W}^{k,p}(\omega, \mathbb{R}^3)$, for $p \geq 2$. This is the reason for the name of the method; however, our analysis motivates to allow $\gamma > 0$, and we refer to $(\mathrm{P})_A^\gamma$ as *scaled projection method.* — The subject of Section 4.2.2 is an error analysis for $(\mathrm{P})_A^\gamma$, by reinterpreting it as a semi-explicit penalization method.

In a next step, we ask for the behavior of finite elements for (LLG) and the interplay of numerical and penalization parameters: Section 5 gives an a priori error analysis for the penalization method (4.13); we select $\phi = \phi_2$ as an example to study main error effects; the proof is sketched for corresponding results valid for $(\mathrm{P})_A^\gamma$.

It is up to this point that we study a simplified version of (LLG), neglecting magnetostatic self-energy contribution, anisotropy and exterior magnetic fields which dynamically act in the ferromagnetic sample. Although this is justified from a mathematical point of view as the main phenomena are already visible here, scaling as well as 'real' dynamics in ferromagnets (like switching processes in nanomagnets) are not properly modeled by the simplified version of (LLG); for this purpose, (nonlocal) demagnetizing effects have to be taken into account, too, and the general Landau-Lifshitz-Gilbert equation for $\omega \subset \mathbb{R}^2$ bounded can be written in the form

$$\mathbf{m}_t = \mathbf{m} \times \mathbf{H}_{\text{eff}} - \alpha\, \mathbf{m} \times \big(\mathbf{m} \times \mathbf{H}_{\text{eff}}\big)\,, \quad \partial_{\mathbf{n}}\mathbf{m}\big|_{\partial\omega} = 0\,, \tag{4.17}$$

for $\mathbf{H}_{\text{eff}} \equiv \mathbf{H}_{\text{eff}}(\mathbf{m}, u)$, and $\mathbf{H}_{\text{eff}} = -\frac{\delta\mathcal{E}}{\delta\mathbf{m}}$, and the free energy $\mathcal{E}$ is given by [17, 39]

$$\mathcal{E}(\mathbf{m}) = \int_\omega \Big\{\frac{1}{2}|\,\nabla\mathbf{m}\,|^2 + \phi(\mathbf{m}) + \langle\nabla u, \mathbf{m}\rangle_{\mathbb{R}^2} - \langle\mathbf{f}, \mathbf{m}\rangle_{\mathbb{R}^3}\Big\}\,\mathrm{d}\mathbf{x}\,. \tag{4.18}$$

The density $\phi$ reflects crystallographic anisotropies of the material ('uniaxial, cubic ferromagnets'), and is supposed to be (Frechet-)differentiable in the sequel, an assumption which is not too restrictive for applications. The magnetic potential $u : \mathbb{R}^2 \to \mathbb{R}$ that gives the stray-field energy contribution in (4.18) is related to $\mathbf{m}$ as follows,

$$\Delta u = \operatorname{div}(\chi_\omega \mathbf{m}) \qquad \text{in } \mathbb{R}^2\,, \tag{4.19}$$

where $\chi_\omega = 1$ on $\omega$, and 0 else. In order to understand basic effects and simplify the setup for our computational experiments, we confine to bounded sets $\Omega \subset \mathbb{R}^2$ that contain the ferromagnet $\omega$, i.e., $\Omega \Supset \omega$, to capture non-local stray-field effects. We mention a corresponding scenario which is considered in Sections 2.1 and 2.2 in static micromagnetism; a generalization to $\Omega = \mathbb{R}^2$ is possible by coupled usage of finite element and boundary element methods; see [51, 78, 49].

For an efficient numerical realization, we split the computation of $\{\mathbf{m}^j, u^j\}$ at each time step. For normalized coefficients, we consider

$$\begin{aligned}\mathcal{E}_d(\mathbf{m}^{j+1}) \;=\; & \int_\omega \Big\{\frac{1}{2}|\,\nabla\mathbf{m}^{j+1}\,|^2 + \phi(\mathbf{m}^{j+1}) \\ & + \langle\nabla u^j, \mathbf{m}^{j+1}\rangle_{\mathbb{R}^2} + \langle\mathbf{f}^{j+1}, \mathbf{m}^{j+1}\rangle_{\mathbb{R}^3}\Big\}\,\mathrm{d}\mathbf{x}\,,\end{aligned} \tag{4.20}$$

where $\mathbf{f}^{j+1} = \mathbf{f}(t_{j+1})$, and

$$\Delta u^{j+1} = \operatorname{div}\left(\chi_\omega \mathbf{m}^{j+1}\right) \quad \text{in } \Omega\,. \tag{4.21}$$

We set $\mathbf{H}_{\text{eff}}^{j+1} = \mathbf{H}_1^{j+1} + \tilde{\mathbf{H}}_2^j$, for

$$\begin{aligned} \mathbf{H}_1^{j+1} &\equiv \mathbf{H}_1^{j+1}(\mathbf{m}^{j+1}) = \Delta \mathbf{m}^{j+1}\,, \\ \tilde{\mathbf{H}}_2^j &\equiv \tilde{\mathbf{H}}_2^j(\mathbf{m}^j, u^j) = -D\phi(\mathbf{m}^j) - \nabla u^j + \mathbf{f}^{j+1}\,. \end{aligned} \tag{4.22}$$

We propose the following **time-splitting scheme $(\mathbf{P})_B^\gamma$**:

1. Given $\left\{\tilde{\mathbf{m}}^j, \mathbf{m}^j, u^j\right\} \in \mathbf{W}^{1,2}(\omega, \mathbb{R}^3) \times \mathbf{W}^{1,2}(\omega, \mathbb{R}^3) \times W_0^{1,2}(\Omega)$, compute $\tilde{\mathbf{m}}^{j+1}$ from

$$\begin{aligned} \frac{1}{k}\left\{\tilde{\mathbf{m}}^{j+1} - \mathbf{m}^{j+1}\right\} - \alpha\,\Delta\tilde{\mathbf{m}}^{j+1} &= \alpha\,|\,\nabla\tilde{\mathbf{m}}^j\,|^2\tilde{\mathbf{m}}^{j+1} + \tilde{\mathbf{m}}^j \times \Delta\tilde{\mathbf{m}}^{j+1} \\ &\quad + \mathbf{m}^j \times \tilde{\mathbf{H}}_2^j - \alpha\,\mathbf{m}^j \times \left(\mathbf{m}^j \times \tilde{\mathbf{H}}_2^j\right). \end{aligned}$$

2. Compute $\mathbf{W}^{1,2}(\omega, \mathbb{R}^3) \ni \mathbf{m}^{j+1} = \frac{\tilde{\mathbf{m}}^{j+1}}{|\,\tilde{\mathbf{m}}^{j+1}\,|^{2-\gamma}}$, for $\gamma \in \mathbb{N}_0$ .

3. Compute $u^{j+1} \in W_0^{1,2}(\Omega)$ from

$$\Delta u^{j+1} = \operatorname{div}\left(\chi_\omega \mathbf{m}^j\right) \quad \text{in } \Omega\,.$$

We discuss convergence behavior of this method in Section 4.4. The proof is sketchy since most arguments that are given in the study of the projection scheme $(\mathrm{P})_A^\gamma$ do apply here as well.

In Section 4.5, we report on computational results for (LLG) in simplified (see (4.6)) and generalized (see (4.17)-(4.19)) versions. In particular, we show the necessity of the stability constraints $F(\varepsilon, k) > 0$ established in Section 4.2 for optimal convergence of the schemes and compare the different penalization techniques as well as the projection scheme $(\mathrm{P})_A^\gamma$, for different values of $\gamma$. A second order projection scheme is also studied there. We conclude this section with a computational study on evolution of magnetization involving defects.

## 4.1 Analysis of the Landau-Lifshitz-Gilbert equation

### 4.1.1 Existence of a smooth solution to (LLG) for small initial energies

We recall results regarding existence of solutions to (LLG) that have been verified by Guo and Hong in [60]. As will turn out, solutions that are studied here are classical, unique and global. However, the analysis relies on the crucial assumption that the initial energy $\| \mathbf{m}_0 \|_{\mathbf{W}^{1,2}}$ is sufficiently small. The intention in this subsection is to get acquainted with technical aspects on how to control nonlinear effects in the given problem. — The main result of this section is stated in the following theorem. In the sequel, let $I = [0, T]$, for $T > 0$ arbitrarily chosen.

**Theorem 4.1** *(from [60]) Let $\omega$ be the flat torus $\mathbb{R}^2/\mathbb{Z}^2$, $\mathbf{m}_0 \in \mathbf{W}^{k,2}(\omega, \mathcal{S}^2)$, for $k \geq 1$. There exists a constant $C_0 = C_0(\alpha)$ such that (LLG) has a smooth unique solution $\mathbf{m} \in L^\infty\big(I; \mathbf{W}^{k,2}(\omega, \mathcal{S}^2)\big) \cap L^2\big(I; \mathbf{W}^{k+1,2}(\omega, \mathcal{S}^2)\big)$, provided that $\| \mathbf{m}_0 \|_{\mathbf{W}^{1,2}(\omega)} \leq C_0$.*

The remainder of the section is organized as follows to verify this theorem: in a first step, sharp a priori bounds for the solution in Sobolev spaces with high index are shown under the assumption that initial energies are small, using Gagliardo-Nirenberg-type inequalities. Then, we benefit from Amann's (local) existence result for quasilinear parabolic systems in combination with a continuation argument to verify existence of a classical, global solution. Uniqueness will then be proved in a final step.

**Lemma 4.1** *Suppose that $\mathbf{m} : I \times \omega \to \mathcal{S}^2$ is a smooth solution to the Cauchy problem (4.2), where $\mathbf{m}_0 \in \mathbf{W}^{1,2}(\omega, \mathcal{S}^2)$. Then*

$$\operatorname*{ess\,sup}_{[0,T]} \| \nabla \mathbf{m} \|_{\mathbf{L}^2} + \Big( \int_0^T \| \mathbf{m}(s) \times \Delta \mathbf{m}(s) \|_{\mathbf{L}^2}^2 \, \mathrm{d}s \Big)^{1/2} \leq \| \nabla \mathbf{m}_0 \|_{\mathbf{L}^2}, \quad \forall T \geq 0.$$

**Proof:**

Multiplication of (4.2) by $-\Delta \mathbf{m}$ leads to

$$\begin{aligned} \frac{1}{2} d_t \| \nabla \mathbf{m} \|_{\mathbf{L}^2}^2 &= \alpha \left( \Delta \mathbf{m}, \mathbf{m} \times (\mathbf{m} \times \Delta \mathbf{m}) \right) \\ &= \alpha \left( \mathbf{m} \times \Delta \mathbf{m}, \Delta \mathbf{m} \times \mathbf{m} \right) = -\alpha \| \mathbf{m} \times \Delta \mathbf{m} \|_{\mathbf{L}^2}^2 . \end{aligned}$$

□

The verification of sharper a priori bounds is possible, in case the initial data satisfy a smallness condition.

**Lemma 4.2** *(from [60]) Suppose that* $\mathbf{m}_0 \in \mathbf{W}^{2,2}(\omega, \mathcal{S}^2)$, *where* $\| \mathbf{m}_0 \|_{\mathbf{W}^{1,2}} \leq C_0$. *Let* $T > 0$ *be a constant, and* $\mathbf{m} : \omega_T \to \mathcal{S}^2$ *a smooth solution to* (4.2). *Then there exists a constant* $C$ *such that*

$$\operatorname*{ess\,sup}_{[0,T]} \| \mathbf{m} \|_{\mathbf{W}^{2,2}} + \Big( \int_0^T \| \mathbf{m}(s) \|^2_{\mathbf{W}^{3,2}} \, \mathrm{d}s \Big)^{1/2} \leq C \, .$$

**Proof:**
In the following, $D^\alpha = \partial^{\alpha_1}_{x_1} \partial^{\alpha_2}_{x_2}$, where $|\,\alpha\,| = \alpha_1 + \alpha_2$, and $\alpha_1, \alpha_2$ are non-negative integers. For simplicity, let $D^k$ denote any kind of $D^\alpha$ satisfying $|\,\alpha\,| = k$.

We apply $D^2$ to (4.6), multiply by $D^2\mathbf{m}$ and integrate with respect to $\mathbf{x}$ to find

$$\begin{aligned} (D^2\mathbf{m}_t, D^2\mathbf{m}) &= \alpha\,(\Delta D^2\mathbf{m}, D^2\mathbf{m}) + \alpha\,(D^2(|\,\nabla\mathbf{m}\,|^2\mathbf{m}), D^2\mathbf{m}) \\ &\quad + (D^2(\mathbf{m} \times \Delta\mathbf{m}), D^2\mathbf{m}) \, . \end{aligned} \tag{4.23}$$

Since $\omega$ has no boundary, we get

$$\begin{aligned} (\Delta D^2\mathbf{m}, D^2\mathbf{m}) &= -\| \nabla D^2\mathbf{m} \|^2_{\mathbf{L}^2} \, , \\ |\,(D^2(|\,\nabla\mathbf{m}\,|^2\mathbf{m}), D^2\mathbf{m})\,| &= |\,(D(|\,\nabla\mathbf{m}\,|^2\mathbf{m}), D^3\mathbf{m})\,| \, , \\ (D^2\mathbf{m}_t, D^2\mathbf{m}) &= \frac{1}{2}\frac{d}{dt}\| D^2\mathbf{m} \|^2_{\mathbf{L}^2} \, . \end{aligned}$$

Because of

$$D(|\,\nabla\mathbf{m}\,|^2\mathbf{m}) = |\,\nabla\mathbf{m}\,|^2 D\mathbf{m} + 2\,\big\langle \mathbf{m}\nabla\mathbf{m}, D\nabla\mathbf{m} \big\rangle_{\mathbb{R}^6} \, ,$$

we infer

$$\| D(|\,\nabla\mathbf{m}\,|^2\mathbf{m}) \|_{\mathbf{L}^2} \leq \| \, |\,\nabla\mathbf{m}\,|^2 \|_{\mathbf{L}^3} \| D\mathbf{m} \|_{\mathbf{L}^6} + 2\,\| \mathbf{m} \|_{\mathbf{L}^\infty} \| \nabla\mathbf{m} \|_{\mathbf{L}^4} \| D\nabla\mathbf{m} \|_{\mathbf{L}^4} \, .$$

In the next step, we make use of the generalized Sobolev inequalities

$$\begin{aligned} \| \nabla\mathbf{m} \|_{\mathbf{L}^4} &\leq C\,\| \nabla\mathbf{m} \|^{1/4}_{\mathbf{W}^{2,2}} \| \nabla\mathbf{m} \|^{3/4}_{\mathbf{L}^2} \, , \\ \| \nabla^2\mathbf{m} \|_{\mathbf{L}^4} &\leq C\,\| \nabla\mathbf{m} \|^{3/4}_{\mathbf{W}^{2,2}} \| \nabla\mathbf{m} \|^{1/4}_{\mathbf{L}^2} \, , \\ \| \nabla\mathbf{m} \|_{\mathbf{L}^6} &\leq C\,\| \nabla\mathbf{m} \|^{1/3}_{\mathbf{W}^{2,2}} \| \nabla\mathbf{m} \|^{2/3}_{\mathbf{L}^2} \, . \end{aligned} \tag{4.24}$$

This allows us to control the second term on the right-hand side of (4.23) as follows, using Lemma 4.1 in the last step.

$$
\begin{aligned}
&|\,(D^2(|\,\nabla\mathbf{m}\,|^2\mathbf{m}), D^2\mathbf{m})\,| \\
&\le \|\,\nabla^3\mathbf{m}\,\|_{\mathbf{L}^2}\Big(2\,\|\,\mathbf{m}\,\|_{\mathbf{L}^\infty}\|\,\nabla\mathbf{m}\,\|_{\mathbf{L}^4}\|\,D\nabla\mathbf{m}\,\|_{\mathbf{L}^4} + \|\,\nabla\mathbf{m}\,\|^3_{\mathbf{L}^6}\Big) \\
&\le C\left(\|\,\nabla\mathbf{m}\,\|_{\mathbf{L}^2} + \|\,\nabla\mathbf{m}\,\|^2_{\mathbf{L}^2}\right)\|\,\nabla\mathbf{m}\,\|^2_{\mathbf{W}^{2,2}} \qquad (4.25) \\
&\le C\left(\|\,\nabla\mathbf{m}_0\,\|_{\mathbf{L}^2} + \|\,\nabla\mathbf{m}_0\,\|^2_{\mathbf{L}^2}\right)\|\,\nabla\mathbf{m}\,\|^2_{\mathbf{W}^{2,2}}\,.
\end{aligned}
$$

We proceed with the control of the last term in (4.23). In the following, we use the shorthand notation $\mathbf{m}\times\nabla\mathbf{m} := (\mathbf{m}\times\partial_1\mathbf{m}, \mathbf{m}\times\partial_2\mathbf{m})$.

$$
\begin{aligned}
&|\,(D^2(\mathbf{m}\times\Delta\mathbf{m}), D^2\mathbf{m})\,| = |\,D^2\mathrm{div}\,(\mathbf{m}\times\nabla\mathbf{m}), D^2\mathbf{m})\,| \\
&= |\,(D^2(\mathbf{m}\times\nabla\mathbf{m}), \nabla D^2\mathbf{m})\,| \\
&= |\,(D^2\mathbf{m}\times\nabla\mathbf{m} + 2D\mathbf{m}\times D\nabla\mathbf{m}, \nabla D^2\mathbf{m}) + (\mathbf{m}\times\nabla D^2\mathbf{m}, \nabla D^2\mathbf{m})\,|\,.
\end{aligned}
$$

Note that the last contribution is zero. We employ again (4.24).

$$
\begin{aligned}
|\,(D^2(\mathbf{m}\times\Delta\mathbf{m}), D^2\mathbf{m})\,| \;&\le\; C\,\|\,\nabla\mathbf{m}\,\|_{\mathbf{L}^4}\,\|\,\nabla^2\mathbf{m}\,\|_{\mathbf{L}^4}\,\|\,\nabla^3\mathbf{m}\,\|_{\mathbf{L}^2} \qquad (4.26) \\
&\le\; C\,\|\,\nabla\mathbf{m}\,\|_{\mathbf{L}^2}\,\|\,\nabla\mathbf{m}\,\|^2_{\mathbf{W}^{2,2}} \\
&\le\; C\,\|\,\nabla\mathbf{m}_0\,\|_{\mathbf{L}^2}\,\|\,\nabla\mathbf{m}\,\|^2_{\mathbf{W}^{2,2}}\,.
\end{aligned}
$$

We may now combine these results to find

$$
\begin{aligned}
&d_t\|\,\nabla^2\mathbf{m}\,\|^2_{\mathbf{L}^2} + \alpha\,\|\,\nabla^3\mathbf{m}\,\|^2_{\mathbf{L}^2} \le C\,\|\,\nabla^2\mathbf{m}\,\|^2_{\mathbf{L}^2} \qquad (4.27) \\
&\quad +\alpha\,\|\,\nabla^3\mathbf{m}\,\|^2_{\mathbf{L}^2}\Big(\|\,\nabla\mathbf{m}_0\,\|^2_{\mathbf{L}^2} + 2\,\|\,\nabla\mathbf{m}_0\,\|_{\mathbf{L}^2} + \frac{1}{\alpha}\|\,\nabla\mathbf{m}_0\,\|_{\mathbf{L}^2}\Big).
\end{aligned}
$$

For $\|\,\nabla\mathbf{m}_0\,\|_{\mathbf{L}^2}$ sufficiently small, the term on the right-hand side may be absorbed on the left-hand side. This settles the proof. □

The following lemma shows that regularity properties will be preserved by (LLG).

**Lemma 4.3** *(from [60]) Suppose that $\mathbf{m}_0 \in \mathbf{W}^{k,2}(\omega, S^2)$, $k \ge 2$, and that the assumptions of Lemma 4.2 are satisfied. Then there exists a constant $C$ such that*

$$
\operatorname*{ess\,sup}_{[0,T]}\|\,\mathbf{m}\,\|_{\mathbf{W}^{k,2}} + \left(\int_0^T \|\,\mathbf{m}(s)\,\|^2_{\mathbf{W}^{k+1,2}}\,\mathrm{d}s\right)^{1/2} \le C\,, \quad \forall T \ge 0\,.
$$

**Proof:**
We proceed by induction, and assume $\mathbf{m}$ to be a smooth solution to (4.6). Again, we take benefit from the fact that $\partial\omega = \emptyset$. We apply $D^3$ to (4.6).

$$D^3\mathbf{m}_t - \alpha\,\Delta D^3\mathbf{m} = \alpha\, D^3\big(|\,\nabla\mathbf{m}\,|^2\mathbf{m}\big) + D^3\big(\mathbf{m}\times\Delta\mathbf{m}\big)\,. \tag{4.28}$$

We have

$$\begin{aligned} D^2\big(|\,\nabla\mathbf{m}\,|^2\mathbf{m}\big) &= |\,\nabla\mathbf{m}\,|^2 D^2\mathbf{m} + 4\,\langle\nabla\mathbf{m}, D\nabla\mathbf{m}\rangle_{\mathbb{R}^6}\nabla\mathbf{m} \\ &\quad +2\,|\,D\nabla\mathbf{m}\,|^2\mathbf{m} + 2\langle D^2\nabla\mathbf{m}, \nabla\mathbf{m}\rangle_{\mathbb{R}^6}\mathbf{m}\,. \end{aligned} \tag{4.29}$$

We test (4.28) by $D^3\mathbf{m}$,

$$\begin{aligned} &\frac{1}{2}d_t\|\,D^3\mathbf{m}\,\|^2_{\mathbf{L}^2} + \alpha\,\|\,\nabla D^3\mathbf{m}\,\|^2_{\mathbf{L}^2} \\ &\quad = -\alpha\,\big(D^2(|\,\nabla\mathbf{m}\,|^2\mathbf{m}), D^4\mathbf{m}\big) - \big(D^3(\mathbf{m}\times\Delta\mathbf{m}), \nabla D^3\mathbf{m}\big)\,. \end{aligned} \tag{4.30}$$

The following estimate will be useful in the sequel (see e.g. [60], Lemma 2.4),

$$\|\,\mathbf{m}\,\|_{\mathbf{L}^\infty} \le C\,\|\,\mathbf{m}\,\|^{3/4}_{\mathbf{W}^{2,2}}\,\|\,\mathbf{m}\,\|^{1/4}_{\mathbf{L}^2}\,, \tag{4.31}$$

Let $0 < \beta = \beta(\alpha)$ be chosen sufficiently small in the sequel. Then, we find the following upper bounds for the nonlinear terms, using (4.29) and (4.24), (4.31).

$$\begin{aligned} |\,\big(D^2(|\,\nabla\mathbf{m}\,|^2\mathbf{m}), D^4\mathbf{m}\big)\,| &\le C\Big(\|\,\nabla\mathbf{m}\,\|^2_{\mathbf{L}^\infty}\|\,D\nabla\mathbf{m}\,\|_{\mathbf{L}^2} + \|\,D\nabla\mathbf{m}\,\|^2_{\mathbf{L}^4} \\ &\quad +\|\,\nabla\mathbf{m}\,\|_{\mathbf{L}^\infty}\|\,D^2\nabla\mathbf{m}\,\|_{\mathbf{L}^2}\Big)\|\,D^4\mathbf{m}\,\|_{\mathbf{L}^2} \\ &\le \frac{\beta}{4}\|\,\nabla\mathbf{m}\,\|^2_{\mathbf{W}^{3,2}} + C_\beta\Big(\|\,\nabla\mathbf{m}\,\|^3_{\mathbf{W}^{2,2}}\,\|\,\nabla\mathbf{m}\,\|^2_{\mathbf{L}^2} + \|\,\nabla\mathbf{m}\,\|^4_{\mathbf{L}^4} \\ &\quad +\|\,\nabla\mathbf{m}\,\|^4_{\mathbf{W}^{2,2}}\|\,\nabla\mathbf{m}\,\|_{\mathbf{L}^2}\Big)\,. \end{aligned} \tag{4.32}$$

Next, we benefit from the fact that $\langle \mathbf{a}\times\mathbf{b},\mathbf{b}\rangle_{\mathbb{R}^3}=0$, and (4.31), (4.24),

$$\begin{aligned}
&|\,(D^3(\mathbf{m}\times\nabla\mathbf{m}),\nabla D^3\mathbf{m})\,|\\
&=|\,(D^3\mathbf{m}\times\nabla\mathbf{m}+5\,D^2\mathbf{m}\times D\nabla\mathbf{m}\\
&\quad+4\,D\mathbf{m}\times D^2\nabla\mathbf{m}+\mathbf{m}\times\nabla D^3\mathbf{m},\nabla D^3\mathbf{m})\,|\\
&\le C\Big(\|\nabla\mathbf{m}\|_{\mathbf{L}^\infty}\|D^3\mathbf{m}\|_{\mathbf{L}^2}\\
&\quad+\|D^2\mathbf{m}\|_{\mathbf{L}^4}\|D\nabla\mathbf{m}\|_{\mathbf{L}^4}+\|D\mathbf{m}\|_{\mathbf{L}^\infty}\Big)\|\nabla D^3\mathbf{m}\|_{\mathbf{L}^2}\\
&\le\frac{\beta}{4}\|\nabla\mathbf{m}\|^2_{\mathbf{W}^{3,2}}+C_\beta\Big(\|\nabla\mathbf{m}\|^{1/2}_{\mathbf{L}^2}\|\mathbf{m}\|^{7/2}_{\mathbf{W}^{3,2}}\\
&\quad+\|\mathbf{m}\|^2_{\mathbf{W}^{3,2}}\|\nabla\mathbf{m}\|^2_{\mathbf{W}^{1,2}}+\|\mathbf{m}\|^{3/2}_{\mathbf{W}^{3,2}}\|D\mathbf{m}\|^{1/2}_{\mathbf{L}^2}\Big)\,.
\end{aligned}$$

Then, Lemma 4.2 together with Gronwall's inequality leads to the result of the lemma for $k=2$.

Let is now suppose

$$\operatorname*{ess\,sup}_{[0,T]}\|\mathbf{m}\|_{\mathbf{W}^{k+1,2}}+\Big(\int_0^T\|\mathbf{m}(s)\|^2_{\mathbf{W}^{k+2,2}}\,\mathrm{d}s\Big)^{1/2}\le C\quad\text{for } k\ge 2. \tag{4.33}$$

We start from

$$D^{k+1}\mathbf{m}_t=\alpha\,\Delta D^{k+1}\mathbf{m}+\alpha\,D^{k+1}(|\nabla\mathbf{m}|^2\mathbf{m})+D^{k+1}(\mathbf{m}\times\Delta\mathbf{m}) \tag{4.34}$$

and employ an iterative argument by employing the following inequality that is due to Kato. For two functions $\mathbf{f},\mathbf{g}$ (see [77]),

$$\|D^s(\mathbf{f}\mathbf{g})-\mathbf{f}D^s\mathbf{g}\|_{\mathbf{L}^p}\le C\Big(\|D\mathbf{f}\|_{\mathbf{L}^\infty}\|D^{s-1}\mathbf{g}\|_{\mathbf{L}^p}+\|D^s\mathbf{f}\|_{\mathbf{L}^p}\|\mathbf{g}\|_{\mathbf{L}^\infty}\Big)\,. \tag{4.35}$$

Setting $\mathbf{f}=\mathbf{m}$, $\mathbf{g}=|\nabla\mathbf{m}|^2$, $s=k$, we find

$$\begin{aligned}
\|D^k(|\nabla\mathbf{m}|^2\mathbf{m})\|_{\mathbf{L}^2}&\le\|\mathbf{m}D^k(|\nabla\mathbf{m}|^2)\|_{\mathbf{L}^2}\\
&\quad+C\Big(\|\nabla\mathbf{m}\|_{\mathbf{L}^\infty}\|D^{k-1}|\nabla\mathbf{m}|^2\|_{\mathbf{L}^2}+\|D^k\mathbf{m}\|_{\mathbf{L}^2}\|\nabla\mathbf{m}\|^2_{\mathbf{L}^\infty}\Big)\\
&\le\|D^k(|\nabla\mathbf{m}|^2)\|_{\mathbf{L}^2}+C\Big(\|D^{k-1}|\nabla\mathbf{m}|^2\|_{\mathbf{L}^2}+\|D^k\mathbf{m}\|_{\mathbf{L}^2}\Big)\\
&\le C\,\|D^{k+1}\mathbf{m}\|_{\mathbf{L}^2}+C\,.
\end{aligned} \tag{4.36}$$

This result may now be employed in the following estimate,

$$\begin{aligned} \left| \left(D^{k+1}(|\nabla \mathbf{m}|^2 \mathbf{m}), D^{k+1}\mathbf{m}\right) \right| &= \left| \left(D^{k}(|\nabla \mathbf{m}|^2 \mathbf{m}), D^{k+2}\mathbf{m}\right) \right| \\ &\leq \frac{\beta}{4} \| \mathbf{m} \|^2_{\mathbf{W}^{k+2,2}} + C_\beta \| \mathbf{m} \|^2_{\mathbf{W}^{k+1,2}} + C_\beta \,. \end{aligned} \tag{4.37}$$

On the other hand, we have

$$\left| \left(D^{k+1}(\mathbf{m} \times \Delta \mathbf{m}), D^{k+1}\mathbf{m}\right) \right| = \left| \left(D^{k+1}(\mathbf{m} \times \nabla \mathbf{m}), \nabla D^{k+1}\mathbf{m}\right) \right| .$$

For numbers $c_\ell \leq C$, $1 \leq \ell \leq k$, we have the identity

$$D^{k+1}(\mathbf{m} \times \nabla \mathbf{m}) = D^{k+1}\mathbf{m} \times \nabla \mathbf{m} + \mathbf{m} \times D^{k+1}\nabla \mathbf{m} + \sum_{\ell=1}^{k} c_\ell D^\ell \mathbf{m} \times D^{k+1-\ell}\nabla \mathbf{m} \,.$$

This leads to

$$\begin{aligned} & \left| \left(D^{k+1}(\mathbf{m} \times \nabla \mathbf{m}), \nabla D^{k+1}\mathbf{m}\right) \right| \\ & \leq \left| \left(D^{k+1}\mathbf{m} \times \nabla \mathbf{m}, \nabla D^{k+1}\mathbf{m}\right) \right| + \sum_{\ell=1}^{k} c_\ell \left| \left(D^\ell \mathbf{m} \times D^{k+1-\ell}\nabla \mathbf{m}, \nabla D^{k+1}\mathbf{m}\right) \right| \\ & \leq \| \nabla \mathbf{m} \|_{\mathbf{L}^\infty} \| D^{k+1}\mathbf{m} \|_{\mathbf{L}^2} + C \| D^k \mathbf{m} \|_{\mathbf{L}^4} \| D^k \nabla \mathbf{m} \|_{\mathbf{L}^4} \| \nabla D^{k+1}\mathbf{m} \|_{\mathbf{L}^2} \\ & \leq \frac{\beta}{4} \| \mathbf{m} \|^2_{\mathbf{W}^{k+2,2}} + C_\beta \| \mathbf{m} \|^2_{\mathbf{W}^{k+1,2}} \end{aligned} \tag{4.38}$$

We may now combine (4.37), (4.38) to control arising terms when multiplying (4.34) with $D^{k+1}\mathbf{m}$,

$$d_t \| D^{k+1}\mathbf{m} \|^2_{\mathbf{L}^2} + \frac{\alpha}{2} \| \nabla D^{k+1}\mathbf{m} \|^2_{\mathbf{L}^2} \leq C \left(1 + \| \mathbf{m} \|^2_{\mathbf{W}^{k+1,2}}\right) . \tag{4.39}$$

Gronwall's inequality then gives the result. □

Lemmata 4.1 to 4.3 provide the basis firstly to verify existence of local smooth solutions to (LLG) in 2D, using results of Amann [5] on quasi-linear parabolic systems. Then, a continuation argument gives global existence of solutions; this program is elaborated in [60].

In a last step, we show uniqueness of smooth global solutions to (LLG). The argumentation presented next is similar to [60].

**Lemma 4.4** *Suppose that the assumptions of Lemma 4.2 are satisfied. Then the solution to (LLG) is unique.*

**Proof:**
Let $\mathbf{m}^i$, $i \in \{1,2\}$ be two solutions that solve (LLG), where $\mathbf{m}^i(0) = \mathbf{m}_0$. In the following, set $\mathbf{e} := \mathbf{m}^1 - \mathbf{m}^2$. Then, we obtain

$$\begin{aligned} \mathbf{e}_t - \alpha\Delta\mathbf{e} &= \alpha\Big(|\nabla\mathbf{m}^1|^2\mathbf{m}^1 \pm |\nabla\mathbf{m}^1|^2\mathbf{m}^2 - |\nabla\mathbf{m}^2|^2\mathbf{m}^2\Big) \\ &\quad +\mathbf{m}^1\times\Delta\mathbf{m}^1 - \mathbf{m}^2\times\Delta\mathbf{m}^2 \pm \mathbf{m}^1\times\Delta\mathbf{m}^2 \\ &= \alpha\Big(|\nabla\mathbf{m}^1|^2\mathbf{e} + \langle\nabla\mathbf{e}, \nabla(\mathbf{m}^1+\mathbf{m}^2)\rangle_{\mathbb{R}^6}\Big) + \mathbf{m}^1\times\Delta\mathbf{e} + \mathbf{e}\times\Delta\mathbf{m}^2 . \end{aligned} \tag{4.40}$$

Multiplication with $\mathbf{e}$ and $\langle\mathbf{a}\times\mathbf{b},\mathbf{c}\rangle_{\mathbb{R}^3} = -\langle\mathbf{a}\times\mathbf{c},\mathbf{b}\rangle_{\mathbb{R}^3}$ then leads to

$$\begin{aligned} \frac{1}{2}d_t\|\mathbf{e}\|^2_{\mathbf{L}^2} + \alpha\|\nabla\mathbf{e}\|^2_{\mathbf{L}^2} &= (\mathbf{e}\times\nabla\mathbf{e}, \nabla\mathbf{m}^1) \\ &\quad +\alpha\Big((|\nabla\mathbf{m}^1|^2\mathbf{e},\mathbf{e}) + (\langle\nabla\mathbf{e},\nabla(\mathbf{m}^1+\mathbf{m}^2)\rangle_{\mathbb{R}^6}\mathbf{m}^2,\mathbf{e})\Big) \\ &\leq C\alpha\Big(\|\nabla\mathbf{m}^1\|^2_{\mathbf{L}^4}\|\mathbf{e}\|^2_{\mathbf{L}^4} + \|\nabla\mathbf{e}\|_{\mathbf{L}^2}\|\nabla(\mathbf{m}^1+\mathbf{m}^2)\|_{\mathbf{L}^4}\|\mathbf{e}\|_{\mathbf{L}^4} \\ &\quad +\|\mathbf{e}\|_{\mathbf{L}^4}\|\nabla\mathbf{e}\|_{\mathbf{L}^2}\|\nabla\mathbf{m}^1\|_{\mathbf{L}^4}\Big) . \end{aligned} \tag{4.41}$$

We may use Gagliardo-Nirenberg's inequality,

$$\|\mathbf{m}\|_{\mathbf{L}^4} \leq C\|\mathbf{m}\|^{1/2}_{\mathbf{L}^2}\|\mathbf{m}\|^{1/2}_{\mathbf{W}^{1,2}} \tag{4.42}$$

to find after absorption

$$\frac{1}{2}d_t\|\mathbf{e}\|^2_{\mathbf{L}^2} + \frac{\alpha}{2}\|\nabla\mathbf{e}\|^2_{\mathbf{L}^2} \leq C\Big(\|\nabla\mathbf{m}^1\|^4_{\mathbf{L}^4}\|\mathbf{e}\|^2_{\mathbf{L}^2} + \|\nabla(\mathbf{m}^1+\mathbf{m}^2)\|^4_{\mathbf{L}^4}\|\mathbf{e}\|^2_{\mathbf{L}^2}\Big) . \tag{4.43}$$

Gronwall's inequality together with Lemma 4.2 then show that $\mathbf{e} \equiv 0$. □

### 4.1.2 Harmonic maps between Riemannian manifolds

The material of this subsection is taken from [120, 122, 121]. — Mathematically, there is a connection between solutions of the Landau-Lifshitz-Gilbert equation and harmonic maps, which gives another motivation to study (LLG): let $\mathcal{M}$ an $m$-dimensional compact Riemannian manifold, with metric tensor $\gamma = (\gamma_{\alpha\beta})_{1\leq\alpha,\beta\leq m}$, and $\mathcal{N}$ an $n$-dimensional compact Riemannian manifold, with metric tensor $g = (g_{ij})_{1\leq i,j\leq n}$. For a $C^1$-map $\mathbf{m} : \mathcal{M} \to \mathcal{N}$, an energy density $e(\mathbf{m})$ is defined in local coordinates $\mathbf{x} = (x^1, .., x^m)^\top$, $\mathbf{m} = (m^1, .., m^n)^\top$ by

$$e(\mathbf{m}) = \frac{1}{2}\gamma^{\alpha\beta}(\mathbf{x})g_{ij}(\mathbf{m})\frac{\partial}{\partial x^\alpha}\mathbf{m}^i\frac{\partial}{\partial x^\beta}\mathbf{m}^j\,, \tag{4.44}$$

where $(\gamma^{\alpha\beta}) = (\gamma_{\alpha\beta})^{-1}$. Then, $\mathbf{m} \in C^1(\mathcal{M}, \mathcal{N})$ is called a harmonic map if it is a stationary point for Dirichlet's energy

$$E(\mathbf{m}) = \int_{\mathcal{M}} e(\mathbf{m})\,\mathrm{d}\mathcal{M}\,. \tag{4.45}$$

Equivalently, harmonic mappings solve the Euler-Lagrange equations

$$-\Delta_{\mathcal{M}}\mathbf{m} = \Gamma(\mathbf{m})\big(\nabla\mathbf{m}, \nabla\mathbf{m}\big)_{\mathcal{M}}\,, \tag{4.46}$$

where $\Delta_{\mathcal{M}}$ denotes the Laplace-Beltrami operator on $\mathcal{M}$, and $\Gamma(\mathbf{m})$ is the second fundamental form of $\mathcal{N}$. Note that for $\mathcal{N} = \mathbb{R}^n$, harmonic maps are harmonic functions.

Subsequently, we present some more details how to derive the Euler-Lagrange equations for harmonic maps, in particular for $\mathcal{N} = \mathcal{S}^{n-1} \subset \mathbb{R}^n$. For this purpose, we introduce the tangent space to the manifold $C^1(\mathcal{M}, \mathcal{N})$, for mappings $\mathbf{m} \in C^1(\mathcal{M}, \mathcal{N})$,

$$T_{\mathbf{m}}C^1(\mathcal{M}, \mathcal{N}) = \big\{\varphi \in C^1(\mathcal{M}, \mathcal{N}) : \ \varphi(\mathbf{x}) \in T_{\mathbf{m}}\mathcal{M}, \ \text{for } \mathbf{x} \in \mathcal{M}\big\}\,.$$

In the sequel, we choose $\mathcal{M} = T^2 = \mathbb{R}^2/\mathbb{Z}^2$, and $\mathcal{N} = \mathcal{S}^{n-1} \subset \mathbb{R}^n$. Then, $\mathbf{m} : T^2 \to \mathcal{S}^{n-1} \subset \mathbb{R}^n$ is harmonic of class $C^2$, if the first variation of $E(\mathbf{m})$ gives

$$0 = \big(\varphi, DE(\mathbf{m})\big) = -\int_{T^2} \Delta\mathbf{m}\,\varphi\,\mathrm{d}\mathbf{x}\,,$$

for all doubly periodic functions $\varphi \in C^1(\mathbb{R}^2, \mathbb{R}^n)$, such that $\varphi(\mathbf{x}) \in T_{\mathbf{m}(\mathbf{x})}\mathcal{N}$, for all $\mathbf{x} \in \mathbb{R}^2$. But this means, that $-\Delta\mathbf{m}(\mathbf{x})$ is orthogonal to the tangent space at $\mathcal{S}^{n-1}$ at any point $\mathbf{x} \in T^2$; in symbols,

$$-\Delta\mathbf{m}(\mathbf{x}) = \lambda(\mathbf{x})\mathbf{n}_{\mathbf{m}}(\mathbf{x}) \ \perp \ T_{\mathbf{m}}(\mathbf{x})\mathcal{N}\,, \tag{4.47}$$

for some normal vector field $\mathbf{n_m} : T^2 \to \mathbb{R}^n$ and a scalar function $\lambda : T^2 \to \mathbb{R}$. Thanks to $|\mathbf{m}| = 1$, testing this relation with $\mathbf{m}$ then gives $\lambda = |\nabla\mathbf{m}|^2$, and hence $-\Delta\mathbf{m} = |\nabla\mathbf{m}|^2\mathbf{m}$.

It is now that we turn to weakly harmonic maps. We denote

$$\mathbf{W}^{1,2}(\mathcal{M},\mathcal{N}) = \{\mathbf{m} \in \mathbf{W}^{1,2}(\mathcal{M},\mathbb{R}^n) : \ \mathbf{m}(\mathbf{x}) \in \mathcal{N}, \text{ for a.e. } \mathbf{x} \in \mathcal{M}\}.$$

This set is weakly closed in $\mathbf{W}^{1,2}(\mathcal{M},\mathbb{R}^n)$, and $C^\infty(\mathcal{M},\mathcal{N})$ is dense in the set $\mathbf{W}^{1,2}(\mathcal{M},\mathcal{N})$, [115]. We call a map $\mathbf{m} \in \mathbf{W}^{1,2}(\mathcal{M},\mathcal{N})$ weakly harmonic if it solves (4.46) in the distributional sense.

A natural generalization of Dirichlet's problem for harmonic maps is the *homotopy problem*: Given a map $\mathbf{m}_0 : \mathcal{M} \to \mathcal{N}$, is there a harmonic map $\mathbf{m}$ homotopic to $\mathbf{m}_0 \in C^0(\mathcal{M},\mathcal{N})$? The notion of homotopy classes of a $\mathbf{W}^{1,2}(\mathcal{M},\mathcal{N})$-map is well-defined but homotopy classes of non-constant maps are not weakly closed, so a direct approach fails to answer this question; in particular, Eells-Wood [48] gave an example where the infimum in a given homotopy class in general need not be attained.

Another strategy to attack the homotopy problem for harmonic maps was proposed by Eells-Sampson [47], where the gradient flow is studied,

$$\mathbf{m}_t - \Delta_{\mathcal{M}}\mathbf{m} - \Gamma(\mathbf{m})(\nabla\mathbf{m},\nabla\mathbf{m})_{\mathcal{M}} = 0\,, \quad \mathbf{m}(0) = \mathbf{m}_0 \in C^\infty(\mathcal{M},\mathcal{N})\,. \tag{4.48}$$

The crucial question they ask is whether the solution $\mathbf{m}(\cdot,t)$ for $t \to \infty$ will come to rest at a critical point of $E$; that is, whether it is a harmonic map. The result of Eells-Sampson [47] furnishes that in case the sectional curvature of $\mathcal{N}$ is non-positive, then there exists a unique, global, smooth solution $\mathbf{m} : \mathcal{M} \times [0,\infty[ \to \mathcal{N}$, which, at $t \to \infty$ suitably, converges smoothly to a harmonic map $\mathbf{m}_\infty \in C^\infty(\mathcal{M},\mathcal{N})$, homotopic to $\mathbf{m}_0$.

Unfortunately, the result of Eells-Sampson [47] is not applicable to the pairing $\{\mathcal{M},\mathcal{N}\} = \{\omega,\mathcal{S}^d\}$, $d \in \{2,3\}$, since the sectional curvature of $\mathcal{S}^d$ is strictly positive. Therefore there remains the question whether in this case the heat flow mapping will develop singularities in finite or infinite time. In fact, an example of a solution to (4.48) is given in [23] for $\dim\mathcal{M} = 2$, and $\mathcal{N} = \mathcal{S}^2$, where finite-time blow-up occurs, provided the initial energy exceeds a certain threshold. This shows that, in general, smooth, global solutions to (4.48) do not exist. However, if we relax the concept of solutions to *weak* ones, there is a result of M. Struwe [120] that establishes global solutions for Riemann surfaces $\mathcal{M}$. They are smooth, apart from finitely many points in space-time.

### 4.1.3 Existence of strong solutions to (LLG) for initial data of finite energy

In [60], Guo and Hong apply techniques that have been developed in [120] to (LLG) to show existence of a unique strong local solution $\mathbf{m} \in L^\infty(I; \mathbf{W}^{1,2}(\mathcal{M}, \mathcal{S}^2)) \cap W^{1,2}(I; \mathbf{L}^2(\mathcal{M}, \mathcal{S}^2)) \cap L^2(I; \mathbf{W}^{2,2}(\mathcal{M}, \mathcal{S}^2))$, for compact Riemannian surfaces $\mathcal{M}$ without boundary, $I = [0,T]$, and some $T > 0$. Its verification proceeds in three steps: (i) Verify a priori bounds for the solution in $L^\infty(I; \mathbf{W}^{1,2}(\mathcal{M}, \mathcal{S}^2)) \cap W^{1,2}(I; \mathbf{L}^2(\mathcal{M}, \mathcal{S}^2)) \cap L^2(I; \mathbf{W}^{2,2}(\mathcal{M}, \mathcal{S}^2))$. This goal is accomplished by means of *local* (in space and time) energy estimates to control its evolution in time. (ii) Exploit isomorphism properties of the linear strongly parabolic operator $\partial_t - \alpha\,\Delta - \mathbf{z} \times \Delta$ to show existence of a unique local (in time) smooth solution to (LLG) for $\mathbf{m}_0 \in C^\infty(\mathcal{M}, \mathcal{S}^2)$ by inverse function theorem. (iii) Approximate $\mathbf{m}_0$ in $\mathbf{W}^{1,2}(\mathcal{M}, \mathcal{S}^2)$ by regular initial data $\mathbf{m}_0^i \in C^\infty(\mathcal{M}, \mathcal{S}^2)$ and identify (weak) limits.

We use this solution concept for our numerical analysis, where we select $\mathcal{M} = \omega = \mathbb{R}^2/\mathbb{Z}^2$. In this section, we show further regularity results for the solution to (LLG) locally in time that are the basis for the numerical analysis presented in Section 4.2; see Theorem 4.3. The arguments are formal but can be made rigorous on the level of standard Faedo-Galerkin approximations.

In the sequel, we consider local strong solutions to (LLG) and provide some useful results from [60, 120].

**Lemma 4.5** *(from [60]) Let $\mathbf{m} : \omega_T \to \mathcal{S}^2$ be a strong solution to (LLG). There holds*

$$\operatorname*{ess\,sup}_{[0,T]} \| \nabla \mathbf{m} \|_{\mathbf{L}^2} + \Big( \int_0^T \| \mathbf{m} \times \Delta \mathbf{m}(s) \|_{\mathbf{L}^2}^2 + \| \mathbf{m}_t(s) \|_{\mathbf{L}^2}^2 \, \mathrm{d}s \Big)^{1/2} \leq \| \nabla \mathbf{m}_0 \|_{\mathbf{L}^2} .$$

**Proof:**

(LLG) is valid in $L^2(I; \mathbf{L}^2(\omega, \mathbb{R}^3))$. Multiply (LLG) by $\boldsymbol{\varphi} = \mathbf{m}_t$ and employ $d_t|\,\mathbf{m}\,|^2 = 0$, almost everywhere,

$$\int_0^T \| \mathbf{m}_t(s) \|_{\mathbf{L}^2}^2 \, \mathrm{d}s + \frac{\alpha}{2} \| \nabla \mathbf{m}(T) \|_{\mathbf{L}^2}^2 \qquad (4.49)$$

$$\leq \frac{\alpha}{2} \| \nabla \mathbf{m}_0 \|_{\mathbf{L}^2}^2 + \int_0^T \big(\mathbf{m}(s) \times \Delta \mathbf{m}(s), \mathbf{m}_t(s)\big) \, \mathrm{d}s .$$

Next, we take the vector product of (LLG) by $\mathbf{m}_t$ and finally multiply by $\mathbf{m}$.

$$\alpha \left\langle \mathbf{m} \times \Delta\mathbf{m}, \mathbf{m}_t \right\rangle_{\mathbb{R}^3} = \left\langle (\mathbf{m} \times \Delta\mathbf{m}) \times \mathbf{m}_t, \mathbf{m} \right\rangle_{\mathbb{R}^3} . \tag{4.50}$$

Let $\theta(s)$ stand for the angle between $\mathbf{m}_t(s)$ and $\mathbf{m}(s) \times \Delta\mathbf{m}(s)$. Then (4.50) can be written

$$|\, \mathbf{m} \times \Delta\mathbf{m} \,||\, \mathbf{m}_t \,||\cos\theta\,| \leq \alpha \,|\, \mathbf{m} \,||\, \mathbf{m} \times \Delta\mathbf{m} \,||\, \mathbf{m}_t \,||\sin\theta\,| ,$$

or, since $|\, \mathbf{m} \,| = 1$ almost everywhere,

$$|\cos\theta\,|^2 \leq \frac{\alpha^2}{1+\alpha^2} . \tag{4.51}$$

We may benefit from this result to control the last term in (4.49): multiplication of (LLG) with $\mathbf{m} \times \Delta\mathbf{m}$ leads to

$$\left\langle \mathbf{m} \times \Delta\mathbf{m}, \mathbf{m}_t \right\rangle_{\mathbb{R}^3} = |\, \mathbf{m} \times \Delta\mathbf{m} \,|^2 , \tag{4.52}$$

or equivalently,

$$|\, \mathbf{m}_t \,|\cos\theta = |\, \mathbf{m} \times \Delta\mathbf{m} \,| , \tag{4.53}$$

which helps to absorb the last term in (4.49) by means of the left-hand side. □

**Theorem 4.2** *(from [60]) There exists $T = T(\mathbf{m}_0) > 0$, such that the strong solution to (LLG) satisfies the following bound,*

$$\int_0^T \|\, \Delta\mathbf{m}(s) \,\|^2_{\mathbf{L}^2} \, \mathrm{d}s \leq C .$$

This result is the gateway to establish further results concerning the solution of (LLG), see Theorem 4.3.

In [120], the author derives a corresponding result for harmonic maps between compact Riemannian surfaces. The proof of Theorem 4.2 (for smooth solutions) is based on a program of auxiliary results that have to be verified, see [120]. For this purpose, we need the following notations: let $B_R(s) = \{\mathbf{y} \in \omega,\, |\, \mathbf{x} - \mathbf{y} \,| < R\}$, and

$$E(\varphi, \omega) = \int_\omega |\, \nabla\varphi \,|^2 \, \mathrm{d}\mathbf{x} \tag{4.54}$$

quantifies local energies created by $\varphi\big|_\omega$. In the sequel, we deal with the function space

$$\mathbf{V}(\omega_T;\mathcal{S}^2) := \Big\{\varphi:\omega_T\to\mathcal{S}^2 \,\big|\, \varphi \text{ measurable and } \operatorname{ess\,sup}_{[0,T]} \|\nabla\varphi\|_{\mathbf{L}^2} + \Big(\int_0^T \big\{\|\nabla^2\varphi(s)\|^2_{\mathbf{L}^2} + \|\mathbf{m}_t(s)\|^2_{\mathbf{L}^2}\big\}\,\mathrm{d}s\Big)^{1/2} < \infty\Big\}. \tag{4.55}$$

The following estimate is a localized version of the inequality $\|\cdot\|_{\mathbf{L}^4} \leq C\|\cdot\|^{1/2}_{\mathbf{L}^2}\|\cdot\|^{1/2}_{\mathbf{W}^{1,2}}$.

**Lemma 4.6** *(from [120]) There exist constants $C, R_0 > 0$ such that for any $\varphi \in \mathbf{V}(\omega_T;\mathcal{S}^2)$ and any $R \in (0, R_0]$ there holds the estimate*

$$\|\nabla\varphi(T)\|^4_{\mathbf{L}^4} \leq C \operatorname{ess}\sup_{(\mathbf{x},T)\in\omega_T} E\big(\varphi(T), B_R(\mathbf{x})\big)\Big(\|\nabla^2\varphi(T)\|^2_{\mathbf{L}^2} + \frac{1}{R^2}E\big(\varphi(0)\big)\Big).$$

**Proof:**
See [120], formula (3.6), and (3.5)'. □

The following result is a localized version of Lemma 4.5.

**Lemma 4.7** *(from [60]) Let $\mathbf{m}:\omega_T\to\mathcal{S}^2$ be a smooth solution to (LLG). Let $\mathbf{x}_0\in\omega$, and $B_R(\mathbf{x}_0)$ be a ball around $\mathbf{x}_0$, for $0 < R \leq 2R_0$. Then, for $R \leq R_0$ there holds*

$$E\big(\mathbf{m}(T); B_R(\mathbf{x}_0)\big) \leq E\big(\mathbf{m}_0; B_{2R}(\mathbf{x}_0)\big) + 2\big(1+\alpha^2\big)^2\frac{T}{R^2}E\big(\mathbf{m}_0\big),$$

*with $C$ independent of $R$.*

**Proof:**
Let $\phi \in C_0^\infty\big(B_{2R}(\mathbf{x}_0)\big)$ satisfy $0 \leq \phi \leq 1$, $\phi \equiv 1$ on $B_r(\mathbf{x}_0)$, with $|\nabla\phi| \leq \frac{2}{R}$, and test (LLG) by $\mathbf{m}_t\phi^2$ to obtain

$$\int_\omega |\mathbf{m}_t|^2\phi^2\,\mathrm{d}\mathbf{x} + \frac{\alpha}{2}d_t\Big(\int_\omega |\nabla\mathbf{m}|^2\phi^2\,\mathrm{d}\mathbf{x}\Big) \leq 2\alpha\int_\omega |\nabla\mathbf{m}||\mathbf{m}_t||\nabla\phi|\phi\,\mathrm{d}\mathbf{x} + (\mathbf{m}\times\Delta\mathbf{m}, \mathbf{m}_t\phi^2). \tag{4.56}$$

Recall from (4.51), (4.52), (4.53), in the proof of Lemma 4.5 that

$$\big|\langle \mathbf{m} \times \Delta\mathbf{m}, \mathbf{m}_t\rangle_{\mathbb{R}^3}\big| \leq |\cos\theta\,|^2\,|\,\mathbf{m}_t\,|^2 \leq \frac{\alpha^2}{1+\alpha^2}|\,\mathbf{m}_t\,|^2 .$$

Hence we find from (4.56),

$$\int_\omega |\,\mathbf{m}_t\,|^2\phi^2\,\mathrm{d}\mathbf{x} + \alpha\, d_t\Big(\int_\omega |\,\nabla\mathbf{m}\,|^2\phi^2\,\mathrm{d}\mathbf{x}\Big) \leq \big(1+\alpha^2\big)^2\int_\omega |\,\nabla\mathbf{m}\,|^2|\,\nabla\phi\,|^2\,\mathrm{d}\mathbf{x}\,,$$

and Lemma 4.5 completes the proof. □

From this lemma, we may draw the following conclusion: given $\varepsilon_1 > 0$, $\mathbf{m}_0 \in \mathbf{W}^{1,2}(\omega, \mathcal{S}^2)$, there exists a number $T_1 > 0$ depending only on a maximal number $R_1 > 0$ and $\alpha > 0$, such that

$$\sup_{\mathbf{x}_0\in\omega} E\big(\mathbf{m}_0; B_{2R_1}(\mathbf{x}_0)\big) < \varepsilon_1\,,$$

with the property that any smooth solution $\mathbf{m}$ to (LLG) satisfies

$$\sup_{\mathbf{x}_0\in\omega,\,0\leq t\leq T_1} E\big(\mathbf{m}(t); B_{R_1}(\mathbf{x}_0)\big) < 2\varepsilon_1\,.$$

We choose $T_1 = \frac{\varepsilon_1 R_1^2}{2E(\mathbf{m}_0)(1+\alpha^2)^2}$ to meet this scenario. In the sequel, we benefit from Lemma 4.6: Let $\varepsilon_1 > 0$ be given and $R_1 > 0$ be determined as above. Let $\{\phi_i\}$ be a set of smooth cut-off functions subordinate to a cover of $\omega$ by balls $\{B_{2R_1}(\mathbf{x}_i)\}$ with finite overlap and such that $0 \leq \phi_i \leq 1$, where $|\,\nabla\phi\,| \leq \frac{2}{R_1}$, and $\sum_i \phi_i^2 = 1$. Then, we infer

$$\begin{aligned} \|\,\nabla\mathbf{m}(t)\,\|_{\mathbf{L}^4}^4 &= \sum_i \int_\omega |\,\nabla\mathbf{m}(t)\,|^4\phi_i^2\,\mathrm{d}\mathbf{x} \qquad (4.57)\\ &\leq C\sup_i E\big(\mathbf{m}(t); B_{2R_1}(\mathbf{x}_i)\big)\Big(\int_\omega |\,\nabla^2\mathbf{m}(t)\,|^2\,\mathrm{d}\mathbf{x} + \frac{1}{R_1^2}E(\mathbf{m}_0)\Big)\\ &\leq C\varepsilon_1\Big(\int_\omega |\,\nabla^2\mathbf{m}(t)\,|^2\,\mathrm{d}\mathbf{x} + \frac{1}{R_1^2}E(\mathbf{m}_0)\Big)\,. \end{aligned}$$

This result is the crucial one to control the term on the right-hand side when multiplying (LLG) with $-\Delta\mathbf{m}$,

$$\frac{1}{2}d_t\|\,\nabla\mathbf{m}\,\|_{\mathbf{L}^2}^2 + \alpha\,\|\,\Delta\mathbf{m}\,\|_{\mathbf{L}^2}^2 = \alpha\,\|\,\nabla\mathbf{m}\,\|_{\mathbf{L}^4}^4\,, \qquad (4.58)$$

where we used $|\mathbf{m}| = 1$, almost everywhere in $\omega$. Time integration then gives

$$\begin{aligned}\alpha\int_0^T \|\Delta\mathbf{m}(s)\|_{\mathbf{L}^2}^2\,\mathrm{d}s &\leq \|\nabla\mathbf{m}_0\|_{\mathbf{L}^2}^2 + \alpha\int_0^T \|\nabla\mathbf{m}(s)\|_{\mathbf{L}^4}^4\,\mathrm{d}s \qquad (4.59)\\ &\leq C\alpha\varepsilon_1\int_0^T \|\Delta\mathbf{m}\|_{\mathbf{L}^2}^2\,\mathrm{d}s + \left(1+\frac{T}{R_1^2}\right)E(\mathbf{m}_0)\,.\end{aligned}$$

Adjusting $\varepsilon_1 > 0$ sufficiently small finally gives the a priori bounds from Theorem 4.2.

We may sum up the results from Section 4.1.1 and Section 4.1.3 as follows: solutions to (LLG) do not blow up in time for small initial energies, but can do so in (in-)finite time for finite initial energies. In the next section, we study numerical schemes for solutions of certain regularity that are valid in both scenarios. The following theorem provides further results that apply to both considered cases.

**Theorem 4.3** *For $0 < T < T_0(\mathbf{m}_0)$, let $\mathbf{m} \in \mathbf{V}(\omega_T; \mathcal{S}^2)$ solve (LLG), and $\mathbf{m}_0 \in \mathbf{W}^{2,2}(\omega, \mathcal{S}^2)$. There exists a constant $C > 0$ such that holds for $\tau(s) := \min\{1, s\}$,*

(a) $$\operatorname*{ess\,sup}_{[0,T]}\Big\{\|\mathbf{m}_t\|_{\mathbf{L}^2} + \|\Delta\mathbf{m}\|_{\mathbf{L}^2}\Big\} + \Big(\int_0^T \|\nabla\mathbf{m}_t\|^2\,\mathrm{d}s\Big)^{\frac12} \leq C\,,$$

(b) $$\operatorname*{ess\,sup}_{[0,T]}\Big\{\sqrt{\tau}\,\|\nabla\mathbf{m}_t\|_{\mathbf{L}^2}\Big\} + \Big(\int_0^T \tau\{\|\mathbf{m}_{tt}\|_{\mathbf{L}^2}^2 + \|\Delta\mathbf{m}_t\|_{\mathbf{L}^2}^2\}\,\mathrm{d}s\Big)^{\frac12} \leq C\,,$$

(c) $$\operatorname*{ess\,sup}_{[0,T]}\tau\Big\{\|\mathbf{m}_{tt}\|_{\mathbf{L}^2} + \|\Delta\mathbf{m}_t\|_{\mathbf{L}^2}\Big\} + \Big(\int_0^T \tau^2\|\nabla\mathbf{m}_{tt}\|^2\,\mathrm{d}s\Big)^{\frac12} \leq C\,,$$

(d) $$\operatorname*{ess\,sup}_{[0,T]}\tau^{3/2}\|\nabla\mathbf{m}_{tt}\|_{\mathbf{L}^2} + \Big(\int_0^T \tau^3\{\|\mathbf{m}_{ttt}\|_{\mathbf{L}^2}^2 + \|\Delta\mathbf{m}_{tt}\|_{\mathbf{L}^2}^2\}\,\mathrm{d}s\Big)^{\frac12} \leq C\,,$$

(e) $$\int_0^T \{\|\mathbf{m}_{tt}\|_{\mathbf{W}^{-1,2}(\omega)}^2 + \tau^2\|\mathbf{m}_{ttt}\|_{\mathbf{W}^{-1,2}(\omega)}^2\}\,\mathrm{d}s \leq C\,.$$

**Remark 4.1** *1. There exist $c_1, c_2 > 0$, such that for every $\boldsymbol{\varphi} \in \mathbf{W}^{-1,2}(\omega, \mathbb{R}^3)$,*

$$c_1\|\boldsymbol{\varphi}\|_{\mathbf{W}^{-1,2}(\omega,\mathbb{R}^3)}^2 \leq (\tilde{\Delta}^{-1}\boldsymbol{\varphi}, \boldsymbol{\varphi}) \leq c_2\|\boldsymbol{\varphi}\|_{\mathbf{W}^{-1,2}(\omega,\mathbb{R}^3)}^2. \qquad (4.60)$$

*To see this, we recall that* $\tilde{\Delta}^{-1} \equiv (\mathrm{Id} - \Delta)^{-1} : \mathbf{W}^{-1,2}(\omega, \mathbb{R}^3) \to \mathbf{W}^{1,2}(\omega, \mathbb{R}^3)$ *is an isomorphism:*

$$\begin{aligned} (\tilde{\Delta}^{-1}\varphi, \varphi) &= \| \tilde{\Delta}^{-1}\varphi \|_{\mathbf{L}^2}^2 + \| \nabla\tilde{\Delta}^{-1}\varphi \|_{\mathbf{L}^2}^2 \\ &= \| \tilde{\Delta}^{-1}\varphi \|_{\mathbf{W}^{1,2}}^2 \leq c_2 \| \varphi \|_{\mathbf{W}^{-1,2}(\omega,\mathbb{R}^3)} . \end{aligned} \tag{4.61}$$

*The first inequality can be shown in a similar way,*

$$\begin{aligned} \| \varphi \|_{\mathbf{W}^{-1,2}(\omega,\mathbb{R}^3)}^2 &= \Big[ \sup_{\xi \in \mathbf{W}^{1,2}} \frac{(\xi, \varphi)}{\| \varphi \|_{\mathbf{W}^{1,2}}} \Big]^2 = \Big[ \sup_{\xi \in \mathbf{W}^{1,2}} \frac{(\xi, \tilde{\Delta}^{-1}\varphi) + (\nabla\xi, \nabla\tilde{\Delta}^{-1}\varphi)}{\| \varphi \|_{\mathbf{W}^{1,2}}} \Big]^2 \\ &\leq \Big[ \sup_{\xi \in \mathbf{W}^{1,2}} \frac{\max\{\| \xi \|_{\mathbf{L}^2}, \| \nabla\xi \|_{\mathbf{L}^2}\}\{\| \tilde{\Delta}^{-1}\varphi \| + \| \nabla\tilde{\Delta}^{-1}\varphi \|\}}{\| \varphi \|_{\mathbf{W}^{1,2}}} \Big]^2 \\ &\leq \| \tilde{\Delta}^{-1}\varphi \|_{\mathbf{W}^{1,2}}^2 . \end{aligned}$$

*Thanks to* (4.61), *this verifies* (4.60).

2. *Stability bound* (e) *will be used when first order time discretization schemes are constructed in Section 4.2. The crucial point here is that these quantities do not carry time weights: as it will turn out in the analysis of the numerical schemes, penalization methods are lacking certain time damping properties that are well-known for linear dissipative systems. This forces us to consider weak spatial norms.*

3. *The arguments given in the proof are formal but can be made rigorous by Faedo-Galerkin approximations.*

**Proof:**
(a) We start from

$$\begin{aligned} \mathbf{m}_{tt} - \alpha\,\Delta\mathbf{m}_t &= 2\,\langle \nabla\mathbf{m}, \nabla\mathbf{m}_t \rangle_{\mathbb{R}^6}\mathbf{m} \\ &\quad +\alpha\,|\,\nabla\mathbf{m}\,|^2\mathbf{m}_t + \mathbf{m}_t \times \Delta\mathbf{m} + \mathbf{m} \times \Delta\mathbf{m}_t . \end{aligned} \tag{4.62}$$

We test with the admissible function $\mathbf{m}_t$ to find

$$\frac{1}{2} d_t \| \mathbf{m}_t \|_{\mathbf{L}^2}^2 + \alpha \| \nabla\mathbf{m}_t \|_{\mathbf{L}^2}^2 \leq C\big(\alpha\{I + II\} + III\big) , \tag{4.63}$$

and deal with the arising terms independently, using the Sobolev inequality

(4.42) repeatedly,

$$
\begin{aligned}
I := 2\alpha \,|\, \big( \langle \nabla \mathbf{m}, \nabla \mathbf{m}_t \rangle_{\mathbb{R}^6} \mathbf{m}, \mathbf{m}_t \big) \,| &\leq C \,\|\, \mathbf{m} \,\|_{\mathbf{W}^{1,4}} \,\|\, \nabla \mathbf{m}_t \,\|_{\mathbf{L}^2} \,\|\, \mathbf{m}_t \,\|_{\mathbf{L}^4} \\
&\leq C_\alpha \,\|\, \mathbf{m} \,\|^4_{\mathbf{W}^{1,4}} \|\, \mathbf{m}_t \,\|^2_{\mathbf{L}^2} + \frac{\alpha}{4} \,\|\, \nabla \mathbf{m}_t \,\|^2_{\mathbf{L}^2} \,, \\
II := \big( |\, \nabla \mathbf{m} \,|^2, |\, \mathbf{m}_t \,|^2 \big) &\leq C_\alpha \,\|\, \mathbf{m} \,\|^4_{\mathbf{W}^{1,4}} \,\|\, \mathbf{m}_t \,\|^2_{\mathbf{L}^2} + \frac{\alpha}{4} \,\|\, \nabla \mathbf{m}_t \,\|^2_{\mathbf{L}^2} \,, \\
III := |\, \big( \mathbf{m} \times \Delta \mathbf{m}_t, \mathbf{m}_t \big) \,| &= |\, \big( \nabla (\mathbf{m} \times \mathbf{m}_t), \nabla \mathbf{m}_t \big) \,| \\
&\leq C_\alpha \,\|\, \mathbf{m} \,\|^4_{\mathbf{W}^{1,4}} \|\, \mathbf{m}_t \,\|^2_{\mathbf{L}^2} + \frac{\alpha}{4} \,\|\, \nabla \mathbf{m}_t \,\|^2_{\mathbf{L}^2} \,.
\end{aligned}
$$

Then, Gronwall's inequality settles the first result.

This result helps to show the bound $\|\, \mathbf{m} \,\|_{L^\infty(I;\mathbf{W}^{2,2})} \leq C$, using Lemma 4.6.

$$
\alpha \,\|\, \Delta \mathbf{m} \,\|^2_{\mathbf{L}^2} \leq \alpha \,\|\, \nabla \mathbf{m} \,\|^4_{\mathbf{L}^4} + \frac{1}{\alpha} \,\|\, \mathbf{m}_t \,\|^2_{\mathbf{L}^2} \,.
$$

In order to show (b), we multiply (4.62) by $-\Delta \mathbf{m}_t$,

$$
\frac{1}{2} d_t \|\, \nabla \mathbf{m}_t \,\|^2_{\mathbf{L}^2} + \alpha \,\|\, \Delta \mathbf{m}_t \,\|^2_{\mathbf{L}^2} \leq \Big( \alpha \{ I + II \} + III + IV \Big) \,, \tag{4.64}
$$

and, by means of the inequalities $\|\cdot\|_{\mathbf{L}^4} \leq C \,\|\cdot\|^{1/2}_{\mathbf{L}^2} \,\|\cdot\|^{1/2}_{\mathbf{W}^{1,2}}$, (4.31), and $\|\cdot\|_{\mathbf{L}^\infty} \leq C \,\|\cdot\|_{\mathbf{W}^{1,4}}$,

$$
\begin{aligned}
I := |\, \big( \langle \nabla \mathbf{m}, \nabla \mathbf{m}_t \rangle_{\mathbb{R}^6} \mathbf{m}, -\Delta \mathbf{m}_t \big) \,| &\leq \|\, \nabla \mathbf{m} \,\|_{\mathbf{L}^4} \,\|\, \nabla \mathbf{m}_t \,\|_{\mathbf{L}^4} \,\|\, \mathbf{m} \,\|_{\mathbf{L}^\infty} \,\|\, \Delta \mathbf{m}_t \,\|_{\mathbf{L}^2} \\
&\leq C_\alpha \,\|\, \nabla \mathbf{m} \,\|^4_{\mathbf{L}^4} \,\|\, \nabla \mathbf{m}_t \,\|^2_{\mathbf{L}^2} + \frac{\alpha}{4} \,\|\, \Delta \mathbf{m}_t \,\|^2_{\mathbf{L}^2} \,, \\
II := |\, \big( |\, \nabla \mathbf{m} \,|^2 \mathbf{m}_t, -\Delta \mathbf{m}_t \big) \,| &\leq C_\alpha \,\|\, \nabla \mathbf{m} \,\|^4_{\mathbf{L}^8} \,\|\, \mathbf{m}_t \,\|^2_{\mathbf{L}^4} + \frac{\alpha}{4} \,\|\, \Delta \mathbf{m}_t \,\|^2_{\mathbf{L}^2} \,, \\
III := |\, \big( \mathbf{m}_t \times \Delta \mathbf{m}, -\Delta \mathbf{m}_t \big) \,| &\leq \|\, \mathbf{m}_t \,\|_{\mathbf{L}^\infty} \,\|\, \Delta \mathbf{m} \,\|_{\mathbf{L}^2} \,\|\, \Delta \mathbf{m}_t \,\|_{\mathbf{L}^2} \,, \\
&\leq C_\alpha \,\|\, \Delta \mathbf{m} \,\|^8_{\mathbf{L}^2} \,\|\, \mathbf{m}_t \,\|^2_{\mathbf{L}^2} + \frac{\alpha}{4} \,\|\, \mathbf{m}_t \,\|^2_{\mathbf{W}^{2,2}} \,, \\
IV := |\, \big( \mathbf{m} \times \Delta \mathbf{m}_t, \Delta \mathbf{m}_t \big) \,| &= 0 \,.
\end{aligned}
$$

Multiplication of (4.64) by the time weight and (a), together with Gronwall's inequality show the first part of (b).

We may benefit from this result to show the remaining part of (b): multiplication of (4.62) by $\mathbf{m}_{tt}$ leads to

$$
\|\, \mathbf{m}_{tt} \,\|^2_{\mathbf{L}^2} + \frac{\alpha}{2} d_t \|\, \nabla \mathbf{m}_t \,\|^2_{\mathbf{L}^2} \leq \Big( \alpha \{ I + II \} + III + IV \Big) \,. \tag{4.65}
$$

$$
\begin{aligned}
I := 2\alpha \,|\, \big(\langle \nabla \mathbf{m}, \nabla \mathbf{m}_t\rangle_{\mathbb{R}^6}\mathbf{m}, \mathbf{m}_{tt}\big) \,| &\leq \| \nabla \mathbf{m} \|_{\mathbf{L}^4} \| \nabla \mathbf{m}_t \|_{\mathbf{L}^4} \| \mathbf{m} \|_{\mathbf{L}^\infty} \| \mathbf{m}_{tt} \|_{\mathbf{L}^2} \\
&\leq C \| \nabla \mathbf{m}_t \|_{\mathbf{L}^2} \| \nabla \mathbf{m}_t \|_{\mathbf{W}^{1,2}} + \frac{1}{8} \| \mathbf{m}_{tt} \|_{\mathbf{L}^2}^2 , \\
II := \alpha \,|\, \big( | \nabla \mathbf{m} |^2 \mathbf{m}_t, \mathbf{m}_{tt}\big) \,| &\leq C \| \mathbf{m}_t \|_{\mathbf{L}^2}^2 \| \nabla \mathbf{m} \|_{\mathbf{L}^8}^8 + \| \nabla \mathbf{m}_t \|_{\mathbf{L}^2}^2 \\
&\quad + \frac{1}{8} \| \mathbf{m}_{tt} \|_{\mathbf{L}^2}^2 , \\
III := |\, \big(\mathbf{m}_t \times \Delta \mathbf{m}, \mathbf{m}_{tt}\big) \,| &\leq \| \mathbf{m}_t \|_{\mathbf{L}^\infty} \| \Delta \mathbf{m} \|_{\mathbf{L}^2} \| \mathbf{m}_{tt} \|_{\mathbf{L}^2} \\
&\leq C \| \mathbf{m}_t \|_{\mathbf{W}^{1,4}} \| \Delta \mathbf{m} \|_{\mathbf{L}^2} \| \mathbf{m}_{tt} \|_{\mathbf{L}^2} \\
&\leq C \| \mathbf{m}_t \|_{\mathbf{W}^{1,2}} \| \mathbf{m}_t \|_{\mathbf{W}^{2,2}} \| \Delta \mathbf{m} \|_{\mathbf{L}^2}^2 \\
&\quad + \frac{1}{8} \| \mathbf{m}_{tt} \|_{\mathbf{L}^2}^2 , \\
IV := |\, \big(\mathbf{m} \times \Delta \mathbf{m}_t, \mathbf{m}_{tt}\big) \,| &\leq \| \mathbf{m} \|_{\mathbf{L}^\infty} \| \Delta \mathbf{m}_t \|_{\mathbf{L}^2} \| \mathbf{m}_{tt} \|_{\mathbf{L}^2} \\
&\leq C \| \Delta \mathbf{m}_t \|_{\mathbf{L}^2}^2 + \frac{1}{8} \| \mathbf{m}_{tt} \|_{\mathbf{L}^2}^2 .
\end{aligned}
$$

Finally, we multiply (4.65) by $\tau(s)$ and integrate with respect to time, employ a result from the previous step and finally make use of Gronwall's inequality. This shows (b).

(c) Formally taking the time derivative two times of (LLG), we find

$$
\begin{aligned}
\mathbf{m}_{ttt} - \alpha \Delta \mathbf{m}_{tt} &= 2\alpha \,|\, \nabla \mathbf{m}_t \,|^2 \mathbf{m} + 2\alpha \langle \nabla \mathbf{m}, \nabla \mathbf{m}_{tt} \rangle_{\mathbb{R}^6} \mathbf{m} \\
&\quad + 4\alpha \langle \nabla \mathbf{m}, \nabla \mathbf{m}_t \rangle_{\mathbb{R}^6} \mathbf{m}_t + \alpha \,|\, \nabla \mathbf{m} \,|^2 \mathbf{m}_{tt} \\
&\quad + 2\, \mathbf{m}_t \times \Delta \mathbf{m}_t + \mathbf{m}_{tt} \times \Delta \mathbf{m} + \mathbf{m} \times \Delta \mathbf{m}_{tt} .
\end{aligned} \tag{4.66}
$$

We start by testing (4.66) with $\tau^2 \mathbf{m}_{tt}$ .

$$
\begin{aligned}
&\frac{1}{2} d_t \big(\tau^2 \| \mathbf{m}_{tt} \|_{\mathbf{L}^2}^2\big) + \alpha \tau^2 \| \nabla \mathbf{m}_{tt} \|_{\mathbf{L}^2}^2 \\
&\qquad = \tau \| \mathbf{m}_{tt} \|_{\mathbf{L}^2}^2 + \alpha \Big(I + .. + IV\Big) + \Big(V + .. + VII\Big) .
\end{aligned} \tag{4.67}
$$

Let $\beta > 0$ be fixed later.

$$
\begin{aligned}
I := 2\tau^2 \,|\, \big(|\, \nabla \mathbf{m}_t \,|^2 \mathbf{m}, \mathbf{m}_{tt}\big) \,| &\leq \tau^2 \,\|\, |\, \nabla \mathbf{m}_t \,|^2 \,\|_{\mathbf{L}^2} \,\|\, \mathbf{m} \,\|_{\mathbf{L}^\infty} \,\|\, \mathbf{m}_{tt} \,\|_{\mathbf{L}^2} \\
&\leq C_\beta \tau^2 \,\|\, \nabla \mathbf{m}_t \,\|_{\mathbf{L}^4}^4 + \beta\tau^2 \,\|\, \mathbf{m}_{tt} \,\|_{\mathbf{L}^2}^2 \\
&\leq C_\beta \big[\tau \|\, \nabla \mathbf{m}_t \,\|_{\mathbf{L}^2}^2\big]\, \tau \,\|\, \mathbf{m}_t \,\|_{\mathbf{W}^{1,2}}^2 + \beta\tau^2 \,\|\, \mathbf{m}_{tt} \,\|_{\mathbf{L}^2}^2 \,, \\
IV := \tau^2 \,|\, \big(|\, \nabla \mathbf{m} \,|^2, |\, \mathbf{m}_{tt} \,|^2\big) \,| &\leq \tau^2 \,\|\, \nabla \mathbf{m} \,\|_{\mathbf{L}^4}^2 \,\|\, \mathbf{m}_{tt} \,\|_{\mathbf{L}^4}^2 \,\|\, \nabla \mathbf{m} \,\|_{\mathbf{L}^4}^4 \\
&\leq C_\beta \tau^2 \,\|\, \mathbf{m}_{tt} \,\|_{\mathbf{L}^2}^2 + \beta\tau^2 \,\|\, \mathbf{m}_{tt} \,\|_{\mathbf{W}^{1,2}}^2 \,, \\
V := \tau^2 \,|\, \big(\mathbf{m}_t \times \Delta \mathbf{m}_t, \mathbf{m}_{tt}\big) \,| &\leq \tau^2 \,\|\, \mathbf{m}_t \,\|_{\mathbf{L}^4} \,\|\, \Delta \mathbf{m}_t \,\|_{\mathbf{L}^2} \,\|\, \mathbf{m}_{tt} \,\|_{\mathbf{L}^4} \\
&\leq C_\beta \big[\tau \,\|\, \mathbf{m}_t \,\|_{\mathbf{L}^4}^2\big] \big[\tau \|\, \Delta \mathbf{m}_t \,\|_{\mathbf{L}^2}^2\big] + \beta\tau^2 \,\|\, \mathbf{m}_{tt} \,\|_{\mathbf{L}^4}^2 \,, \\
VI := \tau^2 \,|\, \big(\mathbf{m}_{tt} \times \Delta \mathbf{m}, \mathbf{m}_{tt}\big) \,| &= 0 \,, \\
VII := \tau^2 \,|\, \big(\mathbf{m}_{tt} \times \Delta \mathbf{m}_{tt}, \mathbf{m}\big) \,| &= \tau^2 \,|\, \big(\operatorname{div}[\mathbf{m}_{tt} \times \nabla \mathbf{m}_{tt}], \mathbf{m}\big) \,| \\
&= \tau^2 \,|\, \big(\mathbf{m}_{tt} \times \nabla \mathbf{m}_{tt}, \nabla \mathbf{m}\big) \,| \\
&\leq \tau^2 \,\|\, \mathbf{m}_{tt} \,\|_{\mathbf{L}^4} \,\|\, \nabla \mathbf{m}_{tt} \,\|_{\mathbf{L}^2} \,\|\, \nabla \mathbf{m} \,\|_{\mathbf{L}^4} \\
&\leq C_\beta \tau^2 \,\|\, \mathbf{m}_{tt} \,\|_{\mathbf{L}^2}^2 \,\|\, \nabla \mathbf{m} \,\|_{\mathbf{L}^4}^4 + \beta\tau^2 \,\|\, \mathbf{m}_{tt} \,\|_{\mathbf{W}^{1,2}}^2 \,,
\end{aligned}
$$

and

$$
\begin{aligned}
II &:= 2\tau^2 \,|\, \big(\langle \nabla \mathbf{m}, \nabla \mathbf{m}_{tt} \rangle_{\mathbb{R}^6}, \langle \mathbf{m}, \mathbf{m}_{tt} \rangle_{\mathbb{R}^3}\big) \,| \\
&\leq 2\tau^2 \,\|\, \nabla \mathbf{m} \,\|_{\mathbf{L}^4} \,\|\, \nabla \mathbf{m}_{tt} \,\|_{\mathbf{L}^2} \,\|\, \mathbf{m}_{tt} \,\|_{\mathbf{L}^4} \,\|\, \mathbf{m} \,\|_{\mathbf{L}^\infty} \\
&\leq C_\beta \tau^2 \,\|\, \nabla \mathbf{m} \,\|_{\mathbf{L}^4}^4 + \beta\tau^2 \,\|\, \mathbf{m}_{tt} \,\|_{\mathbf{W}^{1,2}}^2 \,, \\
III &:= 4\tau^2 \,|\, \big(\langle \nabla \mathbf{m}, \nabla \mathbf{m}_t \rangle_{\mathbb{R}^6}, \langle \mathbf{m}_t, \mathbf{m}_{tt} \rangle_{\mathbb{R}^3}\big) \,| \\
&\leq 4\tau^2 \,\|\, \nabla \mathbf{m} \,\|_{\mathbf{L}^8} \,\|\, \nabla \mathbf{m}_t \,\|_{\mathbf{L}^2} \,\|\, \mathbf{m}_t \,\|_{\mathbf{L}^8} \,\|\, \mathbf{m}_{tt} \,\|_{\mathbf{L}^4} \\
&\leq C_\beta \tau^4 \big[\tau \,\|\, \nabla \mathbf{m}_t \,\|_{\mathbf{L}^2}^2\big]^2 \big[\tau^2 \,\|\, \mathbf{m}_{tt} \,\|_{\mathbf{L}^2}^2\big] + \beta\tau^2 \,\|\, \mathbf{m}_{tt} \,\|_{\mathbf{W}^{1,2}}^2 \,.
\end{aligned}
$$

In order to deal with $VII$, we again used the notation $\mathbf{m} \times \nabla \mathbf{m} := (\mathbf{m} \times \nabla_1 \mathbf{m}, \mathbf{m} \times \nabla_2 \mathbf{m})^\top$.

Now, we integrate in (4.67) and use the results from (a), (b) for sufficiently small $\beta > 0$ to verify one part of (c). The second statement in (c) is now shown from (4.62), which is multiplied by $-\tau^2 \Delta \mathbf{m}_t$.

$$
\begin{aligned}
\tau^2 \|\, \Delta \mathbf{m}_t \,\|_{\mathbf{L}^2}^2 \leq C\tau^2 \Big\{ &\|\, \mathbf{m}_t \,\|_{\mathbf{L}^2}^2 + \|\, \nabla \mathbf{m} \,\|_{\mathbf{L}^4}^2 \,\|\, \nabla \mathbf{m}_t \,\|_{\mathbf{L}^4}^2 \,\|\, \mathbf{m} \,\|_{\mathbf{L}^\infty}^2 \\
&+ \|\, |\, \nabla \mathbf{m} \,|^2 \,\|_{\mathbf{L}^4}^2 \,\|\, \mathbf{m}_t \,\|_{\mathbf{L}^4}^2 + \alpha \,\|\, \mathbf{m}_t \,\|_{\mathbf{L}^\infty}^2 \,\|\, \Delta \mathbf{m} \,\|_{\mathbf{L}^2}^2 \Big\} + \frac{1}{2}\tau^2 \,\|\, \Delta \mathbf{m}_t \,\|_{\mathbf{L}^2}^2 \,.
\end{aligned}
\tag{4.68}
$$

We use the inequality $\|\, \mathbf{m}_t \,\|_{\mathbf{L}^\infty} \leq C \,\|\, \mathbf{m}_t \,\|_{\mathbf{L}^2}^{1/2} \,\|\, \mathbf{m}_t \,\|_{\mathbf{W}^{2,2}}^{1/2}$. Finally the first statement in (b) helps, together with those results already proved in (c) to derive a uniform bound for $\tau \mathbf{m}_t \in L^\infty(I; \mathbf{W}^{2,2})$.

To verify (d), we multiply (4.66) by $-\tau^3\Delta\mathbf{m}_{tt}$.

$$\frac{1}{2}d_t\big(\tau^3\|\nabla\mathbf{m}_{tt}\|^2_{\mathbf{L}^2}\big)+\alpha\tau^3\|\Delta\mathbf{m}_{tt}\|^2_{\mathbf{L}^2} = \frac{3}{2}\tau^2\|\nabla\mathbf{m}_{tt}\|^2_{\mathbf{L}^2}+\alpha\big(I+..+IV\big)+V+..+VII\,.$$

In the sequel, suppose $\beta>0$ to be chosen sufficiently small.

$$\begin{aligned}
I := 2\tau^3\,|\,(|\nabla\mathbf{m}_t|^2\mathbf{m},\Delta\mathbf{m}_{tt})\,| &\leq 2\tau^3\,\|\,|\nabla\mathbf{m}_t|^2\,\|_{\mathbf{L}^2}\,\|\,\mathbf{m}\,\|_{\mathbf{L}^\infty}\,\|\,\Delta\mathbf{m}_{tt}\,\|_{\mathbf{L}^2}\\
&\leq C_\beta\tau^3\,\|\,\nabla\mathbf{m}_t\,\|^4_{\mathbf{L}^4}+\beta\tau^3\,\|\,\Delta\mathbf{m}_{tt}\,\|^2_{\mathbf{L}^2}\\
&\leq C_\beta\big[\tau\|\,\nabla\mathbf{m}_t\,\|^2_{\mathbf{L}^2}\big]\,\tau\,\|\,\mathbf{m}_t\,\|^2_{\mathbf{W}^{2,2}}\\
&\quad+\beta\tau^3\,\|\,\Delta\mathbf{m}_{tt}\,\|^2_{\mathbf{L}^2}\,,\\
&\leq C_\beta\,\big[\tau^{3/2}\,\|\,\nabla\mathbf{m}_t\,\|^3_{\mathbf{L}^2}\big]\,\big[\sqrt{\tau}\,\|\,\Delta\mathbf{m}_t\,\|^2_{\mathbf{L}^2}\big]\\
&\quad+\beta\tau^3\,\|\,\Delta\mathbf{m}_{tt}\,\|^2_{\mathbf{L}^2}\,,\\
V := \tau^3\,|\,(\mathbf{m}_t\times\Delta\mathbf{m}_t,\Delta\mathbf{m}_{tt})\,| &\leq \tau^3\,\|\,\mathbf{m}_t\,\|_{\mathbf{L}^\infty}\,\|\,\Delta\mathbf{m}_t\,\|_{\mathbf{L}^2}\,\|\,\Delta\mathbf{m}_{tt}\,\|_{\mathbf{L}^2}\\
&\leq C_\beta\,\big[\tau^{3/2}\|\,\nabla\mathbf{m}_t\,\|^2_{\mathbf{L}^4}\big]\,\big[\tau\|\,\Delta\mathbf{m}_t\,\|^2_{\mathbf{L}^2}\big]\\
&\quad+\beta\tau^3\,\|\,\Delta\mathbf{m}_{tt}\,\|^2_{\mathbf{L}^2}\,,\\
VI := \tau^3\,|\,(\mathbf{m}_{tt}\times\Delta\mathbf{m},\Delta\mathbf{m}_{tt})\,| &\leq C\tau^3\,\|\,\mathbf{m}_{tt}\,\|_{\mathbf{L}^2}\,\|\,\Delta\mathbf{m}\,\|_{\mathbf{L}^2}\,\|\,\Delta\mathbf{m}_{tt}\,\|_{\mathbf{L}^2}\\
&\leq C\tau^3\,\|\,\mathbf{m}_{tt}\,\|^2_{\mathbf{L}^2}\,\|\,\Delta\mathbf{m}\,\|^2_{\mathbf{L}^2}\\
&\quad+\beta\tau^3\,\|\,\Delta\mathbf{m}_{tt}\,\|^2_{\mathbf{L}^2}\,,\\
VII := \tau^3\,|\,(\mathbf{m}\times\Delta\mathbf{m}_{tt},\Delta\mathbf{m}_{tt})\,| &= 0\,,
\end{aligned}\tag{4.69}$$

and

$$\begin{aligned}
II &:= 2\tau^3\,|\,\big(\langle\nabla\mathbf{m},\nabla\mathbf{m}_{tt}\rangle_{\mathbb{R}^6},\langle\mathbf{m},\Delta\mathbf{m}_{tt}\rangle_{\mathbb{R}^3}\big)\,|\\
&\leq 2\tau^3\,\|\,\nabla\mathbf{m}\,\|_{\mathbf{L}^4}\,\|\,\nabla\mathbf{m}_{tt}\,\|_{\mathbf{L}^4}\,\|\,\mathbf{m}\,\|_{\mathbf{L}^\infty}\,\|\,\Delta\mathbf{m}_{tt}\,\|_{\mathbf{L}^2}\\
&\leq C_\beta\tau^3\,\|\,\nabla\mathbf{m}_{tt}\,\|^2_{\mathbf{L}^2}\,\|\,\nabla\mathbf{m}\,\|^4_{\mathbf{L}^4}+\beta\tau^3\,\|\,\Delta\mathbf{m}_{tt}\,\|^2_{\mathbf{L}^2}\,,\\
III &:= 4\tau^3\,|\,\big(\langle\nabla\mathbf{m},\nabla\mathbf{m}_t\rangle_{\mathbb{R}^6},\langle\mathbf{m}_t,\Delta\mathbf{m}_{tt}\rangle_{\mathbb{R}^3}\big)\,|\\
&\leq 4\tau^3\,\|\,\nabla\mathbf{m}\,\|_{\mathbf{L}^8}\,\|\,\nabla\mathbf{m}_t\,\|_{\mathbf{L}^4}\,\|\,\mathbf{m}_t\,\|_{\mathbf{L}^8}\,\|\,\mathbf{m}_{tt}\,\|_{\mathbf{W}^{2,2}}\\
&\leq C_\beta\,\big[\tau^{3/2}\,\|\,\nabla\mathbf{m}_t\,\|^3_{\mathbf{L}^2}\big]\,\big[\sqrt{\tau}\,\|\,\Delta\mathbf{m}_t\,\|^2_{\mathbf{L}^2}\big]+\beta\tau^3\,\|\,\Delta\mathbf{m}_{tt}\,\|^2_{\mathbf{L}^2}\,,\\
IV &:= \tau^3\,|\,\big(|\nabla\mathbf{m}|^2,\langle\mathbf{m}_{tt},\Delta\mathbf{m}_{tt}\rangle_{\mathbb{R}^3}\big)\,|\\
&\leq \tau^3\,\|\,\nabla\mathbf{m}\,\|^2_{\mathbf{L}^8}\,\|\,\mathbf{m}_{tt}\,\|_{\mathbf{L}^4}\,\|\,\mathbf{m}_{tt}\,\|_{\mathbf{W}^{2,2}}\\
&\leq C_\beta\big[\tau^2\,\|\,\mathbf{m}_{tt}\,\|^2_{\mathbf{L}^2}\big]\,\|\,\nabla\mathbf{m}\,\|^8_{\mathbf{L}^8}+\beta\tau^3\,\|\,\Delta\mathbf{m}_{tt}\,\|^2_{\mathbf{L}^2}\,,
\end{aligned}$$

Gronwall's lemma then helps to prove the first part of (d). The rest of this inequality then follows from multiplication of (4.66) by $\mathbf{m}_{ttt}$ . The arguments are of the same kind as for $I$ through $VI$. Instead of (4.69), we use

$$\tau^3 \,|\, \big(\mathbf{m} \times \Delta\mathbf{m}_{tt}, \mathbf{m}_{ttt}\big) \,| \;\; \leq \;\; C\tau^3 \,\|\, \mathbf{m} \,\|_{\mathbf{L}^\infty} \,\|\, \Delta\mathbf{m}_{tt} \,\|_{\mathbf{L}^2} \,\|\, \mathbf{m}_{ttt} \,\|_{\mathbf{L}^2} \,.$$

This proves part (d).

For the verification of the first statement in (e), we employ (4.62). Since

$$\begin{aligned} &\|\, \mathbf{m}_{tt} \,\|_{L^2(I;\mathbf{W}^{-1,2}(\omega,\mathbb{R}^3))} \\ &\leq C \sup_{\substack{\varphi\in L^2(I;\mathbf{W}^{1,2}(\omega,\mathbb{R}^3)),\\ \|\varphi\|_{L^2(I;\mathbf{W}^{1,2}(\omega,\mathbb{R}^3))}\leq 2}} \int_0^T \Big|\big(\alpha\,\nabla\mathbf{m}_t, \nabla\varphi\big) + I + .. + IV\Big|\,\mathrm{d}s\,, \end{aligned} \tag{4.70}$$

terms $I$ through $IV$ can be bounded as follows,

$$\begin{aligned} I &:= 2\alpha \,|\, \big(\langle\nabla\mathbf{m}, \nabla\mathbf{m}_t\rangle_{\mathbb{R}^6}\mathbf{m}, \varphi\big) \,| \leq C \,\|\, \varphi \,\|_{\mathbf{L}^8} \,\|\, \nabla\mathbf{m} \,\|_{\mathbf{L}^4} \,\|\, \nabla\mathbf{m}_t \,\|_{\mathbf{L}^2} \,\|\, \mathbf{m} \,\|_{\mathbf{L}^\infty} \\ II &:= |\, \big(|\,\nabla\mathbf{m}\,|^2\mathbf{m}_t, \varphi\big) \,| \leq C \,\|\, \varphi \,\|_{\mathbf{L}^4}^4 \,\|\, \mathbf{m}_t \,\|_{\mathbf{L}^2} \,\|\, \nabla\mathbf{m} \,\|_{\mathbf{L}^8}^2 \\ III &:= |\, \big(\nabla(\mathbf{m}_t \times \varphi), \nabla\mathbf{m}\big) \,| \\ &\leq \Big(\|\, \varphi \,\|_{\mathbf{L}^4} \,\|\, \nabla\mathbf{m}_t \,\|_{\mathbf{L}^2} \,\|\, \nabla\mathbf{m} \,\|_{\mathbf{L}^4} + \|\, \mathbf{m}_t \,\|_{\mathbf{L}^4} \,\|\, \nabla\varphi \,\|_{\mathbf{L}^2} \,\|\, \mathbf{m} \,\|_{\mathbf{W}^{1,4}}\Big) \\ IV &:= |\, \big(\nabla(\mathbf{m} \times \varphi), \nabla\mathbf{m}_t\big) \,| \\ &\leq \Big(\|\, \varphi \,\|_{\mathbf{L}^4} \,\|\, \nabla\mathbf{m} \,\|_{\mathbf{L}^4} \,\|\, \nabla\mathbf{m}_t \,\|_{\mathbf{L}^2} + \|\, \nabla\varphi \,\|_{\mathbf{L}^2} \,\|\, \nabla\mathbf{m}_t \,\|_{\mathbf{L}^2} \,\|\, \mathbf{m} \,\|_{\mathbf{L}^\infty}\Big) \end{aligned}$$

This shows (e). □

We end this section with some short comments on existence results for Riemannian manifolds $\mathcal{M}$ of dimension $m \geq 3$, where the analytical approach is different. In [3] and [60], the authors prove existence of weak solutions to (LLG) for mappings $\mathbf{m} : \mathcal{M} \to \mathcal{S}^2$, with $\mathcal{M}$ a Riemannian manifold of dimension $m \geq 3$. This result is obtained by a penalization technique: instead of dealing with mappings with target $\mathcal{S}^2$ directly, one considers for values $\varepsilon > 0$,

$$\mathbf{m}_t^\varepsilon - \alpha\Delta\mathbf{m}^\varepsilon - \mathbf{m}^\varepsilon \times \mathbf{m}_t^\varepsilon - \frac{1}{\varepsilon}\big(|\,\mathbf{m}^\varepsilon\,|^2 - 1\big)\mathbf{m}^\varepsilon = 0\,, \quad \mathbf{m}^\varepsilon\big|_{t=0} = \mathbf{m}_0\,. \tag{4.71}$$

The following results are valid,

$$\operatorname*{ess\,sup}_{[0,T]} \| \nabla \mathbf{m}^\varepsilon \|_{\mathbf{L}^2} + \frac{1}{4\varepsilon}\Big(\int_\omega \big[|\,\mathbf{m}^\varepsilon\,|^2 - 1\big]^2 \,\mathbf{dx}\Big)^{1/2} \leq C\,, \quad \operatorname*{ess\,sup}_{[0,T]} \| \mathbf{m}^\varepsilon \|_{\mathbf{L}^\infty} \leq 1\,.$$

In a second step, the penalization term is removed by taking the wedge product of (4.71) with $\mathbf{m}^\varepsilon$, and (weak-type) convergence of a subsequence $\{\mathbf{m}^\varepsilon\}_{\varepsilon>0}$ can be shown towards a solution $\mathbf{m}$ to (LLG). We refer the reader to [3, 60] for details.

**Remark 4.2** *1. In [24, 25], the authors show uniqueness of weak solutions to (LLG) that start with finite energies, i.e., $\| \mathbf{m}_0 \|_{\mathbf{W}^{1,2}(\omega,\mathcal{S}^2)} \leq C < \infty$.*

*2. The three-dimensional problem is analyzed in [60, 3], where weak solutions are non-unique [3].*

*3. The analysis in [60] to show existence of unique strong local solutions for Riemannian surfaces without boundary can be extended to verify existence and uniqueness of weak local solutions by a continuation argument in time, already used in [120]. It is, however, that the analysis in [60] provides additional information on arising singularities, if compared to [3]: for any $\mathbf{m}_0 \in \mathbf{W}^{1,2}(\omega,\mathcal{S}^2)$, there exists a unique solution of (LLG) which is regular on $\omega \times [0,\infty]$ with the exeption of at most finitely many points $(\mathbf{x}^\ell, T^\ell)$, $1 \leq \ell \leq L$ (Theorem 3.13, [60]).*

## 4.2 Time Discretization of the Landau-Lifshitz-Gilbert equation

Weak solutions to (LLG), $\mathbf{m}_0 \in \mathbf{W}^{2,2}(\omega,\mathcal{S}^2)$ are (i) in $C^\infty\big(0,\infty;\mathbf{W}^{2,2}(\omega,\mathcal{S}^2)\big)$ for small initial energies $E(\mathbf{m}_0;\omega)$, see Lemma 4.2, or (ii) for $T_0 = T_0(\mathbf{m}_0) > 0$ finite in $C^\infty\big(0,T_0,\mathbf{W}^{2,2}(\omega,\mathcal{S}^2)\big)$, see Theorem 4.3. This is the mathematical framework where we study numerical schemes for (LLG) now.

### 4.2.1 Semi-Implicit Discretization of (LLG)

Consider the semidiscretization in time

$$d_t\mathbf{m}^{j+1} - \alpha\,\Delta\mathbf{m}^{j+1} = \alpha\,|\,\nabla\mathbf{m}^j\,|^2\mathbf{m}^{j+1} + \mathbf{m}^{j+1}\times\Delta\mathbf{m}^{j+1}\,, \quad \mathbf{m}^0 = \mathbf{m}_0\,. \tag{4.72}$$

Note that we do not enforce $|\mathbf{m}^{j+1}| = 1$, almost everywhere in $I_k \times \omega$. In [43], severe stability problems are reported by computational experiments, see also Example 4.1.

We prove the following result.

**Theorem 4.4** *Let $0 < t_J < T_0(\mathbf{m})$. Let $\{\mathbf{m}^j\}_{j=0}^J$ be the solution to (4.72), and $\mathbf{m}$ solves (4.6), for $\mathbf{m}_0 \in \mathbf{W}^{2,2}(\omega, \mathcal{S}^2)$. Let $k \leq k_0(t_J)$ be sufficiently small. Then we have*

$$\max_{0\leq j\leq J} \| \mathbf{m}(t_j) - \mathbf{m}^j \|_{\mathbf{L}^2} + \Big(k \sum_{j=0}^{J} \| \nabla\{\mathbf{m}(t_j) - \mathbf{m}^j\} \|_{\mathbf{L}^2}^2\Big)^{1/2} \leq Ck\,.$$

**Proof:**
*1st step:* Consider (4.6), (4.72) in weak form. In the sequel, let $\mathbf{e}^{j+1} := \mathbf{m}(t_{j+1}) - \mathbf{m}^{j+1}$ denote the error governed by

$$\begin{aligned}(d_t\mathbf{e}^{j+1}, \boldsymbol{\varphi}) + \alpha\,(\nabla\mathbf{e}^{j+1}, \nabla\boldsymbol{\varphi}) &= (\boldsymbol{\mathcal{R}}^{j+1}(\mathbf{m}), \boldsymbol{\varphi}) \\ +\alpha\Big[(|\nabla\mathbf{m}(t_{j+1})|^2\mathbf{m}(t_{j+1}), \boldsymbol{\varphi}) &\pm (|\nabla\mathbf{m}(t_j)|^2\mathbf{m}^{j+1}, \boldsymbol{\varphi}) \\ -|\nabla\mathbf{m}^j|^2\mathbf{m}^{j+1}, \boldsymbol{\varphi})\Big] &- \Big[(\mathbf{m}(t_{j+1}) \times \nabla\mathbf{m}(t_{j+1}), \nabla\boldsymbol{\varphi}) \\ \pm(\mathbf{m}^{j+1} \times \nabla\mathbf{m}(t_{j+1}), \nabla\boldsymbol{\varphi}) &- (\mathbf{m}^{j+1} \times \nabla\mathbf{m}^{j+1}, \nabla\boldsymbol{\varphi})\Big]\,,\end{aligned} \tag{4.73}$$

for all $\boldsymbol{\varphi} \in \mathbf{W}^{1,2}(\omega, \mathbb{R}^3)$, and

$$\boldsymbol{\mathcal{R}}^{j+1}(\mathbf{m}) := -\frac{1}{k}\int_{t_j}^{t_{j+1}} (s - t_j)\,\mathbf{m}_{tt}(s)\,\mathrm{d}s\,. \tag{4.74}$$

We use Theorem 4.3, (b) and compute

$$\begin{aligned} k\sum_{j=0}^{J} \|\boldsymbol{\mathcal{R}}(j, \mathbf{m}_{tt})\|_{\mathbf{L}^2(\omega)}^2 & \qquad\qquad (4.75)\\ \leq \frac{1}{k}\sum_{j=0}^{J}\Big(\int_{t_j}^{t_{j+1}} (s-t_j)\,\mathrm{d}s\Big)\Big(\int_{t_j}^{t_{j+1}} s\,\|\mathbf{m}_{tt}(s)\|_{\mathbf{L}^2}^2\,\mathrm{d}s\Big) &\leq Ck\,,\end{aligned}$$

$$\begin{aligned} k\sum_{j=0}^{J} \|\boldsymbol{\mathcal{R}}(j, \mathbf{m}_{tt})\|_{\mathbf{W}^{-1,2}(\omega,\mathbb{R}^3)}^2 & \qquad\qquad (4.76)\\ \leq Ck^2\int_0^{t_{J+1}} \|\mathbf{m}_{tt}(s)\|_{\mathbf{W}^{-1,2}(\omega,\mathbb{R}^3)}^2\,\mathrm{d}s &\leq C\,k^2\,.\end{aligned}$$

We choose $\varphi = \mathbf{e}^{j+1} \in \mathbf{W}^{1,2}(\omega, \mathbb{R}^3)$ in (4.73). Then,

$$
\begin{aligned}
& k \sum_{j=0}^{J} \| \boldsymbol{\mathcal{R}}(j, \mathbf{m}_{tt}) \|_{\mathbf{W}^{-1,2}(\omega,\mathbb{R}^3)} \big( \| \mathbf{e}^{j+1} \|_{\mathbf{L}^2} + \| \nabla \mathbf{e}^{j+1} \|_{\mathbf{L}^2} \big) \\
& \leq k \sum_{j=0}^{J} \Big\{ \frac{\alpha}{16} \| \nabla \mathbf{e}^{j+1} \|_{\mathbf{L}^2}^2 + \frac{1}{t_J} \| \mathbf{e}^{j+1} \|_{\mathbf{L}^2}^2 \Big\} \\
& \qquad + C_\alpha (1 + t_J) k \sum_{j=0}^{J} \| \boldsymbol{\mathcal{R}}(j, \mathbf{m}_{tt}) \|_{\mathbf{W}^{-1,2}(\omega,\mathbb{R}^3)}^2 .
\end{aligned}
$$

In the sequel, we employ the skew-symmetricity property for the terms in the last line of (4.73). We consider the following prototype terms that arise on the right-hand side of (4.73),

$$
\begin{aligned}
\big( | \nabla \mathbf{m}(t_j) |^2 \mathbf{e}^{j+1}, \mathbf{e}^{j+1} \big) &\leq \| \nabla \mathbf{m}(t_j) \|_{\mathbf{L}^4}^2 \| \mathbf{e}^{j+1} \|_{\mathbf{L}^4}^2 \\
&\leq C \| \mathbf{e}^{j+1} \|_{\mathbf{L}^2} \| \mathbf{e}^{j+1} \|_{\mathbf{W}^{1,2}} , \\
\big( \langle \nabla \mathbf{e}^j, \nabla \mathbf{m}(t_j) \rangle_{\mathbb{R}^6} \mathbf{m}(t_{j+1}), \mathbf{e}^{j+1} \big) &\leq \| \nabla \mathbf{e}^j \|_{\mathbf{L}^2} \| \nabla \mathbf{m}(t_j) \|_{\mathbf{L}^4} \\
&\quad \times \| \mathbf{m}(t_{j+1}) \|_{\mathbf{L}^\infty} \| \mathbf{e}^{j+1} \|_{\mathbf{L}^2}^{1/2} \| \mathbf{e}^{j+1} \|_{\mathbf{W}^{1,2}}^{1/2} \\
&\leq \frac{\alpha}{8} \big( \| \nabla \mathbf{e}^j \|_{\mathbf{L}^2}^2 + \| \nabla \mathbf{e}^{j+1} \|_{\mathbf{L}^2}^2 \big) \\
&\quad + C_\alpha \| \nabla \mathbf{m}(t_j) \|_{\mathbf{L}^4}^4 \| \mathbf{e}^{j+1} \|_{\mathbf{L}^2}^2 , \\
\big( \langle \nabla \mathbf{e}^j, \nabla \mathbf{m}(t_j) \rangle_{\mathbb{R}^6}, | \mathbf{e}^{j+1} |^2 \big) &\leq \| \nabla \mathbf{e}^j \|_{\mathbf{L}^2} \| \nabla \mathbf{m}(t_{j+1}) \|_{\mathbf{L}^\infty} \| \mathbf{e}^{j+1} \|_{\mathbf{L}^4}^2 \\
&\leq \frac{\alpha}{8} \| \nabla \mathbf{e}^j \|_{\mathbf{L}^2}^2 \\
&\quad + C_\alpha \| \nabla \mathbf{m}(t_j) \|_{\mathbf{L}^\infty}^4 \| \mathbf{e}^{j+1} \|_{\mathbf{L}^4}^2 , \\
\big( | \nabla \mathbf{e}^j |^2 , | \mathbf{e}^{j+1} |^2 \big) &\leq \frac{\alpha}{8} \| \nabla \mathbf{e}^{j+1} \|_{\mathbf{L}^2}^2 \qquad (4.77) \\
&\quad + C_\alpha \| \nabla \mathbf{e}^j \|_{\mathbf{L}^2}^2 \| \nabla \mathbf{e}^j \|_{\mathbf{W}^{1,2}}^2 \| \mathbf{e}^{j+1} \|_{\mathbf{L}^2}^2 , \\
\big( \big[ \mathbf{m}(t_{j+1}) - \mathbf{e}^{j+1} \big] \times \nabla \mathbf{e}^{j+1}, \nabla \mathbf{e}^{j+1} \big) &= 0 ,
\end{aligned}
$$

and

$$
\begin{aligned}
& k \big( \big\langle \nabla d_t \mathbf{m}(t_{j+1}), \nabla \mathbf{m}(t_{j+1}) \big\rangle_{\mathbb{R}^6} \mathbf{m}(t_{j+1}), \mathbf{e}^{j+1} \big) \\
& \leq k \| \nabla d_t \mathbf{m}(t_{j+1}) \|_{\mathbf{L}^2} \| \nabla \mathbf{m}(t_{j+1}) \|_{\mathbf{L}^4} \| \mathbf{m}(t_{j+1}) \|_{\mathbf{L}^\infty} \| \mathbf{e}^{j+1} \|_{\mathbf{L}^4} \\
& \leq C_\alpha k \int_{t_j}^{t_{j+1}} \| \nabla \mathbf{m}_t(s) \|_{\mathbf{L}^2}^2 \, \mathrm{d}s + \frac{\alpha}{8} \| \mathbf{e}^{j+1} \|_{\mathbf{W}^{1,2}}^2 .
\end{aligned}
$$

*2nd step:* We employ an inductive argument: There exist constants $C_i \equiv C_i(\omega, t_J, \alpha, \mathbf{m}_0)$, $i = 1, 2$, such that for $0 \leq \ell \leq J$, we have

$$\begin{aligned} &\| \mathbf{e}^{\ell+1} \|_{\mathbf{L}^2}^2 + \frac{k^2}{2} \sum_{j=0}^{\ell} \| d_t \mathbf{e}^{j+1} \|_{\mathbf{L}^2}^2 + \frac{\alpha k}{2} \sum_{j=0}^{\ell} \| \nabla \mathbf{e}^{j+1} \|_{\mathbf{L}^2}^2 \\ &\leq C_1 \, k^2 \exp\bigl(C_2 \, t_\ell\bigr) , \end{aligned} \tag{4.78}$$

$$\begin{aligned} &\| \nabla \mathbf{e}^{\ell+1} \|_{\mathbf{L}^2}^2 + \frac{k^2}{2} \sum_{j=0}^{\ell} \| \nabla d_t \mathbf{e}^{j+1} \|_{\mathbf{L}^2}^2 + \frac{\alpha k}{2} \sum_{j=0}^{\ell} \| \Delta \mathbf{e}^{j+1} \|_{\mathbf{L}^2}^2 \\ &\leq C_1 \, k \exp\bigl(C_2 \, t_\ell\bigr) . \end{aligned} \tag{4.79}$$

These statements are valid for $\ell = 0$. — Next, we test (4.73) by $-\Delta \mathbf{e}^{j+1}$ and benefit from (4.31) to control the nonlinear terms:

$$\begin{aligned} \bigl| \bigl(\mathbf{e}^{j+1} \times \Delta \mathbf{m}(t_{j+1}), \Delta \mathbf{e}^{j+1}\bigr) \bigr| &\leq C \| \mathbf{e}^{j+1} \|_{\mathbf{L}^2}^{1/2} \| \mathbf{e}^{j+1} \|_{\mathbf{W}^{2,2}}^{1/2} \\ &\quad \times \| \Delta \mathbf{m}(t_{j+1}) \|_{\mathbf{L}^2} \| \Delta \mathbf{e}^{j+1} \|_{\mathbf{L}^2} , \\ \bigl| \bigl(\mathbf{m}^{j+1} \times \Delta \mathbf{e}^{j+1}, \Delta \mathbf{e}^{j+1}\bigr) \bigr| &= 0 , \\ \bigl| \bigl(| \nabla \mathbf{m}(t_j) |^2 \mathbf{e}^{j+1}, \Delta \mathbf{e}^{j+1}\bigr) \bigr| &\leq \| \nabla \mathbf{m}(t_j) \|_{\mathbf{L}^8}^2 \| \mathbf{e}^{j+1} \|_{\mathbf{L}^4} \| \Delta \mathbf{e}^{j+1} \|_{\mathbf{L}^2} , \end{aligned}$$

$$\begin{aligned} &\bigl| \bigl(\langle \nabla \mathbf{e}^j, \nabla \mathbf{e}^j - 2 \nabla \mathbf{m}(t_j) \rangle_{\mathbb{R}^6} \{ \mathbf{m}(t_j) - \mathbf{e}^{j+1} \}, \Delta \mathbf{e}^{j+1}\bigr) \bigr| \\ &\quad \leq C \| \nabla \mathbf{e}^j \|_{\mathbf{L}^4} \{ \| \nabla \mathbf{e}^j \|_{\mathbf{L}^4} + \| \nabla \mathbf{m}(t_j) \|_{\mathbf{L}^4} \} \\ &\quad\quad \times \bigl\{ \| \mathbf{m}(t_{j+1}) \|_{\mathbf{L}^\infty} + \| \mathbf{e}^{j+1} \|_{\mathbf{L}^2}^{1/2} \| \mathbf{e}^{j+1} \|_{\mathbf{W}^{2,2}}^{1/2} \bigr\} \| \Delta \mathbf{e}^{j+1} \|_{\mathbf{L}^2} \end{aligned} \tag{4.80}$$

and, for $j \geq 1$,

$$\begin{aligned} &k \bigl| \bigl(\langle \nabla d_t \mathbf{m}(t_{j+1}), \nabla \{ \mathbf{m}(t_{j+1}) + \mathbf{m}(t_j) \} \rangle_{\mathbb{R}^6} \mathbf{m}(t_{j+1}), \Delta \mathbf{e}^{j+1}\bigr) \bigr| \\ &\leq k \| \nabla d_t \mathbf{m}(t_{j+1}) \|_{\mathbf{L}^4} \| \nabla \mathbf{m}(t_{j+1}) \|_{\mathbf{L}^4} \| \mathbf{m}(t_{j+1}) \|_{\mathbf{L}^\infty} \| \Delta \mathbf{e}^{j+1} \|_{\mathbf{L}^2} \\ &\leq \frac{\alpha}{8} \| \Delta \mathbf{e}^{j+1} \|_{\mathbf{L}^2}^2 + \sqrt{k} \Bigl( \int_{t_j}^{t_{j+1}} \| \nabla \mathbf{m}_t(s) \|_{\mathbf{L}^2}^2 \, \mathrm{d}s \Bigr)^{\frac{1}{2}} \Bigl( \int_{t_j}^{t_{j+1}} s \, \| \Delta \mathbf{m}_t(s) \|_{\mathbf{L}^2}^2 \, \mathrm{d}s \Bigr)^{\frac{1}{2}} . \end{aligned}$$

The crucial term in (4.80) can be bounded as follows,

$$\begin{aligned} &\| \nabla \mathbf{e}^j \|_{\mathbf{L}^4}^2 \| \mathbf{e}^{j+1} \|_{\mathbf{L}^2}^{1/2} \| \Delta \mathbf{e}^{j+1} \|_{\mathbf{L}^2}^{3/2} \leq \frac{\alpha}{8} \| \Delta \mathbf{e}^{j+1} \|_{\mathbf{L}^2}^2 + C_\alpha \| \nabla \mathbf{e}^j \|_{\mathbf{L}^4}^8 \| \mathbf{e}^{j+1} \|_{\mathbf{L}^2}^2 \\ &\leq \frac{\alpha}{8} \| \Delta \mathbf{e}^{j+1} \|_{\mathbf{L}^2}^2 + C_\alpha \| \nabla \mathbf{e}^j \|_{\mathbf{L}^2}^4 \| \Delta \mathbf{e}^j \|_{\mathbf{L}^2}^4 \| \mathbf{e}^{j+1} \|_{\mathbf{L}^2}^2 \\ &\leq \frac{\alpha}{8} \| \Delta \mathbf{e}^{j+1} \|_{\mathbf{L}^2}^2 + C_\alpha C_1^3 k^3 \exp\bigl(3 C_2 t_J\bigr) \| \Delta \mathbf{e}^j \|_{\mathbf{L}^2}^2 \| \mathbf{e}^{j+1} \|_{\mathbf{L}^2}^2 . \end{aligned} \tag{4.81}$$

For $k \leq k_0(C_1, C_2, t_J)$ sufficiently small, we can do the induction step for (4.79),

$$\| \nabla \mathbf{e}^{\ell+2} \|_{\mathbf{L}^2}^2 + \frac{k^2}{2} \sum_{j=0}^{\ell+1} \| \nabla d_t \mathbf{e}^{j+1} \|_{\mathbf{L}^2}^2 + \frac{\alpha k}{2} \sum_{j=0}^{\ell+1} \| \Delta \mathbf{e}^{j+1} \|_{\mathbf{L}^2}^2 \leq C_1 \, k \exp\bigl(C_2 \, t_{\ell+1}\bigr) . \tag{4.82}$$

As a consequence, we can now control the right-hand side of (4.77), using again Gronwall's lemma. Then, concluding by induction is possible for constants $C_i$, $i = 1, 2$ chosen sufficiently large. □

**Corollary 4.1** *The solution* $\{\mathbf{m}^j\}_{j=0}^J$ *to* (4.72) *enjoys*

$$\text{(a)} \qquad \max_{0 \leq j \leq J} \| 1 - |\mathbf{m}^j|^2 \|_{L^2} \leq C\sqrt{k} \, ,$$

$$\text{(b)} \qquad \max_{1 \leq j \leq J} \bigl\{ \| \mathbf{m}^j \|_{\mathbf{W}^{2,2}} + \| d_t \mathbf{m}^j \|_{\mathbf{L}^2} \bigr\} + \Bigl( k \sum_{j=1}^{J} \| \nabla d_t \mathbf{m}^j \|_{\mathbf{L}^2}^2 \bigr\} \Bigr)^{1/2} \leq C \, ,$$

*for* $k \leq k_0(t_J)$ *sufficiently small and* $C = C(t_J, \omega; \mathbf{m}_0)$ *independent of* $k$.

**Proof:**
Item (a) follows immediately from

$$\begin{aligned} \| \, |\mathbf{m}(t_{j+1})|^2 - |\mathbf{m}^{j+1}|^2 \|_{L^2}^2 &= \| \langle \mathbf{e}^{j+1}, \mathbf{e}^{j+1} + 2\mathbf{m}(t_{j+1}) \rangle_{\mathbb{R}^3} \|_{L^2}^2 \\ &\leq 2 \bigl( \| \mathbf{e}^{j+1} \|_{\mathbf{L}^4}^4 + \| \langle \mathbf{e}^{j+1}, \mathbf{m}(t_{j+1}) \rangle_{\mathbb{R}^3} \|_{L^2}^2 \bigr) . \end{aligned} \tag{4.83}$$

Part (b) is a consequence of (4.78), (4.79). □

### 4.2.2 Penalization Strategies for (LLG)

We study stability as well as convergence behavior of the penalization functions $\phi_i(\cdot)$ given in (4.14). Proofs are shortened, and we omit elaboration of the analysis for the nonlinear terms on the right-hand side of (4.13), since it has already been accomplished in the analysis of (4.72) in the previous subsection.

**Remark 4.3** *1. We discuss well-posedness of* (4.13), *for* $\phi_i$, $i = 2, 3$ *from* (4.14), *and values* $k \leq k_0$*: We consider* (4.13) *in linearized form, and* $i = 2, 3$,

$$d_t\mathbf{m}^{j+1} - \alpha\,\Delta\mathbf{m}^{j+1} + \frac{1}{\varepsilon}\,\phi_i(\mathbf{m}^j, \mathbf{m}^{j+1})\mathbf{m}^{j+1} = \alpha\,|\,\nabla\mathbf{m}^j\,|^2\mathbf{m}^{j+1} + \mathbf{m}^j \times \Delta\mathbf{m}^{j+1}. \tag{4.84}$$

*In abstract form, we have to deal with the following problem at each iteration step: given* $\boldsymbol{\beta} \in \mathbf{W}^{2,2}(\omega, \mathbb{R}^3)$, *such that* $\|\,\boldsymbol{\beta}\,\|_{\mathbf{W}^{2,2}} \leq C$, *and* $\|\,\phi_i(\boldsymbol{\beta}, \boldsymbol{\varphi})\,\|_{\mathbf{L}^\infty} \leq C\sqrt{k}$, *and* $i = 2, 3$, *define*

$$\mathcal{L}_{i;\boldsymbol{\beta}}\boldsymbol{\varphi} \equiv \boldsymbol{\varphi} - k\,\Delta\boldsymbol{\varphi} - \boldsymbol{\beta} - k\Big\{-\frac{1}{\varepsilon}\phi_i(\boldsymbol{\beta}, \boldsymbol{\varphi})\boldsymbol{\varphi} + \alpha\,|\,\nabla\boldsymbol{\beta}\,|^2\boldsymbol{\varphi} + \boldsymbol{\beta} \times \Delta\boldsymbol{\varphi}\Big\}.$$

*The linear mapping* $\mathcal{L}_{i;\boldsymbol{\beta}} : \mathbf{W}^{2,2}(\omega, \mathbb{R}^3) \to \mathbf{L}^2(\omega, \mathbb{R}^3)$, *with* $\mathcal{L}_{i;\boldsymbol{\beta}}\boldsymbol{\psi} = \mathbf{g}$ *has a (unique) solution, for* $\varepsilon = \mathcal{O}(k)$, *and* $k \leq k_0$ *sufficiently small. This follows from continuity and* $\mathbf{W}^{1,2}(\omega, \mathbb{R}^3)$*-coercivity of the associated bilinear form* $a(\cdot,\cdot) : \mathbf{W}^{1,2} \times \mathbf{W}^{1,2} \to \mathbb{R}$, *acting through* $a(\boldsymbol{\psi}, \boldsymbol{\chi}) := (\mathcal{L}_{i;\boldsymbol{\beta}}\boldsymbol{\psi}, \boldsymbol{\chi})$, *together with a standard regularity result for elliptic operators. We confine to showing coercivity, taking* $\boldsymbol{\chi} = \boldsymbol{\psi}$. *In what follows, we use interpolation of* $\mathbf{L}^4$ *between* $\mathbf{L}^2$ *and* $\mathbf{W}^{1,2}$, *and the inequality* $\|\,\boldsymbol{\varphi}\boldsymbol{\chi}\,\|_{\mathbf{L}^{4/3}} \leq C\,\|\,\boldsymbol{\varphi}\,\|_{\mathbf{L}^2}\,\|\,\boldsymbol{\chi}\,\|_{\mathbf{L}^4}$.

$$\begin{aligned}(\mathcal{L}_{i;\boldsymbol{\beta}}\boldsymbol{\psi}, \boldsymbol{\psi}) &= \|\,\boldsymbol{\psi}\,\|^2_{\mathbf{L}^2} + k\,\|\,\nabla\boldsymbol{\psi}\,\|^2_{\mathbf{L}^2} \\ &\quad -k\Big\{\frac{1}{\varepsilon}|\,(\phi_i(\boldsymbol{\beta}, \boldsymbol{\beta})\boldsymbol{\psi}, \boldsymbol{\psi})\,| + \alpha\,|\,(|\,\nabla\boldsymbol{\beta}\,|^2\boldsymbol{\psi}, \boldsymbol{\psi})\,| + |\,(\boldsymbol{\beta} \times \Delta\boldsymbol{\psi}, \boldsymbol{\psi})\,|\Big\}.\end{aligned} \tag{4.85}$$

*We derive upper bounds for the last three terms separately.*

$$\begin{aligned}&\leq k\,\Big\{\|\,\frac{1}{\varepsilon}\phi_i(\boldsymbol{\beta}, \boldsymbol{\varphi})\boldsymbol{\psi}\,\|_{\mathbf{L}^2}\,\|\,\boldsymbol{\psi}\,\|_{\mathbf{L}^2} + \alpha\,\|\,|\,\nabla\boldsymbol{\beta}\,|^2\boldsymbol{\psi}\,\|_{\mathbf{L}^{4/3}}\|\,\boldsymbol{\psi}\,\|_{\mathbf{L}^4} \\ &\qquad + \|\,|\,\nabla\boldsymbol{\beta}\,|\,|\,\boldsymbol{\psi}\,|\,\|_{\mathbf{L}^2}\,\|\,\nabla\boldsymbol{\psi}\,\|_{\mathbf{L}^2}\Big\} \\ &\leq k\Big\{\|\,\frac{1}{\varepsilon}\phi_i(\boldsymbol{\beta}, \boldsymbol{\varphi})\boldsymbol{\psi}\,\|_{\mathbf{L}^2}\,\|\,\boldsymbol{\psi}\,\|_{\mathbf{L}^2} + \alpha\,\|\,\nabla\boldsymbol{\beta}\,\|^2_{\mathbf{L}^4}\|\,\boldsymbol{\psi}\,\|_{\mathbf{L}^2}\,\|\,\nabla\boldsymbol{\psi}\,\|_{\mathbf{L}^2} \\ &\qquad + \|\,\nabla\boldsymbol{\beta}\,\|_{\mathbf{L}^4}\,\|\,\boldsymbol{\psi}\,\|^{1/2}_{\mathbf{L}^2}\,\|\,\nabla\boldsymbol{\psi}\,\|^{3/2}_{\mathbf{L}^2}\Big\}.\end{aligned} \tag{4.86}$$

*We now come back to* (4.85) *and absorb corresponding terms, which is e.g. possible for* $k \leq k_0$, *and values* $\varepsilon = \mathcal{O}(k)$. *— The requirement* $\|\,\phi_i(\boldsymbol{\beta}, \boldsymbol{\beta})\,\|_{\mathbf{L}^\infty} \leq C\sqrt{k}$ *can be realized in the semidiscretization scheme of (LLG) by interpolation of error statements in* $\ell^\infty(I_k; \mathbf{L}^2)$ *and* $\ell^\infty(I_k; \mathbf{W}^{2,2})$; *see the error analysis below.*

*2. The semidiscretization scheme*

$$d_t\mathbf{m}^{j+1} - \alpha\,\Delta\mathbf{m}^{j+1} + \frac{1}{\varepsilon}\,\phi_1(\mathbf{m}^j,\mathbf{m}^{j+1})\mathbf{m}^{j+1} = \alpha\,|\,\nabla\mathbf{m}^j\,|^2\mathbf{m}^{j+1} + \mathbf{m}^j \times \Delta\mathbf{m}^{j+1}$$

*is well-posed, for* $\varepsilon^{-1} = o(\frac{1}{k})$, *and* $k \le k_0$ *sufficiently small. In an abstract setting, we introduce the nonlinear problem* $\mathcal{N}_{j+1;\beta}\varphi = 0$, *where*

$$\mathcal{N}_{j+1;\beta}\varphi \equiv \varphi - k\,\Delta\varphi - \beta - k\Big\{-\frac{1}{\varepsilon}\phi_1(\varphi,\varphi)\varphi + \alpha\,|\,\nabla\beta\,|^2\varphi + \beta \times \Delta\varphi\Big\}\,, \tag{4.87}$$

*and we show existence of a unique solution* $\varphi \in \mathbf{W}^{1,2}(\omega,\mathbb{R}^3)$, *for* $\beta \in \mathbf{W}^{2,2}(\omega,\mathbb{R}^3)$, *such that* $\|\,\beta\,\|_{\mathbf{W}^{2,2}} \le C$, *and* $\varepsilon^{-1} = o(\frac{1}{k})$, *provided* $k \le k_0$.

*To see this, consider the linear map* $\mathcal{N}^{\mathbf{U}}_{j+1;\beta}$ *on* $\mathbf{W}^{1,2}(\omega,\mathbb{R}^3)$,

$$\mathcal{N}^{\mathbf{U}}_{j+1;\beta}\varphi \equiv \varphi - k\,\Delta\varphi - \beta - k\Big\{-\frac{1}{\varepsilon}\phi_1(\mathbf{U},\mathbf{U})\varphi + \alpha\,|\,\nabla\beta\,|^2\varphi + \beta \times \Delta\varphi\Big\}\,, \tag{4.88}$$

*for given* $\beta \in \mathbf{W}^{2,2}(\omega,\mathbb{R}^3) \cap \mathcal{A}^{C_1/4}_{j+1}$, $\mathbf{U} \in \mathcal{A}^{C_1}_{j+1} \equiv \mathcal{A}^{C_1}_{j+1}(\mathbf{m}(t_{j+1}))$, *and* $C_1 \equiv C_1(t_{j+1},\alpha) > 0$. *Here,*

$$\begin{aligned}\mathcal{A}^{C_1}_{j+1} \;:=\; &\Big\{\varphi \in \mathbf{W}^{1,2}(\omega,\mathbb{R}^3) : \|\,\varphi - \mathbf{m}(t_{j+1})\,\|_{\mathbf{L}^2}\\ &+\sqrt{\alpha k}\,\|\,\varphi - \mathbf{m}(t_{j+1})\,\|_{\mathbf{W}^{1,2}} \le C_1\,k\Big\}\,.\end{aligned}$$

*Solve* $\mathcal{N}^{\varphi^\ell}_{j+1;\beta}\varphi^{\ell+1} = 0$, *for* $\varphi^{\ell+1} \in \mathbf{W}^{1,2}(\omega,\mathbb{R}^3)$, *and* $\ell \in \mathbb{N}_0$. *We make the following observations for* $\mathcal{N}^{\mathbf{U}}_{j+1;\beta}\varphi = 0$, *for* $k \le k_0$.

*(i)* $\mathbf{U} \in \mathcal{A}^{C_1}_{j+1}$ *implies* $\varphi \in \mathcal{A}^{C_1}_{j+1}$: *We proceed as in the proof of Theorem 4.4, for* $\mathbf{e}^{j+1} := \mathbf{m}(t_{j+1}) - \varphi$ *and* $\mathbf{e}^j := \mathbf{m}(t_j) - \beta$. *'Nonlinear effects' from the last two contributions in (4.88) as well as the consistency error term (4.74) are gathered in the term 'rest' in the sequel, which is non-critical in the following argument: it can be handled for* $k \le k_0(t_J)$ *sufficiently small, see also (4.77), and we skip the elaboration of this argument at this place.*

$$\begin{aligned}&\|\,\mathbf{e}^{j+1}\,\|^2_{\mathbf{L}^2} - \|\,\mathbf{e}^j\,\|^2_{\mathbf{L}^2} + 2\alpha k\,\|\,\nabla\mathbf{e}^{j+1}\,\|^2_{\mathbf{L}^2}\\ &\le \frac{2k}{\varepsilon}\,|\,(\{|\,\mathbf{U}\,|^2 - 1\}\{\mathbf{e}^{j+1} - \mathbf{m}(t_{j+1})\},\mathbf{e}^{j+1})\,| + 2k\,\mathit{rest}\\ &\le \frac{2k}{\varepsilon}\,\|\,|\,\mathbf{U}\,|^2 - 1\,\|_{L^2}\Big(\|\,\mathbf{e}^{j+1}\,\|^2_{\mathbf{L}^4} + \|\,\mathbf{e}^{j+1}\,\|_{\mathbf{L}^2}\Big) + 2k\,\mathit{rest}\,.\end{aligned} \tag{4.89}$$

*We interpolate $L^4$ between $L^2$ and $W^{1,2}$ and employ*

$$\begin{aligned} \| \, |\mathbf{U}|^2 - 1 \, \|_{L^2}^2 &= \| \langle \mathbf{m}(t_{j+1}) - \mathbf{U}, \mathbf{m}(t_{j+1}) - \mathbf{U} - 2\mathbf{m}(t_{j+1}) \rangle_{\mathbb{R}^3} \|_{L^2}^2 \\ &\leq 2 \Big( \| \mathbf{m}(t_{j+1}) - \mathbf{U} \|_{\mathbf{L}^4}^4 + \| \langle \mathbf{m}(t_{j+1}) - \mathbf{U}, \mathbf{m}(t_{j+1}) \rangle_{\mathbb{R}^3} \|_{L^2}^2 \Big) \qquad (4.90) \\ &\leq C \, \| \mathbf{m}(t_{j+1}) - \mathbf{U} \|_{\mathbf{L}^2}^2 \Big( \| \mathbf{m}(t_{j+1}) - \mathbf{U} \|_{\mathbf{W}^{1,2}}^2 + 1 \Big) \leq C_1^2 \, k^2 . \end{aligned}$$

*For $\varepsilon^{-1} = o(\frac{1}{k})$, the assertion then follows from (4.89).*

*(ii) There holds*

$$\begin{aligned} &\| \varphi_1 - \varphi_2 \|_{\mathbf{L}^2} + \sqrt{\alpha k} \, \| \nabla(\varphi_1 - \varphi_2) \|_{\mathbf{L}^2} \qquad (4.91) \\ &\quad \leq \frac{3}{4} \Big\{ \| \mathbf{U}_1 - \mathbf{U}_2 \|_{\mathbf{L}^2} + \sqrt{\alpha k} \, \| \nabla(\mathbf{U}_1 - \mathbf{U}_2) \|_{\mathbf{L}^2} \Big\} , \end{aligned}$$

*for $\mathcal{N}^{\mathbf{U}_\kappa}_{j+1;\beta}\varphi_\kappa = 0$, with $\mathbf{U}_\kappa \in \mathcal{A}^{C_1}_{j+1}$, $\kappa = 1,2$, and $\varepsilon^{-1} = o(\frac{1}{k})$, $k \leq k_0$ sufficiently small. Again, we confine to the most critical terms in the analysis. Let $\mathbf{E} := \mathbf{U}_1 - \mathbf{U}_2$, and $\mathbf{e} := \varphi_1 - \varphi_2$.*

$$\begin{aligned} &\| \mathbf{e} \|_{\mathbf{L}^2}^2 + \alpha k \, \| \nabla \mathbf{e} \|_{\mathbf{L}^2}^2 \leq \frac{k}{\varepsilon} \Big| \Big( (|\mathbf{U}_1|^2 - 1)\varphi_1 - |\mathbf{U}_2|^2 - 1)\varphi_2 \Big) \Big| + \overline{rest} \\ &\leq \frac{k}{\varepsilon} \Big( | ((|\mathbf{U}_1|^2 - 1)\mathbf{e}, \mathbf{e}) | + | (| (\mathbf{U}_1|^2 - |\mathbf{U}_2|^2) \{ \varphi_2 \pm \mathbf{m}(t_{j+1}) \}, \mathbf{e}) | + \overline{rest} \\ &\leq \frac{k}{\varepsilon} \Big( C \, \| \, |\mathbf{U}_1|^2 - 1 \, \|_{L^2} \, \| \mathbf{e} \|_{\mathbf{L}^2} \| \mathbf{e} \|_{\mathbf{W}^{1,2}} \qquad (4.92) \\ &\quad + \| \, |\mathbf{U}_1|^2 - |\mathbf{U}_2|^2 \, \|_{L^2} \big( \| \mathbf{e} \|_{\mathbf{L}^2} + \| \varphi_2 - \mathbf{m}(t_{j+1}) \|_{\mathbf{L}^4} \, \| \mathbf{e} \|_{\mathbf{L}^4} \big) \Big) + \overline{rest} . \end{aligned}$$

*According to (4.90), $\| \, |\mathbf{U}_1|^2 - 1 \, \|_{L^2} \leq C_1 k$, such that the first term on the right-hand side of (4.92) can be absorbed on the left-hand side. For the second term, we notice*

$$\begin{aligned} &\| \langle \mathbf{U}_1 - \mathbf{U}_2, \pm 2\mathbf{m}(t_{j+1}) + \mathbf{U}_1 + \mathbf{U}_2 \rangle_{\mathbb{R}^3} \|_{L^2} \\ &\quad \leq 2 \Big( \| \mathbf{U}_1 - \mathbf{U}_2 \|_{\mathbf{L}^2} + \max_{\kappa=1,2} \| \mathbf{m}(t_{j+1}) - \mathbf{U}_\kappa \|_{\mathbf{L}^4} \, \| \mathbf{U}_1 - \mathbf{U}_2 \|_{\mathbf{L}^4} \Big) \\ &\qquad + \| \mathbf{U}_1 - \mathbf{U}_2 \|_{\mathbf{L}^2} \\ &\quad \leq 3 \Big( \| \mathbf{U}_1 - \mathbf{U}_2 \|_{\mathbf{L}^2} + C\alpha^{-1/4} k^{3/4} \, \| \mathbf{U}_1 - \mathbf{U}_2 \|_{\mathbf{W}^{1,2}} \Big) , \end{aligned}$$

*which follows from interpolation. We may conclude (4.91) from (4.92) then.*

*From (i), (ii), solvability of $\mathcal{N}_{j+1;\beta}\varphi = 0$ then follows from Banach contraction mapping principle.*

**Analysis of the penalization function $\phi_1(\cdot)$ in (4.13)**

The penalty term $\phi_1(\mathbf{m}^j,\mathbf{m}^{j+1}) = \frac{1}{\varepsilon}\big(|\,\mathbf{m}^{j+1}\,|^2 - 1\big)$ was already motivated to show existence of solutions in a three-dimensional setting; see (4.71). At this place, we discuss the convergence behavior of the method with respect to the parameters $k, \varepsilon > 0$.

**Theorem 4.5** *Let $\{\mathbf{m}^j\}_{j=0}^J$, for $0 < t_J < T_0(\mathbf{m}_0)$ be the solution to (4.13), with $\phi(\cdot) = \phi_1(\cdot)$ and $\varepsilon^{-1} = o(\frac{1}{k})$. Let further $\mathbf{m}$ solve (4.6), for $\mathbf{m}_0 \in \mathbf{W}^{2,2}(\omega, \mathcal{S}^2)$. For $k \leq k_0(t_J;\varepsilon)$ sufficiently small, there holds*

$$\max_{0\leq j\leq J} \|\,\mathbf{m}(t_j) - \mathbf{m}^j\,\|_{\mathbf{L}^2} + \Big(k\sum_{j=0}^J \|\,\nabla\{\mathbf{m}(t_j) - \mathbf{m}^j\}\,\|_{\mathbf{L}^2}^2\Big)^{1/2}$$

$$+\frac{1}{\sqrt{\varepsilon}}\Big(k\sum_{j=0}^J\Big\{\|\,\mathbf{m}(t_j) - \mathbf{m}^j\,\|_{\mathbf{L}^4}^4 + \|\,\big\langle \mathbf{m}(t_j) - \mathbf{m}^j, \mathbf{m}(t_j)\big\rangle_{\mathbb{R}^3}\,\|_{L^2}^2\Big\}\Big)^{1/2} \leq C\,k\,.$$

*The positive number $C = C(\omega, t_J, \mathbf{m}_0)$ does not depend on $k$ or $\varepsilon$.*

**Remark 4.4** *The constraint $\varepsilon^{-1} = o(\frac{1}{k})$ is motivated from Remark 4.3, 2. to ensure existence of solutions. The subsequent analysis holds for values $\varepsilon > 1.8k$.*

**Proof:**
We introduce the shorthand notation $\mathbf{e}^j := \mathbf{m}(t_j) - \mathbf{m}^j$, where $\mathbf{m}(t_j)$ solves (4.6), and $\mathbf{m}^j$ is a solution to (4.13), with $\phi(\cdot) = \phi_1(\cdot)$ defined in (4.14).

*1st step:* Let $\boldsymbol{\mathcal{E}}^{j+1}$ denote the error caused by the term $\alpha\,|\,\nabla\mathbf{m}^j\,|^2\mathbf{m}^{j+1} + \mathbf{m}^{j+1}\times\Delta\mathbf{m}^{j+1}$. It can be bounded like in the proof of Theorem 4.4. Moreover, from (4.75), (4.76) we have

$$k\sum_{j=0}^J\Big\{\|\,\boldsymbol{\mathcal{R}}(j;\mathbf{m}_{tt})\,\|_{\mathbf{W}^{-1,2}(\omega,\mathbb{R}^3)}^2 + k\,\|\,\boldsymbol{\mathcal{R}}(j;\mathbf{m}_{tt})\,\|_{\mathbf{L}^2}^2\Big\} \leq C\,k^2\,.$$

We obtain in a similar way to the first step of in the proof of Theorem 4.4,

$$\frac{1}{2}\|\,\mathbf{e}^{J+1}\,\|_{\mathbf{L}^2}^2 + k\sum_{j=0}^J\Big\{\frac{k}{2}\|\,d_t\mathbf{e}^{j+1}\,\|_{\mathbf{L}^2}^2$$
$$+\|\,\nabla\mathbf{e}^{j+1}\,\|_{\mathbf{L}^2}^2 + \frac{1}{\varepsilon}\big(1 - |\,\mathbf{m}^{j+1}\,|^2, \langle\mathbf{m}^{j+1},\mathbf{e}^{j+1}\rangle_{\mathbb{R}^3}\big)\Big\}$$
$$\leq Ck^2 + k\sum_{j=0}^J \|\,\boldsymbol{\mathcal{E}}^{j+1}\,\|_{\mathbf{W}^{-1,2}(\omega,\mathbb{R}^3)}^2\,. \tag{4.93}$$

The last term in the second line of (4.93) can be rewritten in the form

$$
\begin{aligned}
&-\frac{1}{\varepsilon}\Big(|\,\mathbf{m}(t_{j+1})\,|^2 - |\,\mathbf{m}^{j+1}\,|^2, |\,\mathbf{e}^{j+1}\,|^2\Big) \\
&+\frac{1}{\varepsilon}\Big(|\,\mathbf{m}(t_{j+1})\,|^2 - |\,\mathbf{m}^{j+1}\,|^2, \langle \mathbf{m}(t_{j+1}), \mathbf{e}^{j+1}\rangle_{\mathbb{R}^3}\Big) \qquad (4.94)\\
&= \frac{1}{\varepsilon}\Big(\langle \mathbf{e}^{j+1}, 2\mathbf{m}(t_{j+1}) - \mathbf{e}^{j+1}\rangle_{\mathbb{R}^3}, \langle \mathbf{m}(t_{j+1}), \mathbf{e}^{j+1}\rangle_{\mathbb{R}^3} - |\,\mathbf{e}^{j+1}\,|^2\Big)\,.
\end{aligned}
$$

We can now combine (4.93), (4.94) to yield

$$
\begin{aligned}
&\frac{1}{2}\,\|\,\mathbf{e}^{J+1}\,\|^2_{\mathbf{L}^2} + k\sum_{j=0}^{J}\Big\{\frac{k}{2}\,\|\,d_t\mathbf{e}^{j+1}\,\|^2_{\mathbf{L}^2} + \alpha\,\|\,\nabla\mathbf{e}^{j+1}\,\|^2_{\mathbf{L}^2} \\
&\quad+\frac{1}{\varepsilon}\Big(2\|\,\langle \mathbf{e}^{j+1}, \mathbf{m}(t_{j+1})\rangle_{\mathbb{R}^3}\,\|^2_{L^2} + \|\,\mathbf{e}^{j+1}\,\|^4_{\mathbf{L}^4}\Big)\Big\} \qquad (4.95)\\
&\le Ck^2 + k\sum_{j=0}^{J+1}\|\,\boldsymbol{\mathcal{E}}^j\,\|^2_{\mathbf{W}^{-1,2}(\omega,\mathbb{R}^3)} + \frac{3}{\varepsilon}k\sum_{j=0}^{J+1}\big|\,\big(\langle \mathbf{m}(t_j), \mathbf{e}^j\rangle_{\mathbb{R}^3}, |\,\mathbf{e}^j\,|^2\big)\,\big|\,.
\end{aligned}
$$

We confine to the last term. By means of Young's inequality, we have for $\beta, \delta_1 > 0$,

$$
\begin{aligned}
&3\beta\,\big|\,\big(\langle \mathbf{m}(t_{j+1}), \mathbf{e}^{j+1}\rangle_{\mathbb{R}^3}, |\,\mathbf{e}^{j+1}\,|^2\big)\,\big| \qquad (4.96)\\
&\quad\le \frac{3\beta}{4\delta_1}\,\|\,\langle \mathbf{m}(t_{j+1}), \mathbf{e}^{j+1}\rangle_{\mathbb{R}^3}\,\|^2_{L^2} + 3\beta\delta_1\,\|\,\mathbf{e}^{j+1}\,\|^4_{\mathbf{L}^4}\,.
\end{aligned}
$$

For choices $3\beta\delta_1 < \frac{7}{8}$, and $\beta < \left(\frac{3}{7\delta_1}\right)^{-1}$ we can absorb (4.96) on the left-hand side of (4.95). Since $\beta < \frac{7}{\sqrt{72}} \approx 0.82$, we have to bound $(1-\beta)$ times the last term in (4.95).

$$
\begin{aligned}
&\big|\,\big(\langle \mathbf{m}(t_{j+1}), \mathbf{e}^{j+1}\rangle_{\mathbb{R}^3}, |\,\mathbf{e}^{j+1}\,|^2\big)\,\big| \qquad (4.97)\\
&\le \big|\,\big(\langle \mathbf{m}(t_j), \mathbf{e}^j\rangle_{\mathbb{R}^3}, |\,\mathbf{e}^{j+1}\,|^2\big)\,\big| + k\,\big|\,\big(\langle d_t\mathbf{m}(t_{j+1}), \mathbf{e}^j\rangle_{\mathbb{R}^3}, |\,\mathbf{e}^{j+1}\,|^2\big)\,\big| \\
&\quad+k\,\big|\,\big(\langle \mathbf{m}(t_{j+1}), d_t\mathbf{e}^{j+1}\rangle_{\mathbb{R}^3}, |\,\mathbf{e}^{j+1}\,|^2\big)\,\big|\,.
\end{aligned}
$$

The first contribution on the right-hand side can be controlled by interpolating $L^4$ between $L^2$ and $W^{1,2}$, together with Young's inequality,

$$
\begin{aligned}
&\frac{1}{\varepsilon}\,\big(\langle \mathbf{m}(t_j), \mathbf{e}^j\rangle_{\mathbb{R}^3}, |\,\mathbf{e}^{j+1}\,|^2\big) \le \frac{C}{\varepsilon}\,\|\,\langle \mathbf{m}(t_j), \mathbf{e}^j\rangle_{\mathbb{R}^3}\,\|_{L^2}\,\|\,\mathbf{e}^{j+1}\,\|_{\mathbf{L}^2}\,\|\,\nabla\mathbf{e}^{j+1}\,\|_{\mathbf{L}^2} \\
&\le \frac{C}{\varepsilon^2}\,\|\,\langle \mathbf{m}(t_j), \mathbf{e}^j\rangle_{\mathbb{R}^3}\,\|^2_{L^2}\,\|\,\mathbf{e}^{j+1}\,\|^2_{\mathbf{L}^2} + \frac{\alpha}{2}\,\|\,\nabla\mathbf{e}^{j+1}\,\|^2_{\mathbf{L}^2}\,. \qquad (4.98)
\end{aligned}
$$

What regards the second and (scaled) third term on the right-hand side of (4.97), we conclude for $\delta_2 > 0$,

$$\begin{aligned}
&k \,\|\, d_t\mathbf{m}(t_{j+1}) \,\|_{\mathbf{L}^4} \,\|\, \mathbf{e}^j \,\|_{\mathbf{L}^4} \,\|\, \mathbf{e}^{j+1} \,\|_{\mathbf{L}^4}^2 \\
&\quad\leq C_\alpha k^2 \,\|\, d_t\mathbf{m}(t_{j+1}) \,\|_{\mathbf{L}^2} \,\|\, \nabla d_t\mathbf{m}(t_{j+1}) \,\|_{\mathbf{L}^2} \,\|\, \mathbf{e}^j \,\|_{\mathbf{L}^2} \,\|\, \nabla\mathbf{e}^j \,\|_{\mathbf{L}^2} \,\|\, \mathbf{e}^{j+1} \,\|_{\mathbf{L}^2}^2 \\
&\qquad+\frac{\alpha}{8} \,\|\, \nabla\mathbf{e}^{j+1} \,\|_{\mathbf{L}^2}^2 , \\
&3(1-\beta)\frac{k}{\varepsilon} \,\|\, d_t\mathbf{e}^{j+1} \,\|_{\mathbf{L}^2} \,\|\, \mathbf{e}^{j+1} \,\|_{\mathbf{L}^4}^2 \qquad\qquad (4.99)\\
&\quad\leq 3(1-\beta)\frac{1}{4\delta_2}\frac{k^2}{\varepsilon} \,\|\, d_t\mathbf{e}^{j+1} \,\|_{\mathbf{L}^2}^2 + 3(1-\beta)\frac{\delta_2}{\varepsilon} \,\|\, \mathbf{e}^{j+1} \,\|_{\mathbf{L}^4}^4 .
\end{aligned}$$

For absorption on the left-hand side of (4.95), we need to satisfy

$$3(1-\beta)\delta_2 < \frac{1}{8}, \qquad \frac{3(1-\beta)}{4\delta_2}\frac{k}{\varepsilon} < \frac{1}{2}, \tag{4.100}$$

where choices $\varepsilon > 1.8k$ are sufficient.

*2nd step:* We proceed by induction to verify existence of constants $C_i \equiv C_i(\omega, t_J, \alpha, \mathbf{m}_0)$, $i = 1, 2$, for $0 \leq \ell \leq J$, such that

$$\begin{aligned}
&\|\, \mathbf{e}^{\ell+1} \,\|_{\mathbf{L}^2}^2 + \frac{k^2}{2}\sum_{j=0}^{\ell} \|\, d_t\mathbf{e}^{j+1} \,\|_{\mathbf{L}^2}^2 + \frac{\alpha k}{2}\sum_{j=0}^{\ell} \|\, \nabla\mathbf{e}^{j+1} \,\|_{\mathbf{L}^2}^2 \qquad\qquad (4.101)\\
&\quad+\frac{k}{\varepsilon}\sum_{j=0}^{\ell}\big\{\|\, \mathbf{e}^{j+1} \,\|_{\mathbf{L}^4}^4 + \|\, \langle \mathbf{m}(t_{j+1}), \mathbf{e}^{j+1}\rangle_{\mathbb{R}^3} \,\|_{L^2}^2\big\} \leq C_1\, k^2 \exp\big(C_2\, t_\ell\big) ,
\end{aligned}$$

$$\begin{aligned}
&\|\, \nabla\mathbf{e}^{\ell+1} \,\|_{\mathbf{L}^2}^2 + \frac{k^2}{2}\sum_{j=0}^{\ell} \|\, \nabla d_t\mathbf{e}^{j+1} \,\|_{\mathbf{L}^2}^2 \qquad\qquad (4.102)\\
&\quad+\frac{\alpha k}{2}\sum_{j=0}^{\ell} \|\, \Delta\mathbf{e}^{j+1} \,\|_{\mathbf{L}^2}^2 \leq C_1\, k \exp\big(C_2\, t_\ell\big) ,
\end{aligned}$$

for $F(\varepsilon, k) > 0$ as stated in the theorem, and $k \leq k_0(t_J; \varepsilon)$. These statements are obviously valid for $\ell = 0$. — We may now collect the statements in the first step to absorb portions that stem from the penalty term by the second term on the left-hand side of (4.93). The contributions $\{\boldsymbol{\mathcal{E}}^j\}_{j=0}^{J+1}$ are dealt with as in in the proof of Theorem 4.4. For (4.98), we need the constraint $\varepsilon^{-1/2} = o(\frac{1}{k})$ to successfully apply Gronwall's inequality in the induction step and verify $\frac{k}{\varepsilon^2}\sum_{j=0}^{\ell+1} \| \langle \mathbf{e}^j, \mathbf{m}(t_j)\rangle_{\mathbb{R}^3} \|_{L^2}^2 \leq \frac{C_2 t_{J+1}}{2}$, where $k \leq k_0(t_J;\varepsilon)$; for (4.99), $\varepsilon > 1.8k$ is sufficient in that respect.

To show (4.102) at $(\ell+1)$, we can copy the argumentation from the proof of Theorem 4.4 to control $\{\mathcal{E}^j\}_{j=0}^{J+1}$ in the norm $\ell^2(I_k;\mathbf{L}^2)$. The additional term comes from the penalty term; we exploit (4.83) in the following inequality, together with $\|\cdot\|_{\mathbf{L}^\infty} \leq C\|\cdot\|_{\mathbf{L}^2}^{1/2}\|\cdot\|_{\mathbf{W}^{2,2}}^{1/2}$.

$$
\begin{aligned}
&\frac{k}{\varepsilon^2}\sum_{j=0}^{\ell+2}\Big\{\| (|\mathbf{m}(t_j)|^2 - |\mathbf{m}^j|^2)\mathbf{m}(t_j)\|_{\mathbf{L}^2}^2 + \|(|\mathbf{m}(t_j)|^2 - |\mathbf{m}^j|^2)\mathbf{e}^j\|_{\mathbf{L}^2}^2\Big\} \\
&\leq \frac{2k}{\varepsilon^2}\sum_{j=0}^{\ell+2}\Big\{\|\mathbf{e}^j\|_{\mathbf{L}^4}^4 + \|\langle \mathbf{e}^j, \mathbf{m}(t_j)\rangle_{\mathbb{R}^3}\|_{L^2}^2\Big\}\Big\{1 + C\|\mathbf{e}^j\|_{\mathbf{L}^2}\|\Delta \mathbf{e}^j\|_{\mathbf{L}^2}\Big\} \\
&\leq \frac{2k}{\varepsilon^2}\sum_{j=0}^{\ell+2}\Big\{\|\mathbf{e}^j\|_{\mathbf{L}^4}^4 + \|\langle \mathbf{e}^j, \mathbf{m}(t_j)\rangle_{\mathbb{R}^3}\|_{L^2}^2\Big\} \qquad (4.103)\\
&\quad + C_\alpha \max_{0\leq j\leq \ell+1}\|\mathbf{e}^j\|_{\mathbf{L}^2}^4\Big[\frac{k}{\varepsilon^4}\sum_{j=0}^{\ell+1}\|\mathbf{e}^j\|_{\mathbf{L}^4}^4\|\nabla \mathbf{e}^j\|_{\mathbf{L}^2}^2 + \frac{\alpha k}{2}\sum_{j=0}^{\ell+2}\|\Delta \mathbf{e}^j\|_{\mathbf{L}^2}^2\Big] \\
&\quad + C_\alpha \max_{0\leq j\leq \ell+2}\Big\{\|\mathbf{e}^j\|_{\mathbf{L}^2}^2\|\langle \mathbf{e}^j, \mathbf{m}(t_j)\rangle_{\mathbb{R}^3}\|_{L^2}^2\Big\}\frac{k}{\varepsilon^4}\sum_{j=0}^{\ell+2}\|\langle \mathbf{e}^j, \mathbf{m}(t_j)\rangle_{\mathbb{R}^3}\|_{\mathbf{L}^2}^2 .
\end{aligned}
$$

The crucial term is the first one next to the inequality sign. Again, we exploit the constraint $F(\varepsilon, k) > 0$ to settle the induction step. □

We can draw further conclusions from this theorem.

**Corollary 4.2** *Let the assumptions of Theorem 4.5 be valid. Then,* $\{\mathbf{m}^j\}_{j=0}^J$ *enjoys*

(a) $$\max_{0\leq j\leq J}\|1 - |\mathbf{m}^j|^2\|_{L^2} \leq C\sqrt{k\varepsilon},$$

(b) $$\max_{1\leq j\leq J}\Big\{\|\mathbf{m}^j\|_{\mathbf{W}^{2,2}} + \|d_t\mathbf{m}^j\|_{\mathbf{L}^2}\Big\} + \Big(k\sum_{j=1}^J\|\nabla d_t\mathbf{m}^j\|_{\mathbf{L}^2}^2\Big)^{1/2} \leq C,$$

*for* $C = C(t_J, \omega, \mathbf{m}_0)$ *independent of* $k$.

**Proof:**
Result (a) comes from (4.83) and Theorem 4.5; (b) is a consequence of (4.101), (4.102). □

**Analysis of the penalization function $\phi_2(\cdot)$ in (4.13)**

For $\phi_2(\mathbf{m}^j, \mathbf{m}^{j+1}) = \frac{1}{\varepsilon}\big(1 - \frac{1}{|\mathbf{m}^j|^2}\big)$, we make use of iterates from the previous time-step. In order to avoid degeneracy, i.e., $|\mathbf{m}^{j+1}| = 0$ on a set $\omega' \subset \omega$ of positive Lebesgue measure, we choose time-steps $k \leq k_0(t_J)$ sufficiently small. If compared to $F(\varepsilon, k) > 0$ from Remark 4.4, we need a slightly more restrictive constraint since some more phenomena enter the picture.

**Theorem 4.6** *Let $\{\mathbf{m}^j\}_{j=0}^J$, for $0 < t_J < T_0(\mathbf{m}_0)$ be the solution to (4.13), with $\phi(\cdot) = \phi_2(\cdot)$, for $\varepsilon > 1.9k$, and $k \leq k_0(t_J)$ sufficiently small. Let $\mathbf{m}$ solve (4.6), for $\mathbf{m}_0 \in \mathbf{W}^{2,2}(\omega, \mathcal{S}^2)$. Then*

$$\max_{0\leq j\leq J} \| \mathbf{m}(t_j) - \mathbf{m}^j \|_{\mathbf{L}^2} + \Big(k \sum_{j=0}^{J} \| \nabla\{\mathbf{m}(t_j) - \mathbf{m}^j\} \|_{\mathbf{L}^2}^2\Big)^{1/2}$$

$$+\frac{1}{\sqrt{\varepsilon}}\Big(k \sum_{j=0}^{J}\Big\{\| \mathbf{m}(t_j) - \mathbf{m}^j \|_{\mathbf{L}^4}^4 + \| \langle \mathbf{m}(t_j) - \mathbf{m}^j, \mathbf{m}(t_j)\rangle_{\mathbb{R}^3} \|_{L^2}^2\Big\}\Big)^{1/2} \leq C\,k.$$

*Here, $C = C(\omega, t_J, \mathbf{m}_0)$ is a positive number that does not depend on $k$.*

**Proof:**
*1st step:* Let $\mathbf{e}^j := \mathbf{m}(t_j) - \mathbf{m}^j$, where $\mathbf{m}(t_j)$ and $\mathbf{m}^j$ solve (4.6) and (4.13), respectively, and $\phi(\cdot) = \phi_2(\cdot)$ from (4.14). We focus on the effect of penalization since discretization and amplification mechanisms from the nonlinear terms on the right-hand of (4.6), (4.13) work the same way as studied in Section 4.2.1 for (4.72). The error identity reads as follows,

$$d_t\mathbf{e}^{j+1} - \alpha\Delta\mathbf{e}^{j+1} - \frac{1}{\varepsilon}\big(1 - \frac{1}{|\mathbf{m}^j|^2}\big)\mathbf{m}^{j+1} = \boldsymbol{\mathcal{E}}^{j+1}\,, \qquad (4.104)$$

and $\boldsymbol{\mathcal{E}}^{j+1} \in \ell^2(I_k; \mathbf{L}^2)$ summarizes terms due to consistency error and nonlinear effects in the scheme. — After testing (4.104) with $\mathbf{e}^{j+1}$, we may reformulate the penalty part as

$$
\begin{aligned}
&\frac{1}{\varepsilon}\Big(\frac{1}{|\mathbf{m}^j|^2}\{|\mathbf{m}(t_j)|^2 - |\mathbf{m}^j|^2\}, \langle \mathbf{e}^{j+1}, \mathbf{m}^{j+1}\rangle_{\mathbb{R}^3}\Big) \\
&= \frac{1}{\varepsilon}\Big(\frac{1}{|\mathbf{m}^j|^2}\langle \mathbf{e}^j, -\mathbf{e}^j + 2\,\mathbf{m}(t_j)\rangle_{\mathbb{R}^3}, \langle \mathbf{e}^{j+1}, \mathbf{m}^{j+1}\rangle_{\mathbb{R}^3}\Big) \qquad (4.105)\\
&= \frac{1}{\varepsilon}\Big\{\|\frac{|\mathbf{e}^j|^2}{|\mathbf{m}^j|}\|^2_{L^2} + 2\,\|\frac{\langle \mathbf{e}^j, \mathbf{m}(t_j)\rangle_{\mathbb{R}^3}}{|\mathbf{m}^j|}\|^2_{L^2} - 3\Big(\frac{|\mathbf{e}^j|^2}{|\mathbf{m}^j|}, \frac{\langle \mathbf{m}(t_j), \mathbf{e}^j\rangle_{\mathbb{R}^3}}{|\mathbf{m}^j|}\Big)\Big\} \\
&\quad + \frac{k^2}{\varepsilon}\Big(\frac{\langle \mathbf{e}^j, -\mathbf{e}^j + 2\,\mathbf{m}(t_j)\rangle_{\mathbb{R}^3}}{|\mathbf{m}^j|}, \frac{\langle d_t\mathbf{e}^{j+1}, d_t\mathbf{m}(t_{j+1}) - d_t\mathbf{e}^{j+1}\rangle_{\mathbb{R}^3}}{|\mathbf{m}^j|}\Big) \\
&\quad + \frac{k}{\varepsilon}\Big(\frac{\langle \mathbf{e}^j, -\mathbf{e}^j + 2\,\mathbf{m}(t_j)\rangle_{\mathbb{R}^3}}{|\mathbf{m}^j|}, \frac{\langle d_t\mathbf{e}^{j+1}, \mathbf{m}^j\rangle_{\mathbb{R}^3} + \langle \mathbf{e}^j, d_t\mathbf{m}(t_{j+1}) - d_t\mathbf{e}^{j+1}\rangle_{\mathbb{R}^3}}{|\mathbf{m}^j|}\Big) \\
&= \frac{1}{\varepsilon}\Big\{\|\frac{|\mathbf{e}^j|^2}{|\mathbf{m}^j|}\|^2_{L^2} + 2\,\|\frac{\langle \mathbf{e}^j, \mathbf{m}(t_j)\rangle_{\mathbb{R}^3}}{|\mathbf{m}^j|}\|^2_{L^2} - 3\,\Big(\frac{|\mathbf{e}^j|^2}{|\mathbf{m}^j|}, \frac{\langle \mathbf{m}(t_j), \mathbf{e}^j\rangle_{\mathbb{R}^3}}{|\mathbf{m}^j|}\Big) \\
&\quad + k^2\,\|\frac{|\mathbf{e}^j|\,|d_t\mathbf{e}^{j+1}|}{|\mathbf{m}^j|}\|^2_{L^2} + \frac{k^2}{\varepsilon}\Big(\frac{|\mathbf{e}^j|^2}{|\mathbf{m}^j|}, \frac{\langle d_t\mathbf{e}^{j+1}, d_t\mathbf{m}(t_{j+1})\rangle_{\mathbb{R}^3}}{}\Big)\Big\} \\
&\quad + \frac{k}{\varepsilon}\Big\{\Big(-\frac{|\mathbf{e}^j|^2}{|\mathbf{m}^j|}, \frac{\langle d_t\mathbf{e}^{j+1}, \mathbf{m}^j\rangle_{\mathbb{R}^3} + \langle \mathbf{e}^j, d_t\mathbf{m}(t_{j+1})\rangle_{\mathbb{R}^3}}{|\mathbf{m}^j|}\Big) \\
&\quad + 2k\,\Big(\frac{\langle \mathbf{m}(t_j), \mathbf{e}^j\rangle_{\mathbb{R}^3}}{|\mathbf{m}^j|}, \frac{-|d_t\mathbf{e}^{j+1}|^2 + \langle d_t\mathbf{e}^{j+1}, d_t\mathbf{m}(t_{j+1})\rangle_{\mathbb{R}^3}}{|\mathbf{m}^j|}\Big) \\
&\quad + \Big(\frac{\langle \mathbf{e}^j, -\mathbf{e}^j + 2\,\mathbf{m}(t_j)\rangle_{\mathbb{R}^3}}{|\mathbf{m}^j|}, \frac{\langle d_t\mathbf{e}^{j+1}, \mathbf{m}^j\rangle_{\mathbb{R}^3} + \langle \mathbf{e}^j, d_t\mathbf{m}(t_{j+1}) - d_t\mathbf{e}^{j+1}\rangle_{\mathbb{R}^3}}{|\mathbf{m}^j|}\Big)\Big\}.
\end{aligned}
$$

The terms in the second line next to the last equality sign can be controlled through ($\delta_3 > 0$)

$$
\geq \frac{k^2}{2\varepsilon}\,\|\frac{|\mathbf{e}^j|\,|d_t\mathbf{e}^{j+1}|}{|\mathbf{m}^j|}\|^2_{\mathbf{L}^2} - \frac{Ck^2}{\varepsilon}\,\|\frac{|\mathbf{e}^j|\,|d_t\mathbf{m}(t_{j+1})|}{|\mathbf{m}^j|}\|^2_{\mathbf{L}^2},
$$

and

$$
\begin{aligned}
\frac{Ck^2}{\varepsilon}\,\|\mathbf{e}^j\|^2_{\mathbf{L}^\infty}\,\|d_t\mathbf{m}(t_{j+1})\|^2_{\mathbf{L}^2} &\leq \frac{Ck^2}{\varepsilon}\,\|\mathbf{e}^j\|_{\mathbf{L}^2}\,\|\mathbf{e}^j\|_{\mathbf{W}^{2,2}}\,\|d_t\mathbf{m}(t_{j+1})\|^2_{\mathbf{L}^2} \\
&\leq \delta_3\,\|\mathbf{e}^j\|^2_{\mathbf{L}^2}\,\|\mathbf{e}^j\|^2_{\mathbf{W}^{2,2}} + C_{\delta_3}\frac{k^4}{\varepsilon^2}\,\|d_t\mathbf{m}(t_{j+1})\|^2_{\mathbf{L}^2}.
\end{aligned}
$$

The terms in the last three lines of (4.105) can be bounded by

$$\frac{\delta_4}{\varepsilon}\Big\{ \| \frac{|\mathbf{e}^j|^2}{|\mathbf{m}^j|} \|_{L^2}^2 + 2 \| \frac{\langle \mathbf{e}^j, \mathbf{m}(t_j)\rangle_{\mathbb{R}^3}}{|\mathbf{m}^j|} \|_{L^2}^2 \Big\} \tag{4.106}$$
$$+\frac{1}{4\delta_4}\frac{k^2}{\varepsilon}\Big\{ \| \mathbf{e}^j \|_{\mathbf{L}^4}^2 \| \frac{d_t\mathbf{m}(t_{j+1})}{|\mathbf{m}^j|} \|_{\mathbf{L}^4}^2 + \| \frac{\langle \mathbf{e}^j, d_t\mathbf{e}^{j+1}\rangle_{\mathbb{R}^3}}{|\mathbf{m}^j|} \|_{L^2}^2 + \| \frac{d_t\mathbf{e}^{j+1}}{|\mathbf{m}^j|} \|_{\mathbf{L}^2}^2$$
$$+2k \| \frac{d_t\mathbf{e}^{j+1}}{\sqrt{|\mathbf{m}^j|}} \|_{\mathbf{L}^2} \{ \| \nabla \mathbf{e}^j \|_{\mathbf{L}^2} + \| \nabla \mathbf{e}^{j+1} \|_{\mathbf{L}^2} \}$$
$$+k^2 \| \frac{d_t\mathbf{m}(t_{j+1})}{\sqrt{|\mathbf{m}^j|}} \|_{\mathbf{L}^4}^4 + \| \mathbf{e}^j \|_{\mathbf{L}^\infty}^2 \| \frac{d_t\mathbf{e}^{j+1}}{\sqrt{|\mathbf{m}^j|}} \|_{\mathbf{L}^2}^2 \Big\},$$

where $\delta_4 = \delta_4(\delta_1, \delta_2, \beta) > 0$, with $\delta_1, \delta_2, \beta$ from the proof of Theorem 4.5, is chosen sufficiently small to allow for absorption of the leading term in (4.106) by the fourth and fifth term on the left-hand side of an error inequality analogous to (4.95). The last term in (4.106) can be controlled as follows,

$$\| \mathbf{e}^j \|_{\mathbf{L}^\infty}^2 \| d_t\mathbf{e}^{j+1} \|_{\mathbf{L}^2}^2 \le C \| \mathbf{e}^j \|_{\mathbf{L}^2} \| \mathbf{e}^j \|_{\mathbf{W}^{2,2}} \| d_t\mathbf{e}^{j+1} \|_{\mathbf{L}^2}^2 .$$

and the last but one term (in summarized form)

$$\begin{aligned} k\sum_{j=0}^{J} \| d_t\mathbf{m}(t_{j+1}) \|_{\mathbf{L}^4}^4 &\le k^{-3}\sum_{j=0}^{J}\Big( \int_{t_j}^{t_{j+1}} \| \mathbf{m}_t(s) \|_{\mathbf{L}^4} \, \mathrm{d}s \Big)^4 \\ &\le C \sup_{[0,t_{J+1}]} \| \mathbf{m}_t \|_{\mathbf{L}^2}^2 \int_0^{t_{J+1}} \| \mathbf{m}_t(s) \|_{\mathbf{W}^{1,2}}^2 \, \mathrm{d}s \le C . \end{aligned}$$

Here, we made use of Theorem 4.3.

*2nd step:* We verify by induction the statements that correspond to (4.101), (4.102). We can follow the argumentation of step 2 in the proof of Theorem 4.5 once we make sure that

$$\max_{0\le j\le \ell} \| \mathbf{m}^j \|_{\mathbf{L}^\infty} \ge c_0 , \tag{4.107}$$

for $0 \le \ell \le J$, and $|1 - c_0| < 1$ sufficiently small. This is a consequence of the induction argument, since

$$\| \mathbf{e}^j \|_{\mathbf{L}^\infty} \le C \| \mathbf{e}^j \|_{\mathbf{L}^2}^{1/2} \| \mathbf{e}^j \|_{\mathbf{W}^{2,2}}^{1/2} \le C_1 \exp\big(C_2 t_\ell\big)\, k. \tag{4.108}$$

We require $k \le k_0(t_J)$ to enforce (4.107). □

**Corollary 4.3** *Let the assumptions of Theorem 4.6 be valid. Then* $\{\mathbf{m}^j\}_{j=0}^J$ *enjoys*

$$\text{(a)} \qquad \max_{0\leq j\leq J} \| 1 - |\mathbf{m}^j|^2 \|_{L^2} \leq C\sqrt{k\varepsilon},$$

$$\text{(b)} \qquad \max_{1\leq j\leq J} \{ \| \mathbf{m}^j \|_{\mathbf{W}^{2,2}} + \| d_t \mathbf{m}^j \|_{\mathbf{L}^2} \} + \Big( k \sum_{j=1}^J \| \nabla d_t \mathbf{m}^j \|_{\mathbf{L}^2}^2 \Big)^{1/2} \leq C,$$

*where* $C = C(t_J, \omega, \mathbf{m}_0)$ *does not dependent of* $k$.

**Analysis of the penalization function $\phi_3(\cdot)$ in (4.13) and some modifications of it**

The convergence analysis for $\phi_3(\mathbf{m}^j, \mathbf{m}^{j+1}) = \frac{1}{\varepsilon}\big(1 - \frac{1}{|\mathbf{m}^j|}\big)$ is given next, and it turns out that this penalization method leads to increased stability properties with respect to choices of tuples $\varepsilon, k$, if compared to $\phi_2(\cdot)$. The proof of the subsequent theorem essentially follows the lines of the proof of Theorem 4.6, and is based on an inductive argumentation and a combination of stability features of the scheme due to (implicit) time discretization and (semi-explicit in time) penalization. In the following, we focus on deviating new aspects of the penalty function $\phi_3(\cdot)$ if compared to $\phi_2(\cdot)$.

**Theorem 4.7** *Let* $\{\mathbf{m}^j\}_{j=0}^J$, *for* $0 < t_J < T_0(\mathbf{m}_0)$ *be the solution to* (4.13), *with* $\phi(\cdot) = \phi_3(\cdot)$, *and* $\mathbf{m}_0 \in \mathbf{W}^{2,2}(\omega, \mathcal{S}^2)$. *Suppose* $\varepsilon \geq k$, *for* $k \leq k_0(t_J)$ *sufficiently small. Let further* $\mathbf{m}$ *solve* (4.6). *Then*

$$\max_{0\leq j\leq J} \| \mathbf{m}(t_j) - \mathbf{m}^j \|_{\mathbf{L}^2} + \Big( k \sum_{j=0}^J \| \nabla \{ \mathbf{m}(t_j) - \mathbf{m}^j \} \|_{\mathbf{L}^2}^2 \Big)^{1/2}$$

$$+ \frac{1}{\sqrt{\varepsilon}} \Big( k \sum_{j=0}^J \Big\{ \| \mathbf{m}(t_j) - \mathbf{m}^j \|_{\mathbf{L}^4}^4 + \| \langle \mathbf{m}(t_j) - \mathbf{m}^j, \mathbf{m}(t_j) \rangle_{\mathbb{R}^3} \|_{L^2}^2 \Big\} \Big)^{1/2} \leq C k,$$

*with* $C = C(\omega, t_J, \mathbf{m}_0)$ *a positive number that does not depend on* $k$.

**Remark 4.5** *The crucial observation to see increased freedom in the choice of* $\varepsilon, k > 0$ *for this method is the larger denominator in the fractions given in* (4.112) *below. This, in turn, delivers a constant* $\frac{3}{8\delta_2}(1-\beta)$ *rather than* $\frac{3}{4\delta_2}(1-\beta)$ *in* (4.99), *for* $k \leq k(t_J)$ *sufficiently small.*

**Proof:**
*1st step:* We set $\mathbf{e}^j := \mathbf{m}(t_j) - \mathbf{m}^j$, where $\mathbf{m}(t_j)$ and $\mathbf{m}^j$ solve (4.6)) and (4.13), respectively, and $\phi(\cdot) = \phi_3(\cdot)$ from (4.14). Again, we focus on the effect of penalization, and omit to study further error contributions that are summarized by $\boldsymbol{\mathcal{E}}^j$ in the next formula and that we already analyzed before (see proofs of Theorems 4.4 to 4.6),

$$d_t\mathbf{e}^{j+1} - \alpha\Delta\mathbf{e}^{j+1} - \frac{1}{\varepsilon}\Big(1 - \frac{1}{|\,\mathbf{m}^j\,|}\Big)\mathbf{m}^{j+1} = \boldsymbol{\mathcal{E}}^{j+1}\,. \tag{4.109}$$

We carry out some algebraic manipulations of the penalty term.

$$\begin{aligned}
|\,\mathbf{m}(t_j)\,| - |\,\mathbf{m}^j\,| &= \Big\langle \frac{-\mathbf{m}^j}{\sqrt{|\,\mathbf{m}^j\,|}} + \mathbf{m}(t_j), \frac{\mathbf{m}^j}{\sqrt{|\,\mathbf{m}^j\,|}} + \mathbf{m}(t_j)\Big\rangle_{\mathbb{R}^3} \\
&= \Big\langle \frac{\mathbf{e}^j}{\sqrt{|\,\mathbf{m}^j\,|}} + \Big(1 - \frac{1}{\sqrt{|\,\mathbf{m}^j\,|}}\Big)\mathbf{m}(t_j), \Big(1 + \frac{1}{\sqrt{|\,\mathbf{m}^j\,|}}\Big)\mathbf{m}(t_j) - \frac{\mathbf{e}^j}{\sqrt{|\,\mathbf{m}^j\,|}}\Big\rangle_{\mathbb{R}^3} \\
&= -\frac{|\,\mathbf{e}^j\,|^2}{|\,\mathbf{m}^j\,|} + \Big(1 + \frac{1}{\sqrt{|\,\mathbf{m}^j\,|}}\Big)\frac{1}{\sqrt{|\,\mathbf{m}^j\,|}}\langle \mathbf{e}^j, \mathbf{m}(t_j)\rangle_{\mathbb{R}^3} \qquad (4.110) \\
&\quad + \Big(\frac{1}{\sqrt{|\,\mathbf{m}^j\,|}} - 1\Big)\frac{1}{\sqrt{|\,\mathbf{m}^j\,|}}\langle \mathbf{e}^j, \mathbf{m}(t_j)\rangle_{\mathbb{R}^3} + |\,\mathbf{m}(t_j)\,|^2\Big(1 - \frac{1}{|\,\mathbf{m}^j\,|}\Big) \\
&= -\frac{|\,\mathbf{e}^j\,|^2}{|\,\mathbf{m}^j\,|} + \frac{2}{|\,\mathbf{m}^j\,|}\langle \mathbf{e}^j, \mathbf{m}(t_j)\rangle_{\mathbb{R}^3} + \frac{|\,\mathbf{m}^j\,| - |\,\mathbf{m}(t_j)\,|}{|\,\mathbf{m}^j\,|}\,.
\end{aligned}$$

Thus, we obtain

$$\big(1 + |\,\mathbf{m}^j\,|\big)\big\{|\,\mathbf{m}(t_j)\,| - |\,\mathbf{m}^j\,|\big\} = -|\,\mathbf{e}^j\,|^2 + 2\,\langle \mathbf{e}^j, \mathbf{m}(t_j)\rangle_{\mathbb{R}^3}\,. \tag{4.111}$$

We may now benefit from (4.111) to rewrite the penalty term in (4.115), if tested by $\mathbf{e}^{j+1}$. We discuss arising terms separately.

The subsequent argument is adopted from (4.105), (4.106) to the present situation.

$$\frac{1}{\varepsilon}\Big(\frac{-|\,\mathbf{e}^j\,|^2+2\langle \mathbf{e}^j,\mathbf{m}(t_j)\rangle_{\mathbb{R}^3}}{|\,\mathbf{m}^j\,|(1+|\,\mathbf{m}^j\,|)},\langle \mathbf{m}^{j+1},\mathbf{e}^{j+1}\rangle_{\mathbb{R}^3}\Big)\geq \tag{4.112}$$
$$\frac{1}{\varepsilon}\Big\{\|\,\frac{|\,\mathbf{e}^j\,|^2}{\sqrt{|\,\mathbf{m}^j\,|(1+|\,\mathbf{m}^j\,|)}}\,\|_{L^2}^2+2\,\|\,\frac{\langle \mathbf{e}^j,\mathbf{m}(t_j)\rangle_{\mathbb{R}^3}}{\sqrt{|\,\mathbf{m}^j\,|(1+|\,\mathbf{m}^j\,|)}}\,\|_{L^2}^2$$
$$+k^2\,\|\,\frac{|\,\mathbf{e}^j\,|\,|\,d_t\mathbf{e}^{j+1}\,|}{\sqrt{|\,\mathbf{m}^j\,|\,(1+|\,\mathbf{m}^j\,|)}}\,\|_{L^2}^2-3\,\Big(\frac{|\,\mathbf{e}^j\,|^2,\langle \mathbf{e}^j,\mathbf{m}(t_j)\rangle_{\mathbb{R}^3}}{|\,\mathbf{m}^j\,|(1+|\,\mathbf{m}^j\,|)}\Big)\Big\}$$
$$-\frac{\delta_3}{\varepsilon}\Big\{\|\,\frac{|\,\mathbf{e}^j\,|^2}{\sqrt{|\,\mathbf{m}^j\,|\,(1+|\,\mathbf{m}^j\,|)}}\,\|_{L^2}^2+2\,\|\,\frac{\langle \mathbf{e}^j,\mathbf{m}(t_j)\rangle_{\mathbb{R}^3}}{\sqrt{|\,\mathbf{m}^j\,|\,(1+|\,\mathbf{m}^j\,|)}}\,\|_{L^2}^2\Big\}$$
$$-\frac{1}{4\delta_3}\frac{k^2}{\varepsilon}\Big\{\|\,\mathbf{e}^j\,\|_{\mathbf{L}^4}^2\,\|\,\frac{d_t\mathbf{m}(t_{j+1})}{\sqrt{|\,\mathbf{m}^j\,|\,(1+|\,\mathbf{m}^j\,|)}}\,\|_{\mathbf{L}^4}^2+\|\,\frac{\langle \mathbf{e}^j,d_t\mathbf{e}^{j+1}\rangle_{\mathbb{R}^3}}{\sqrt{|\,\mathbf{m}^j\,|\,(1+|\,\mathbf{m}^j\,|)}}\,\|_{L^2}^2$$
$$+\|\,\frac{d_t\mathbf{e}^{j+1}}{\sqrt{|\,\mathbf{m}^j\,|\,(1+|\,\mathbf{m}^j\,|)}}\,\|_{\mathbf{L}^2}^2$$
$$+2k\,\|\,\frac{d_t\mathbf{e}^{j+1}}{\sqrt[4]{|\,\mathbf{m}^j\,|\,(1+|\,\mathbf{m}^j\,|)}}\,\|_{\mathbf{L}^2}\,\{\|\,\nabla\mathbf{e}^j\,\|_{\mathbf{L}^2}+\|\,\nabla\mathbf{e}^{j+1}\,\|_{\mathbf{L}^2}\}$$
$$+k^2\,\|\,\frac{d_t\mathbf{m}(t_{j+1})}{\sqrt[4]{|\,\mathbf{m}^j\,|\,(1+|\,\mathbf{m}^j\,|)}}\,\|_{\mathbf{L}^4}^4+\|\,\mathbf{e}^j\,\|_{\mathbf{L}^\infty}^2\,\|\,\frac{d_t\mathbf{e}^{j+1}}{\sqrt{|\,\mathbf{m}^j\,|\,(1+|\,\mathbf{m}^j\,|)}}\,\|_{\mathbf{L}^2}^2\Big\},$$

for $\delta_3=\delta_3(\delta_1,\delta_2,\beta)>0$ chosen sufficiently small. We refer to the proof of Theorem 4.6 for notation.

*2nd step:* This part of the proof of Theorem 4.6 applies here as well. □

**Corollary 4.4** *Let the assumptions of Theorem 4.7 be valid. Then* $\{\mathbf{m}^j\}_{j=0}^J$ *enjoys*

(a) $$\max_{0\leq j\leq J}\|\,1-|\,\mathbf{m}^j\,|\,\|_{L^2}\leq C\sqrt{k\varepsilon}\,,$$

(b) $$\max_{0\leq j\leq J}\Big\{\|\,\mathbf{m}^j\,\|_{\mathbf{W}^{2,2}}+\|\,d_t\mathbf{m}^j\,\|_{\mathbf{L}^2}\Big\}+\Big(k\sum_{j=1}^J\|\,\nabla d_t\mathbf{m}^j\,\|_{\mathbf{L}^2}^2\Big)^{1/2}\leq C\,,$$

*where* $C=C(\omega,t_J,\mathbf{m}_0)$ *does not dependent of* $k$.

**Remark 4.6** *Part* (a) *even holds for the squared modulus of the iterates.*

As can be seen from the proof, there are competing effects of penalization caused by $\phi_3(\cdot)$ (on scale $\varepsilon$) and by time discretization, using the implicit Euler scheme (on scale $k$). In order to increase upon stability with respect to larger values of $k$ and allow for more flexible choices $F(\varepsilon, k) > 0$, we introduce the penalization $\phi_3^\gamma(\mathbf{m}^j, \mathbf{m}^{j+1}) = \frac{1}{\varepsilon}\{1 - \frac{1}{|\mathbf{m}^j|^{2-\gamma}}\}\mathbf{m}^{j+1}$, for $\gamma \in \mathbb{N}_0$. Herefore, note that

$$|\mathbf{m}(t_j)|^{2-\gamma} - |\mathbf{m}^j|^{2-\gamma} = \frac{|\mathbf{m}(t_j)| - |\mathbf{m}^j|}{\prod_{\kappa=1}^{\gamma}\{|\mathbf{m}(t_j)|^{2-\kappa} + |\mathbf{m}^j|^{2-\kappa}\}}. \tag{4.113}$$

The last term can again be reformulated according to (4.110), (4.111). The well-posedness of this strategy can be assured for sufficiently small time-steps $k \leq k_0(t_J)$, using the bound $\| \mathbf{e}^j \|_{\mathbf{L}^\infty} \leq C \| \mathbf{e}^j \|_{\mathbf{L}^2} \| \mathbf{e}^j \|_{\mathbf{W}^{2,2}}$. Note that the product in (4.113) scales the penalization contribution relative to the time discretization effect, such that we can choose $\varepsilon = \mathcal{O}(k)$ in a corresponding error analysis: again, we can mimic the steps given in the previous proofs to find a result that corresponds to (4.95) (resp. (4.105), (4.106), and (4.113)). To run an inductive argumentation successfully, we found (4.98) to be the crucial inequality; however, this relation is now scaled by $\prod_{\kappa=0}^{\gamma}\{1 + |\mathbf{m}^j|^{2-\kappa}\}^{-1}$, and the argument works for $\gamma$ sufficiently large such that (see (4.99), (4.100))

$$3(1-\beta)\delta_2 < \frac{1}{8}, \quad \prod_{\kappa=0}^{\gamma}\{1 + |\mathbf{m}^j|^{2-\kappa}\}^{-1}\frac{3(1-\beta)k}{4\delta_2\varepsilon} < \frac{1}{2},$$

and $k \leq k_0(t_J; \gamma)$ sufficiently small to make sure that for $|1 - c_0|$ sufficiently small,

$$\max_{0 \leq j \leq \ell+1} \| \mathbf{m}^{j+1} \|_{\mathbf{L}^\infty} \geq c_0, \quad 0 \leq \ell \leq J.$$

So far, we gave the sketch of a proof for the following theorem. We emphasize that the dependency $\gamma = \gamma(\varepsilon)$ stated in the following theorem is non-crucial.

**Theorem 4.8** *Let $\{\mathbf{m}^j\}_{j=0}^J$ be the solution to* (4.13), *with*

$$\phi(\mathbf{m}^{j+1}) = \phi_3^\gamma(\mathbf{m}^{j+1}) = \frac{1}{\varepsilon}\{1 - \frac{1}{|\mathbf{m}^j|^{2-\gamma}}\}\mathbf{m}^{j+1},$$

$\mathbf{m}_0 \in \mathbf{W}^{2,2}(\omega, \mathcal{S}^2)$, *and* $0 < t_J < T_0(\mathbf{m}_0)$. *Let* $\varepsilon = \mathcal{O}(k)$, *for* $\mathbb{N}_0 \ni \gamma \equiv \gamma(\varepsilon)$ *sufficiently large, and* $k \leq k_0(t_J; \gamma)$ *sufficiently small. Let further* $\mathbf{m}$ *solve*

(4.6). *Then*

$$\text{(a)}\quad \max_{0\leq j\leq J} \| \mathbf{m}(t_j) - \mathbf{m}^j \|_{\mathbf{L}^2} + \Big( k \sum_{j=0}^{J} \| \nabla \{ \mathbf{m}(t_j) - \mathbf{m}^j \} \|_{\mathbf{L}^2}^2 \Big)^{1/2}$$

$$+ \Big( \frac{k}{\varepsilon} \sum_{j=0}^{J} \Big\{ \| \mathbf{m}(t_j) - \mathbf{m}^j \|_{\mathbf{L}^4}^4 + \| \langle \mathbf{m}(t_j) - \mathbf{m}^j, \mathbf{m}(t_j) \rangle_{\mathbb{R}^3} \|_{L^2}^2 \Big\} \Big)^{\frac{1}{2}} \leq Ck\,,$$

$$\text{(b)}\quad \max_{0\leq j\leq J} \| 1 - | \mathbf{m}^j | \|_{L^2} \leq C \sqrt{k\varepsilon}\,,$$

$$\text{(c)}\quad \max_{0\leq j\leq J} \Big\{ \| \mathbf{m}^j \|_{\mathbf{W}^{2,2}} + \| d_t \mathbf{m}^j \|_{\mathbf{L}^2} \Big\} + \Big( k \sum_{j=1}^{J} \| \nabla d_t \mathbf{m}^j \|_{\mathbf{L}^2}^2 \Big)^{1/2} \leq C\,,$$

*where $C = C(\omega, t_J, \mathbf{m}_0)$ is a positive number that does not depend on $k$.*

### Analysis of the Projection Scheme

In order to understand the error effects of the method $(\mathrm{P})_A^\gamma$, for $\gamma \in \mathbb{N}_0$ (see p. 139), we substitute the quantity without a twiddle by the formula from the second step, and end up with

$$d_t \tilde{\mathbf{m}}^{j+1} - \alpha \Delta \tilde{\mathbf{m}}^j + \frac{1}{k} \{ 1 - \frac{1}{| \tilde{\mathbf{m}}^j |^{2-\gamma}} \} \tilde{\mathbf{m}}^j = \alpha | \nabla \tilde{\mathbf{m}}^j |^2 \tilde{\mathbf{m}}^{j+1} + \tilde{\mathbf{m}}^{j+1} \times \Delta \tilde{\mathbf{m}}^{j+1}\,. \tag{4.114}$$

This is a semi-explicit penalty method of the third form given in (4.14), which establishes the close connection between projection methods and penalization strategies, choosing $\varepsilon = k$.

The goal of this section is to verify the following theorem by using results from the previous section.

**Theorem 4.9** *Let $\{\tilde{\mathbf{m}}^j\}_{j=0}^J$, for $0 < t_J < T_0(\mathbf{m}_0)$ be the solution to (4.114), and $\mathbf{m}$ solves (4.6), for $\mathbf{m}_0 \in \mathbf{W}^{2,2}(\omega, \mathcal{S}^2)$. For $\gamma \in \mathbb{N}_0$, and sufficiently small time-steps $k \leq k_0(t_J; \gamma)$, there exists a constant $C = C(t_J, \omega, \mathbf{m}_0; \gamma)$ that is* *independent of $k$, such that*

$$\text{(a)} \quad \max_{0\le j\le J} \| \mathbf{m}(t_j) - \mathbf{m}^j \|_{\mathbf{L}^2} + \Big(k\sum_{j=0}^{J} \| \nabla\{\mathbf{m}(t_j) - \mathbf{m}^j\} \|_{\mathbf{L}^2}^2\Big)^{1/2}$$

$$+ \Big(\sum_{j=0}^{J}\Big\{\| \mathbf{m}(t_j) - \mathbf{m}^j \|_{\mathbf{L}^4}^4 + \| \langle \mathbf{m}(t_j) - \mathbf{m}^{j+1}, \mathbf{m}(t_j)\rangle_{\mathbb{R}^3} \|_{L^2}^2\Big\}\Big)^{1/2} \le C\,k\,,$$

$$\text{(b)} \quad \max_{1\le j\le J}\Big\{\frac{1}{k} \| 1 - |\mathbf{m}^j| \|_{L^2} + \| \mathbf{m}^j \|_{\mathbf{W}^{2,2}} + \| d_t\mathbf{m}^j \|_{\mathbf{L}^2}\Big\}$$

$$+\Big(k\sum_{j=1}^{J} \| \nabla d_t\mathbf{m}^j \|_{\mathbf{L}^2}^2\Big)^{1/2} \le C\,.$$

**Remark 4.7** *Choosing $\gamma > 0$ improves stability of the method. However, the case $\gamma = 0$ is covered herewith as well.*

**Proof:**

The subsequent arguments exploit Theorem 4.8. — Let $\mathbf{m}_A^j$ denote the solution to (4.13), with $\phi(\cdot) = \phi_3^\gamma(\cdot)$, $\mathbf{m}_B^j \equiv \tilde{\mathbf{m}}^j$ solves (4.114), and $\mathbf{e}^j := \mathbf{m}_A^j - \mathbf{m}_B^j$.

$$d_t\mathbf{e}^{j+1} - \alpha\,\Delta\mathbf{e}^{j+1} + \frac{1}{k}\Big\{\{1 - \frac{1}{|\mathbf{m}_A^j|^{2-\gamma}}\}\mathbf{m}_A^{j+1} - \{1 - \frac{1}{|\mathbf{m}_B^j|^{2-\gamma}}\}\mathbf{m}_B^j\Big\} = \boldsymbol{\mathcal{E}}^{j+1}\,, \tag{4.115}$$

where $\boldsymbol{\mathcal{E}}^{j+1}$ collects contributions from the nonlinear terms on the right-hand sides of (4.13), (4.114). We again skip its analysis since arguments given in the proofs of Theorems 4.4 to 4.6 do apply here as well.

In order to deal with the explicit character of the penalization effectively, we have to modify the argument from the proof of Theorem 4.6. At first, we follow the argument in (4.110) to conclude

$$\begin{aligned}
|\mathbf{m}_A^j| - |\mathbf{m}_B^j| &= -\frac{|\mathbf{e}^j|^2}{|\mathbf{m}_B^j|} + (1 + \frac{1}{\sqrt{|\mathbf{m}_B^j|}})\frac{1}{\sqrt{|\mathbf{m}_A^j|}}\langle \mathbf{e}^j, \mathbf{m}_B^j\rangle_{\mathbb{R}^3} \\
&\quad +(\frac{1}{\sqrt{|\mathbf{m}_B^j|}} - 1)\frac{1}{\sqrt{|\mathbf{m}_B^j|}}\langle \mathbf{e}^j, \mathbf{m}_A^j\rangle_{\mathbb{R}^3} + |\mathbf{m}_A^j|^2(1 - \frac{1}{|\mathbf{m}_B^j|}) \\
&= -\frac{|\mathbf{e}^j|^2}{|\mathbf{m}_B^j|} + \frac{2}{|\mathbf{m}_B^j|}\langle \mathbf{e}^j, \mathbf{m}_A^j\rangle_{\mathbb{R}^3} + \frac{|\mathbf{m}_B^j| - |\mathbf{m}_A^j|}{|\mathbf{m}_B^j|}|\mathbf{m}_A^j|^2,
\end{aligned}$$

or

$$|\mathbf{m}_A^j| - |\mathbf{m}_B^j| = \frac{-|\mathbf{e}^j|^2 + 2\langle \mathbf{e}^j, \mathbf{m}_A^j\rangle_{\mathbb{R}^3}}{|\mathbf{m}_B^j| + |\mathbf{m}_A^j|^2}. \tag{4.116}$$

Secondly, the difference in (4.115) can be restated as follows, using (4.113),

$$\begin{aligned}
&\frac{1}{k}\Big\{\frac{1}{|\mathbf{m}_A^j|^{2-\gamma}} - \frac{1}{|\mathbf{m}_B^j|^{2-\gamma}}\Big\}\{\mathbf{e}^j - \mathbf{m}_A^j\} + \frac{1}{k}\Big\{1 - \frac{1}{|\mathbf{m}_A^j|^{2-\gamma}}\Big\}\{\mathbf{e}^j + k\, d_t\mathbf{m}_A^{j+1}\} \\
&= \prod_{\kappa=0}^{\gamma} \frac{1}{|\mathbf{m}_A^j|^{2-\kappa} + |\mathbf{m}_B^j|^{2-\kappa}} \frac{|\mathbf{e}^j|^2 - 2\langle \mathbf{e}^j, \mathbf{m}_A^j\rangle_{\mathbb{R}^3}}{(|\mathbf{m}_A^j|^2 + |\mathbf{m}_B^j|^2)|\mathbf{m}_A^j|^{2-\gamma}|\mathbf{m}_B^j|^{2-\gamma}}\{\mathbf{e}^j - \mathbf{m}_A^j\} \\
&\quad + \frac{1}{k}\Big\{1 - \frac{1}{|\mathbf{m}_A^j|^{2-\gamma}}\Big\}\{\mathbf{e}^j + k\, d_t\mathbf{m}_A^{j+1}\}.
\end{aligned} \tag{4.117}$$

We deal with the first term as in (4.112). The last term in (4.117), if tested by $\mathbf{e}^{j+1}$, is bounded by

$$\begin{aligned}
&\frac{1}{k}\Big\| 1 - \frac{1}{|\mathbf{m}_A^j|^{2-\gamma}}\Big\|_{\mathbf{L}^2} \|\mathbf{e}^j\|_{\mathbf{L}^4} \|\mathbf{e}^{j+1}\|_{\mathbf{L}^4} \\
&\quad + \Big(\Big\| 1 - \frac{1}{|\mathbf{m}_A^j|^{2-\gamma}}\Big\|_{L^2} \|d_t\mathbf{m}_A^j\|_{\mathbf{W}^{1,2}}^{1/2}\Big) \|d_t\mathbf{m}_A^j\|_{\mathbf{L}^2}^{1/2} \|\mathbf{e}^{j+1}\|_{\mathbf{L}^2}^{1/2} \|\mathbf{e}^{j+1}\|_{\mathbf{W}^{1,2}}^{1/2} \\
&\leq \frac{C_\alpha \delta_4}{k^2}\Big\| 1 - \frac{1}{|\mathbf{m}_A^j|^{2-\gamma}}\Big\|_{L^2}^2 \|\mathbf{e}^j\|_{\mathbf{L}^2} \|\mathbf{e}^{j+1}\|_{\mathbf{L}^2} \\
&\quad + C\delta_4\Big\{\Big(\Big\| 1 - \frac{1}{|\mathbf{m}_A^j|^{2-\gamma}}\Big\|_{L^2} \|\nabla d_t\mathbf{m}_A^j\|_{\mathbf{L}^2}^{1/2}\Big)^4 + \Big(\|d_t\mathbf{m}_A^j\|_{\mathbf{L}^2}^{1/2} \|\mathbf{e}^{j+1}\|_{\mathbf{L}^2}^{1/2}\Big)^4\Big\} \\
&\quad + \frac{\alpha}{4\delta_4}\Big\{\|\nabla\mathbf{e}^j\|_{\mathbf{L}^2}^2 + \|\nabla\mathbf{e}^{j+1}\|_{\mathbf{L}^2}^2\Big\},
\end{aligned}$$

thanks to Young's generalized inequality, and $\delta_4 > 0$ sufficiently large. Because of Theorem 4.8, we can bound these terms by Gronwall's inequality. We may now follow the argumentation via induction that was given to show Theorem 4.8. □

## 4.3 Finite Elements for the Landau-Lifshitz-Gilbert equation

In Section 4.2, different time discretization methods are studied, and it is now our goal to deal with their finite element formulations. We discuss a $\mathbf{W}^{1,2}(\omega, \mathbb{R}^3)$-conforming **P**1-finite element formulation of the penalized schemes (4.13)-(4.15) and projection methods $(\mathrm{P})_A^\gamma$ and study the interplay of time discretization, spatial discretization, and penalization effects in the arising schemes: Section 4.3.1 gives an error analysis of the penalization method (4.13) for $\phi = \phi_2$ which we consider the prototype algorithm that gathers principal difficulties in a finite element analysis of all considered penalization strategies. Section 4.3.2 is then concerned with a finite element analysis for $(\mathrm{P})_A^\gamma$.

As will turn out, the interplay of numerical discretization and penalization strategies to satisfy the saturation constraint in (LLG) is primarily concentrated on time-discretization effects, but a slight dependency — stated in the form of a mild constraint $F(k,h) > 0$ — is observed.

### 4.3.1 Finite element error analysis for the penalization method (4.13) using $\phi(\cdot) = \phi_2(\cdot)$

The finite element formulation of method (4.13) for $\phi(\cdot) = \phi_2(\cdot)$ reads as follows: Given $\mathbf{m}_h^j \in \mathcal{S}^1(\omega)$, compute $\mathbf{m}_h^{j+1} \in \mathcal{S}^1(Q)$ from

$$\begin{aligned}(d_t\mathbf{m}_h^{j+1}, \varphi_h) + \alpha(\nabla\mathbf{m}_h^{j+1}, \nabla\varphi_h) + \frac{1}{\varepsilon}(\{1 - \frac{1}{|\mathbf{m}_h^j|^2}\}\mathbf{m}_h^{j+1}, \varphi_h) \qquad (4.118)\\ = \alpha\left(|\nabla\mathbf{m}_h^j|^2\mathbf{m}_h^{j+1}, \varphi_h\right) - \left(\mathbf{m}_h^{j+1} \times \nabla\mathbf{m}_h^{j+1}, \nabla\varphi_h\right) \quad \forall\varphi_h \in \mathcal{S}^1(Q),\end{aligned}$$

and $\mathbf{m}^0 = \mathbf{P}_h\mathbf{m}_0$. We again make use of the notation $\left(\mathbf{m}_h^{j+1} \times \nabla\mathbf{m}_h^{j+1}, \nabla\varphi_h\right) = \sum_{\ell=1}^2\left(\mathbf{m}_h^{j+1} \times \partial_\ell\mathbf{m}_h^{j+1}, \partial_\ell\varphi_h\right)$. — The main result in this subsection is stated in the following theorem, valid for quasi-uniform meshes $\mathcal{T}_h$.

**Theorem 4.10** *Let $\{\mathbf{m}_h^j\}_{j=0}^J$, $0 < t_J < T_0(\mathbf{m}_0)$ be the solution to (4.118), and $\mathbf{m}$ solves (4.6), for $\mathbf{m}_0 \in \mathbf{W}^{2,2}(\omega, \mathcal{S}^2)$. Suppose that $\varepsilon > 1.9k$, $k^{-1/2} = o(\frac{1}{h})$, and $k \le k_0(t_J)$, $h \le h_0(t_J)$ are chosen sufficiently small,*

$$\max_{0\le j\le J} \|\mathbf{m}(t_j) - \mathbf{m}_h^j\|_{\mathbf{L}^2} + \Big(k\sum_{j=0}^{J}\|\mathbf{m}(t_j) - \mathbf{m}^j\|_{\mathbf{W}^{1,2}}^2\Big)^{\frac{1}{2}}$$

$$+\sqrt{\frac{k}{\varepsilon}}\Big(\sum_{j=0}^{J}\{\|\mathbf{m}(t_j) - \mathbf{m}_h^j\|_{\mathbf{L}^4}^4 + \|\langle\mathbf{m}(t_j) - \mathbf{m}_h^j, \mathbf{m}(t_j)\rangle_{\mathbb{R}^3}\|_{L^2}^2\}\Big)^{\frac{1}{2}} \le C\{k + h$$

*with $C = C(\omega, t_J, \mathbf{m}_0)$ a positive number that does not depend on $k, h$. In addition, there holds*

$$\max_{0 \le j \le J} \| 1 - |\mathbf{m}_h^j|^2 \|_{L^2} \le C\{k + h\}.$$

**Remark 4.8** *In contrast to prior argumentations, we run the inductive argument in the 2nd step of the proof only for the given inequalities and skip another one that involves higher order spatial derivatives; the reason for this procedure is the limited regularity of used (affine) finite elements, and 'problematic' terms are controlled by means of inverse inequalities and parameter constraints $F(h,k;\varepsilon) > 0$. Hence, the analysis differs for finite elements that show increased regularity properties, like e.g. $\mathbf{W}^{2,2}(\omega, \mathbb{R}^3)$-conforming Hermite-type elements, where we expect no restriction $F(h,k) > 0$.*

**Proof:**

*1st step:* Thanks to conformity of the finite element method, $\mathbf{e}^j := \mathbf{m}(t_j) - \mathbf{m}_h^j \in \mathbf{W}^{1,2}(\omega, \mathbb{R}^3)$, for $j \ge 0$. We subtract the equations (4.118), (4.6) from each other and then test with the admissible function $\mathbf{P}_{h,\mathbf{W}^{1,2}}\mathbf{e}^{j+1}$.

$$\begin{aligned}
&\frac{1}{2}\| \mathbf{e}^{J+1} \|_{\mathbf{L}^2}^2 + \frac{k}{2}\sum_{j=1}^{J+1}\{k \| d_t\mathbf{e}^j \|_{\mathbf{L}^2}^2 + \frac{\alpha}{2}\| \nabla\mathbf{e}^j \|_{\mathbf{L}^2}^2\} + k\sum_{j=1}^{J+1}(\mathcal{P}^j, \mathbf{P}_{h,\mathbf{W}^{1,2}}\mathbf{e}^j) \\
&\le \frac{1}{2}\| \mathbf{m}^{J+1} - \mathbf{P}_{h,\mathbf{W}^{1,2}}\mathbf{m}^{J+1} \|_{\mathbf{L}^2}^2 + Ck\sum_{j=1}^{J+1}\Big\{k \| d_t\{\mathbf{m}^j - \mathbf{P}_{h,\mathbf{W}^{1,2}}\mathbf{m}^j\} \|_{\mathbf{L}^2}^2 \\
&\quad + \| \nabla\{\mathbf{m}^j - \mathbf{P}_{h,\mathbf{W}^{1,2}}\mathbf{m}^j\} \|_{\mathbf{L}^2}^2\Big\} + k\sum_{j=1}^{J+1}|(\mathcal{F}^j, \mathbf{P}_{h,\mathbf{W}^{1,2}}\mathbf{e}^j)|, \qquad (4.119)
\end{aligned}$$

where we used the fact that $(\nabla\{\mathbf{m} - \mathbf{P}_{h,\mathbf{W}^{1,2}}\mathbf{m}\}, \nabla\varphi_h) = 0$, for all $\varphi_h \in \mathcal{S}^1(Q)$. $\mathcal{F}^{j+1} := \alpha\mathcal{F}_1^{j+1} - \mathcal{F}_2^{j+1}$ gathers the terms

$$\begin{aligned}
(\mathcal{F}_1^{j+1}, \varphi_h) &:= (|\nabla\mathbf{m}^j|^2\mathbf{m}^{j+1} - |\nabla\mathbf{m}_h^j|^2\mathbf{m}_h^{j+1}, \varphi_h), &(4.120)\\
(\mathcal{F}_2^{j+1}, \varphi_h) &:= (\mathbf{m}^{j+1} \times \nabla\mathbf{m}^{j+1} - \mathbf{m}_h^{j+1} \times \nabla\mathbf{m}_h^{j+1}, \nabla\varphi_h). &(4.121)
\end{aligned}$$

We first consider the penalization part $\big(\mathcal{P}^{j+1}, \mathbf{P}_{h,\mathbf{W}^{1,2}}\mathbf{e}^{j+1}\big)$ in (4.119),

$$\begin{aligned}
&\big|\,\big(\mathcal{P}^{j+1}, \mathbf{P}_{h,\mathbf{W}^{1,2}}\mathbf{e}^{j+1}\big)\,\big| \qquad (4.122)\\
&\equiv \frac{1}{\varepsilon}\Big|\Big(\{1-\frac{1}{|\,\mathbf{m}^j\,|^2}\}\mathbf{m}^{j+1} - \{1-\frac{1}{|\,\mathbf{m}_h^j\,|^2}\}\mathbf{m}_h^{j+1}, \mathbf{P}_{h,\mathbf{W}^{1,2}}\mathbf{e}^{j+1}\Big)\Big|\\
&= \frac{1}{\varepsilon}\Big|\Big(\{\frac{1}{|\,\mathbf{m}_h^j\,|^2} - \frac{1}{|\,\mathbf{m}^j\,|^2}\}\mathbf{m}^{j+1} + \{1-\frac{1}{|\,\mathbf{m}_h^j\,|^2}\}\mathbf{e}^{j+1}, \mathbf{P}_{h,\mathbf{W}^{1,2}}\mathbf{e}^{j+1}\Big)\Big|\\
&:= I + II\,.
\end{aligned}$$

We proceed with the terms $I, II$ independently.

$$\begin{aligned}
I \;&=\; \frac{1}{\varepsilon}\Big(\frac{|\,\mathbf{m}^j\,|^2 - |\,\mathbf{m}_h^j\,|^2}{|\,\mathbf{m}_h^j\,|^2\,|\,\mathbf{m}^j\,|^2}\big\{\mathbf{m}^j + k\,d_t\mathbf{m}^{j+1}\big\}, \mathbf{P}_{h,\mathbf{W}^{1,2}}\mathbf{e}^{j+1}\Big)\\
&\geq \frac{1}{\varepsilon}\Big(\frac{|\,\mathbf{m}^j\,|^2 - |\,\mathbf{m}_h^j\,|^2}{|\,\mathbf{m}_h^j\,|^2\,|\,\mathbf{m}^j\,|^2}\mathbf{m}^j, \mathbf{P}_{h,\mathbf{W}^{1,2}}\mathbf{e}^j\Big)\\
&\quad + \frac{k}{\varepsilon}\Big(\frac{|\,\mathbf{m}^j\,|^2 - |\,\mathbf{m}_h^j\,|^2}{|\,\mathbf{m}_h^j\,|^2\,|\,\mathbf{m}^j\,|^2}\mathbf{m}^j, \mathbf{P}_{h,\mathbf{W}^{1,2}}d_t\mathbf{e}^{j+1}\Big)\\
&\quad - \frac{k}{\varepsilon}\,\Big\|\,\frac{|\,\mathbf{m}^j\,|^2 - |\,\mathbf{m}_h^j\,|^2}{|\,\mathbf{m}_h^j\,|^2\,|\,\mathbf{m}^j\,|^2}\,\Big\|_{L^2}\,\|\,d_t\mathbf{m}^{j+1}\,\|_{\mathbf{L}^4}\,\|\,\mathbf{P}_{h,\mathbf{W}^{1,2}}\mathbf{e}^{j+1}\,\|_{\mathbf{L}^4}\\
&\geq \frac{1}{\varepsilon}\Big(\frac{|\,\mathbf{m}^j\,|^2 - |\,\mathbf{m}_h^j\,|^2}{|\,\mathbf{m}_h^j\,|^2\,|\,\mathbf{m}^j\,|^2}\mathbf{m}^j, \mathbf{e}^j\Big) \qquad (4.123)\\
&\quad - \frac{C}{\varepsilon}\,\Big\|\,\frac{\mathbf{m}^j + \mathbf{m}_h^j}{|\,\mathbf{m}_h^j\,|\,|\,\mathbf{m}^j\,|}\,\Big\|_{\mathbf{L}^\infty}^2\,\|\,\mathbf{e}^j\,\|_{\mathbf{L}^2}\,\|\,\mathbf{m}^j - \mathbf{P}_{h,\mathbf{W}^{1,2}}\mathbf{m}^j\,\|_{\mathbf{L}^2}\\
&\quad - C\delta_1\,\frac{k^2}{\varepsilon}\,\Big\|\,\frac{\mathbf{m}^j + \mathbf{m}_h^j}{|\,\mathbf{m}_h^j\,|^2\,|\,\mathbf{m}^j\,|^2}\,\Big\|_{\mathbf{L}^\infty}^2\,\|\,\mathbf{P}_{h,\mathbf{W}^{1,2}}d_t\mathbf{e}^{j+1}\,\|_{\mathbf{L}^2}^2\\
&\quad - C\delta_1\,\frac{k^2}{\varepsilon^2}\Big\|\,\frac{\mathbf{m}^j + \mathbf{m}_h^j}{|\,\mathbf{m}_h^j\,|^2\,|\,\mathbf{m}^j\,|^2}\,\Big\|_{\mathbf{L}^\infty}^2\,\|\,\mathbf{e}^j\,\|_{\mathbf{L}^2}^2\|\,d_t\mathbf{m}^{j+1}\,\|_{\mathbf{L}^4}^2\\
&\quad - \frac{1}{\delta_1\varepsilon}\,\Big\|\,\frac{|\,\mathbf{m}^j\,|^2 - |\,\mathbf{m}_h^j\,|^2}{|\,\mathbf{m}_h^j\,|^2\,|\,\mathbf{m}^j\,|^2}\,\Big\|_{L^2}^2 - \frac{\alpha}{8}\,\|\,\nabla\mathbf{P}_{h,\mathbf{W}^{1,2}}\mathbf{e}^{j+1}\,\|_{\mathbf{L}^2}^2\,,
\end{aligned}$$

for $\delta_1 > 0$. Term $II$ from (4.122) can be bounded in the following way.

$$\begin{aligned} II &= \frac{1}{\varepsilon}\Big(\frac{|\mathbf{m}_h^j|^2 - |\mathbf{m}(t_j)|^2}{|\mathbf{m}_h^j|^2}\mathbf{e}^{j+1}, \mathbf{P}_{h,\mathbf{W}^{1,2}}\mathbf{e}^{j+1}\Big) \qquad (4.124) \\ &\leq \frac{C}{\varepsilon} \| \mathbf{m}_h^j - \mathbf{m}(t_j) \|_{\mathbf{L}^2} \| \frac{\mathbf{m}_h^j + \mathbf{m}(t_j)}{|\mathbf{m}_h^j|^2} \|_{\mathbf{L}^2} \\ &\quad \times \big\{ \| \mathbf{P}_{h,\mathbf{W}^{1,2}}\mathbf{e}^{j+1} \|_{\mathbf{L}^4}^2 + \| \mathbf{m}^{j+1} - \mathbf{P}_{h,\mathbf{W}^{1,2}}\mathbf{m}^{j+1} \|_{\mathbf{L}^4}^2 \big\} \\ &\leq \frac{C}{\varepsilon^2} \| \mathbf{m}_h^j - \mathbf{m}(t_j) \|_{\mathbf{L}^2}^2 \| \frac{\mathbf{m}_h^j + \mathbf{m}(t_j)}{|\mathbf{m}_h^j|^2} \|_{\mathbf{L}^2} \\ &\quad \times \big\{ \| \mathbf{P}_{h,\mathbf{W}^{1,2}}\mathbf{e}^{j+1} \|_{\mathbf{L}^2}^2 + \| \mathbf{m}^{j+1} - \mathbf{P}_{h,\mathbf{W}^{1,2}}\mathbf{m}^{j+1} \|_{\mathbf{L}^4}^4 \big\} \\ &\quad + \frac{\alpha}{8} \{ \| \nabla \mathbf{P}_{h,\mathbf{W}^{1,2}}\mathbf{e}^{j+1} \|_{\mathbf{L}^2}^2 + \| \nabla \{ \mathbf{m}^{j+1} - \mathbf{P}_{h,\mathbf{W}^{1,2}} \} \mathbf{m}^{j+1} \|_{\mathbf{L}^2}^2 \}. \end{aligned}$$

We use certain continuity properties of $\mathbf{P}_{h,\mathbf{W}^{1,2}}$ and inverse inequalities in the following to bound the last term in (4.119) as follows,

$$\begin{aligned} &(\mathcal{F}_1^{j+1}, \mathbf{P}_{h,\mathbf{W}^{1,2}}\mathbf{e}^{j+1}) \\ &= \big(\langle \nabla(\mathbf{m}^j - \mathbf{P}_{h,\mathbf{W}^{1,2}}\mathbf{m}^j), \nabla(\mathbf{m}^j + \mathbf{P}_{h,\mathbf{W}^{1,2}}\mathbf{m}^j)\rangle_{\mathbb{R}^6} \mathbf{m}^{j+1}, \mathbf{P}_{h,\mathbf{W}^{1,2}}\mathbf{e}^{j+1}\big) \\ &\quad + \big(|\nabla \mathbf{P}_{h,\mathbf{W}^{1,2}}\mathbf{m}^j|^2(\mathbf{m}^{j+1} - \mathbf{P}_{h,\mathbf{W}^{1,2}}\mathbf{m}^{j+1} + \mathbf{P}_{h,\mathbf{W}^{1,2}}\mathbf{e}^{j+1}), \mathbf{P}_{h,\mathbf{W}^{1,2}}\mathbf{e}^{j+1}\big) \\ &\quad + \big(\langle \nabla \mathbf{P}_{h,\mathbf{W}^{1,2}}\mathbf{e}^j, \nabla\{-\mathbf{P}_{h,\mathbf{W}^{1,2}}\mathbf{e}^j + 2\mathbf{P}_{h,\mathbf{W}^{1,2}}\mathbf{m}^j\}\rangle_{\mathbb{R}^6}, \\ &\quad \langle \mathbf{P}_{h,\mathbf{W}^{1,2}}\mathbf{m}^{j+1} - \mathbf{P}_{h,\mathbf{W}^{1,2}}\mathbf{e}^{j+1}, \mathbf{P}_{h,\mathbf{W}^{1,2}}\mathbf{e}^{j+1}\rangle_{\mathbb{R}^6}\big) \\ &\leq C\Big\{ \| \mathbf{m}^{j+1} \|_{\mathbf{L}^\infty} \| \mathbf{m}^j - \mathbf{P}_{h,\mathbf{W}^{1,2}}\mathbf{m}^j \|_{\mathbf{W}^{1,2}} \| \nabla \mathbf{m}^j \|_{\mathbf{L}^4} \| \mathbf{P}_{h,\mathbf{W}^{1,2}}\mathbf{e}^{j+1} \|_{\mathbf{L}^4} \\ &\quad + \| |\nabla \mathbf{P}_{h,\mathbf{W}^{1,2}}\mathbf{m}^j|^2 \|_{\mathbf{L}^4} \{ \| \mathbf{P}_{h,\mathbf{W}^{1,2}}\mathbf{e}^{j+1} \|_{\mathbf{L}^2} \qquad (4.125) \\ &\quad + \| \mathbf{m}^{j+1} - \mathbf{P}_{h,\mathbf{W}^{1,2}}\mathbf{m}^{j+1} \|_{\mathbf{L}^2} \} \| \mathbf{P}_{h,\mathbf{W}^{1,2}}\mathbf{e}^{j+1} \|_{\mathbf{L}^4} \\ &\quad + \| \nabla \mathbf{P}_{h,\mathbf{W}^{1,2}}\mathbf{e}^j \|_{\mathbf{L}^2} \| \nabla \mathbf{P}_{h,\mathbf{W}^{1,2}}\mathbf{e}^j \|_{\mathbf{L}^{\frac{2q}{q-2}}} \| \mathbf{m}^{j+1} \|_{\mathbf{L}^\infty} \| \mathbf{P}_{h,\mathbf{W}^{1,2}}\mathbf{e}^{j+1} \|_{\mathbf{L}^q} \\ &\quad + \| \nabla \mathbf{P}_{h,\mathbf{W}^{1,2}}\mathbf{e}^j \|_{\mathbf{L}^2} \| \nabla \mathbf{P}_{h,\mathbf{W}^{1,2}}\mathbf{e}^j \|_{\mathbf{L}^q} \| \mathbf{e}^{j+1} \|_{\mathbf{L}^{\frac{4q}{q-2}}}^2 \\ &\quad + \| \nabla \mathbf{P}_{h,\mathbf{W}^{1,2}}\mathbf{e}^j \|_{\mathbf{L}^4} \| \nabla \mathbf{m}^{j+1} \|_{\mathbf{L}^4} \| \mathbf{m}^{j+1} - \mathbf{e}^{j+1} \|_{\mathbf{L}^4} \| \mathbf{P}_{h,\mathbf{W}^{1,2}}\mathbf{e}^{j+1} \|_{\mathbf{L}^4} \Big\} \\ &\leq C_\alpha \Big\{ \| \mathbf{m}^j - \mathbf{P}_{h,\mathbf{W}^{1,2}}\mathbf{m}^j \|_{\mathbf{W}^{1,2}}^2 \\ &\quad + \| \mathbf{m}^{j+1} - \mathbf{P}_{h,\mathbf{W}^{1,2}}\mathbf{m}^{j+1} \|_{\mathbf{L}^2}^2 + \| \mathbf{P}_{h,\mathbf{W}^{1,2}}\mathbf{e}^{j+1} \|_{\mathbf{L}^2}^2 \Big\} \\ &\quad + \frac{\alpha}{8}\Big\{ \| \nabla \mathbf{P}_{h,\mathbf{W}^{1,2}}\mathbf{e}^j \|_{\mathbf{L}^2}^2 + \| \nabla \mathbf{P}_{h,\mathbf{W}^{1,2}}\mathbf{e}^{j+1} \|_{\mathbf{L}^2}^2 + \delta_2 \| \mathbf{P}_{h,\mathbf{W}^{1,2}}\mathbf{e}^{j+1} \|_{\mathbf{L}^q}^2 \Big\} \\ &\quad + C_{\alpha,\delta_2} h^{2\frac{2-q}{q}} \| \nabla \mathbf{e}^j \|_{\mathbf{L}^2}^4 + C\, h^{\frac{2-q}{q}} \| \nabla \mathbf{P}_{h,\mathbf{W}^{1,2}}\mathbf{e}^j \|_{\mathbf{L}^2}^2 \| \mathbf{e}^{j+1} \|_{\mathbf{L}^{\frac{4q}{2-q}}}^2 \end{aligned}$$

for $q > 2$, $\delta_2 = \delta_2(q) > 0$. — Furthermore,

$$\begin{aligned}
&\big(\boldsymbol{\mathcal{F}}_2^{j+1}, \mathbf{P}_{h,\mathbf{W}^{1,2}}\mathbf{e}^{j+1}\big) = \big(\mathbf{e}^{j+1} \times \nabla \mathbf{m}^{j+1}, \nabla \mathbf{P}_{h,\mathbf{W}^{1,2}}\mathbf{e}^{j+1}\big) \\
&\quad + \big(\mathbf{m}_h^{j+1} \times \nabla \mathbf{e}^{j+1}, \nabla\{\mathbf{m}^{j+1} - \mathbf{P}_{h,\mathbf{W}^{1,2}}\mathbf{m}^{j+1}\}\big) \\
&\quad \leq C_\alpha \Big\{ \{\| \mathbf{P}_{h,\mathbf{W}^{1,2}}\mathbf{e}^{j+1} \|_{\mathbf{L}^2}^2 + \| \mathbf{m}^{j+1} - \mathbf{P}_{h,\mathbf{W}^{1,2}}\mathbf{m}^{j+1} \|_{\mathbf{L}^2}^2 \} \| \nabla \mathbf{m}^{j+1} \|_{\mathbf{L}^4}^4 \\
&\quad + \| \mathbf{m}^{j+1} - \mathbf{P}_{h,\mathbf{W}^{1,2}}\mathbf{m}^{j+1} \|_{\mathbf{W}^{1,2}}^2 \| \mathbf{m}^{j+1} \|_{\mathbf{L}^\infty}^2 \Big\} \\
&\quad + \frac{\alpha}{8} \| \nabla \mathbf{P}_{h,\mathbf{W}^{1,2}}\mathbf{e}^{j+1} \|_{\mathbf{L}^2}^2 \qquad\qquad (4.126)\\
&\quad + \| \mathbf{e}^{j+1} \|_{\mathbf{L}^4} \| \nabla \mathbf{e}^{j+1} \|_{\mathbf{L}^2} \| \nabla \{\mathbf{m}^{j+1} - \mathbf{P}_{h,\mathbf{W}^{1,2}}\mathbf{m}^{j+1}\} \|_{\mathbf{L}^4} .
\end{aligned}$$

*2nd step:* We show the following result by induction, for $C_i \equiv C_i(\omega, t_J, \alpha, \mathbf{m}_0)$, $i = 1, 2$, for $0 \leq \ell \leq J$,

$$\begin{aligned}
&\| \mathbf{e}^{\ell+1} \|_{\mathbf{L}^2}^2 + \frac{k^2}{2} \sum_{j=0}^{\ell} \| d_t \mathbf{e}^{j+1} \|_{\mathbf{L}^2}^2 + \frac{\alpha k}{2} \sum_{j=0}^{\ell} \| \nabla \mathbf{e}^{j+1} \|_{\mathbf{L}^2}^2 \qquad (4.127)\\
&+ \frac{k}{\varepsilon} \sum_{j=0}^{\ell+1} \{ \| \mathbf{e}^j \|_{\mathbf{L}^4}^4 + \| \langle \mathbf{m}(t_j), \mathbf{e}^j \rangle_{\mathbb{R}^2} \|_{L^2}^2 \} \leq C_1 \{k^2 + h^2\} \exp(C_2 t_\ell) ,
\end{aligned}$$

$$\max_{0 \leq j \leq \ell} \| \, | \mathbf{m}_h^{j+1} |^2 - 1 \|_{L^\infty} \leq | 1 - c_0 | , \qquad (4.128)$$

for $| 1 - c_0 |$ arbitrarily small.

By induction, we can establish (4.128) at every iteration step, thanks to the embedding $W^{1,q} \hookrightarrow L^\infty$, for $q > 2$, an inverse inequality, (4.127), and the requirement of restricted values $k^{-1/2} = o(\frac{1}{h})$, for $k \leq k_0(t_J)$, $h \leq h_0(t_J)$ sufficiently small.

We continue with the arising terms (4.120), (4.121), (4.122): thanks to the embedding $W^{1,2} \hookrightarrow L^{\frac{4p}{2-p}}$ and the restriction $k^{-1/2} = o(\frac{1}{h})$, the last two terms in (4.125) can be absorbed on the left-hand side of (4.119), for $k \leq k_0(t_J)$, $h \leq h_0(t_J)$ sufficiently small.

As to (4.123), the first term next to the inequality sign can now be dealt with as in the proof of Theorem 4.6, cf. inequality (4.105). Using the constraint $\varepsilon > 1.9k$, for $k \leq k_0(t_J)$, $h \leq h_0(t_J)$, is then sufficient to run an inductive argument, and deal with the crucial term effectively that corresponds to (4.98). □

### 4.3.2 Finite element error analysis for $(\mathrm{P})^{\gamma}_{A}$

The following result is shown in a way that is similar to the proof of Theorem 4.10, using Theorem 4.9.

**Theorem 4.11** *Let $\{\tilde{\mathbf{m}}_h^j\}_{j=0}^J$, $0 < t_J < T_0(\mathbf{m}_0)$ be the solution to $(P)^{\gamma}_{A}$, for $\gamma \in \mathbb{N}_0$, and $\mathbf{m}$ solves (4.6), for $\mathbf{m}_0 \in \mathbf{W}^{2,2}(\omega, \mathcal{S}^2)$. Suppose that $k^{-1/2} = o(\frac{1}{h})$, and $k \le k_0(t_J)$, $h \le h_0(t_J)$ are chosen sufficiently small, then*

$$\max_{0\le j\le J} \| \mathbf{m}(t_j) - \tilde{\mathbf{m}}_h^j \|_{\mathbf{L}^2} + \Big( k \sum_{j=0}^{J} \| \mathbf{m}(t_j) - \tilde{\mathbf{m}}_h^j \|^2_{\mathbf{W}^{1,2}} \Big)^{1/2}$$

$$+ \sqrt{\frac{k}{\varepsilon}} \Big( \sum_{j=0}^{J} \{ \| \mathbf{m}(t_j) - \tilde{\mathbf{m}}_h^j \|^4_{\mathbf{L}^4} + \| \langle \mathbf{m}(t_j) - \tilde{\mathbf{m}}_h^j, \mathbf{m}(t_j) \rangle_{\mathbb{R}^3} \|^2_{L^2} \} \Big)^{\frac{1}{2}} \le C\{k+h\},$$

*with $C = C(\omega, t_J, \mathbf{m}_0)$ a positive number that does not depend on $k, h$. In addition, there holds*

$$\max_{0\le j\le J} \| 1 - |\tilde{\mathbf{m}}_h^j|^2 \|_{L^2} \le C\,\{k+h\}.$$

**Proof:**
(Sketch) The proof is a simplified version of the one to show Theorem 4.10: in (4.122) all iterates are used in an explicit way. As a consequence, it is only the first line next to the second inequality sign in formula (4.123) that is relevant; in (4.124), $\mathbf{e}^j$ replaces $\mathbf{e}^{j+1}$, a manipulation which does not affect the argumentation. □

## 4.4 Generalization of the physical model

We include surface energy contribution, anisotropy and exterior magnetic fields into the Landau-Lifshitz-Gilbert equation. The study of the simplified model is justified from a mathematical point of view, but scaling as well as actual dynamics in ferromagnets (like switching processes in nanomagnets) are not modeled reliably.

**Remark 4.9** *1. Note that in physical experiments of thin films or nanomagnets, it is the competition between surface energy ('implicit treatment* *in the algorithm') and stray-field minimization ('explicit treatment in the algorithm') that is crucial for driving the dynamics.*

*2. The algorithm $(P)_B^\gamma$ splits the numerical difficulties connected to satisfying the saturation constraint and non-local stray-field energy contributions.*

**Theorem 4.12** *Let $\mathcal{V}^j \equiv \{\chi_\omega \mathbf{m}^j, \nabla u^j\}_{j=0}^J$, for $0 < t_J \le T_0(\mathbf{m}_0)$ be the solution to $(P)_B^\gamma$ at the $j$-th iteration step, $1 \le j \le J$, and let $\mathcal{V} \equiv \{\chi_\omega \mathbf{m}, \nabla u\}$ solve (4.17)-(4.19), for $\mathbf{m}_0 \in \mathbf{W}^{2,2}(\omega, \mathcal{S}^2)$, and $\gamma \in \mathbb{N}_0$. Suppose $\mathbf{f} \in W^{2,2}(I, \mathbf{L}^2(\omega, \mathbb{R}^3))$. For $k \le k_0(t_J)$ chosen to be sufficiently small,*

$$\max_{0 \le j \le J} \| \mathcal{V}(t_j) - \mathcal{V}^j \|_{\mathbf{L}^2(\Omega)} + \Big( k \sum_{j=0}^{J} \| \nabla \{ \mathbf{m}(t_j) - \mathbf{m}^j \} \|_{\mathbf{L}^2}^2 \Big)^{1/2}$$

$$+ \frac{1}{\sqrt{\varepsilon}} \Big( k \sum_{j=0}^{J} \Big\{ \| \mathbf{m}(t_j) - \mathbf{m}^j \|_{\mathbf{L}^4}^4 + \| \langle \mathbf{m}(t_j) - \mathbf{m}^j, \mathbf{m}(t_j) \rangle_{\mathbb{R}^3} \|_{L^2}^2 \Big\} \Big)^{1/2} \le C\, k\,,$$

*with $C = C(\omega, t_J, \mathbf{m}_0)$ a positive number that does not depend on $k$.*

**Proof:**

(Sketch) To verify this result, we may benefit from the analysis of the 'projection part' which has been performed in Chapter 4.2. As to the explicit update of $u^j$, we may rely on (stability) properties of the linear operator $\mathbf{C} \equiv \nabla \Delta_D^{-1} \mathrm{div} \chi_\omega$; for example, $\mathbf{C} : \mathbf{L}^q(\Omega) \to \mathbf{L}^q(\Omega)$ is a continuous mapping, $q > 1$. In particular, we need to convince ourselves of the following facts, by exploiting that the further contribution in (4.17), (4.18), is 'of lower order':

1. Lemma 4.2 (resp. Theorem 4.2) and Theorem 4.3 remain valid.

2. Theorems 4.7 and 4.9 remain valid.

□

**Remark 4.10** *Results pertaining to finite element discretization that correspond to Theorem 4.11 can also be verified. We skip the verification of this assertion here.*

# 4.5 Computational Experiments

In the first part, we study the penalization strategies (4.13), (4.14), (4.15) and the projection scheme $(\mathrm{P})_A^\gamma$ computationally; in particular, we see deteriorate convergence in case that constraint $F(\varepsilon, k) > 0$ is violated. In a next step, we present a second order time discretization scheme based on the projection idea and report on convergence behavior for an academic example. In the last example, we present computational results for the physically relevant model (4.17)-(4.19) for a starting magnetization that shows two defects, using $(\mathrm{P})_B^0$.

**Example 4.3** *We report on computational results for the magnetization* $\mathbf{m}^{j+1} : \mathbb{R}^2 \supset \omega \to \mathbb{R}^2$ *governed by*

$$d_t\mathbf{m}^{j+1} - \Delta\mathbf{m}^{j+1} + \frac{1}{\varepsilon}\phi_i(\mathbf{m}^j, \mathbf{m}^{j+1})\mathbf{m}^{j+1} = |\nabla\mathbf{m}^j|^2\mathbf{m}^{j+1} + \mathbf{f}^{j+1}, \quad \mathbf{m}^0 = \mathbf{m}_0, \tag{4.129}$$

*for* $i = 2, 3$, $\varepsilon = f(k)$, *and* $\omega = (0,1)^2$. *Continuous* **P**1*-elements are used for spatial discretization. The right-hand side* $\mathbf{f}^{j+1} \equiv \mathbf{f}(t_{j+1})$ *is computed from the exact solution*

$$\mathbf{m}(x, y, t) = (xt, \sqrt{1 - x^2t^2})^\top, \quad \mathbf{m}_0(x, y) = \mathbf{m}(x, y, 0), \tag{4.130}$$

*of the continuous version of* (4.129), *and we choose boundary data* $\mathbf{m}^{j+1}\big|_{\partial\omega} = \mathbf{m}(x, y, t_{j+1})\big|_{\partial\omega}$ *in* (4.129). *Tables 4.1 and 4.2 compare different penalization strategies* $\phi_2$ *and* $\phi_3$ *for* $\varepsilon = 10k$ *and* $\varepsilon = k$, *respectively. We observe that the constraint* $F(\varepsilon, k) > 0$ *stated in Theorems 4.6 and 4.7 is indeed necessary for optimal convergence behavior of the methods.*

*In Table 4.3, convergence studies for* $(P)_A^\gamma$, $\gamma \in \{-1, 0, 1\}$ *are reported for the harmonic heat flow problem, i.e., the torque is removed from* $(P)_A^\gamma$. *We did not observe qualitative changes in the convergence behavior of method* $(P)_A^\gamma$ *for different values of* $\gamma$ *in our example.*

| $\ell$ | $m_2^{j+1}$ | $\nabla m_2^{j+1}$ | $\lvert \mathbf{m}^{j+1} \rvert^2 - 1$ | $m_2^{j+1}$ | $\nabla m_2^{j+1}$ | $\lvert \mathbf{m}^{j+1} \rvert - 1$ |
|---|---|---|---|---|---|---|
| 1 | 4.49 − 3 | 2.01 − 2 | 1.57 − 2 | 4.34 − 3 | 2.03 − 2 | 7.89 − 3 |
| 2 | 2.43 − 3 | 1.13 − 2 | 8.20 − 3 | 2.36 − 3 | 1.11 − 2 | 4.11 − 3 |
| 3 | 1.27 − 3 | 6.37 − 3 | 4.21 − 3 | 1.24 − 3 | 6.24 − 3 | 2.11 − 3 |
| 4 | 6.85 − 4 | 4.06 − 3 | 2.12 − 3 | 6.55 − 4 | 3.95 − 3 | 1.07 − 3 |
| 5 | 3.84 − 4 | 3.13 − 3 | 1.06 − 3 | 3.84 − 4 | 3.13 − 4 | 5.28 − 4 |
| 6 | 2.31 − 4 | 2.81 − 3 | 5.15 − 4 | 2.01 − 4 | 2.76 − 4 | 2.62 − 4 |
| 7 | 1.40 − 4 | 2.70 − 3 | 2.44 − 4 | 1.24 − 4 | 2.67 − 4 | 1.26 − 4 |
| order | 0.84 | 0.59 | 1.00 | 0.86 | 0.55 | 0.99 |

Table 4.1: $L^2$-errors at $t = 0.5$ for $\phi_2$ (left) and $\phi_3$ (right), with $\varepsilon = 10k$ and time-steps $k = 0.1 \cdot 2^{-\ell+1}$ ($h = \frac{1}{32}$).

| $\ell$ | $m_2^{j+1}$ | $\nabla m_2^{j+1}$ | $\lvert \mathbf{m}^{j+1} \rvert^2 - 1$ | $m_2^{j+1}$ | $\nabla m_2^{j+1}$ | $\lvert \mathbf{m}^{j+1} \rvert - 1$ |
|---|---|---|---|---|---|---|
| 1 | 4.81 − 3 | 2.24 − 2 | 1.56 − 2 | 4.61 − 3 | 2.13 − 2 | 7.94 − 3 |
| 2 | 2.93 − 3 | 1.35 − 2 | 8.17 − 3 | 2.60 − 3 | 1.23 − 2 | 4.08 − 3 |
| 3 | 1.78 − 3 | 8.77 − 3 | 4.09 − 3 | 1.56 − 3 | 7.43 − 3 | 2.17 − 3 |
| 4 | 1.20 − 3 | 6.32 − 3 | 1.99 − 3 | 9.81 − 4 | 5.08 − 3 | 1.04 − 3 |
| 5 | 8.76 − 4 | 5.00 − 3 | 9.25 − 4 | 6.17 − 4 | 3.91 − 3 | 4.97 − 4 |
| 6 | 6.51 − 4 | 4.14 − 3 | 4.02 − 4 | 4.45 − 4 | 3.37 − 3 | 2.29 − 4 |
| 7 | 4.66 − 4 | 3.49 − 3 | 1.59 − 4 | 3.28 − 4 | 3.06 − 3 | 9.79 − 5 |
| order | 0.56 | 0.45 | 1.10 | 0.64 | 0.47 | 1.06 |

Table 4.2: $L^2$-errors at $t = 0.5$ for $\phi_2$ (left) and $\phi_3$ (right), with $\varepsilon = k$ and time-steps $k = 0.1 \cdot 2^{-\ell+1}$ ($h = \frac{1}{32}$).

**Example 4.4** *So far, we considered time discretization schemes of first order to solve (LLG). The construction of robust second order schemes is more involved, and we only report on computational results for the following projection scheme to solve the heat flow problem; this scheme was already proposed in [43] (in modified form). Let* $\overline{\mathbf{f}}^{j+1/2} = \frac{1}{2}\{\mathbf{f}^j + \mathbf{f}^{j+1}\}$ *in the following.*

| $\ell$ | $m_2^j$ | $\mid \mathbf{m}^j \mid -1$ | $m_2^j$ | $\mid \mathbf{m}^j \mid -1$ | $m_2^j$ | $\mid \mathbf{m}^j \mid -1$ |
|---|---|---|---|---|---|---|
| 1 | 2.46 − 3 | 4.56 − 3 | 3.17 − 3 | 5.81 − 3 | 3.67 − 3 | 6.69 − 3 |
| 2 | 9.34 − 4 | 1.62 − 3 | 1.32 − 3 | 2.32 − 3 | 1.69 − 3 | 2.96 − 3 |
| 3 | 3.13 − 4 | 5.04 − 4 | 4.80 − 4 | 8.84 − 4 | 6.78 − 4 | 1.16 − 3 |
| 4 | 1.64 − 4 | 1.55 − 4 | 1.68 − 4 | 2.49 − 4 | 2.50 − 4 | 4.03 − 4 |
| 5 | 6.83 − 5 | 4.23 − 5 | 6.85 − 5 | 7.15 − 5 | 9.52 − 5 | 1.23 − 4 |
| 6 | 3.44 − 5 | 1.34 − 5 | 3.88 − 5 | 1.95 − 5 | 4.66 − 5 | 3.52 − 5 |
| order | 1.23 | 1.68 | 1.36 | 1.64 | 1.27 | 1.51 |

Table 4.3: $L^2$-errors at $t = 0.5$ for $(\mathrm{P})_A^\gamma$, for $\gamma = -1$ (left), $\gamma = 0$ (middle), and $\gamma = 1$ (right) ($h = \frac{1}{32}$).

1. *Compute* $\tilde{\mathbf{m}}^{j+1}$ *from*

$$\frac{1}{k}\{\tilde{\mathbf{m}}^{j+1} - \mathbf{m}^j\} - \alpha\,\Delta\overline{\tilde{\mathbf{m}}}^{j+1/2} = \frac{3\alpha}{2} \mid \nabla\tilde{\mathbf{m}}^j \mid^2 \tilde{\mathbf{m}}^j - \frac{\alpha}{2} \mid \nabla\tilde{\mathbf{m}}^{j-1} \mid^2 \tilde{\mathbf{m}}^{j-1} + \overline{\mathbf{f}}^{j+1/2}\,. \tag{4.131}$$

2. *Compute* $\mathbf{m}^{j+1} = \frac{\tilde{\mathbf{m}}^{j+1}}{|\tilde{\mathbf{m}}^{j+1}|}$.

*Table 4.4 displays computational results that we obtained with this scheme for the academic example (4.130), evidencing 2nd order of accuracy at least for smooth solutions.*

| $\ell$ | $m_2^{j+1}$ | $\nabla m_2^{j+1}$ | $\mid \mathbf{m}^{j+1} \mid^2 -1$ |
|---|---|---|---|
| 1 | 2.68 − 3 | 1.28 − 2 | 1.08 − 2 |
| 2 | 5.66 − 4 | 3.44 − 3 | 2.42 − 3 |
| 3 | 1.03 − 4 | 2.14 − 3 | 4.61 − 4 |
| 4 | 2.21 − 5 | 2.07 − 3 | 7.60 − 5 |
| order | 2.31 | 0.89 | 2.38 |

Table 4.4: $L^2$-errors at $t = 0.6$ for the 2nd order projection scheme, for time-steps $k = 0.2 \cdot 2^{-\ell+1}$ ($h = \frac{1}{64}$).

In the last example, we present computations for the physically more relevant model (4.17)-(4.19). For this purpose, we implemented the time-splitting/projection scheme $(\mathrm{P})^0_B$. We exploit (4.3) to state step 1. in the following weak form: Find $\tilde{\mathbf{m}}_h^{j+1} \in \boldsymbol{\mathcal{S}}^1(\omega)$, such that for $u_h^j \in \mathcal{S}_0^1(\Omega)$ and all $\varphi_h \in \mathcal{S}^1(\omega)$,

$$\begin{aligned}
&\frac{1}{k}(\tilde{\mathbf{m}}_h^{j+1} - \mathbf{m}_h^j, \varphi_h) + \alpha\,(\nabla\tilde{\mathbf{m}}_h^{j+1}, \nabla\varphi_h) = \alpha\,(|\nabla\tilde{\mathbf{m}}_h^j|^2 \tilde{\mathbf{m}}_h^{j+1}, \varphi_h) \qquad (4.132)\\
&-(\nabla[\varphi_h \times \tilde{\mathbf{m}}_h^j], \nabla\tilde{\mathbf{m}}_h^{j+1}) + \int_{\partial\omega} \left\langle \partial_{\mathbf{n}}\tilde{\mathbf{m}}_h^{j+1}, \alpha\,\varphi_h + \varphi_h \times \tilde{\mathbf{m}}_h^j \right\rangle_{\mathbb{R}^3} ds\\
&-\left(\mathbf{m}_h^j \times \{D\phi(\mathbf{m}_h^j) + \nabla u_h^j - \mathbf{f}_h^{j+1}\}, \varphi_h\right) - \alpha\left(D\phi(\mathbf{m}_h^j) + \nabla u_h^j - \mathbf{f}_h^{j+1}, \varphi_h\right)\\
&+\alpha\left(\{\langle \mathbf{m}_h^j, \nabla u_h^j\rangle_{\mathbb{R}^2} + \langle \mathbf{m}_h^j, D\phi(\mathbf{m}_h^j)\rangle_{\mathbb{R}^3} - \langle \mathbf{m}_h^j, \mathbf{f}^{j+1}\rangle_{\mathbb{R}^3}\}\mathbf{m}_h^j, \varphi_h\right).
\end{aligned}$$

Parts of the operator that contain $\partial_3(\cdot)$ are set equal to zero. — Next, we study the evolution of initial singularities for the generalized problem (LLG).

**Example 4.5** *Let $\omega = (-1,1) \times (-.5,.5)$, $\mathbf{f} = (-100,0,0)^\top$, for $\phi(\mathbf{m}) = \frac{1}{2}\langle \mathbf{m}, \mathbf{e}_\perp\rangle^2_{\mathbb{R}^3}$, with $\mathbf{e} = (-\frac{1}{\sqrt{2}}, -\frac{1}{\sqrt{2}}, 0)^\top$, $\partial_{\mathbf{n}}\mathbf{m} = 0$ on $\partial\omega$, and $\mathbf{m}_0(x,y) = \hat{\mathbf{m}}/\sqrt{|\hat{\mathbf{m}}|^2 + \eta^2}$, where $\hat{\mathbf{m}}(x,y) = (x^2 + y^2 - \frac{1}{4}, y, 0)^\top$, and $\alpha = 1$. For the mesh-size $h = \frac{1}{16}$, we choose $\eta = 0.2$ to regularize the singularities (i.e., point defects of degrees $\pm 1$). Problem (4.19) is solved on $\Omega = (-4,4)^2$, for $u_h\big|_{\partial\Omega} = 0$. We use P1 conforming elements for both, $\mathbf{m}_h^{j+1}$, and $u_h^j$. Figure 4.1 displays $(\tilde{m}_{h,1}^j, \tilde{m}_{h,2}^j)^\top$ and its modulus at different times $t_j$, for $k = 10^{-4}$, where the singularities move to the boundary $\partial\omega$ to finally leave $\omega$ and give way to a uniform magnetization in direction of $\mathbf{f}$. We also tried $\mathbf{f} = 0$ which also moved singularities, but without cancellation during the time of the simulation. Note that the constraint $F(h,k) > 0$ is not satisfied for our computation. We found sensitivity of the scheme with respect to choices of the time-step (indicated by blow-up of the modulus of magnetization for too large values of $k$) but not with respect to the grid-size $h$.*

As a summary, we draw the following conclusions from our experiments:

1. Penalization methods have to satisfy the derived stability constraints in order to show optimal convergence behavior. In compuational experiments, we observe more sensitivity with respect to proper choices for $k$ than for $h$.

2. Second order projection schemes seem to be a good alternative for smooth magnetizations, at least; an analysis of this method will be presented elsewhere.

3. Algorithm $(\mathrm{P})_B^\gamma$ using P1-elements for $\{u_h^j, \tilde{\mathbf{m}}_h^j\}$ works well even for the case of existing point defects. However, although time-steps are known to be chosen quite small at certain times (in particular close to $t = 0$), [117], adaptively chosen time-steps should give way to significantly reduce computational effort at this place.

4. At each time-step, we solve linear problems in our computational experiments; however, we expect nonlinear strategies to be more appropriate for complex flows.

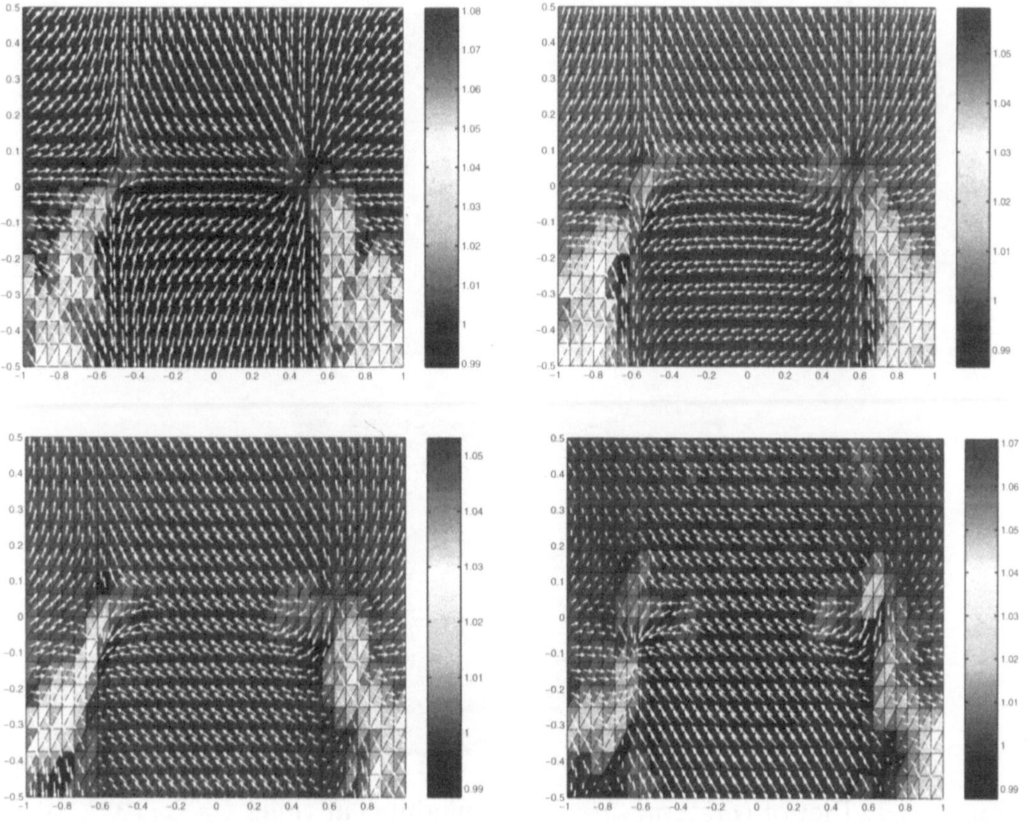

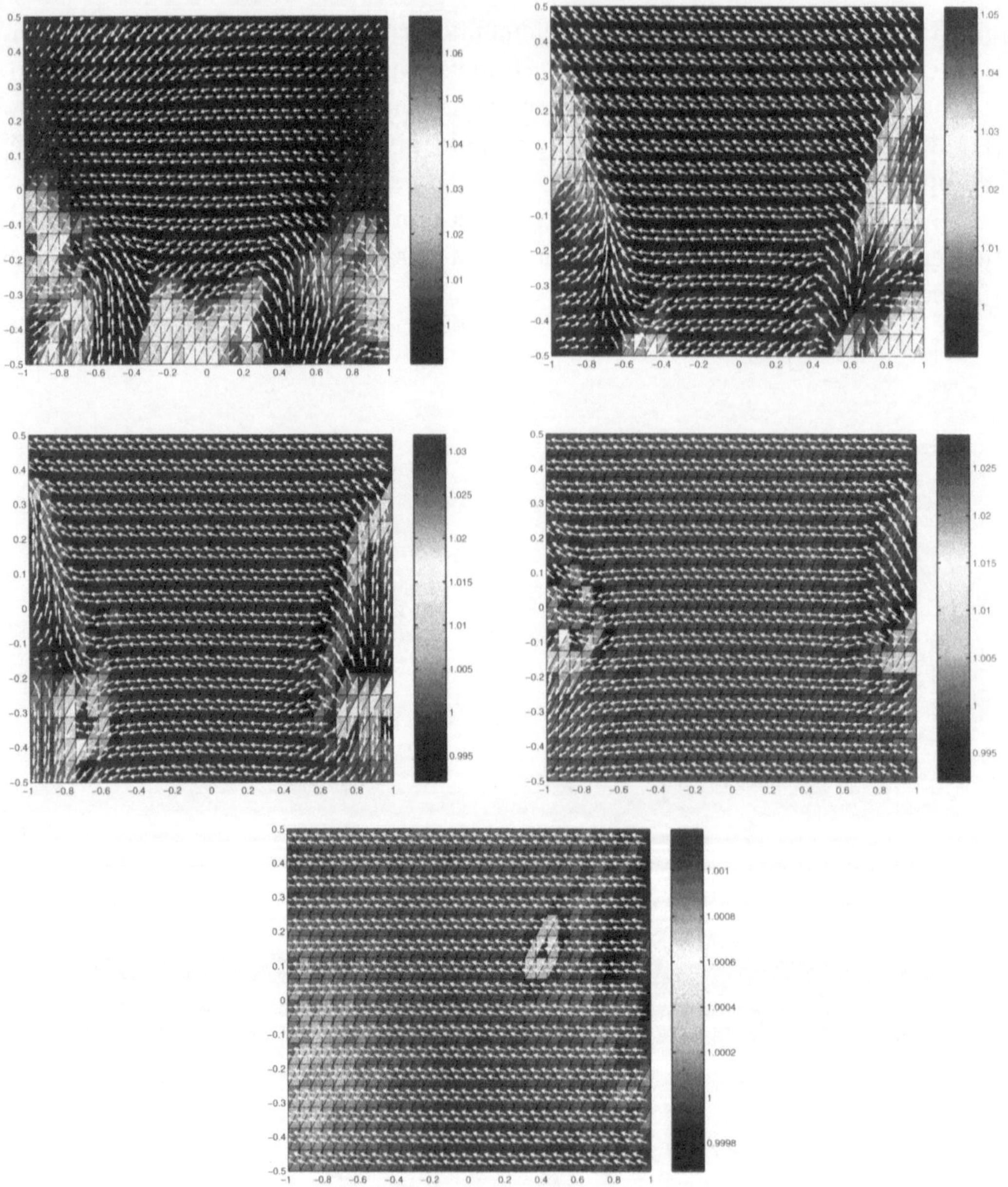

Figure 4.1: Plot of $(\tilde{m}_{h,1}^j, \tilde{m}_{h,2}^j)^\top$ and its modulus at times $t = 5 \cdot 10^{-3}$, $10^{-2}$, $1.5 \cdot 10^{-2}$, $2 \cdot 10^{-2}$, $5.5 \cdot 10^{-2}$, $7.5 \cdot 10^{-2}$, $9 \cdot 10^{-2}$, $10^{-1}$, $1.65 \cdot 10^{-1}$ ($h = \frac{1}{16}$, $k = 10^{-4}$) (note different scaling).

# Chapter 5

# The Maxwell-Landau-Lifshitz-Gilbert Equations

The behavior of electromagnetic dynamic phenomena on small scales is described by the Maxwell-Landau-Lifshitz-Gilbert equations, shortly (MLLG). For a magnetization $\mathbf{m} : \mathbb{R}^d \supset \omega \to \mathbb{R}^3$, a magnetic field $\mathbf{H} : \mathbb{R}^d \supset \omega \to \mathbb{R}^3$, and an electric field $\mathbf{E} : \mathbb{R}^d \supset \omega \to \mathbb{R}^3$, the equations read as follows,

$$\mathbf{m}_t = \mathbf{m} \times (\Delta\mathbf{m} + \mathbf{H}) - \kappa\,\mathbf{m} \times (\mathbf{m} \times (\Delta\mathbf{m} + \mathbf{H}))\,, \tag{5.1}$$

$$\nabla \times \mathbf{H} = \mathbf{E}_t + \sigma\,\mathbf{E}\,, \tag{5.2}$$

$$\nabla \times \mathbf{E} = -(\mathbf{H} + \beta\,\mathbf{m})_t\,, \tag{5.3}$$

$$\operatorname{div}(\mathbf{H} + \beta\,\mathbf{m}) = 0\,, \quad \operatorname{div}\mathbf{E} = 0\,. \tag{5.4}$$

with constants $\sigma, \beta, \kappa > 0$. The goal in this chapter is twofold: firstly, we show optimal convergence behavior for two penalized implicit Euler methods which discretize (MLLG) in time, for $\omega = (0, 2D)^2$ and space-periodic functions, see Section 5.2; a finite element discretization is discussed in Section 5.4. We mention the work [98], where a spatial semidiscretization of (MLLG) is analyzed under the assumption of smooth solutions in the case of absent exchange interaction.

To reach our goals, we need stability bounds for the solution of (5.1)-(5.4) that have partly been verified in [62, 63]. It is because of their crucial importance for the construction of time discretization schemes that we recall the proof of known results in Section 5.1 before we verify new stability bounds for the solution to (MLLG).

Secondly, we propose and analyze two (stabilized) time-splitting schemes in Section 5.3 that combine ideas of (scaled) projection schemes to effectively treat the Landau-Lifshitz-Gilbert equation and splitting techniques to compute $\{\mathbf{E}^j, \mathbf{H}^j\}$ independently from each other at each time-step. A finite element analysis is provided in Section 5.4.

## 5.1 Existence and Regularity of Solutions to (MLLG); a priori bounds

In [62], the existence of global weak solutions on $\omega = (0, 2D)^2$ is shown for (MLLG) with periodic boundary conditions (in 3D), i.e., for $\mathbf{x} \in \omega \subset \mathbb{R}^2$, and $t \geq 0$,

$$\begin{aligned} &\mathbf{m}(\mathbf{x} + 2D\mathbf{e}_i, t) = \mathbf{m}(\mathbf{x}, t)\,, \quad \mathbf{H}(\mathbf{x} + 2D\mathbf{e}_i, t) = \mathbf{H}(\mathbf{x}, t)\,, \qquad (5.5)\\ &\mathbf{E}(\mathbf{x} + 2D\mathbf{e}_i, t) = \mathbf{E}(\mathbf{x}, t)\,,\\ &\mathbf{m}(\mathbf{x}, 0) = \mathbf{m}_0(\mathbf{x})\,, \quad \mathbf{H}(\mathbf{x}, 0) = \mathbf{H}_0(\mathbf{x})\,, \quad \mathbf{E}(\mathbf{x}, 0) = \mathbf{E}_0(\mathbf{x})\,, \qquad (5.6) \end{aligned}$$

where $\mathbf{x} + 2D\mathbf{e}_i = (x_1, .., x_i + 2D, .., x_2)$, $i = 1, 2$, and $D > 0$. — We start with an elementary observation.

**Lemma 5.1** *(from [62]) Let* $|\,\mathbf{m}_0\,| = 1$. *Then for any smooth solution of the periodic initial value problem* (5.1)-(5.6) *there holds*

$$|\,\mathbf{m}(\mathbf{x}, t)\,| = 1, \quad \mathbf{x} \in \omega\,, \quad t \geq 0\,.$$

**Proof:**
Multiply (5.1) by $\mathbf{m}$ to obtain $\frac{d}{dt}|\,\mathbf{m}(\mathbf{x}, t)\,|^2 = 0$. □

We adopt the notion of weak solutions to (MLLG) from [62]: the triple $\{\mathbf{m}, \mathbf{H}, \mathbf{E}\} \in L^\infty(I; \mathbf{W}^{1,2}_{\text{per}}(\omega, \mathbb{R}^3)) \cap W^{1,2}(I; \mathbf{W}^{0,2}_{\text{per}}(\omega, \mathbb{R}^3)) \times L^\infty(I; \mathbf{W}^{0,2}_{\text{per}}(\omega, \mathbb{R}^3)) \times L^\infty(I; \mathbf{W}^{0,2}_{\text{per}}(\omega, \mathbb{R}^3))$ is called a weak solution of problem (MLLG) stated in (5.1)-(5.6), if $|\,\mathbf{m}\,| = 1$, almost everywhere in $\omega_T$, $\mathbf{m}(\mathbf{x}, 0) = \mathbf{m}_0$ holds in the sense of traces, and for any vector-valued test function $\boldsymbol{\psi}(\mathbf{x}, t)$, and test function $\xi(\mathbf{x}, t) \in C^\infty(I; C^\infty_{\text{per}}(\omega))$, with $\boldsymbol{\psi}(\mathbf{x}, T) = 0$, $\boldsymbol{\psi}(\mathbf{x}+D\mathbf{e}_i, t) = \boldsymbol{\psi}(\mathbf{x}-D\mathbf{e}_i, t)$, for $i = 1, 2$, the following equalities hold,

$$\int_{\omega_T} \langle \mathbf{m}_t, \boldsymbol{\psi} \rangle_{\mathbb{R}^3} \, \mathrm{d}\mathbf{x}\mathrm{d}s + \kappa \int_I (\nabla \mathbf{m}, \nabla \boldsymbol{\psi}) \, \mathrm{d}s = \kappa \int_I (| \nabla \mathbf{m} |^2 \mathbf{m}, \boldsymbol{\psi}) \, \mathrm{d}s$$
$$- \int_I (\mathbf{m} \times \nabla \mathbf{m}, \nabla \boldsymbol{\psi}) \, \mathrm{d}s + \int_I (\mathbf{m} \times \mathbf{H}, \boldsymbol{\psi}) \, \mathrm{d}s \tag{5.7}$$
$$+\kappa \int_I (\mathbf{m} \times \mathbf{H}, \mathbf{m} \times \boldsymbol{\psi}) \, \mathrm{d}s = 0 \,,$$
$$\int_I e^{\sigma s} (\mathbf{E}, \boldsymbol{\psi}_t) \, \mathrm{d}s + \int_I e^{\sigma s} (\nabla \times \boldsymbol{\psi}, \mathbf{H}) \, \mathrm{d}s + (\mathbf{E}_0, \boldsymbol{\psi}(\mathbf{x}, 0)) = 0, \tag{5.8}$$
$$\int_I (\mathbf{H} + \beta \, \mathbf{m}, \boldsymbol{\psi}_t) \, \mathrm{d}s - \int_I (\nabla \times \boldsymbol{\psi}, \mathbf{E}) \, \mathrm{d}s \tag{5.9}$$
$$+ (\mathbf{H}_0 + \beta \, \mathbf{m}_0, \boldsymbol{\psi}(\mathbf{x}, 0)) = 0 \,,$$
$$\int_I (\nabla \xi, \mathbf{H} + \beta \, \mathbf{m}) \, \mathrm{d}s = 0 \,, \quad \int_I (\nabla \xi, \mathbf{E}) \, \mathrm{d}s = 0 \,, \tag{5.10}$$
$$\mathbf{m}(\mathbf{x}, 0) = \mathbf{m}_0(\mathbf{x}), \quad \text{a.e. } \mathbf{x} \in \omega \,.$$

The proof of the following result can be found in [62].

**Lemma 5.2** *(from [62]) If the initial value vector functions* $\mathbf{E}_0, \mathbf{H}_0$ *satisfy the condition*

$$(\nabla \xi_0, \mathbf{E}_0) = 0 \,, \quad (\nabla \xi_0, \mathbf{H}_0 + \beta \, \mathbf{m}_0) = 0 \,, \tag{5.11}$$

*where* $\xi \in C^1(\omega_T)$, $\xi(\mathbf{x}, T) = 0$, $\xi_0 = \xi(\mathbf{x}, 0)$, *then from* (5.8), (5.10) *it follows that*

$$\int_I (\nabla \xi, \mathbf{E}) \, \mathrm{d}s = 0 \,, \quad \int_I (\nabla \xi, \mathbf{H} + \beta \, \mathbf{m}) \, \mathrm{d}s = 0 \,. \tag{5.12}$$

**Theorem 5.1** *(from [62], valid for $d = 3$) Assume that* $\{\mathbf{m}_0, \mathbf{H}_0, \mathbf{E}_0\} \in \mathbf{W}^{1,2}_{\mathrm{per}}(\omega, \mathcal{S}^2) \times \left[\mathbf{W}^{0,2}_{\mathrm{per}}(\omega, \mathbb{R}^3)\right]^2$, *for* $\omega = (0, 2D)^2$ *satisfy* (5.11). *Then* (5.1)-(5.6) *has at least one weak solution* $\{\mathbf{m}, \mathbf{H}, \mathbf{E}\}$, *such that*

$$\mathbf{m} \in L^\infty(I; \mathbf{W}^{1,2}_{\mathrm{per}}(\omega, \mathcal{S}^2)) \cap C^{0,2/3}(I; \mathbf{W}^{0,3}_{\mathrm{per}}(\omega, \mathbb{R}^3)) \,,$$
$$\mathbf{E}, \mathbf{H} \in L^\infty(I; \mathbf{W}^{0,2}_{\mathrm{per}}(\omega, \mathbb{R}^3)) \cap C^{0,1/2}(I; \mathbf{W}^{-1,2}_{\mathrm{per}}(\omega, \mathbb{R}^3)) \,.$$

Under certain assumptions on the given data of the problem, the solution to (MLLG) is smooth. We have the following result due to Su and Guo that is proved by means of Galerkin method, using sharp a priori bounds for the solution to (5.7)-(5.10), (5.5)-(5.6), under the assumption (5.11), and standard compactness arguments.

**Theorem 5.2** *(from [62]) Assume constants $\kappa > 0$, $\beta \geq 0$, $\sigma \geq 0$, periodic initial value functions $\{\mathbf{m}_0, \mathbf{H}_0, \mathbf{E}_0\} \in \mathbf{W}^{k,2}_{\text{per}}(\omega, \mathcal{S}^2) \times [\mathbf{W}^{k-1,2}_{\text{per}}(\omega, \mathbb{R}^3)]^2$, $k \geq 3$, $\omega = (0, 2D)^2$, and $\operatorname{div}(\mathbf{H}_0 + \beta\, \mathbf{m}_0) = 0$, $\operatorname{div} \mathbf{E}_0 = 0$. Moreover, let*

$$\| \nabla \mathbf{m}_0 \|_{\mathbf{L}^2} + \| \mathbf{E}_0 \|_{\mathbf{L}^2} + \| \mathbf{H}_0 \|_{\mathbf{L}^2} \leq \delta\,,$$

*where $\delta$ is a suitably small constant. Then, (MLLG) has a unique global smooth solution such that $|\, \mathbf{m}(\mathbf{x}, t)\, | = 1$, $\mathbf{x} \in \omega$, $t \geq 0$, and*

$$\mathbf{m} \in \bigcap_{s=0}^{[\frac{k}{2}]} W^{s,\infty}\big(I; \mathbf{W}^{k-2s,2}_{\text{per}}(\omega, \mathcal{S}^2)\big)\,, \quad \mathbf{H}, \mathbf{E} \in \bigcap_{s=0}^{k-1} W^{s,\infty}\big(I; \mathbf{W}^{k-1-s,2}_{\text{per}}(\omega, \mathbb{R}^3)\big)\,.$$

We sketch the proof of this theorem here that corresponds to the one for (LLG); see Lemmata 4.1, 4.2, and 4.3. The bootstrapping arguments from the proof of Lemma 4.3 will not be explicitly given here since we are mainly interested in the dynamics for initial data with limited regularity.

**Lemma 5.3** *Let $\{\mathbf{m}_0, \mathbf{H}_0, \mathbf{E}_0\} \in \mathbf{W}^{1,2}_{\text{per}}(\omega, \mathcal{S}^2) \times [\mathbf{W}^{0,2}_{\text{per}}(\omega, \mathbb{R}^3)]^2$ satisfy (5.11) and $\{\mathbf{m}, \mathbf{H}, \mathbf{E}\}$ be a smooth solution to (MLLG). Then there exists a constant $C$ such that*

$$\text{(a)} \qquad \operatorname*{ess\,sup}_{[0,T]} \{ \| \nabla \mathbf{m} \|_{\mathbf{L}^2} + \| \mathbf{H} \|_{\mathbf{L}^2} + \| \mathbf{E} \|_{\mathbf{L}^2} \}$$
$$+ \frac{1}{C} \Big( \int_0^T \{ \| \mathbf{m}(s) \times (\Delta \mathbf{m}(s) + \mathbf{H}(s)) \|^2_{\mathbf{L}^2} + \| \mathbf{m}_t(s) \|^2_{\mathbf{L}^2} \}\, \mathrm{d}s \Big)^{\frac{1}{2}} \leq E_0(\omega)\,,$$

$$\text{(b)} \qquad \int_0^T \| \Delta \mathbf{m} \|^2_{\mathbf{L}^2}\, \mathrm{d}s \leq C\,.$$

*Statement (b) is only valid for $E_0(\omega) := \| \nabla \mathbf{m}_0 \|_{\mathbf{L}^2} + \| \mathbf{H}_0 \|_{\mathbf{L}^2} + \| \mathbf{E}_0 \|_{\mathbf{L}^2} \leq \delta$ sufficiently small.*

**Proof:**
(a) (partly from [63]): We multiply (5.2) by $\mathbf{E}$ and (5.3) with $-\mathbf{H}$ and add both equalities.

$$\begin{aligned}&\langle \nabla \times \mathbf{H}, \mathbf{E}\rangle_{\mathbb{R}^3} - \langle \nabla \times \mathbf{E}, \mathbf{H}\rangle_{\mathbb{R}^3} \\ &\quad = \frac{1}{2}\frac{d}{dt}|\,\mathbf{E}\,|^2 + \sigma\,|\,\mathbf{E}\,|^2 + \frac{1}{2}\frac{d}{dt}|\,\mathbf{H}\,|^2 + \beta\,\langle \mathbf{m}_t, \mathbf{H}\rangle_{\mathbb{R}^3}\,.\end{aligned} \tag{5.13}$$

We make use of the formula

$$\langle \nabla \times \mathbf{H}, \mathbf{E}\rangle_{\mathbb{R}^3} - \langle \nabla \times \mathbf{E}, \mathbf{H}\rangle_{\mathbb{R}^3} = \operatorname{div}(\mathbf{H} \times \mathbf{E})\,, \tag{5.14}$$

when integrating (5.13) over $\mathbf{x} \in \omega$ to find

$$\frac{1}{2}\frac{d}{dt}\{\|\,\mathbf{E}\,\|^2_{\mathbf{L}^2} + \|\,\mathbf{H}\,\|^2_{\mathbf{L}^2}\} + \sigma\,\|\,\mathbf{E}\,\|^2_{\mathbf{L}^2} + \beta\,\big(\mathbf{m}_t, \mathbf{H}\big) = 0\,. \tag{5.15}$$

Next, we multiply (5.1) by $\Delta\mathbf{m} + \mathbf{H}$ to get

$$\langle \Delta\mathbf{m} + \mathbf{H}, \mathbf{m}_t\rangle_{\mathbb{R}^3} = -\kappa\,\langle \Delta\mathbf{m} + \mathbf{H}, \mathbf{m} \times \big(\mathbf{m} \times (\Delta\mathbf{m} + \mathbf{H})\big)\rangle_{\mathbb{R}^3}\,. \tag{5.16}$$

We use the identity

$$-\langle \Delta\mathbf{m} + \mathbf{H}, \mathbf{m} \times \big(\mathbf{m} \times (\Delta\mathbf{m} + \mathbf{H})\big)\rangle_{\mathbb{R}^3} = |\,\mathbf{m} \times \big(\Delta\mathbf{m} \times \mathbf{H}\big)\,|^2$$

to obtain

$$\big(\Delta\mathbf{m} + \mathbf{H}, \mathbf{m}_t\big) = \kappa\,\|\,\mathbf{m} \times \big(\Delta\mathbf{m} + \mathbf{H}\big)\,\|^2_{\mathbf{L}^2}\,. \tag{5.17}$$

Combining (5.15) and (5.17) leads to

$$\begin{aligned}&\frac{1}{2}\frac{d}{dt}\{\|\,\mathbf{E}\,\|^2_{\mathbf{L}^2} + \|\,\mathbf{H}\,\|^2_{\mathbf{L}^2}\} + \frac{\beta}{2}\frac{d}{dt}\|\,\nabla\mathbf{m}\,\|^2_{\mathbf{L}^2} \\ &\qquad + \sigma\,\|\,\mathbf{E}\,\|^2_{\mathbf{L}^2} + \beta\kappa\,\|\,\mathbf{m} \times \big(\Delta\mathbf{m} + \mathbf{H}\big)\,\|^2_{\mathbf{L}^2} = 0\,.\end{aligned}$$

We verify the remaining result in (a). Therefore, we multiply (5.1) by $\mathbf{m}_t$ and exploit $d_t|\,\mathbf{m}\,|^2 = 0$.

$$\begin{aligned}&\int_0^T \|\,\mathbf{m}_t(s)\,\|^2\,\mathrm{d}s + \frac{\kappa}{2}\,\|\,\nabla\mathbf{m}(T)\,\|^2_{\mathbf{L}^2} \leq \frac{\kappa}{2}\,\|\,\nabla\mathbf{m}(0)\,\|^2_{\mathbf{L}^2} \\ &\quad + \int_0^T \big(\mathbf{m}(s) \times (\Delta\mathbf{m}(s) + \mathbf{H}(s)), \mathbf{m}_t(s)\big)\,\mathrm{d}s + \kappa \int_0^T \big(\mathbf{H}(s), \mathbf{m}_t(s)\big)\,\mathrm{d}s\,.\end{aligned} \tag{5.18}$$

We deal with $\kappa \int_0^T \big(\mathbf{m}(s) \times [\Delta \mathbf{m}(s) + \mathbf{H}(s)], \mathbf{m}_t(s)\big)\, \mathrm{d}s$ separately. We take the vector product of (5.1) by $\mathbf{m}_t$ and finally multiply by $\mathbf{m}$.

$$\kappa \big\langle \mathbf{m} \times \big(\Delta \mathbf{m} + \mathbf{H}\big), \mathbf{m}_t \big\rangle_{\mathbb{R}^3} = \big\langle \big(\mathbf{m} \times \big[\Delta \mathbf{m} + \mathbf{H}\big]\big) \times \mathbf{m}_t, \mathbf{m} \big\rangle_{\mathbb{R}^3} \tag{5.19}$$

Let $\theta(s)$ stand for the angle between $\mathbf{m}_t(s)$ and $\mathbf{m}(s) \times \big(\Delta \mathbf{m}(s) + \mathbf{H}(s)\big)$. Then (5.19) reads

$$\kappa \,|\, \mathbf{m} \times \big(\Delta \mathbf{m} + \mathbf{H}\big) \,|\,|\, \mathbf{m}_t \,|\,|\cos\theta\,| \leq |\, \mathbf{m} \times \big(\Delta \mathbf{m} + \mathbf{H}\big) \,|\,|\, \mathbf{m}_t \,|\,|\sin\theta\,| \tag{5.20}$$

or

$$|\cos\theta\,|^2 \leq \frac{\kappa^2}{1+\kappa^2}\,. \tag{5.21}$$

Then, we multiply (5.1) by $\mathbf{m} \times \big(\Delta \mathbf{m} + \mathbf{H}\big)$.

$$\big\langle \mathbf{m} \times \big(\Delta \mathbf{m} + \mathbf{H}\big), \mathbf{m}_t \big\rangle_{\mathbb{R}^3} = |\, \mathbf{m} \times \big(\Delta \mathbf{m} + \mathbf{H}\big) \,|^2\,,$$

or equivalently

$$|\, \mathbf{m}_t \,|\cos\theta = |\, \mathbf{m} \times \big(\Delta \mathbf{m} + \mathbf{H}\big) \,|\,. \tag{5.22}$$

From this consideration, we find

$$|\,\langle \mathbf{m} \times \big(\Delta \mathbf{m} + \mathbf{H}\big), \mathbf{m}_t \rangle_{\mathbb{R}^3} \,| \leq |\, \mathbf{m}_t \,|^2 |\cos\theta\,|^2 \leq \frac{\kappa^2}{1+\kappa^2}\, |\, \mathbf{m}_t \,|^2, \tag{5.23}$$

which allows to absorb this term on the left-hand side of (5.18).

(b) To see this, we multiply (5.1) by $-\Delta \mathbf{m}$, and make use of the relation

$$\|\, \nabla \mathbf{m} \,\|_{\mathbf{L}^4}^4 \leq C \,\|\, \nabla \mathbf{m} \,\|_{\mathbf{L}^2}^2 \,\|\, \nabla^2 \mathbf{m} \,\|_{\mathbf{L}^2}^2 \leq C\, E_0(\omega)\, \|\, \nabla^2 \mathbf{m} \,\|_{\mathbf{L}^2}^2\,.$$

□

**Lemma 5.4** *Let* $\big\{\mathbf{m}_0, \mathbf{H}_0, \mathbf{E}_0\big\} \in \mathbf{W}^{2,2}_{\mathrm{per}}(\omega, \mathcal{S}^2) \times \big[\mathbf{W}^{1,2}_{\mathrm{per}}(\omega, \mathbb{R}^3)\big]^2$ *satisfy* (5.11), *and*

$$\|\, \nabla \mathbf{m}_0 \,\|_{\mathbf{L}^2} + \|\, \mathbf{H}_0 \,\|_{\mathbf{L}^2} + \|\, \mathbf{E}_0 \,\|_{\mathbf{L}^2} \leq \delta\,,$$

*for* $\delta > 0$ *suitably small. Let* $\{\mathbf{m}, \mathbf{H}, \mathbf{E}\}$ *be a smooth solution to (MLLG) that satisfies periodic boundary conditions. Then there exists a constant* $C$ *such that*

$$\operatorname*{ess\,sup}_{[0,T]} \Big\{ \|\, \nabla^2 \mathbf{m} \,\|_{\mathbf{L}^2} + \|\, \nabla \mathbf{H} \,\|_{\mathbf{L}^2} + \|\, \nabla \mathbf{E} \,\|_{\mathbf{L}^2} \Big\} + \Big( \int_0^T \|\, \mathbf{m}(s) \,\|_{\mathbf{W}^{3,2}}^2\, \mathrm{d}s \Big)^{1/2} \leq C\,.$$

**Proof:**
In a next step, we make the scalar products of $-\Delta\mathbf{E}$ with (5.2) and $\Delta\mathbf{H}$ with (5.3), resp. Thanks to the facts $\mathbf{curl}\,\nabla\xi = 0$ and $-\Delta\boldsymbol{\xi}+\nabla\,\mathrm{div}\,\boldsymbol{\xi} = \mathbf{curl}\,\mathbf{curl}\,\boldsymbol{\xi}$, we may employ periodicity of functions to conclude

$$\begin{aligned}&(\nabla\times\Delta\mathbf{H},\mathbf{E}) - (\nabla\times\Delta\mathbf{E},\mathbf{H})\\ &\quad = \big(\mathbf{curl}\,\mathbf{curl}\,\mathbf{curl}\,\mathbf{H},\mathbf{E}\big) - \big(\mathbf{curl}\,\mathbf{curl}\,\mathbf{curl}\,\mathbf{E},\mathbf{H}\big) = 0\,.\end{aligned}$$

Therefore

$$\frac{d}{dt}\{\|\nabla\mathbf{E}\|_{\mathbf{L}^2}^2 + \|\nabla\mathbf{H}\|_{\mathbf{L}^2}^2\} + \sigma\|\nabla\mathbf{E}\|_{\mathbf{L}^2}^2 = 2\beta\big(\nabla\mathbf{m}_t,\nabla\mathbf{H}\big)\,. \tag{5.24}$$

On the other hand, we can proceed as in the proof of Lemma 4.2, with those terms in (5.1) that do not involve $\mathbf{H}$, see (4.27),

$$\begin{aligned}d_t\|\nabla^2\mathbf{m}\|_{\mathbf{L}^2}^2 + \alpha\|\nabla^3\mathbf{m}\|_{\mathbf{L}^2}^2 &\le C\|\nabla^2\mathbf{m}\|_{\mathbf{L}^2}^2\\ &+\alpha\|\nabla^3\mathbf{m}\|_{\mathbf{L}^2}^2\big(\|\nabla\mathbf{m}_0\|_{\mathbf{L}^2}^2 + 2\|\nabla\mathbf{m}_0\|_{\mathbf{L}^2} + \frac{1}{\alpha}\|\nabla\mathbf{m}_0\|_{\mathbf{L}^2}\big)\\ &+|\big(D(\mathbf{m}\times\mathbf{H} - \kappa\,\mathbf{m}\times\mathbf{m}\times\mathbf{H}\big)|\,.\end{aligned} \tag{5.25}$$

The crucial terms are

$$\begin{aligned}\|D\mathbf{m}\|_{\mathbf{L}^4}\|\mathbf{H}\|_{\mathbf{L}^4}\|D^3\mathbf{m}\|_{\mathbf{L}^2} &\le C\|D\mathbf{m}\|_{\mathbf{L}^2}^2\|D\mathbf{m}\|_{\mathbf{W}^{1,2}}^2\|\mathbf{H}\|_{\mathbf{L}^2}^2\\ &\qquad + \frac{\alpha}{8}\|\nabla^3\mathbf{m}\|_{\mathbf{L}^2}^2 + \frac{1}{2}\|\nabla\mathbf{H}\|_{\mathbf{L}^2}^2\,,\\ \|\mathbf{m}\|_{\mathbf{L}^\infty}\|\nabla\mathbf{H}\|_{\mathbf{L}^2}\|D^3\mathbf{m}\|_{\mathbf{L}^2} &\le C\|\mathbf{m}\|_{\mathbf{L}^2}\|\mathbf{m}\|_{\mathbf{W}^{2,2}}\|\nabla\mathbf{H}\|_{\mathbf{L}^2}^2 + \frac{\alpha}{8}\|\nabla^3\mathbf{m}\|_{\mathbf{L}^2}^2\,.\end{aligned}$$

Then, Lemma 5.3 helps to control the second contribution. Finally Gronwall's inequality settles the result. □

Theorem 5.2 states that solutions to (MLLG) are smooth, provided that the initial energy $E_0(\omega)$ is suitably small. Similarly to (LLG), one can also show local existence of smooth solutions to (MLLG), with degree of regularity and length of the time interval dependent on $\mathbf{m}_0$. — We employ the technique of M. Struwe to show the following result.

**Lemma 5.5** *Let $\{\mathbf{m},\mathbf{H},\mathbf{E}\}$ be a space-periodic smooth solution to (MLLG), for $\{\mathbf{m}_0,\mathbf{H}_0,\mathbf{E}_0\} \in \mathbf{W}^{2,2}_{\mathrm{per}}(\omega,S^2)\times\big[\mathbf{W}^{1,2}_{\mathrm{per}}(\omega,\mathbb{R}^3)\big]^2$ satisfying (5.11). For*

*$\mathbf{x}_0 \in \omega$, let $B_R(\mathbf{x}_0)$ be a ball around $\mathbf{x}_0$, for $0 < R \leq 2R_0$. Then, for $R \leq R_0$ there exists $T > 0$ and $C = C(\mathbf{m}_0, R, \kappa)$, such that holds*

$$\text{(a)} \quad \| \nabla \mathbf{m}(T) \|^2_{\mathbf{L}^2(B_R(\mathbf{x}_0))} \leq E_0\big(B_{2R}(\mathbf{x}_0)\big) + 2\,(1+\kappa^2)^2 \frac{T}{R^2} E_0(\omega)\,,$$

$$\text{(b)} \quad \int_0^T \| \Delta \mathbf{m}(s) \|^2 \, \mathrm{d}s \leq C\,.$$

**Proof:**
(a) Let $\phi \in C_0^\infty\big(B_{2R}(\mathbf{x}_0)\big)$ satisfy $0 \leq \phi \leq 1$, $\phi \equiv 1$ on $B_R(\mathbf{x}_0)$, with $|\nabla \phi| \leq \frac{2}{R}$, and multiply (5.1) by $\mathbf{m}_t \phi^2$ to obtain

$$\begin{aligned} \int_\omega |\mathbf{m}_t|^2 \phi^2 \, \mathrm{d}\mathbf{x} + \kappa\, d_t\Big(\int_\omega |\nabla \mathbf{m}|^2 \phi^2 \, \mathrm{d}\mathbf{x}\Big) &\leq 2\kappa \int_\omega |\nabla \mathbf{m}|\,|\mathbf{m}_t|\,|\nabla\phi|\,|\phi| \, \mathrm{d}\mathbf{x} \\ -\Big(\mathbf{m} \times (\Delta \mathbf{m} + \mathbf{H}), \mathbf{m}_t \phi^2\Big) - \kappa \Big(\mathbf{m} \times (\mathbf{m} \times \mathbf{H}), \mathbf{m}_t \phi^2\Big) \end{aligned} \qquad (5.26)$$

It is thanks to (5.22), (5.23), that we can control the second term on the right-hand side by the leading one on the left-hand side. Together with Lemma 5.3, (a), this settles the proof.

(b) From this local energy estimate, we may now prove another bound for the magnetization, according to the proof of Lemma 4.6: given $\varepsilon_1 > 0$, $\mathbf{m}_0 \in \mathbf{W}^{1,2}_{\text{per}}(\omega, \mathcal{S}^2)$, there exists a number $T_1 > 0$ depending only on a maximal number $R_1 > 0$ and $\alpha > 0$, such that

$$\sup_{\mathbf{x}_0 \in \omega, 0 \leq s \leq T_1} \| \nabla \mathbf{m}(s) \|_{\mathbf{L}^2(B_R(\mathbf{x}_0))} < 2\varepsilon_1\,. \qquad (5.27)$$

Let $\varepsilon_1 > 0$ and $R_1 > 0$ be given, and $\{\phi_i\}$ a set of smooth cut-off functions subordinate to a cover of $\omega$ of balls $\{B_{2R_1}(\mathbf{x}_i)\}$ with finite overlap and such that $0 \leq \phi_i \leq 1$, such that $|\nabla \phi_i| \leq \frac{2}{R_1}$, and $\sum_i \phi_i^2 = 1$. Then, owing to (4.57),

$$\begin{aligned} \| \nabla \mathbf{m} \|^4_{\mathbf{L}^4} &\leq C \sup_i \| \nabla \mathbf{m}(s) \|_{\mathbf{L}^2(B_R(\mathbf{x}_i))} \Big( \int_\Omega |\nabla^2 \mathbf{m}|^2 \, \mathrm{d}\mathbf{x} + \frac{1}{R_1^2} \| \nabla \mathbf{m}_0 \|^2_{\mathbf{L}^2} \Big) \\ &\leq C\varepsilon_1 \Big( \int_\Omega |\nabla \mathbf{m}|^2 \, \mathrm{d}\mathbf{x} + \frac{1}{R_1^2} \| \nabla \mathbf{m}_0 \|^2_{\mathbf{L}^2} \Big). \end{aligned} \qquad (5.28)$$

We may now multiply (5.1) by $-\Delta \mathbf{m}$ and employ a (local in time) argument similar to the one in the proof of Lemma 4.6, formula (4.59); the further terms $\| \mathbf{m} \times (\Delta \mathbf{m} + \mathbf{H}) \|_{L^2(I,\mathbf{L}^2)}$ and $\| \mathbf{m} \times (\mathbf{m} \times \mathbf{H}) \|_{L^2(I,\mathbf{L}^2)}$ do not cause further difficulties, using Lemma 5.3, (a). $\square$

We are interested in further a priori estimates for the solution $\{\mathbf{m}, \mathbf{H}, \mathbf{E}\}$ to (MLLG). The following results apply to both considered scenarios involving global/local smooth solutions.

**Theorem 5.3** *Let* $\{\mathbf{m}_0, \mathbf{H}_0, \mathbf{E}_0\} \in \mathbf{W}^{2,2}_{\text{per}}(\omega, \mathcal{S}^2) \times [\mathbf{W}^{1,2}_{\text{per}}(\omega, \mathbb{R}^3)]^2$ *satisfy (5.11), and the solution* $\{\mathbf{m}, \mathbf{H}, \mathbf{E}\}$ *to (MLLG) satisfies periodic boundary conditions. There exists* $T = T(\mathbf{m}_0) > 0$, *such that*

$$\text{(a) } \operatorname*{ess\,sup}_{[0,T]} \{ \| \mathbf{m}_t \|_{\mathbf{L}^2} + \| \mathbf{H}_t \|_{\mathbf{L}^2} + \| \mathbf{E}_t \|_{\mathbf{L}^2} \} + \Big( \int_0^T \| \nabla \mathbf{m}_t(s) \|^2_{\mathbf{L}^2} \, ds \Big)^{1/2}$$
$$+ \operatorname*{ess\,sup}_{[0,T]} \{ \| \Delta \mathbf{m} \|_{\mathbf{L}^2} + \| \nabla \mathbf{H} \|_{\mathbf{L}^2} + \| \nabla \mathbf{E} \|_{\mathbf{L}^2} \} \leq C \,,$$

$$\text{(b) } \Big( \int_0^T \{ \| \mathbf{m}_{tt}(s) \|_{\mathbf{W}^{-1,2}_{\text{per}}} + \| \mathbf{H}_{tt}(s) \|_{\mathbf{W}^{-1,2}_{\text{per}}} + \| \mathbf{E}_{tt}(s) \|_{\mathbf{W}^{-1,2}_{\text{per}}} \}^2 \, ds \Big)^{1/2} \leq C \,,$$

$$\text{(c) } \operatorname*{ess\,sup}_{[0,T]} \sqrt{\tau} \, \| \nabla \mathbf{m}_t \|_{\mathbf{L}^2} + \Big( \int_0^T \tau \{ \| \mathbf{m}_{tt} \|_{\mathbf{L}^2} + \| \Delta \mathbf{m}_t \|_{\mathbf{L}^2}$$
$$+ \| \mathbf{H}_{tt} \|_{\mathbf{L}^2} + \| \mathbf{E}_{tt} \|_{\mathbf{L}^2} \}^2 (s) \, ds \Big)^{1/2} \leq C \,,$$

*where* $\tau(s) := \min\{1, s\}$.

**Proof:**
(a) Differentiating (5.1) with respect to time gives

$$\begin{aligned} \mathbf{m}_{tt} &= \mathbf{m}_t \times (\Delta \mathbf{m} + \mathbf{H}) + \mathbf{m} \times (\Delta \mathbf{m}_t + \mathbf{H}_t) + \kappa \Delta \mathbf{m}_t + \kappa | \nabla \mathbf{m} |^2 \mathbf{m}_t \\ &\quad + 2\kappa \langle \nabla \mathbf{m}, \nabla \mathbf{m}_t \rangle_{\mathbb{R}^3} \mathbf{m} - \kappa \{ \langle \mathbf{m}_t, \mathbf{H} \rangle_{\mathbb{R}^3} + \langle \mathbf{m}, \mathbf{H}_t \rangle_{\mathbb{R}^3} \} \mathbf{m} \\ &\quad - \kappa \langle \mathbf{m}, \mathbf{H} \rangle_{\mathbb{R}^3} \mathbf{m}_t + \kappa \mathbf{H}_t \,. \end{aligned}$$

We multiply this identity with $\mathbf{m}_t$ and integrate with respect to $\mathbf{x} \in \omega$.

$$\begin{aligned} \frac{1}{2} \frac{d}{dt} \| \mathbf{m}_t \|^2_{\mathbf{L}^2} + \kappa \| \nabla \mathbf{m}_t \|^2_{\mathbf{L}^2} &\leq C \Big( \{ | (\mathbf{m} \times \Delta \mathbf{m}_t, \mathbf{m}_t) | + | (\mathbf{m} \times \mathbf{H}_t, \mathbf{m}_t) | \} \\ &\quad + \kappa \{ | (\mathbf{H}_t, \mathbf{m}_t) | + | (| \nabla \mathbf{m} |^2, | \mathbf{m}_t |^2) | + | (| \mathbf{m}_t |^2 \mathbf{m}, \mathbf{H}) | \} \Big) . \end{aligned} \tag{5.29}$$

The first and fourth term can be controlled by using (c) and the property $(\mathbf{a} \times \nabla \mathbf{b}, \nabla \mathbf{b}) = 0$.

$$
\begin{aligned}
(\mathbf{m} \times \Delta \mathbf{m}_t, \mathbf{m}_t) &= (\nabla \mathbf{m} \times \mathbf{m}_t, \nabla \mathbf{m}_t) \\
&\leq \frac{\kappa}{8} \| \nabla \mathbf{m}_t \|_{\mathbf{L}^2}^2 + C_\kappa \| \nabla \mathbf{m} \|_{\mathbf{L}^4}^4 \| \mathbf{m}_t \|_{\mathbf{L}^2}^2 , \\
(| \nabla \mathbf{m} |^2, | \mathbf{m}_t |^2) &\leq C \| \nabla \mathbf{m} \|_{\mathbf{L}^4}^2 \| \mathbf{m}_t \|_{\mathbf{L}^4}^2 \\
&\leq \frac{\kappa}{8} \| \nabla \mathbf{m}_t \|_{\mathbf{L}^2}^2 + C_\kappa \| \nabla \mathbf{m} \|_{\mathbf{L}^4}^4 \| \mathbf{m}_t \|_{\mathbf{L}^2}^2 .
\end{aligned}
$$

By (a), we get

$$
| (| \mathbf{m}_t |^2 \mathbf{m}, \mathbf{H}) | \leq C \| \mathbf{H} \|_{\mathbf{L}^2} \| \mathbf{m}_t \|_{\mathbf{L}^4}^2 \leq C \| \mathbf{m}_t \|_{\mathbf{L}^2} \| \nabla \mathbf{m}_t \|_{\mathbf{L}^2} . \tag{5.30}
$$

In order to bound the remaining two terms in (5.29), we benefit from (5.3) twice,

$$
\begin{aligned}
| (\mathbf{m} \times \mathbf{H}_t, \mathbf{m}_t) | &= | (\mathbf{m} \times (\nabla \times \mathbf{E}), \mathbf{m}_t) | \leq C \| \nabla \mathbf{E} \|_{\mathbf{L}^2} \| \mathbf{m}_t \|_{\mathbf{L}^2} , & (5.31) \\
| (\mathbf{H}_t, \mathbf{m}_t) | &= | \beta \| \mathbf{m}_t \|_{\mathbf{L}^2}^2 - (\nabla \times \mathbf{E}, \mathbf{m}_t) | & (5.32) \\
&\leq C \left( \| \nabla \mathbf{E} \|_{\mathbf{L}^2}^2 + \| \mathbf{m}_t \|_{\mathbf{L}^2}^2 \right) .
\end{aligned}
$$

By Gronwall's lemma, this shows that $\mathbf{m}_t \in L^\infty(I, \mathbf{W}^{0,2}_{\text{per}}(\omega, \mathbb{R}^3))$ is bounded.

If we multiply (5.1) by $-\Delta \mathbf{m}$, the last result and (a) imply

$$
\kappa \| \Delta \mathbf{m} \|_{\mathbf{L}^2}^2 \leq \kappa \| \nabla \mathbf{m} \|_{\mathbf{L}^4}^4 + \frac{1}{\kappa} \{ \| \mathbf{m}_t \|_{\mathbf{L}^2}^2 + \| \mathbf{H} \|_{\mathbf{L}^2}^2 \} . \tag{5.33}
$$

Moreover, we already have (5.24) which is

$$
\frac{d}{dt} \{ \| \nabla \mathbf{E} \|_{\mathbf{L}^2}^2 + \| \nabla \mathbf{H} \|_{\mathbf{L}^2}^2 \} + \sigma \| \nabla \mathbf{E} \|_{\mathbf{L}^2}^2 = 2\beta \left( \nabla \mathbf{m}_t, \nabla \mathbf{H} \right) .
$$

The right-hand side can be bounded thanks to (5.29).

The verification of the upper bounds for $\mathbf{E}_t, \mathbf{H}_t \in L^\infty(I; \mathbf{W}^{0,2}_{\text{per}}(\omega, \mathbb{R}^3))$ is now immediate.

(b) The first result follows from (5.29), since upper bounds next to (4.70) are valid here as well. Additional terms do not cause further difficulties. From $\nabla \times \mathbf{H}_t = -\mathbf{E}_{tt} + \sigma \, \mathbf{m}_{tt}$, we infer

$$
\| \mathbf{E}_{tt} \|_{L^2(I; \mathbf{W}^{-1,2}_{\text{per}})} \leq \sigma \| \mathbf{m}_{tt} \|_{L^2(I; \mathbf{W}^{-1,2}_{\text{per}})} + \| \mathbf{H}_t \|_{L^2(I; \mathbf{L}^2)} , \tag{5.34}
$$

and similarly for the bound $\| \mathbf{H}_{tt} \|_{L^2\left(I;\mathbf{W}^{-1,2}_{\mathrm{per}}\right)} \leq C$ through (5.4).

(c) We multiply (5.29) by $-\Delta \mathbf{m}_t$. Terms listed next to (4.64) are summarized in $|\,(\mathcal{Z}, -\Delta \mathbf{m}_t)\,|$.

$$\begin{aligned}
&\frac{1}{2}\frac{d}{dt}\| \nabla \mathbf{m}_t \|^2_{\mathbf{L}^2} + \kappa \| \Delta \mathbf{m}_t \|^2_{\mathbf{L}^2} \\
&\leq |\,(\mathcal{Z}, -\Delta \mathbf{m}_t)\,| + \Big\{ |\,(\mathbf{m}_t \times \mathbf{H} + \mathbf{m} \times \mathbf{H}_t, -\Delta \mathbf{m}_t)\,| \Big\} \\
&\quad + \kappa \Big\{ |\,(\langle \mathbf{m}_t, \mathbf{H}\rangle_{\mathbb{R}^3} + \langle \mathbf{m}, \mathbf{H}_t\rangle_{\mathbb{R}^3}, \langle \mathbf{m}, -\Delta \mathbf{m}_t\rangle_{\mathbb{R}^3})\,| \\
&\quad + |\,(\langle \mathbf{m}, \mathbf{H}\rangle_{\mathbb{R}^3}, \langle \mathbf{m}_t, -\Delta \mathbf{m}_t\rangle_{\mathbb{R}^3})\,| + (\mathbf{H}_t, -\Delta \mathbf{m}_t) \Big\}.
\end{aligned} \tag{5.35}$$

Due to (d), bounding these terms is straightforward. For example, the last but one term can be controlled through

$$\| \mathbf{m} \|_{\mathbf{L}^\infty} \| \mathbf{H} \|_{\mathbf{L}^4} \| \mathbf{m}_t \|_{\mathbf{L}^4} \| \Delta \mathbf{m}_t \|_{\mathbf{L}^2}\,.$$

Multiplication by $\tau(s)$, integrating over $I$ and finally (d) and Gronwall's inequality lead to the first bounds in (f).

Next, we multiply (5.29) by $\mathbf{m}_{tt}$. Terms that are already treated next to (4.65) are collected in $|\,(\mathcal{Y}, \mathbf{m}_{tt})\,|$.

$$\begin{aligned}
&\| \mathbf{m}_{tt} \|^2_{\mathbf{L}^2} + \frac{\kappa}{2}\frac{d}{dt}\| \nabla \mathbf{m}_t \|^2_{\mathbf{L}^2} \leq |\,(\mathcal{Y}, \mathbf{m}_{tt})\,| + |\,(\mathbf{m}_t \times \mathbf{H} + \mathbf{m} \times \mathbf{H}_t, \mathbf{m}_{tt})\,| \\
&\quad + \kappa \Big\{ |\,(\langle \mathbf{m}_t, \mathbf{H}\rangle_{\mathbb{R}^3} + \langle \mathbf{m}, \mathbf{H}_t\rangle_{\mathbb{R}^3}, \langle \mathbf{m}, \mathbf{m}_{tt}\rangle_{\mathbb{R}^3})\,| \\
&\quad + |\,(\langle \mathbf{m}, \mathbf{H}\rangle_{\mathbb{R}^3}, \langle \mathbf{m}_t, \mathbf{m}_{tt}\rangle_{\mathbb{R}^3})\,| + (\mathbf{H}_t, \mathbf{m}_{tt}) \Big\}.
\end{aligned} \tag{5.36}$$

The argumentation for (5.35) then immediately applies to (5.36).

Now, we may verify upper bounds for $\| \sqrt{\tau}\mathbf{H}_{tt} \|_{L^2(I;\mathbf{L}^2)}$, $\| \sqrt{\tau}\mathbf{E}_{tt} \|_{L^2(I;\mathbf{L}^2)}$: Time derivatives of (5.2), (5.3) and multiplication by $\mathbf{E}_{tt}$ and $\mathbf{H}_{tt}$, resp., further using (5.36), (5.14) and integration by parts lead to

$$\begin{aligned}
&\| \mathbf{E}_{tt} \|^2_{\mathbf{L}^2} + \frac{3}{4} \| \mathbf{H}_{tt} \|^2_{\mathbf{L}^2} + \frac{\sigma}{2}\frac{d}{dt}\| \mathbf{E}_t \|^2_{\mathbf{L}^2} \\
&\leq \| \mathbf{m}_{tt} \|^2_{\mathbf{L}^2} - (\nabla \times \mathbf{H}_t, \mathbf{E}_{tt}) + (\nabla \times \mathbf{E}_{tt}, \mathbf{H}_t) \\
&= \| \mathbf{m}_{tt} \|^2_{\mathbf{L}^2}\,.
\end{aligned} \tag{5.37}$$

Together with (d), this shows the remaining part in (f), if we multiply by $\tau(s)$ and integrate over the interval $I$. □

## 5.2 The implicit penalized Euler scheme

The mathematical setting for (MLLG) is similar for (LLG): either we are provided with (i) global existence of smooth solutions for small initial energies, or (ii) local existence of smooth solutions for finite initial energies.

We discuss two schemes $(\text{MLLG})_{k,\varepsilon}^{E_1}$ and $(\text{MLLG})_{k,\varepsilon}^{E_2}$. The first one is kind of natural, whereas the construction of $(\text{MLLG})_{k,\varepsilon}^{E_2}$ is based on the shortcomings of the first strategy.

### 5.2.1 Semi-Discretization in time: the scheme $(\text{MLLG})_{k,\varepsilon}^{E_1}$

We consider the following penalized time discretization scheme of (MLLG), which is in the sequel addressed as $(\text{MLLG})_{k,\varepsilon}^{E_1}$. Find $\{\mathbf{m}^j, \mathbf{H}^j, \mathbf{E}^j\}_{j=0}^{J} \in \ell^2(I_k; \mathbf{W}^{1,2}_{\text{per}}) \times [\ell^2(I_k; \mathbf{H}(\mathbf{curl}, \omega))]^2$ satisfying (5.5) and assume initial data $\{\mathbf{m}^0, \mathbf{H}^0, \mathbf{E}^0\} = \{\mathbf{m}_0, \mathbf{H}_0, \mathbf{E}_0\}$ from (5.6), that solve

$$
\begin{aligned}
&d_t\mathbf{m}^{j+1} - \kappa\,\Delta\mathbf{m}^{j+1} - \mathbf{l}_\varepsilon(\mathbf{m}^{j+1})\mathbf{m}^{j+1} = \kappa\,|\,\nabla\mathbf{m}^j\,|^2\mathbf{m}^{j+1} &(5.38)\\
&\quad -\gamma'\,\mathbf{m}^j \times \left(\Delta\mathbf{m}^{j+1} + \mathbf{H}^{j+1}\right) + \kappa\left(\mathbf{H}^{j+1} - \langle\mathbf{m}^j, \mathbf{H}^j\rangle_{\mathbb{R}^3}\mathbf{m}^{j+1}\right), &\\
&\nabla \times \mathbf{H}^{j+1} = d_t\mathbf{E}^{j+1} + \sigma\,\mathbf{E}^{j+1}, &(5.39)\\
&\nabla \times \mathbf{E}^{j+1} = -d_t\mathbf{H}^{j+1} - \beta\left(d_t\mathbf{m}^{j+1} - \mathbf{l}_\varepsilon(\mathbf{m}^{j+1})\mathbf{m}^{j+1}\right), &(5.40)\\
&\operatorname{div}\mathbf{E}^{j+1} = 0, &(5.41)
\end{aligned}
$$

where $\mathbf{l}_\varepsilon(\varphi) = \frac{1}{\varepsilon}\left(|\,\varphi\,|^2 - 1\right)$. At each time-step, we are given a well-posed problem. The subsequent theorem collects the main results in this section.

**Theorem 5.4** *Let* $\{\mathbf{m}, \mathbf{H}, \mathbf{E}\}$ *solve* (5.1)-(5.6), *for* $0 < t_J < T_0(\mathbf{m}_0)$, *where* $\{\mathbf{m}_0, \mathbf{H}_0, \mathbf{E}_0\} \in \mathbf{W}^{2,2}_{\text{per}}(\omega, \mathcal{S}^2) \times [\mathbf{W}^{1,2}_{\text{per}}(\omega, \mathbb{R}^3)]^2$ *satisfies* (5.11). *Let the triple* $\{\mathbf{m}^j, \mathbf{H}^j, \mathbf{E}^j\}_{j=0}^{J}$ *solve* $(\text{MLLG})_{k,\varepsilon}^{E_1}$ *using periodic boundary conditions, and*

$\varepsilon^{-1} = o(\frac{1}{k})$. *For sufficiently small time-steps* $k \leq k_0(t_J)$, *there holds*

$$\text{(a)} \max_{0\leq j\leq J}\Big\{\| \mathbf{m}(t_j) - \mathbf{m}^j \|_{\mathbf{L}^2} + \| \mathbf{H}(t_j) - \mathbf{H}^j \|_{\mathbf{W}^{-1,2}_{\text{per}}} + \| \mathbf{E}(t_j) - \mathbf{E}^j \|_{\mathbf{W}^{-1,2}_{\text{per}}}\Big\}$$

$$+\Big(k\sum_{j=0}^{J}\{\| \mathbf{m}(t_j) - \mathbf{m}^j \|^2_{\mathbf{W}^{1,2}} + \frac{\beta}{\varepsilon}[\| \langle \mathbf{m}(t_j) - \mathbf{m}^j, \mathbf{m}(t_j)\rangle_{\mathbb{R}^3} \|^2_{L^2}$$

$$+\| \mathbf{m}(t_j) - \mathbf{m}^j \|^4_{\mathbf{L}^4}]\}\Big)^{1/2} \leq C\,k\,,$$

$$\text{(b)} \max_{0\leq j\leq J}\Big\{\| \mathbf{H}(t_j) - \mathbf{H}^j \|_{\mathbf{L}^2} + \| \mathbf{E}(t_j) - \mathbf{E}^j \|_{\mathbf{L}^2}\Big\} \leq C\sqrt{k}\,,$$

$$\text{(c)} \Big(k\sum_{j=1}^{J}\| 1 - |\mathbf{m}^j|^2 \|^2_{L^2}\Big)^{1/2} \leq C\sqrt{\varepsilon k^2}\,,$$

$$\text{(d)} \max_{0\leq j\leq J}\Big\{\| \operatorname{div}(\mathbf{H}^j + \beta\,\mathbf{m}^j) \|_{W^{-1,2}_{\text{per}}} + \sqrt{\frac{k}{\varepsilon}}\,\| \operatorname{div}(\mathbf{H}^j + \beta\,\mathbf{m}^j) \|_{L^2}\Big\} \leq C\sqrt{\frac{k^2}{\varepsilon}}\,,$$

*with* $C = C(\mathbf{m}_0, \mathbf{H}_0, \mathbf{E}_0, \omega, t_J)$ *independent of* $k$. *Moreover, the solution to* $(MLLG)^{E_1}_{k,\varepsilon}$ *satisfies the a priori bounds, with* $C$ *independent of* $k$,

$$\text{(e)} \max_{0\leq j\leq J}\Big\{\| d_t\mathbf{m}^j \|_{\mathbf{L}^2} + \| d_t\mathbf{H}^j \|_{\mathbf{L}^2} + \| d_t\mathbf{E}^j \|_{\mathbf{L}^2} + \| \mathbf{m}^j \|_{\mathbf{W}^{2,2}}$$

$$+\| \mathbf{H}^j \|_{\mathbf{W}^{1,2}} + \| \mathbf{E}^j \|_{\mathbf{W}^{1,2}}\Big\} + \Big(k\sum_{j=1}^{J}\| \nabla d_t\mathbf{m}^j \|^2_{\mathbf{L}^2}\Big)^{1/2} \leq C\,.$$

**Remark 5.1** *1. Note the penalization term in both equations* (5.38), (5.40).

*2. As can be seen from (c), (d), increased penalization with respect to the constraint* $|\mathbf{m}| = 1$ *leads to larger violation of the incompressibility requirement of* $\mathbf{H}^j + \beta\,\mathbf{m}^j$ *in the scheme. Notice that the approximation property as well as a priori bounds are not affected by this coupling effect.*

**Proof:**
*1st step:* The error $\{\mathbf{e}^j, \boldsymbol{\eta}^j, \boldsymbol{\pi}^j\} := \{\mathbf{m}(t_j) - \mathbf{m}^j, \mathbf{H}(t_j) - \mathbf{H}^j, \mathbf{E}(t_j) - \mathbf{E}^j\}$ is

governed by (see (4.94), (4.95), and (4.74) for further details and notation)

$$
\begin{aligned}
& d_t \mathbf{e}^{j+1} - \kappa\big(\Delta \mathbf{e}^{j+1} + \boldsymbol{\eta}^{j+1}\big) + \mathbf{l}_\varepsilon(\mathbf{m}^{j+1})\mathbf{m}^{j+1} && (5.42)\\
& \quad = \boldsymbol{\mathcal{R}}^{j+1}(\mathbf{m}) + \boldsymbol{\mathcal{N}}^{j+1}, \\
& \nabla \times \boldsymbol{\eta}^{j+1} = d_t \boldsymbol{\pi}^{j+1} + \sigma \boldsymbol{\pi}^{j+1} - \boldsymbol{\mathcal{R}}^{j+1}(\mathbf{E}), && (5.43)\\
& \nabla \times \boldsymbol{\pi}^{j+1} = -d_t \boldsymbol{\eta}^{j+1} - \beta\big(d_t \mathbf{e}^{j+1} + \mathbf{l}_\varepsilon(\mathbf{m}^{j+1})\mathbf{m}^{j+1}\big) && (5.44)\\
& \qquad + \boldsymbol{\mathcal{R}}^{j+1}(\mathbf{H}) + \beta \boldsymbol{\mathcal{R}}^{j+1}(\mathbf{m}), \\
& \operatorname{div} \boldsymbol{\pi}^{j+1} = 0, && (5.45)
\end{aligned}
$$

where

$$
\begin{aligned}
\boldsymbol{\mathcal{N}}^{j+1} &= \kappa\Big(|\nabla \mathbf{m}(t_j)|^2 \mathbf{e}^{j+1} + k^2 |\nabla d_t \mathbf{m}(t_{j+1})|^2 \mathbf{m}(t_{j+1}) \\
&\quad + 2k \langle \nabla d_t \mathbf{m}(t_{j+1}), \nabla \mathbf{m}(t_j)\rangle_{\mathbb{R}^3} \mathbf{m}(t_{j+1}) \\
&\quad + \langle \nabla \mathbf{e}^j, \nabla\{2\mathbf{m}(t_j) - \mathbf{e}^j\}\rangle_{\mathbb{R}^3} \{\mathbf{m}(t_{j+1}) - \mathbf{e}^{j+1}\}\Big) \\
&\quad - \gamma'\Big(k\, d_t \mathbf{m}(t_{j+1}) \times (\Delta \mathbf{m}(t_{j+1}) + \mathbf{H}(t_{j+1})) \\
&\quad + \mathbf{e}^j \times (\Delta \mathbf{m}(t_{j+1}) + \mathbf{H}(t_{j+1})) \qquad (5.46)\\
&\quad + \mathbf{m}(t_j) \times \big(\Delta \mathbf{e}^{j+1} + \boldsymbol{\eta}^{j+1}\big) + \mathbf{e}^j \times \big(\Delta \mathbf{e}^{j+1} + \boldsymbol{\eta}^{j+1}\big)\Big) + \kappa \boldsymbol{\eta}^{j+1} \\
&\quad + \kappa\Big(k \langle d_t \mathbf{m}(t_{j+1}), \mathbf{H}(t_{j+1})\rangle_{\mathbb{R}^3} \mathbf{m}(t_{j+1}) + k \big\langle \mathbf{m}(t_j), d_t \mathbf{H}(t_{j+1})\big\rangle_{\mathbb{R}^3} \mathbf{m}(t_{j+1}) \\
&\quad + \langle \mathbf{e}^j, \mathbf{H}(t_j)\rangle_{\mathbb{R}^3} \mathbf{m}(t_{j+1}) + \langle \mathbf{m}^j, \boldsymbol{\eta}^j\rangle_{\mathbb{R}^3} \mathbf{m}(t_{j+1}) + \langle \mathbf{m}^j, \mathbf{H}(t_j) - \boldsymbol{\eta}^j\rangle_{\mathbb{R}^3} \mathbf{e}^{j+1}\Big) \\
&= \sum_{\ell=1}^{14} \boldsymbol{\mathcal{N}}_\ell^{j+1}.
\end{aligned}
$$

We multiply (5.43) by $\tilde{\Delta}^{-1}\boldsymbol{\pi}^{j+1}$ and (5.44) by $\tilde{\Delta}^{-1}\boldsymbol{\eta}^{j+1}$, respectively, and observe

$$
\begin{aligned}
& \big(\nabla \times \tilde{\Delta}[\tilde{\Delta}^{-1}\boldsymbol{\eta}^j], \tilde{\Delta}^{-1}\boldsymbol{\pi}^j\big) - \big(\nabla \times \tilde{\Delta}[\tilde{\Delta}^{-1}\boldsymbol{\pi}^j], \tilde{\Delta}^{-1}\boldsymbol{\eta}^j\big) \\
& \quad = \big(\mathbf{curl\,curl\,curl}\, \tilde{\Delta}^{-1}\boldsymbol{\eta}^j, \tilde{\Delta}^{-1}\boldsymbol{\pi}^j\big) - \big(\mathbf{curl\,curl\,curl}\, \tilde{\Delta}^{-1}\boldsymbol{\pi}^j, \tilde{\Delta}^{-1}\boldsymbol{\eta}^j\big) \\
& \qquad + \big(\nabla \times \tilde{\Delta}^{-1}\boldsymbol{\eta}^j, \tilde{\Delta}^{-1}\boldsymbol{\pi}^j\big) - \big(\nabla \times \tilde{\Delta}^{-1}\boldsymbol{\pi}^j, \tilde{\Delta}^{-1}\boldsymbol{\eta}^j\big) = 0.
\end{aligned}
$$

We obtain

$$
\begin{aligned}
&\frac{1}{2}d_t\{\| \boldsymbol{\pi}^j \|^2_{\mathbf{W}^{-1,2}_{\mathrm{per}}} + \| \boldsymbol{\eta}^j \|^2_{\mathbf{W}^{-1,2}_{\mathrm{per}}}\} + \frac{k}{2}\{\| d_t\boldsymbol{\pi}^j \|^2_{\mathbf{W}^{-1,2}_{\mathrm{per}}} + \| d_t\boldsymbol{\eta}^j \|^2_{\mathbf{W}^{-1,2}_{\mathrm{per}}}\} \\
&\quad +\sigma \| \boldsymbol{\pi}^j \|^2_{\mathbf{W}^{-1,2}_{\mathrm{per}}} - \beta\left(d_t\mathbf{e}^{j+1} + \mathbf{l}_\varepsilon(\mathbf{m}^j)\mathbf{m}^j, \tilde{\Delta}^{-1}\boldsymbol{\eta}^j\right) \\
&\quad \le C\left\{\| \boldsymbol{\mathcal{R}}^j(\mathbf{E}) \|^2_{\mathbf{W}^{-1,2}_{\mathrm{per}}} + \| \boldsymbol{\mathcal{R}}^j(\mathbf{H}) \|^2_{\mathbf{W}^{-1,2}_{\mathrm{per}}} + \| \boldsymbol{\mathcal{R}}^j(\mathbf{m}) \|^2_{\mathbf{W}^{-1,2}_{\mathrm{per}}}\right\} \\
&\quad +\frac{1}{4}\left\{\sigma \| \boldsymbol{\pi}^j \|^2_{\mathbf{W}^{-1,2}_{\mathrm{per}}} + (1+\beta)\| \boldsymbol{\eta}^j \|^2_{\mathbf{W}^{-1,2}_{\mathrm{per}}}\right\}. \qquad (5.47)
\end{aligned}
$$

Next, multiplication of (5.42) by $\mathbf{e}^{j+1} + \tilde{\Delta}^{-1}\boldsymbol{\eta}^{j+1}$ gives

$$
\begin{aligned}
&-\left\langle \tilde{\Delta}^{-1}\boldsymbol{\eta}^{j+1}, d_t\mathbf{e}^{j+1} + \mathbf{l}_\varepsilon(\mathbf{m}^{j+1})\mathbf{m}^{j+1}\right\rangle_{\mathbb{R}^3} \qquad (5.48) \\
&\quad = \left\langle \mathbf{e}^{j+1}, d_t\mathbf{e}^{j+1} + \mathbf{l}_\varepsilon(\mathbf{m}^{j+1})\mathbf{m}^{j+1}\right\rangle_{\mathbb{R}^3} \\
&\quad -\kappa\left\langle \mathbf{e}^{j+1} + \tilde{\Delta}^{-1}\boldsymbol{\eta}^{j+1}, \Delta\mathbf{e}^{j+1} + \boldsymbol{\eta}^{j+1}\right\rangle_{\mathbb{R}^3} \\
&\quad -\left\langle \boldsymbol{\mathcal{R}}^{j+1}(\mathbf{m}) + \boldsymbol{\mathcal{N}}^{j+1}, \mathbf{e}^{j+1} + \tilde{\Delta}^{-1}\boldsymbol{\eta}^{j+1}\right\rangle_{\mathbb{R}^3}.
\end{aligned}
$$

We insert (5.48) into (5.47). The first term on the right-hand side of (5.48) can be reformulated by using (4.94).

$$
\begin{aligned}
&\frac{1}{2}d_t\left\{\beta \| \mathbf{e}^{j+1} \|^2_{\mathbf{L}^2} + \| \boldsymbol{\pi}^{j+1} \|^2_{\mathbf{W}^{-1,2}_{\mathrm{per}}} + \| \boldsymbol{\eta}^{j+1} \|^2_{\mathbf{W}^{-1,2}_{\mathrm{per}}}\right\} \qquad (5.49) \\
&\quad +\frac{k}{2}\left\{\beta \| d_t\mathbf{e}^{j+1} \|^2_{\mathbf{L}^2} + \| d_t\boldsymbol{\pi}^{j+1} \|^2_{\mathbf{W}^{-1,2}_{\mathrm{per}}} + \| d_t\boldsymbol{\eta}^{j+1} \|^2_{\mathbf{W}^{-1,2}_{\mathrm{per}}}\right\} \\
&\quad +\sigma \| \boldsymbol{\pi}^{j+1} \|^2_{\mathbf{W}^{-1,2}_{\mathrm{per}}} + \frac{\beta\kappa}{2} \| \nabla\mathbf{e}^{j+1} \|^2_{\mathbf{L}^2} \\
&\quad +\frac{\beta}{\varepsilon}\left\{2 \| \langle \mathbf{e}^{j+1}, \mathbf{m}(t_{j+1})\rangle_{\mathbb{R}^3} \|^2_{L^2} + \| \mathbf{e}^{j+1} \|^4_{\mathbf{L}^4}\right\} \\
&\le C\left\{\| \boldsymbol{\mathcal{R}}^{j+1}(\mathbf{E}) \|^2_{\mathbf{W}^{-1,2}_{\mathrm{per}}} + \| \boldsymbol{\mathcal{R}}^{j+1}(\mathbf{H}) \|^2_{\mathbf{W}^{-1,2}_{\mathrm{per}}} + \| \boldsymbol{\mathcal{R}}^{j+1}(\mathbf{m}) \|^2_{\mathbf{W}^{-1,2}_{\mathrm{per}}}\right\} \\
&\quad +C_\beta\left\{\| \boldsymbol{\pi}^{j+1} \|^2_{\mathbf{W}^{-1,2}_{\mathrm{per}}} + \| \boldsymbol{\eta}^{j+1} \|^2_{\mathbf{W}^{-1,2}_{\mathrm{per}}} + \left| \left(\boldsymbol{\mathcal{N}}^{j+1}, \mathbf{e}^{j+1} + \Delta^{-1}\boldsymbol{\eta}^{j+1}\right) \right|\right\} \\
&\quad +\frac{3\beta}{\varepsilon} \left| \left(\langle \mathbf{m}(t_{j+1}), \mathbf{e}^{j+1}\rangle_{\mathbb{R}^3}, | \mathbf{e}^{j+1} |^2\right) \right|.
\end{aligned}
$$

Hence, it remains to control the last term. We skip the elaboration of this part for the terms $\left| \left(\sum_{\ell=1}^6 \boldsymbol{\mathcal{N}}_\ell^{j+1}, \mathbf{e}^{j+1} + \tilde{\Delta}^{-1}\boldsymbol{\eta}^{j+1}\right) \right|$, see also (4.77). We continue with the (most crucial) terms $|\left(\sum_{\ell=7,8}\mathcal{N}_\ell^{j+1}, \mathbf{e}^{j+1}+\tilde{\Delta}^{-1}\boldsymbol{\eta}^{j+1}\right)|$

$$\begin{aligned}
&|\Big(\nabla\mathbf{m}(t_j)\times\nabla\tilde{\Delta}^{-1}\boldsymbol{\eta}^{j+1}, \mathbf{e}^{j+1}+\tilde{\Delta}^{-1}\boldsymbol{\eta}^{j+1}\Big) \qquad (5.50)\\
&\quad+\Big(\mathbf{m}(t_j)\times\nabla\tilde{\Delta}^{-1}\boldsymbol{\eta}^{j+1}, \nabla\big[\mathbf{e}^{j+1}+\tilde{\Delta}^{-1}\boldsymbol{\eta}^{j+1}\big]\Big)|\\
&\leq C\|\nabla\mathbf{m}(t_j)\|_{\mathbf{L}^4}\|\nabla\tilde{\Delta}^{-1}\boldsymbol{\eta}^{j+1}\|_{\mathbf{L}^2}\\
&\quad\times\Big\{\|\mathbf{e}^{j+1}+\tilde{\Delta}^{-1}\boldsymbol{\eta}^{j+1}\|_{\mathbf{L}^4}+\|\nabla\mathbf{e}^{j+1}\|_{\mathbf{L}^2}+\|\nabla\tilde{\Delta}^{-1}\boldsymbol{\eta}^{j+1}\|_{\mathbf{L}^2}\Big\},\\
&|\Big(\nabla\mathbf{e}^j\times\nabla\big[\mathbf{e}^{j+1}+\tilde{\Delta}^{-1}\boldsymbol{\eta}^{j+1}\big], \mathbf{e}^{j+1}+\tilde{\Delta}^{-1}\boldsymbol{\eta}^{j+1}\Big) \qquad (5.51)\\
&\quad+\Big(\mathbf{e}^j\times\nabla\big[\mathbf{e}^{j+1}+\tilde{\Delta}^{-1}\boldsymbol{\eta}^{j+1}\big], \nabla\big[\mathbf{e}^{j+1}+\tilde{\Delta}^{-1}\boldsymbol{\eta}^{j+1}\big]\Big)|\\
&\leq C\|\nabla\mathbf{e}^j\|_{\mathbf{L}^4}\|\nabla\big[\mathbf{e}^{j+1}+\tilde{\Delta}^{-1}\boldsymbol{\eta}^{j+1}\big]\|_{\mathbf{L}^2}\|\mathbf{e}^{j+1}+\tilde{\Delta}^{-1}\boldsymbol{\eta}^{j+1}\|_{\mathbf{L}^4}.
\end{aligned}$$

*2nd step:* By induction, we prove for constants $C_i = C_i(\mathbf{m}_0, \mathbf{E}_0, \mathbf{H}_0; t_J, \omega)$, $i = 1, 2$, that

$$\begin{aligned}
&\Big\{\|\mathbf{e}^{\ell+1}\|_{\mathbf{L}^2}^2+\|\boldsymbol{\pi}^{\ell+1}\|_{\mathbf{W}_{\text{per}}^{-1,2}}^2+\|\boldsymbol{\eta}^{\ell+1}\|_{\mathbf{W}_{\text{per}}^{-1,2}}^2\Big\}+\frac{k^2}{2}\sum_{j=1}^{\ell+1}\Big\{\|d_t\mathbf{e}^j\|_{\mathbf{L}^2}^2\\
&+\|d_t\boldsymbol{\pi}^j\|_{\mathbf{W}_{\text{per}}^{-1,2}}^2+\|d_t\boldsymbol{\eta}^j\|_{\mathbf{W}_{\text{per}}^{-1,2}}^2\Big\}+\frac{\kappa k}{2}\sum_{j=1}^{\ell+1}\Big\{\|\nabla\mathbf{e}^j\|_{\mathbf{L}^2}^2 \qquad (5.52)\\
&+\frac{\beta}{\varepsilon\kappa}\{\|\langle\mathbf{e}^j, \mathbf{m}(t_j)\rangle_{\mathbb{R}^3}\|_{L^2}^2+\|\mathbf{e}^j\|_{\mathbf{L}^4}^4\}\Big\}\leq C_1 k^2\exp\big(C_2 t_\ell\big),\\
&\Big\{\|\nabla\mathbf{e}^{\ell+1}\|_{\mathbf{L}^2}^2+\|\boldsymbol{\pi}^{\ell+1}\|_{\mathbf{L}^2}^2+\|\boldsymbol{\eta}^{\ell+1}\|_{\mathbf{L}^2}^2\Big\}+\frac{k^2}{2}\sum_{j=1}^{\ell+1}\Big\{\|\nabla d_t\mathbf{e}^j\|_{\mathbf{L}^2}^2+\|d_t\boldsymbol{\pi}^j\|_{\mathbf{L}^2}^2\\
&+\|d_t\boldsymbol{\eta}^j\|_{\mathbf{L}^2}^2+\frac{\kappa}{2k}\|\Delta\mathbf{e}^j\|_{\mathbf{L}^2}^2\Big\}\leq C_1 k\exp\big(C_2 t_\ell\big), \qquad (5.53)
\end{aligned}$$

for $F(\varepsilon, k) > 0$ as stated in the theorem, and $k \leq k_0(t_J; \varepsilon)$. These statements are obviously valid for $\ell = 0$. To verify (5.52) at $\ell + 1$, we first employ (5.53) to control contributions from $\{\mathcal{N}^j\}_{j=0}^{\ell+1}$, due to (5.46), (5.50)-(5.51). The crucial contribution that comes from penalization is again (4.98), which requires the restriction $F(\varepsilon, k) > 0$. Finally, we may employ Theorem 5.3,

(b) to bound

$$k\sum_{j=1}^{\ell+2}\Big\{\| \mathcal{R}(\mathbf{m}) \|^2_{\mathbf{W}^{-1,2}_{\text{per}}} + \| \mathcal{R}(\mathbf{H}) \|^2_{\mathbf{W}^{-1,2}_{\text{per}}} + \| \mathcal{R}(\mathbf{E}) \|^2_{\mathbf{W}^{-1,2}_{\text{per}}}\Big\} \leq \frac{C_1}{2}\, k^2 \exp\big(\frac{C_2}{2}\, t_{\ell+1}\big)\,.$$

To see (5.53), we test (5.42)-(5.44) by $\{-\Delta \mathbf{e}^{j+1} - \boldsymbol{\eta}^{j+1}, \boldsymbol{\pi}^{j+1}, -\boldsymbol{\eta}^{j+1}\}$. We start with (5.43), (5.44) and use Theorem 5.3, (a).

$$\begin{aligned}
&\frac{1}{2}\, d_t \| \boldsymbol{\pi}^{j+1} \|^2_{\mathbf{L}^2} + \frac{k}{2} \| d_t\boldsymbol{\pi}^{j+1} \|^2_{\mathbf{L}^2} + \frac{\sigma}{2} \| \boldsymbol{\pi}^{j+1} \|^2_{\mathbf{L}^2} \qquad (5.54)\\
&\quad \leq (\nabla \times \boldsymbol{\eta}^{j+1}, \boldsymbol{\pi}^{j+1}) + \frac{1}{2\sigma} \| \mathcal{R}(\mathbf{E}) \|^2_{\mathbf{L}^2}\,,\\
&\frac{1}{2}\, d_t \| \boldsymbol{\eta}^{j+1} \|^2_{\mathbf{L}^2} + \frac{k}{2} \| d_t\boldsymbol{\eta}^{j+1} \|^2_{\mathbf{L}^2} \qquad (5.55)\\
&\quad \leq (\nabla \times \boldsymbol{\pi}^{j+1}, -\boldsymbol{\eta}^{j+1}) + \frac{1}{4} \| \mathcal{R}(\mathbf{H}) + \mathcal{R}(\mathbf{m}) \|^2_{\mathbf{L}^2}\\
&\qquad + \frac{1}{4} \| \boldsymbol{\eta}^{j+1} \|^2_{\mathbf{L}^2} - \beta\left(d_t \mathbf{e}^{j+1} + \mathbf{l}_\varepsilon(\mathbf{m}^{j+1})\mathbf{m}^{j+1}, \boldsymbol{\eta}^{j+1}\right).
\end{aligned}$$

Notice that if we add these inequalities, the leading terms on the right-hand sides of the inequality signs annihilate, see (5.13). The last term in (5.55) cancels out if we add (5.42) (tested by $-\Delta \mathbf{e}^{j+1} - \boldsymbol{\eta}^{j+1}$). Hence, we obtain

$$\begin{aligned}
&\frac{1}{2}\, d_t\Big\{\| \nabla \mathbf{e}^{j+1} \|^2_{\mathbf{L}^2} + \| \boldsymbol{\pi}^{j+1} \|^2_{\mathbf{L}^2} + \| \boldsymbol{\eta}^{j+1} \|^2_{\mathbf{L}^2}\Big\} \qquad (5.56)\\
&\qquad + \frac{k}{2}\Big\{\| \nabla d_t\mathbf{e}^{j+1} \|^2_{\mathbf{L}^2} + \| d_t\boldsymbol{\pi}^{j+1} \|^2_{\mathbf{L}^2} + \| d_t\boldsymbol{\eta}^{j+1} \|^2_{\mathbf{L}^2}\Big\} + \frac{\kappa}{2} \| \Delta \mathbf{e}^{j+1} \|^2_{\mathbf{L}^2}\\
&\quad \leq \frac{\kappa}{8} \| \mathcal{R}(\mathbf{m}) \|^2_{\mathbf{L}^2} + \frac{1}{2\sigma} \| \mathcal{R}(\mathbf{E}) \|^2_{\mathbf{L}^2} + \frac{1}{4} \| \mathcal{R}(\mathbf{H}) + \mathcal{R}(\mathbf{m}) \|^2_{\mathbf{L}^2}\\
&\qquad + \frac{\kappa}{8} \| \boldsymbol{\mathcal{N}}^{j+1} \|^2_{\mathbf{L}^2} + C_\kappa \| \boldsymbol{\eta}^{j+1} \|^2_{\mathbf{L}^2} + \frac{\kappa}{4} \| \mathbf{l}_\varepsilon(\mathbf{m}^{j+1})\mathbf{m}^{j+1} \|^2_{\mathbf{L}^2}\,.
\end{aligned}$$

The same argument as in (4.103) applies to control the last term in (5.56). As to $\| \boldsymbol{\mathcal{N}}^{j+1} \|^2_{\mathbf{L}^2}$, we confine to the contributions $\| \boldsymbol{\mathcal{N}}^{j+1}_\ell \|_{\mathbf{L}^2}$, $\ell \geq 12$. Again, we benefit from Theorem 5.3, (d) and the inequalities $\| \cdot \|_{\mathbf{L}^4} \leq C \| \cdot \|^{1/2}_{\mathbf{L}^2} \| \cdot \|^{1/2}_{\mathbf{W}^{1,2}}$, $\| \cdot \|_{\mathbf{L}^\infty} \leq C \| \cdot \|^{1/2}_{\mathbf{L}^2} \| \cdot \|^{1/2}_{\mathbf{W}^{2,2}}$, and $\| \cdot \|_{\mathbf{L}^8} \leq C \| \cdot \|^{1/4}_{\mathbf{L}^2} \| \cdot \|^{3/4}_{\mathbf{W}^{1,2}}$.

$$\begin{aligned}
&\| \langle \mathbf{e}^j, \mathbf{H}(t_j)\rangle_{\mathbb{R}^3} \|^2_{L^2} \leq \| \mathbf{m}(t_{j+1}) \|^2_{\mathbf{L}^\infty} \| \mathbf{e}^j \|^2_{\mathbf{L}^4} \| \mathbf{H}(t_j) \|^2_{\mathbf{L}^4} \qquad (5.57)\\
&\quad \leq C \| \mathbf{e}^j \|^2_{\mathbf{L}^2} + \| \nabla \mathbf{e}^j \|^2_{\mathbf{L}^2}\,,
\end{aligned}$$

$$
\begin{aligned}
&\| \langle \mathbf{m}(t_j) - \mathbf{e}^j, \boldsymbol{\eta}^j \rangle_{\mathbb{R}^3} \mathbf{m}(t_{j+1}) \|_{\mathbf{L}^2}^2 \\
&\quad \leq \{ \| \mathbf{m}(t_j) \|_{\mathbf{L}^\infty}^2 + \| \mathbf{e}^j \|_{\mathbf{L}^\infty}^2 \} \| \mathbf{m}(t_{j+1}) \|_{\mathbf{L}^\infty}^2 \| \boldsymbol{\eta}^j \|_{\mathbf{L}^2}^2 \\
&\quad \leq 2 \| \boldsymbol{\eta}^j \|_{\mathbf{L}^2}^2 , && (5.58) \\
&\| \langle \mathbf{m}(t_j) - \mathbf{e}^j, \mathbf{H}(t_{j+1}) \rangle_{\mathbb{R}^3} \mathbf{e}^{j+1} \|_{\mathbf{L}^2}^2 && (5.59) \\
&\quad \leq \{ \| \mathbf{m}(t_j) \|_{\mathbf{L}^\infty}^2 + \| \mathbf{e}^j \|_{\mathbf{L}^\infty}^2 \} \| \mathbf{H}(t_{j+1}) \|_{\mathbf{L}^4}^2 \| \mathbf{e}^{j+1} \|_{\mathbf{L}^2} \| \mathbf{e}^{j+1} \|_{\mathbf{W}^{1,2}} , \\
&\| \langle \mathbf{m}(t_j) - \mathbf{e}^j, \boldsymbol{\eta}^j \rangle_{\mathbb{R}^3} \mathbf{e}^{j+1} \|_{\mathbf{L}^2}^2 \\
&\quad \leq C \| \mathbf{m}(t_j) - \mathbf{e}^j \|_{\mathbf{L}^\infty}^2 \| \boldsymbol{\eta}^j \|_{\mathbf{L}^2}^2 \| \mathbf{e}^{j+1} \|_{\mathbf{L}^\infty}^2 \\
&\quad \leq C \| \mathbf{m}(t_j) - \mathbf{e}^{j+1} \|_{\mathbf{L}^\infty}^2 \| \boldsymbol{\eta}^j \|_{\mathbf{L}^2}^2 \| \mathbf{e}^{j+1} \|_{\mathbf{L}^2} \| \mathbf{e}^{j+1} \|_{\mathbf{W}^{2,2}} . && (5.60)
\end{aligned}
$$

Then the inductive argument also works for (5.53), thanks to (5.52), and for sufficiently small time-steps $k \leq k_0(t_J, \varepsilon)$, and $F(\varepsilon, k) > 0$.

Verification of (c) is now an easy consequence from (a), (b). Most statements in (e) do also follow from (a), (b) and we only have to verify $\| \mathbf{H}^j \|_{\mathbf{W}^{1,2}} + \| \mathbf{E}^j \|_{\mathbf{W}^{1,2}} \leq C$. Similarly to (5.24), we make scalar products of (5.39) with $-\Delta \mathbf{E}^{j+1}$ and of (5.40) with $\Delta \mathbf{H}^{j+1}$.

$$
\begin{aligned}
&d_t \{ \| \nabla \mathbf{E}^{j+1} \|_{\mathbf{L}^2}^2 + \| \nabla \mathbf{H}^{j+1} \|_{\mathbf{L}^2}^2 \} \\
&\qquad + k \{ \| \nabla d_t \mathbf{E}^{j+1} \|_{\mathbf{L}^2}^2 + \| \nabla d_t \mathbf{H}^{j+1} \|_{\mathbf{L}^2}^2 \} + \sigma \| \nabla \mathbf{E}^{j+1} \|_{\mathbf{L}^2}^2 \\
&\quad = 2\beta (\nabla d_t \mathbf{m}^{j+1}, \nabla \mathbf{H}^{j+1}) + (\nabla [\mathbf{l}_\varepsilon(\mathbf{m}^{j+1}) \mathbf{m}^{j+1}], \nabla \mathbf{H}^{j+1}) \\
&\quad \leq C_\beta \| \nabla d_t \mathbf{m}^{j+1} \|_{\mathbf{L}^2}^2 + 2 \| \nabla [\mathbf{l}_\varepsilon(\mathbf{m}^{j+1}) \mathbf{m}^{j+1}] \|_{\mathbf{L}^2}^2 + \| \nabla \mathbf{H}^{j+1} \|_{\mathbf{L}^2}^2
\end{aligned}
$$

The most crucial term to deal with is the last but one term. We make use of Theorem 5.3, (a), and part (a) in the sequel.

$$
\begin{aligned}
&\| \mathbf{l}_\varepsilon(\mathbf{m}^j) \nabla \mathbf{m}^j \|_{\mathbf{L}^2} \\
&\quad \leq \| \mathbf{l}_\varepsilon(\mathbf{m}^j) \|_{\mathbf{L}^4} \| \nabla \mathbf{m}^j \|_{\mathbf{L}^4} \leq \frac{1}{\varepsilon} \| \langle \mathbf{e}^j, \mathbf{e}^j + 2\, \mathbf{m}(t_j) \rangle_{\mathbb{R}^3} \|_{\mathbf{L}^4} \\
&\quad \leq \frac{1}{\varepsilon} \| \langle \mathbf{e}^j, \mathbf{e}^j + 2\, \mathbf{m}(t_j) \rangle_{\mathbb{R}^3} \|_{\mathbf{L}^2}^{1/2} \| \langle \mathbf{e}^j, \mathbf{e}^j + 2\, \mathbf{m}(t_j) \rangle_{\mathbb{R}^3} \|_{\mathbf{W}^{1,2}}^{1/2} \\
&\quad \leq \frac{1}{\varepsilon} \| \langle \mathbf{e}^j, \mathbf{e}^j + 2\, \mathbf{m}(t_j) \rangle_{\mathbb{R}^3} \|_{\mathbf{L}^2}^{1/2} \Big\{ \| \mathbf{e}^j \|_{\mathbf{W}^{1,2}}^{1/2} \{ \| \mathbf{m}(t_j) \|_{\mathbf{L}^\infty} + \| \mathbf{e}^j \|_{\mathbf{L}^\infty} \}^{1/2} \\
&\qquad + \| \mathbf{e}^j \|_{\mathbf{L}^4}^{1/2} \{ \| \nabla \mathbf{e}^j \|_{\mathbf{L}^4} + \| \nabla \mathbf{m}(t_j) \|_{\mathbf{L}^4} \}^{1/2} \Big\} , \\
&\| \nabla \mathbf{l}_\varepsilon(\mathbf{m}^j) \|_{\mathbf{L}^2} = \frac{1}{\varepsilon} \| \nabla \langle \mathbf{e}^j, \mathbf{e}^j + 2\, \mathbf{m}(t_j) \rangle_{\mathbb{R}^3} \|_{\mathbf{L}^2} \\
&\quad \leq \frac{C}{\varepsilon} \{ \| \nabla \mathbf{e}^j \|_{L^2} + \| \mathbf{e}^j \|_{\mathbf{L}^4} \} .
\end{aligned}
$$

Hence,

$$k \sum_{j=1}^{J+1} \| \mathbf{l}_\varepsilon(\mathbf{m}^j) \nabla \mathbf{m}^j \|_{\mathbf{L}^2}^2 \tag{5.61}$$

$$\leq k \sum_{j=1}^{J+1} \frac{1}{\varepsilon^2} \| \langle \mathbf{e}^j, \mathbf{e}^j + 2\, \mathbf{m}(t_j) \rangle_{\mathbb{R}^3} \|_{L^2} \{ \| \nabla \mathbf{e}^j \|_{\mathbf{L}^2} + \| \mathbf{e}^j \|_{\mathbf{L}^2}^{1/2} \| \mathbf{e}^j \|_{\mathbf{W}^{1,2}}^{1/2} \}$$

$$\leq \frac{k}{\varepsilon^{2.5}} \sum_{j=1}^{J+1} \| \langle \mathbf{e}^j, \mathbf{e}^j + 2\, \mathbf{m}(t_j) \rangle_{\mathbb{R}^3} \|_{L^2}^2 + \frac{k}{\varepsilon^{1.5}} \sum_{j=1}^{J+1} \| \mathbf{e}^j \|_{\mathbf{W}^{1,2}}^2 \leq C \sqrt{k}\,,$$

$$k \sum_{j=1}^{J+1} \| \nabla \mathbf{l}_\varepsilon(\mathbf{m}^j) \|_{\mathbf{L}^2}^2 \leq C\,. \tag{5.62}$$

To see the second part of (d), we may benefit from (5.61), (5.62). We confine to showing the first part of it here. We start from (5.40), and employ $\operatorname{div}(\mathbf{H}^0 + \beta\, \mathbf{m}^0) = 0$.

$$\begin{aligned} \| \operatorname{div}(\mathbf{H}^J + \beta\, \mathbf{m}^J) \|_{W_{\rm per}^{-1,2}}^2 &\leq C_{t_J}\, k \sum_{j=1}^{J+1} \| \operatorname{div} d_t(\mathbf{H}^j + \beta\, \mathbf{m}^j) \|_{W_{\rm per}^{-1,2}}^2 \\ &\leq C_{t_J}\, k \sum_{j=1}^{J+1} \| \mathbf{l}_\varepsilon(\mathbf{m}^j) \mathbf{m}^j \|_{\mathbf{L}^2}^2 \leq C\,, \end{aligned} \tag{5.63}$$

thanks to an argument that is similar to (4.103). This settles the verification of the theorem. □

### 5.2.2 Semi-Discretization in time: the scheme $(\mathrm{MLLG})_{k,\varepsilon}^{E_2}$

Optimal statements of convergence for $\{\mathbf{H}^j, \mathbf{E}^j\}_{j=0}^J$ for method $(\mathrm{MLLG})_{k,\varepsilon}^{E_1}$, $\varepsilon^{-1} = o(\frac{1}{k})$ are available in $\ell^\infty\big(I; \mathbf{W}_{\rm per}^{-1,2}(\omega, \mathbb{R}^3)\big)$, rather than $\ell^\infty\big(I; \mathbf{L}^2(\omega, \mathbb{R}^3)\big)$. The reason for this is the balancing of the different nonlinearities: basically, in order to control the last term (or either a modification, where $\tilde{\Delta}^{-1}$ is deleted from the second argument) in the second line of (5.47), we have to test (5.42) by $\mathbf{e}^{j+1} + \tilde{\Delta}^{-1} \boldsymbol{\eta}^{j+1}$; for the suggested modification, this would require to use the test function $\Delta \mathbf{e}^{j+1} + \boldsymbol{\eta}^{j+1}$ — which interferes with the penalization term that cannot be handled efficiently in this scenario.

It is based on this consideration, that we are motivated to study a following modified Euler scheme, where the penalty functions $\mathbf{l}_\varepsilon(\mathbf{m}^j)\mathbf{m}^j$ in (5.38), (5.40) are replaced by $\tilde{\Delta}^{-1}\big[\mathbf{l}_\varepsilon(\mathbf{m}^j)\mathbf{m}^j\big]$. For this scheme, we introduce a stretched time-grid $\mathcal{G}_2(k_{j+1})$ from [107], where the grid function $k_{j+1}$ is given as follows,

$$k : j \mapsto k_{j+1} \equiv \begin{cases} (j+1)k_0^2 \,, & \text{for } 0 \le t_{j+1} \le 1, \\ \gamma k_0 \,, & \text{for } t_{j+1} \ge 1, \end{cases} \tag{5.64}$$

with $k_0$ the basic grid size and $\gamma = \mathcal{O}(1)$. Obviously, this grid structure is very fine near the origin, with increasing mesh-size for increasing times. Note that asymptotically $\mathcal{O}(\frac{\sqrt{2}}{k_0})$ steps are necessary to bridge the time interval $[0,1]$, and that time-steps $k_{j+1}$ at time $t = 1$ are of magnitude $k_0$. At this time, the mesh is then replaced by an equi-distant one of size $k_0$. — As will be seen in the subsequent theorem, this construction enables to control residual terms in stronger norms, which gives rise to improved convergence and stability results with respect to chosen norms; cf. Theorem 5.4. However, this discretization is impractical since it would lead to a pseudo-differential problem. Instead, we consider the following semi-explicit method, referred to as $(\mathrm{MLLG})_{k,\varepsilon}^{E_2}$ in the sequel, with $\varepsilon = \{\varepsilon_j\}_{j\ge 0}$: Let $\mathcal{G}_2(k_{j+1})$ from (5.64), and $d_t\varphi^{j+1} := \frac{1}{k_{j+1}}\{\varphi^{j+1} - \varphi^j\}$. Find $\{\mathbf{m}^j, \mathbf{H}^j, \mathbf{E}^j\}_{j\ge 0} \in \ell^2(I_k; \mathbf{W}^{1,2}_{\mathrm{per}}) \times \big[\ell^2(I_k; \mathbf{W}^{0,p}_{\mathrm{per}}(\omega,\mathbb{R}^3))\big]^2$ for $\{\mathbf{m}^0, \mathbf{H}^0, \mathbf{E}^0\} = \{\mathbf{m}_0, \mathbf{H}_0, \mathbf{E}_0\}$ from (5.6) that solve

$$\begin{aligned} & d_t\mathbf{m}^{j+1} - \kappa\,\Delta\mathbf{m}^{j+1} + \tilde{\Delta}^{-1}\big[\mathbf{l}_{\varepsilon_j}(\mathbf{m}^j)\mathbf{m}^j\big] = \kappa\,|\,\nabla\mathbf{m}^j\,|^2\mathbf{m}^{j+1} && (5.65)\\ & \qquad -\gamma'\,\mathbf{m}^j \times \big(\Delta\mathbf{m}^{j+1} + \mathbf{H}^{j+1}\big) + \kappa\,\big(\mathbf{H}^{j+1} - \langle \mathbf{m}^j, \mathbf{H}^j\rangle_{\mathbb{R}^3}\mathbf{m}^{j+1}\big)\,, && \\ & \nabla \times \mathbf{H}^{j+1} = d_t\mathbf{E}^{j+1} + \sigma\,\mathbf{E}^{j+1}\,, && (5.66)\\ & \nabla \times \mathbf{E}^{j+1} = -d_t\mathbf{H}^{j+1} - \beta\,\big(d_t\mathbf{m}^{j+1} + \tilde{\Delta}^{-1}\big[\mathbf{l}_{\varepsilon_j}(\mathbf{m}^j)\mathbf{m}^j\big]\big)\,, && (5.67)\\ & \operatorname{div}\mathbf{E}^{j+1} = 0\,. && (5.68) \end{aligned}$$

The following convergence statements are valid for iterates of $(\mathrm{MLLG})_{k,\varepsilon}^{E_2}$.

**Theorem 5.5** *Let $\{\mathbf{m}, \mathbf{H}, \mathbf{E}\}$ solve (5.1)-(5.6), $\omega = (0, 2D)^2$ and suppose periodic boundary data, for $0 < t_J < T_0(\mathbf{m}_0)$, and let the given initial data $\{\mathbf{m}_0, \mathbf{H}_0, \mathbf{E}_0\} \in \mathbf{W}^{2,2}_{\mathrm{per}}(\omega, \mathcal{S}^2) \times \big[\mathbf{W}^{1,2}_{\mathrm{per}}(\omega, \mathbb{R}^3)\big]^2$ satisfy (5.11). We denote by $\{\mathbf{m}^j, \mathbf{H}^j, \mathbf{E}^j\}_{j=0}^J$ the solution of $(\mathrm{MLLG})_{k,\varepsilon}^{E_2}$, $\varepsilon_{j+1}^{-1} = o(\frac{1}{k_{j+1}})$, on $\mathcal{G}_2(k_{j+1})$ from*

(5.64). *For sufficiently small $k_0 \leq k_A(t_J)$ and $\varepsilon_j$, there exists a constant $C = C(\mathbf{m}_0, \mathbf{H}_0, \mathbf{E}_0, \omega, t_J)$ independent of $k$, such that*

$$\text{(a)} \quad \max_{1\leq j\leq J}\Big\{\| \mathbf{m}(t_j) - \mathbf{m}^j \|_{\mathbf{W}^{1,2}} + \| \mathbf{H}(t_j) - \mathbf{H}^j \|_{\mathbf{L}^2} + \| \mathbf{E}(t_j) - \mathbf{E}^j \|_{\mathbf{L}^2}\Big\}$$
$$+\Big(\sum_{j=1}^J k_j \| \mathbf{m}(t_j) - \mathbf{m}^j \|^2_{\mathbf{W}^{2,2}} + \frac{\beta k_j}{\varepsilon_j}\big[\| \langle \mathbf{m}(t_j) - \mathbf{m}^j, \mathbf{m}(t_j)\rangle_{\mathbb{R}^3} \|^2_{L^2}$$
$$+\| \mathbf{m}(t_j) - \mathbf{m}^j \|^4_{\mathbf{L}^4}\big]\Big)^{1/2} \leq C\, k_0\,,$$

$$\text{(b)} \quad \Big(\sum_{j=1}^J \frac{k_j}{\varepsilon_j} \| 1 - |\mathbf{m}^j|^2 \|^2_{L^2}\Big)^{1/2} \leq C\, k_0\,,$$

$$\text{(c)} \quad \max_{1\leq j\leq J}\Big\{\Big(\sum_{\ell=1}^j \frac{k_\ell}{\varepsilon_\ell}\Big)^{-1/2} \| \operatorname{div}(\mathbf{H}^j + \beta\, \mathbf{m}^j) \|_{W^{1,2}}\Big\} \leq C k_0\,,$$

*with $C = C(\mathbf{m}_0, \mathbf{H}_0, \mathbf{E}_0, \omega, t_J)$ independent of $k_j$.*

**Remark 5.2** *Note the different scaling of nonstationary effects* (5.65)-(5.67).

**Proof:**
Let $\{\mathbf{e}^j, \boldsymbol{\eta}^j, \boldsymbol{\pi}^j\} := \{\mathbf{m}(t_j) - \mathbf{m}^j, \mathbf{H}(t_j) - \mathbf{H}^j, \mathbf{E}(t_j) - \mathbf{E}^j\}$. Then

$$d_t\mathbf{e}^{j+1} - \kappa\big(\Delta\mathbf{e}^{j+1} + \boldsymbol{\eta}^{j+1}\big) - \tilde{\Delta}^{-1}\mathbf{l}_{\varepsilon_j}(\mathbf{m}^j)\mathbf{m}^j = \boldsymbol{\mathcal{R}}^{j+1}(\mathbf{m}) + \boldsymbol{\mathcal{N}}^{j+1} \tag{5.69}$$
$$\nabla\times\boldsymbol{\eta}^{j+1} = d_t\boldsymbol{\pi}^{j+1} + \sigma\,\boldsymbol{\pi}^{j+1} - \boldsymbol{\mathcal{R}}^{j+1}(\mathbf{E})\,, \tag{5.70}$$
$$\nabla\times\boldsymbol{\pi}^{j+1} = -d_t\boldsymbol{\eta}^{j+1} - \beta\big(d_t\mathbf{e}^{j+1} - \tilde{\Delta}^{-1}\mathbf{l}_{\varepsilon_j}(\mathbf{m}^j)\mathbf{m}^j\big) \tag{5.71}$$
$$+\boldsymbol{\mathcal{R}}^{j+1}(\mathbf{H}) + \beta\,\boldsymbol{\mathcal{R}}^{j+1}(\mathbf{m})\,,$$
$$\operatorname{div}\boldsymbol{\pi}^{j+1} = 0\,, \tag{5.72}$$

with $\boldsymbol{\mathcal{N}}^{j+1}$ from (5.46) and

$$\boldsymbol{\mathcal{R}}^{j+1}(\varphi) := -\frac{1}{k_{j+1}}\int_{t_j}^{t_{j+1}} (s - t_j)\,\varphi_{tt}(s)\,\mathrm{d}s\,. \tag{5.73}$$

Next, we confine to pointing out the differences in the error analysis if compared to the proof of Theorem 5.4.

*1st step:* We show that there holds

$$\Big(\sum_{j=1}^J k_j\{\| \boldsymbol{\mathcal{R}}^j(\mathbf{m}) \|^2_{\mathbf{L}^2} + \| \boldsymbol{\mathcal{R}}^j(\mathbf{H}) \|^2_{\mathbf{L}^2} + \| \boldsymbol{\mathcal{R}}^j(\mathbf{E}) \|^2_{\mathbf{L}^2}\}\Big)^{1/2} \leq C\, k_0\,. \tag{5.74}$$

We exploit Theorem 5.3, (c).

$$\sum_{j=0}^{J} k_{j+1}\| \mathcal{R}^{j+1}(\mathbf{m}) \|_{\mathbf{L}^2}^2 \le Ck_0^4 + \sum_{j=0}^{J} \frac{1}{k_{j+1}} \Big\| \int_{t_j}^{t_{j+1}} (s-t_j)\mathbf{m}_{tt}(s)\,ds \Big\|_{\mathbf{L}^2}^2$$

$$\le C\,k_0^2 + \sum_{j=1}^{J} \frac{1}{k_{j+1}} \Big( \int_{t_j}^{t_{j+1}} \frac{1}{t_j}(s-t_j)^2\,\mathrm{d}s \Big) \Big( \int_{t_j}^{t_{j+1}} s\,\| \mathbf{m}_{tt}(s) \|_{\mathbf{L}^2}^2\,\mathrm{d}s \Big)$$

$$\le C\,k_0^2 + C \max_{1\le j\le J} \Big\{ \frac{1}{k_{j+1}} \int_{t_j}^{t_{j+1}} \frac{1}{t_j}(s-t_j)^2\,\mathrm{d}s \Big\} \tag{5.75}$$

$$\le C\,k_0^2 + C \max_{1\le j\le J} \frac{k_{j+1}^2}{t_j} = C\,k_0^2 + C\,k_0^4 \max_{1\le j\le J} \frac{(j+1)^2}{t_j} \le C\,k_0^2\,.$$

*2nd step:* We test (5.69) by $\tilde{\Delta}\mathbf{e}^{j+1} + \boldsymbol{\eta}^{j+1}$, (5.73) by $\boldsymbol{\pi}^{j+1}$, and (5.71) by $\boldsymbol{\eta}^{j+1}$; we proceed as in the proof of Theorem 5.4 to find the following result that corresponds to (5.49).

$$\begin{aligned}
&\frac{1}{2} d_t \big\{ \beta \| \nabla \mathbf{e}^{j+1} \|_{\mathbf{L}^2}^2 + \| \boldsymbol{\pi}^{j+1} \|_{\mathbf{L}^2}^2 + \| \boldsymbol{\eta}^{j+1} \|_{\mathbf{L}^2}^2 \big\} \qquad (5.76)\\
&\quad + \frac{k}{2} \big\{ \beta \| \nabla d_t \mathbf{e}^{j+1} \|_{\mathbf{L}^2}^2 + \| d_t \boldsymbol{\pi}^{j+1} \|_{\mathbf{L}^2}^2 + \| d_t \boldsymbol{\eta}^{j+1} \|_{\mathbf{L}^2}^2 \big\}\\
&\quad + \sigma \| \boldsymbol{\pi}^{j+1} \|_{\mathbf{L}^2}^2 + \frac{\beta\kappa}{2} \| \mathbf{e}^{j+1} \|_{\mathbf{W}^{2,2}}^2 + \frac{\beta}{\varepsilon_j} \big\{ 2 \| \langle \mathbf{e}^j, \mathbf{m}(t_j) \rangle \|_{\mathbf{L}^2}^2 + \| \mathbf{e}^j \|_{\mathbf{L}^4}^4 \big\}\\
&\le C \big\{ \| \mathcal{R}^{j+1}(\mathbf{E}) \|_{\mathbf{L}^2}^2 + \| \mathcal{R}^{j+1}(\mathbf{H}) \|_{\mathbf{L}^2}^2 + \| \mathcal{R}^{j+1}(\mathbf{m}) \|_{\mathbf{L}^2}^2 \big\}\\
&\quad + C \big\{ \| \boldsymbol{\pi}^{j+1} \|_{\mathbf{L}^2}^2 + \| \boldsymbol{\eta}^{j+1} \|_{\mathbf{L}^2}^2 + | \big( \boldsymbol{\mathcal{N}}^{j+1}, \tilde{\Delta}\mathbf{e}^{j+1} + \boldsymbol{\eta}^{j+1} \big) | \big\} + | \mathcal{X}^{j+1} |\\
&\quad + \frac{3\beta}{\varepsilon_j} | \big( \langle \mathbf{m}(t_{j+1}), \mathbf{e}^{j+1} \rangle_{\mathbb{R}^3}, | \mathbf{e}^{j+1} |^2 \big) |
\end{aligned}$$

Note that $\big(\boldsymbol{\mathcal{N}}_8^{j+1}, \tilde{\Delta}\mathbf{e}^{j+1} + \boldsymbol{\eta}^{j+1}\big) = 0$ in (5.46). We discuss the crucial remaining terms.

$$\| \boldsymbol{\mathcal{N}}_1^{j+1} \|_{\mathbf{L}^2} \;\le\; C \| \nabla \mathbf{m}(t_j) \|_{\mathbf{L}^8}^2 \| \mathbf{e}^{j+1} \|_{\mathbf{L}^4}\,, \tag{5.77}$$

$$\| \boldsymbol{\mathcal{N}}_2^{j+1} \|_{\mathbf{L}^2} \;\le\; C\,k_0^2\, t_{j+1} \| \nabla d_t \mathbf{m}(t_{j+1}) \|_{\mathbf{L}^4}^2 \| \mathbf{m}(t_{j+1}) \|_{\mathbf{L}^\infty}\,. \tag{5.78}$$

Theorem 5.3, (f) then helps to bound the right-hand side of the last inequality after integration in (5.76). We continue with $\| \boldsymbol{\mathcal{N}}_{4,a}^{j+1} \|_{\mathbf{L}^4}$, which requires $k_0 \le k_A(t_J)$ for absorption. The most interesting term here is

$$\| \boldsymbol{\mathcal{N}}_{4,a}^{j+1} \|_{\mathbf{L}^2} \;\le\; C \| \nabla \mathbf{e}^j \|_{\mathbf{L}^4} \| \nabla \mathbf{e}^j \|_{\mathbf{L}^4} \| \mathbf{e}^{j+1} \|_{\mathbf{L}^\infty}\,, \tag{5.79}$$

$$\| \boldsymbol{\mathcal{N}}_6^{j+1} \|_{\mathbf{L}^2} \;\le\; C \| \mathbf{e}^j \|_{\mathbf{L}^\infty} \| \tilde{\Delta}\mathbf{m}(t_{j+1}) + \mathbf{H}(t_{j+1}) \|_{\mathbf{L}^2}\,. \tag{5.80}$$

We use the inequality $\| \boldsymbol{\varphi} \|_{\mathbf{L}^\infty} \leq C \| \boldsymbol{\varphi} \|_{\mathbf{W}^{2,2}}^{1/2} \| \boldsymbol{\varphi} \|_{\mathbf{W}^{1,2}}^{1/2}$ to bound this contribution. Similarly,

$$\begin{aligned} \| \mathcal{N}_{11,a}^{j+1} \|_{\mathbf{L}^2} &\leq C \| \mathbf{e}^j \|_{\mathbf{W}^{2,2}}^{1/2} \| \mathbf{e}^j \|_{\mathbf{L}^2}^{1/2} \| \boldsymbol{\eta}^{j+1} \|_{\mathbf{L}^2} \leq C \| \boldsymbol{\eta}^{j+1} \|_{\mathbf{L}^2} , &(5.81) \\ \| \mathcal{N}_{12,a}^{j+1} \|_{\mathbf{L}^2} &\leq \| \boldsymbol{\eta}^{j+1} \|_{\mathbf{L}^2}^2 + \| \mathbf{e}^j \|_{\mathbf{L}^2} \| \mathbf{e}^{j+1} \|_{\mathbf{L}^2} \| \mathbf{e}^j \|_{\mathbf{W}^{2,2}} \| \Delta \mathbf{e}^{j+1} \|_{\mathbf{L}^2} &(5.82) \end{aligned}$$

We come to $| \mathcal{X}^{j+1} |$ in (5.76): this term takes the effect of the explicit penalization term into account.

$$\begin{aligned} \mathcal{X}^{j+1} &= k_{j+1} \left( \mathbf{l}_\varepsilon(\mathbf{m}^j)\mathbf{m}^j, d_t \mathbf{e}^{j+1} \right) \\ &= \frac{k_{j+1}}{\varepsilon} \left( \langle \mathbf{e}^j, 2\mathbf{m}(t_j) - \mathbf{e}^j \rangle_{\mathbb{R}^3}, \langle \mathbf{m}(t_j) - \mathbf{e}^j, d_t \mathbf{e}^{j+1} \rangle_{\mathbb{R}^3} \right) \qquad (5.83) \\ &\leq \frac{\beta}{5} \{ 2 \| \langle \mathbf{e}^j, \mathbf{m}(t_j) \rangle_{\mathbb{R}^3} \|_{L^2} \| d_t \mathbf{e}^{j+1} \|_{\mathbf{L}^2} \\ &\quad + 3 \| \mathbf{e}^j \|_{\mathbf{L}^4}^2 \| d_t \mathbf{e}^{j+1} \|_{\mathbf{L}^2} + \| \mathbf{e}^j \|_{\mathbf{L}^6}^3 \| d_t \mathbf{e}^{j+1} \|_{\mathbf{L}^2} \}. \end{aligned}$$

*3rd step:* We argue by induction, similar to the proof of Theorem 5.4, step 2: there exist constants $C_i = C_i(\mathbf{m}_0, \mathbf{E}_0, \mathbf{H}_0; t_J, \omega)$, $i = 1, 2$, such that

$$\begin{aligned} &\| \nabla \mathbf{e}^{\ell+1} \|_{\mathbf{L}^2}^2 + \| \boldsymbol{\pi}^{\ell+1} \|_{\mathbf{L}^2}^2 + \| \boldsymbol{\eta}^{\ell+1} \|_{\mathbf{L}^2}^2 \qquad (5.84) \\ &+ \sum_{j=1}^{\ell+1} \frac{\beta k_j}{\varepsilon_j} \{ \| \langle \mathbf{e}^j, \mathbf{m}(t_j) \rangle_{\mathbb{R}^3} \|_{L^2}^2 + \| \mathbf{e}^j \|_{\mathbf{L}^4}^4 \} + \sum_{j=1}^{\ell+1} \frac{k_j^2}{2} \{ \| \nabla d_t \mathbf{e}^j \|_{\mathbf{L}^2}^2 \\ &+ \| d_t \boldsymbol{\pi}^j \|_{\mathbf{L}^2}^2 + \| d_t \boldsymbol{\eta}^j \|_{\mathbf{L}^2}^2 + \frac{\kappa}{2 k_j} \| \mathbf{e}^j \|_{\mathbf{W}^{2,2}}^2 \} \leq C_1 \, k_0^2 \exp\left( C_2 t_\ell \right) , \end{aligned}$$

for $F(\varepsilon_j, k_j) > 0$ as stated in the theorem, and $k_0 \leq k_A(t_J; \varepsilon)$. This statement is valid for $\ell = 0$. It is stronger than the statements in the proof of Theorem 5.4, step 2. Putting together (5.74), (5.77)-(5.81), and (5.83) in (5.76), we can then accomplish the induction step, using stability features from (5.84) for the solution of the previous step.

In order to see (c), we apply $\operatorname{div}(\cdot)$ to (5.67), and make use of (b) and an upper bound for $\mathbf{m}^j \in \mathbf{L}^\infty$, $1 \leq j \leq J$, which we obtain from interpolation of the approximation results in (a).

$$\begin{aligned}
\|\operatorname{div}(\mathbf{H}^J + \beta\,\mathbf{m}^J)\|_{W^{1,2}} &\leq \sum_{j=1}^{J} k_j \,\|\, d_t\operatorname{div}(\mathbf{H}^j + \beta\,\mathbf{m}^j)\,\|_{W^{1,2}} \\
&\leq \sum_{j=1}^{J} k_j \,\|\, \operatorname{div}\tilde{\Delta}^{-1}\big[\mathbf{l}_{\varepsilon_j}(\mathbf{m}^j)\mathbf{m}^j\big]\big)\,\|_{W^{1,2}} \\
&\leq \Big(\sum_{j=1}^{J} \frac{k_j}{\varepsilon_j}\Big)^{1/2} \Big(\sum_{j=1}^{J} \frac{k_j}{\varepsilon_j} \|\,|\mathbf{m}^j|^2 - 1\,\|^2_{\mathbf{L}^2}\Big)^{1/2}.
\end{aligned}$$

□

## 5.3 Time-Splitting Schemes to solve (MLLG)

Let us come back to uniform time-grids, for the sake of simplicity. In Section 5.2.2, we discussed a scheme in which the side constraint is implemented in an explicit way. The analysis is strongly related to the following projection based time splitting scheme $(\text{MLLG})_k^{P_1}$:

1. Given $\{\tilde{\mathbf{m}}^j, \mathbf{m}^j, \mathbf{H}^j, \mathbf{E}^j\}$, find $\tilde{\mathbf{m}}^{j+1}$ that solves

$$\begin{aligned}
\frac{1}{k}\{\tilde{\mathbf{m}}^{j+1} - \mathbf{m}^j\} - \kappa\,\Delta\tilde{\mathbf{m}}^{j+1} &= \kappa\,|\,\nabla\tilde{\mathbf{m}}^j\,|^2\tilde{\mathbf{m}}^{j+1} \\
&+\tilde{\mathbf{m}}^j \times \big(\Delta\tilde{\mathbf{m}}^{j+1} + \mathbf{H}^j\big) + \kappa\,\big(\mathbf{H}^j - \langle\tilde{\mathbf{m}}^j, \mathbf{H}^j\rangle_{\mathbb{R}^3}\tilde{\mathbf{m}}^{j+1}\big)\,.
\end{aligned} \tag{5.85}$$

2. Given $\{\tilde{\mathbf{m}}^{j+1}, \mathbf{m}^j, \mathbf{H}^j, \mathbf{E}^j\}$, compute $\{\mathbf{H}^{j+1}, \mathbf{E}^{j+1}\}$ from

$$\nabla \times \mathbf{H}^{j+1} = d_t\mathbf{E}^{j+1} + \sigma\,\mathbf{E}^{j+1}\,, \tag{5.86}$$

$$\nabla \times \mathbf{E}^{j+1} = -d_t\mathbf{H}^{j+1} - \frac{\beta}{k}\{\tilde{\mathbf{m}}^{j+1} - \mathbf{m}^j\}\,. \tag{5.87}$$

3. Compute $\mathbf{m}^{j+1} = \frac{\tilde{\mathbf{m}}^{j+1}}{|\,\tilde{\mathbf{m}}^{j+1}\,|^{2-\gamma}}$, for $\gamma \in \mathbb{N}_0$.

In this scheme, the magnetization $\tilde{\mathbf{m}}^j$ is computed independently from $\{\mathbf{H}^j, \mathbf{E}^j\}$ for each $j \geq 0$, whereas the computation of iterates for magnetic and electric field is coupled with each other. A strategy of further simplification would be then to replace (5.86) by

$$\nabla \times \mathbf{H}^j = d_t \mathbf{E}^{j+1} + \sigma\, \mathbf{E}^{j+1}\,, \tag{5.88}$$

in $(\text{MLLG})_k^{P_1}$. Unfortunately, we are not able to verify 'satisfying' convergence results for this approach. Instead, we consider the following stabilized version $(\text{MLLG})_k^{P_2}$ thereof: Given $\{\tilde{\mathbf{m}}^j, \mathbf{m}^j, \mathbf{H}^j, \mathbf{E}^j\}$, and $\delta \geq 1$.

1. Given $\{\tilde{\mathbf{m}}^j, \mathbf{m}^j, \mathbf{H}^j, \mathbf{E}^j\}$, find $\tilde{\mathbf{m}}^{j+1}$ that solves

$$\begin{aligned} \frac{1}{k}\{\tilde{\mathbf{m}}^{j+1} - \mathbf{m}^j\} - \kappa\, \Delta \tilde{\mathbf{m}}^{j+1} = \kappa\, |\, \nabla \tilde{\mathbf{m}}^j\, |^2 \tilde{\mathbf{m}}^{j+1} \\ \tilde{\mathbf{m}}^j \times \left(\Delta \tilde{\mathbf{m}}^{j+1} + \mathbf{H}^j\right) + \kappa \left(\mathbf{H}^j - \langle \tilde{\mathbf{m}}^j, \mathbf{H}^j \rangle_{\mathbb{R}^3} \tilde{\mathbf{m}}^{j+1}\right), \end{aligned} \tag{5.89}$$

2. Given $\{\mathbf{H}^j, \mathbf{E}^j\}$, determine $\mathbf{E}^{j+1}$ from

$$d_t \mathbf{E}^{j+1} - \delta k\, \tilde{\Delta} \mathbf{E}^{j+1} + \sigma\, \mathbf{E}^{j+1} = \nabla \times \mathbf{H}^j\,. \tag{5.90}$$

3. Given $\{\mathbf{m}^j, \tilde{\mathbf{m}}^{j+1}, \mathbf{H}^j\}$, update $\mathbf{H}^{j+1}$ and $\mathbf{m}^{j+1}$, for $\gamma \in \mathbb{N}_0$, via

$$d_t \mathbf{H}^{j+1} = -\nabla \times \mathbf{E}^{j+1} - \frac{\beta}{k}\{\tilde{\mathbf{m}}^{j+1} - \mathbf{m}^j\}\,, \qquad \mathbf{m}^{j+1} = \frac{\tilde{\mathbf{m}}^{j+1}}{|\, \tilde{\mathbf{m}}^{j+1}\, |^{2-\gamma}}\,. \tag{5.91}$$

The following results describe the convergence behavior of both, the schemes $(\text{MLLG})_k^{P_1}$ and $(\text{MLLG})_k^{P_2}$. As a conclusion of the subsequent analysis, successful splitting in the computation of $\mathbf{E}^j$ and $\mathbf{H}^j$ relies on strong regularity properties of the electric field, cf. Theorem 5.3, (a).

**Theorem 5.6** *Suppose* $\{\mathbf{m}, \mathbf{H}, \mathbf{E}\}$ *to solve* (5.1)-(5.6), *for* $\omega = (0, 2D)^2$ *and periodic boundary data,* $0 < t_J < T_0(\mathbf{m}_0)$. *Let* $\{\mathbf{m}_0, \mathbf{H}_0, \mathbf{E}_0\} \in \mathbf{W}^{2,2}_{\text{per}}(\omega, \mathcal{S}^2) \times [\mathbf{W}^{1,2}_{\text{per}}(\omega, \mathbb{R}^3)]^2$ *satisfy* (5.11). *Let* $\{\tilde{\mathbf{m}}^j, \mathbf{H}^j, \mathbf{E}^j\}_{j=0}^J$ *solve* $(MLLG)_k^{P_\ell}$, $\ell = 1, 2$, *with periodic boundary data, for* $\gamma \in \mathbb{N}_0$. *For sufficiently small time-steps*

$k \leq k_0(t_J)$, *there holds*

$$\text{(a)} \quad \max_{0\leq j\leq J}\Big\{\| \mathbf{m}(t_j) - \tilde{\mathbf{m}}^j \|_{\mathbf{L}^2} + \| \mathbf{H}(t_j) - \mathbf{H}^j \|_{\mathbf{W}^{-1,2}_{\text{per}}} + \| \mathbf{E}(t_j) - \mathbf{E}^j \|_{\mathbf{W}^{-1,2}_{\text{per}}}\Big\}$$
$$+\Big(k \sum_{j=0}^{J} \| \mathbf{m}(t_j) - \tilde{\mathbf{m}}^j \|^2_{\mathbf{W}^{1,2}} + \frac{\beta}{k}\big[\| \langle \mathbf{m}(t_j) - \tilde{\mathbf{m}}^j, \mathbf{m}(t_j)\rangle_{\mathbb{R}^3} \|^2_{L^2}$$
$$+\| \mathbf{m}(t_j) - \tilde{\mathbf{m}}^j \|^4_{\mathbf{L}^4}\big]\Big)^{1/2} \leq C\, k\,,$$

$$\text{(b)} \quad \Big(\sum_{j=0}^{J} \| 1 - |\tilde{\mathbf{m}}^j|^2 \|^2_{\mathbf{L}^2}\Big)^{1/2} \leq C\, k\,,$$

*with* $C = C(\mathbf{m}_0, \mathbf{H}_0, \mathbf{E}_0, \omega, t_J, \gamma_0)$ *independent of* $k$. *Moreover, the solution to* $(MLLG)^{E_1}_{k,\varepsilon}$ *satisfies the a priori bounds, with* $C$ *independent of* $k$,

$$\text{(c)} \quad \max_{0\leq j\leq J}\Big\{\| d_t\tilde{\mathbf{m}}^j \|_{\mathbf{L}^2} + \| d_t\mathbf{H}^j \|_{\mathbf{L}^2} + \| d_t\mathbf{E}^j \|_{\mathbf{L}^2} + \| \tilde{\mathbf{m}}^j \|_{\mathbf{W}^{2,2}}$$
$$+\| \mathbf{H}^j \|_{\mathbf{W}^{1,2}} + \| \mathbf{E}^j \|_{\mathbf{W}^{1,2}}\Big\} + \Big(k \sum_{j=1}^{J} \| \nabla d_t\tilde{\mathbf{m}}^j \|^2_{\mathbf{L}^2}\Big)^{1/2} \leq C\,.$$

**Remark 5.3** *Note that no statement is made concerning approximation of* $\operatorname{div}(\mathbf{m} + \beta\mathbf{H}) = 0$. *In order to obtain a result in this direction, we have to weaken the projection step; see also Remark 5.1, item 2. We suggest two scenarios:*

*1. Replace the projection steps in* $(MLLG)^{P_\ell}_k$, $\ell \in \{1,2\}$ *by* $\mathbf{m}^{j+1} = \tilde{\mathbf{m}}^{j+1} + k^\alpha\big(1 - \frac{1}{|\tilde{\mathbf{m}}^{j+1}|}\big)\tilde{\mathbf{m}}^{j+1}$, *for* $0 \leq \alpha \leq 1$. *If we substitute this into (5.85), say, we find for* $\overline{\mathbf{l}}^{j+1}_\varepsilon(\varphi) = \frac{1}{\varepsilon}\big(1 - \frac{1}{|\varphi|}\big)$,

$$d_t\tilde{\mathbf{m}}^{j+1} - \kappa\,\Delta\tilde{\mathbf{m}}^{j+1} + \overline{\mathbf{l}}^{j+1}_{k^{1-\alpha}}(\tilde{\mathbf{m}}^j)\tilde{\mathbf{m}}^j = \kappa\,|\nabla\tilde{\mathbf{m}}^j|^2\tilde{\mathbf{m}}^{j+1} \qquad (5.92)$$
$$-\gamma'\,\tilde{\mathbf{m}}^j \times \big(\Delta\tilde{\mathbf{m}}^{j+1} + \mathbf{H}^j\big) + \kappa\big(\mathbf{H}^j - \langle\tilde{\mathbf{m}}^j, \mathbf{H}^j\rangle_{\mathbb{R}^3}\tilde{\mathbf{m}}^{j+1}\big)\,.$$

*2. We replace the projection steps in* $(MLLG)^{P_\ell}_k$, $\ell \in \{1,2\}$ *by* $\mathbf{m}^{j+1} = \tilde{\mathbf{m}}^{j+1} - k^\alpha\,\tilde{\Delta}^{-1}\big[\big(1 - \frac{1}{|\tilde{\mathbf{m}}^{j+1}|}\big)\tilde{\mathbf{m}}^{j+1}\big]$. *Similarly to item 1., we are lead to (5.92), for* $\hat{\mathbf{l}}^{j+1}_\varepsilon(\varphi) = -\tilde{\Delta}^{-1}\big[\overline{\mathbf{l}}^{j+1}_\varepsilon(\varphi)\varphi\big]$.

*For the first scenario, we can follow the subsequent proof of Theorem 5.6, where* $\tilde{\mathbf{l}}^{j+1}_k(\tilde{\mathbf{m}}^j)\tilde{\mathbf{m}}^j$ *in (5.93), (5.95) is replaced by* $\overline{\mathbf{l}}^{j+1}_{k^{1-\alpha}}(\tilde{\mathbf{m}}^j)\tilde{\mathbf{m}}^j$. *Showing Theorem 5.4, (d) requires argument (5.63).*

*In case we make use of the time-grid strategy $\mathcal{G}_2(k_{j+1})$, the statements of Theorem 5.5 apply; in particular, we find a result that corresponds to statement (c), for $\varepsilon = k^{1-\alpha}$.*

**Proof:**
We keep the proof sketchy, since we can apply many arguments already worked out before in Section 4.2.2 for (LLG). We start with the analysis of $(\mathrm{MLLG})_k^{P_1}$.

*1st step:* The reformulation of the scheme as penalization scheme leads to the identities, for $\{\mathbf{e}^j, \boldsymbol{\eta}^j, \boldsymbol{\pi}^j\} := \{\mathbf{m}(t_j) - \tilde{\mathbf{m}}^j, \mathbf{H}(t_j) - \mathbf{H}^j, \mathbf{E}(t_j) - \mathbf{E}^j\}$,

$$
\begin{aligned}
& d_t\mathbf{e}^{j+1} - \kappa\big(\Delta\mathbf{e}^{j+1} + \boldsymbol{\eta}^{j+1}\big) + \tilde{\mathbf{l}}_k^{j+1}(\tilde{\mathbf{m}}^j)\tilde{\mathbf{m}}^j && (5.93)\\
&\quad = \boldsymbol{\mathcal{R}}^{j+1}(\mathbf{m}) + \boldsymbol{\mathcal{N}}^{j+1} - k\big(\kappa\, d_t\boldsymbol{\eta}^{j+1} + \gamma'\,[\mathbf{m}(t_j) - \mathbf{e}^j] \times d_t\mathbf{H}^{j+1}\big)\,, \\
& \nabla \times \boldsymbol{\eta}^{j+1} = d_t\boldsymbol{\pi}^{j+1} + \sigma\,\boldsymbol{\pi}^{j+1} - \boldsymbol{\mathcal{R}}^{j+1}(\mathbf{E})\,, && (5.94)\\
& \nabla \times \boldsymbol{\pi}^{j+1} = -d_t\boldsymbol{\eta}^{j+1} - \beta\left(d_t\mathbf{e}^{j+1} + \tilde{\mathbf{l}}_k^{j+1}(\tilde{\mathbf{m}}^j)\tilde{\mathbf{m}}^j\right) && (5.95)\\
&\quad + \boldsymbol{\mathcal{R}}^{j+1}(\mathbf{H}) + \beta\,\boldsymbol{\mathcal{R}}^{j+1}(\mathbf{m})\,, \\
& \operatorname{div}\boldsymbol{\pi}^{j+1} = 0\,, && (5.96)
\end{aligned}
$$

where $\tilde{\mathbf{l}}_k^{j+1}(\tilde{\mathbf{m}}^j) = \frac{1}{k}\big(1 - \frac{1}{|\,\tilde{\mathbf{m}}^j\,|^{2-\gamma}}\big)$, and $\boldsymbol{\mathcal{N}}^{j+1}$ from (5.46).

We can proceed as in the proof of Theorem 5.4, and test (5.93)-(5.96) by the function $\{\mathbf{e}^{j+1} + \tilde{\Delta}^{-1}\boldsymbol{\eta}^{j+1}, \tilde{\Delta}^{-1}\boldsymbol{\pi}^{j+1}, \tilde{\Delta}^{-1}\boldsymbol{\eta}^{j+1}\}$. We obtain result (5.49) for $k \le k_0(t_j)$ sufficiently small, with the following modifications:

1. The last two equations in front of the inequality sign in (5.49) are shifted by $-1$ (see also (5.76)), and new contributions arise that reflect usage of explicit penalization.

2. Two new contributions from (5.93) enter the analysis.

We address these items in the next two steps independently.

*2nd step:* We follow (5.83), with $\mathcal{X}^{j+1} = k\left(\tilde{\mathbf{l}}_k^{j+1}(\tilde{\mathbf{m}}^j)\tilde{\mathbf{m}}^j, d_t\mathbf{e}^{j+1}\right)$.

*3rd step:*

$$
\begin{aligned}
& |\,\big(d_t\boldsymbol{\eta}^{j+1}, \mathbf{e}^{j+1} + \tilde{\Delta}^{-1}\boldsymbol{\eta}^{j+1}\big)\,| && (5.97)\\
&\quad \le C\,\|\, d_t\boldsymbol{\eta}^{j+1}\,\|_{\mathbf{W}^{-1,2}_{\mathrm{per}}}\,\|\,\nabla\big[\mathbf{e}^{j+1} + \tilde{\Delta}^{-1}\boldsymbol{\eta}^{j+1}\big]\,\|_{\mathbf{L}^2}\,, \\
& |\,\big(\mathbf{e}^j \times d_t\mathbf{H}^{j+1}, \mathbf{e}^{j+1} + \tilde{\Delta}^{-1}\boldsymbol{\eta}^{j+1}\big)\,| && (5.98)\\
&\quad \le C\,\|\,\mathbf{e}^j\,\|_{\mathbf{L}^4}\,\|\, d_t\mathbf{H}^{j+1}\,\|_{\mathbf{L}^2}\,\|\,\mathbf{e}^{j+1} + \tilde{\Delta}^{-1}\boldsymbol{\eta}^{j+1}\,\|_{\mathbf{L}^4}\,,
\end{aligned}
$$

thanks to Theorem 5.3, (d).

Hence, we find a modified inequality (5.49), with additional terms that do not reduce order of convergence. Gronwall's lemma then establishes the result for $(\mathrm{MLLG})_{k,\varepsilon}^{P_1}$.

Let us consider $(\mathrm{MLLG})_{k,\varepsilon}^{P_2}$: The error identities (5.93) and (5.95) remain valid, and (5.94) is replaced by

$$\begin{aligned}\nabla \times \boldsymbol{\eta}^{j+1} &= d_t\boldsymbol{\pi}^{j+1} - \delta k\,\tilde{\Delta}\boldsymbol{\pi}^{j+1} + \sigma\,\boldsymbol{\pi}^{j+1} - \boldsymbol{\mathcal{R}}^{j+1}(\mathbf{E}) \\ &\quad + k\big(\delta\,\tilde{\Delta}\mathbf{E}(t_{j+1}) + \nabla\times d_t\boldsymbol{\eta}^{j+1} - \nabla\times d_t\mathbf{H}(t_{j+1})\big)\,. \end{aligned} \tag{5.99}$$

We can test (5.93), (5.99), (5.95) with $\{\mathbf{e}^{j+1} + \tilde{\Delta}^{-1}\boldsymbol{\eta}^{j+1}, \tilde{\Delta}^{-1}\boldsymbol{\pi}^{j+1}, \tilde{\Delta}^{-1}\boldsymbol{\eta}^{j+1}\}$ and proceed as before, getting a modified inequality (5.49) like in the scenario of $(\mathrm{MLLG})_{k,\varepsilon}^{P_1}$. However, there are further modifications of it.

1. There is an additional term $\delta k\,\|\,\boldsymbol{\pi}^{j+1}\,\|_{\mathbf{L}^2}^2$ on the left-hand side.

2. We have to deal with further contributions

$$k\,\big|\big(\delta\,\Delta\mathbf{E}(t_{j+1}) + \nabla\times d_t\boldsymbol{\eta}^{j+1} - \nabla\times d_t\mathbf{H}(t_{j+1}), \tilde{\Delta}^{-1}\boldsymbol{\pi}^{j+1}\big)\big|\,. \tag{5.100}$$

Thanks to Theorem 5.3, (d), we get the upper bound

$$\begin{aligned} &\leq Ck\,\big\{\delta\,\|\,\nabla\mathbf{E}(t_{j+1})\,\|_{\mathbf{L}^2} + \|\,d_t\mathbf{H}(t_{j+1})\,\|_{\mathbf{L}^2}\big\}\|\,\boldsymbol{\pi}^{j+1}\,\|_{\mathbf{W}^{-1,2}_{\mathrm{per}}} \\ &\quad + k\,\|\,d_t\boldsymbol{\eta}^{j+1}\,\|_{\mathbf{W}^{-1,2}_{\mathrm{per}}}\|\,\boldsymbol{\pi}^{j+1}\,\|_{\mathbf{L}^2} \\ &\leq Ck^2 + \|\,\boldsymbol{\pi}^{j+1}\,\|^2_{\mathbf{W}^{-1,2}_{\mathrm{per}}} + \frac{k}{2\delta}\|\,d_t\boldsymbol{\eta}^{j+1}\,\|^2_{\mathbf{W}^{-1,2}_{\mathrm{per}}} + \frac{k\delta}{2}\|\,\boldsymbol{\pi}^{j+1}\,\|^2_{\mathbf{L}^2}\,. \end{aligned} \tag{5.101}$$

The latter two terms can be absorbed on the left-hand side of the modified inequality (5.49). □

Next, we consider a finite element discretization of the considered time discretization schemes. For $(\mathrm{MLLG})_{k,\varepsilon}^{E_\ell}$ (without damping term), $\ell = 1, 2$, Nédélec spaces for $\mathbf{H}_h^j, \mathbf{E}_h^j$, and $\mathbf{m}_h^j \in \boldsymbol{\mathcal{S}}^1(\omega)$ lead to conforming discretizations, [100, 98]. In the following section, we address a conforming finite element discretization of $(\mathrm{MLLG})_k^{P_2}$, using $\mathbf{m}_h^j, \mathbf{H}_h^j, \mathbf{E}_h^j \in \boldsymbol{\mathcal{S}}^1(\omega)$.

## 5.4 Stabilized Finite Element Realization of $(\mathrm{MLLG})_k^{P_2}$

We consider a stabilized conforming finite element discretization $(\mathrm{MLLG})_{k;h}^{P_2}$ of (5.90)-(5.91):

1. Given $\tilde{\mathbf{m}}_h^j, \mathbf{m}_h^j, \mathbf{H}_h^j, \mathbf{E}_h^j \in \boldsymbol{\mathcal{S}}^1(\omega)$, find $\tilde{\mathbf{m}}_h^{j+1} \in \boldsymbol{\mathcal{S}}^1(\omega)$ that solves

$$
\begin{aligned}
\frac{1}{k}(\tilde{\mathbf{m}}_h^{j+1} - \mathbf{m}_h^j, \varphi_h)_h + \kappa\,(\nabla \tilde{\mathbf{m}}_h^{j+1}, \nabla \varphi_h)_h &= \kappa\,(|\nabla \tilde{\mathbf{m}}_h^j|^2 \tilde{\mathbf{m}}_h^{j+1}, \varphi_h)_h \\
&\quad -(\nabla[\varphi_h \times \tilde{\mathbf{m}}_h^j], \nabla \tilde{\mathbf{m}}_h^{j+1})_h \qquad (5.102)\\
&\quad +\kappa\,(\mathbf{H}_h^j - \langle \tilde{\mathbf{m}}_h^j, \mathbf{H}_h^j\rangle_{\mathbb{R}^3} \tilde{\mathbf{m}}_h^j, \varphi_h)_h\,.
\end{aligned}
$$

2. Given $\mathbf{H}_h^j, \mathbf{E}_h^j$, compute $\mathbf{E}_h^{j+1} \in \boldsymbol{\mathcal{S}}^1(\omega)$ from

$$
\begin{aligned}
\frac{1}{k}(\mathbf{E}_h^{j+1} - \mathbf{E}_h^j, \boldsymbol{\psi}_h)_h + \delta k\,(\nabla \mathbf{E}_h^{j+1}, \nabla \boldsymbol{\psi}_h)_h & \qquad (5.103)\\
+\sigma\,(\mathbf{E}_h^{j+1}, \boldsymbol{\psi}_h)_h = (\nabla \times \mathbf{H}_h^j, \boldsymbol{\psi}_h)_h\,. &
\end{aligned}
$$

3. For $\gamma \in \mathbb{N}_0$, and given $\mathbf{m}_h^j, \tilde{\mathbf{m}}_h^j, \mathbf{H}_h^j$, determine $\mathbf{H}_h^{j+1}, \mathbf{m}_h^{j+1} \in \boldsymbol{\mathcal{S}}^1(\omega)$ from

$$
\begin{aligned}
\frac{1}{k}(\mathbf{H}_h^{j+1} - \mathbf{H}_h^j, \boldsymbol{\psi}_h)_h + h\,(\nabla \mathbf{H}_h^{j+1}, \nabla \boldsymbol{\psi}_h)_h & \qquad (5.104)\\
= -(\nabla \times \mathbf{E}_h^{j+1}, \boldsymbol{\psi}_h)_h + \frac{\beta}{k}(\tilde{\mathbf{m}}_h^{j+1} - \mathbf{m}_h^j, \boldsymbol{\psi}_h)_h\,, &\\
\mathbf{m}_h^{j+1} = \frac{\tilde{\mathbf{m}}_h^{j+1}}{|\tilde{\mathbf{m}}_h^{j+1}|^{2-\gamma}}\,, &
\end{aligned}
$$

for all $\varphi_h, \boldsymbol{\psi}_h \in \boldsymbol{\mathcal{S}}^1(\omega)$. — The following convergence result is valid for $(\mathrm{MLLG})_{k;h}^{P_2}$.

**Theorem 5.7** *Let* $\{\mathbf{m}, \mathbf{H}, \mathbf{E}\}$ *solve* (5.1)-(5.6), *for* $0 < t_J < T_0(\mathbf{m}_0)$, *and* $\{\mathbf{m}_0, \mathbf{H}_0, \mathbf{E}_0\} \in \mathbf{W}^{2,2}(\omega, \mathcal{S}^2) \times [\mathbf{W}^{1,2}(\omega, \mathbb{R}^3)]^2$ *satisfies* (5.11), *for* $\omega = (0, 2D)^2$. *Let* $\{\tilde{\mathbf{m}}_h^j, \mathbf{H}_h^j, \mathbf{E}_h^j\}_{j=0}^J$ *solve* $(MLLG)_{k,h}^{P_2}$, *and let* $\gamma \in \mathbb{N}_0$. *Let* $\delta \geq \delta_0(\omega)$ *be suf-*

*ficiently large. For $h = \mathcal{O}(k)$ and $k \leq k_0(t_J)$ sufficiently small, there holds*

$$\max_{0\leq j\leq J}\Big\{\| \mathbf{m}(t_j) - \tilde{\mathbf{m}}_h^j \|_{\mathbf{L}^2} + \| \mathbf{H}(t_j) - \mathbf{H}_h^j \|_{\mathbf{W}_{\rm per}^{-1,2}} + \| \mathbf{E}(t_j) - \mathbf{E}_h^j \|_{\mathbf{W}_{\rm per}^{-1,2}}\Big\}$$
$$+\Big(k\sum_{j=0}^{J} \| \mathbf{m}(t_j) - \tilde{\mathbf{m}}_h^j \|_{\mathbf{W}^{1,2}}^2 + \frac{1}{k} \| \mathbf{m}(t_j) - \tilde{\mathbf{m}}_h^j \|_{\mathbf{L}^4}^4$$
$$+\frac{1}{k} \| 1 - |\tilde{\mathbf{m}}^j| \|_{\mathbf{L}^2}^2\Big)^{1/2} \leq C\{k+h\},$$

*with $C = C(\mathbf{m}_0, \mathbf{H}_0, \mathbf{E}_0, \omega, t_J, \gamma_0, \delta_0)$ independent of $k, h$.*

**Remark 5.4** *1. The restriction $F(k,h) > 0$ is essentially due to Theorem 4.10, where we required $\varepsilon^{-1/2} = o(\frac{1}{k})$, and control of the rotational parts by the stabilizing terms which are scaled in both, $h$ and $k$. Actually, computations in Section 5.5 show that the choice of values $k, h$ seems to be more flexible.*

*2. Alternatively, we can use the modified projection schemes from Remark 5.3, where $\varepsilon = k^{1-\alpha}$ also keeps track of the incompressibility of $\mathbf{m}+\beta\,\mathbf{H}$. The following analysis can be straightforwardly applied to the modified projection schemes.*

**Proof:**
For $\{\mathbf{e}^j, \boldsymbol{\eta}^j, \boldsymbol{\pi}^j\} := \{\tilde{\mathbf{m}}^j - \tilde{\mathbf{m}}_h^j, \mathbf{H}^j - \mathbf{H}_h^j, \mathbf{E}^j - \mathbf{E}_h^j\}$, and $\mathbf{l}_k(\varphi) = \frac{1}{k}\,(1 - \frac{1}{|\varphi|^{2-\gamma}})$, we find

$$(d_t\mathbf{e}^{j+1}, \boldsymbol{\varphi}_h)_h + \kappa\,(\nabla\mathbf{e}^{j+1}, \nabla\boldsymbol{\varphi}_h)_h - \kappa\,(\boldsymbol{\eta}^{j+1}, \boldsymbol{\varphi}_h)_h$$
$$-(\mathbf{l}_k(\tilde{\mathbf{m}}^j)\tilde{\mathbf{m}}^j - \mathbf{l}_k(\tilde{\mathbf{m}}_h^j)\tilde{\mathbf{m}}_h^j, \boldsymbol{\varphi}_h)_h = (\boldsymbol{\mathcal{N}}^{j+1}, \boldsymbol{\varphi}_h)_h\,,$$
$$(\nabla\times\boldsymbol{\eta}^{j+1}, \boldsymbol{\psi}_h)_h = (d_t\boldsymbol{\pi}^{j+1}, \boldsymbol{\psi}_h)_h + \sigma\,(\boldsymbol{\pi}^{j+1}, \boldsymbol{\psi}_h)_h +$$
$$\delta k\,(\nabla\boldsymbol{\pi}^{j+1}, \nabla\boldsymbol{\psi}_h)_h + k\,(\nabla\times d_t\boldsymbol{\eta}^{j+1}, \boldsymbol{\psi}_h)_h\,,$$
$$(\nabla\times\boldsymbol{\pi}^{j+1}, \boldsymbol{\psi}_h)_h = -(d_t\boldsymbol{\eta}^{j+1}, \boldsymbol{\psi}_h)_h - h\big[(\boldsymbol{\eta}^{j+1}, \boldsymbol{\psi}_h)_h + (\nabla\boldsymbol{\eta}^{j+1}, \nabla\boldsymbol{\psi}_h)_h\big]$$
$$-h\,(\nabla\mathbf{E}^{j+1}, \nabla\boldsymbol{\psi}_h)_h - \beta\,(d_t\mathbf{e}^{j+1} - \mathbf{l}_k(\tilde{\mathbf{m}}^j)\tilde{\mathbf{m}}^j + \mathbf{l}_k(\tilde{\mathbf{m}}_h^j)\tilde{\mathbf{m}}_h^j, \boldsymbol{\psi}_h)_h\,,$$

for all $\boldsymbol{\varphi}, \boldsymbol{\psi} \in \boldsymbol{\mathcal{S}}^1(\omega)$. — We test by $\{\mathbf{P}^c_{h,\mathbf{L}^2}\mathbf{e}^{j+1} + \tilde{\Delta}_h^{-1}\boldsymbol{\eta}^{j+1}, \tilde{\Delta}_T^{-1}\boldsymbol{\pi}^{j+1}, \tilde{\Delta}_T^{-1}\boldsymbol{\eta}^{j+1}\}$ and make the following observations:

$$\begin{aligned}
&-(\nabla\boldsymbol{\eta}^{j+1}, \nabla\tilde{\Delta}_T^{-1}\boldsymbol{\eta}^{j+1})_h = -(\nabla\tilde{\Delta}_T^{-1}\boldsymbol{\eta}^{j+1}, \nabla\{\boldsymbol{\eta}^{j+1} \pm \mathbf{P}_{h;\mathbf{L}^2}\mathbf{H}^j\})_h \qquad (5.105)\\
&\quad = -(\nabla\tilde{\Delta}_T^{-1}\boldsymbol{\eta}^{j+1}, \nabla\{\mathbf{H}^{j+1} - \mathbf{P}_{h;\mathbf{L}^2}\mathbf{H}^{j+1}\})_h + \|\,\boldsymbol{\eta}^{j+1}\,\|^2_{\mathbf{L}^2}\\
&\qquad +(\boldsymbol{\eta}^{j+1}, \mathbf{H}^{j+1} - \mathbf{P}_{h;\mathbf{L}^2}\mathbf{H}^{j+1})_h\\
&(\nabla\times\boldsymbol{\pi}^{j+1}, [\tilde{\Delta}_h^{-1} - \tilde{\Delta}^{-1}]\boldsymbol{\eta}^{j+1})_h \qquad (5.106)\\
&\quad = (\mathbf{curl}\,\mathbf{curl}\,\tilde{\Delta}^{-1}\boldsymbol{\pi}^{j+1}, \mathbf{curl}\,[\tilde{\Delta}_h^{-1} - \tilde{\Delta}^{-1}]\boldsymbol{\eta}^{j+1})_h\\
&\quad \leq Ch\,\|\,\boldsymbol{\eta}^{j+1}\,\|_{\mathbf{L}^2}\,\|\,\boldsymbol{\pi}^{j+1}\,\|_{\mathbf{L}^2}\,,\\
&(\nabla\times\boldsymbol{\eta}^{j+1}, [\tilde{\Delta}_h^{-1} - \tilde{\Delta}^{-1}]\boldsymbol{\pi}^{j+1})_h \leq Ch\,\|\,\boldsymbol{\eta}^{j+1}\,\|_{\mathbf{L}^2}\,\|\,\boldsymbol{\pi}^{j+1}\,\|_{\mathbf{L}^2}\,,\\
&(\nabla\times\boldsymbol{\pi}^{j+1}, \tilde{\Delta}^{-1}\boldsymbol{\eta}^{j+1})_h - (\nabla\times\boldsymbol{\eta}^{j+1}, \tilde{\Delta}^{-1}\boldsymbol{\pi}^{j+1})_h = 0\,,\\
&k\,(d_t\boldsymbol{\eta}^{j+1}, \nabla\times\tilde{\Delta}_h^{-1}\boldsymbol{\pi}^{j+1}) \qquad (5.107)\\
&\quad = k\,(d_t[\mathbf{E}^{j+1} - \mathbf{P}^c_{h,\mathbf{L}^2}\mathbf{E}^{j+1} + \mathbf{P}^c_{h,\mathbf{L}^2}\boldsymbol{\eta}^{j+1}], \nabla\times\tilde{\Delta}_h^{-1}\boldsymbol{\pi}^{j+1})\\
&\quad \leq k\,\|\,d_t\mathbf{E}^{j+1}\,\|_{\mathbf{L}^2}\,\|\,\nabla\tilde{\Delta}_h^{-1}\boldsymbol{\pi}^{j+1}\,\|_{\mathbf{L}^2} + k\,\|\,\nabla\tilde{\Delta}_h^{-1}d_t\boldsymbol{\eta}^{j+1}\,\|_{\mathbf{L}^2}\,\|\,\nabla\tilde{\Delta}_h^{-1}\boldsymbol{\pi}^{j+1}\,\|_{\mathbf{L}^2}\,.
\end{aligned}$$

We exploit $\mathbf{W}^{1,2}$-stability of the $\mathbf{L}^2$-projection operator next.

$$\begin{aligned}
&k\,(d_t\boldsymbol{\eta}^{j+1}, \mathbf{P}^c_{h,\mathbf{L}^2}\nabla\times\tilde{\Delta}^{-1}\boldsymbol{\pi}^{j+1})\\
&\quad = k\,(\nabla\tilde{\Delta}_h^{-1}d_t\boldsymbol{\eta}^{j+1}, \nabla\mathbf{P}^c_{h,\mathbf{L}^2}\nabla\times\tilde{\Delta}^{-1}\boldsymbol{\pi}^{j+1})\\
&\quad \leq Ck\,\|\,\nabla\tilde{\Delta}_h^{-1}d_t\boldsymbol{\eta}^{j+1}\,\|_{\mathbf{L}^2}\,\|\,\boldsymbol{\pi}^{j+1}\,\|_{\mathbf{L}^2}\,, \qquad (5.108)\\
&k\,(\mathbf{P}^c_{h,\mathbf{L}^2}d_t\boldsymbol{\eta}^{j+1}, \nabla\times[\tilde{\Delta}_h^{-1} - \tilde{\Delta}]\boldsymbol{\pi}^{j+1})\\
&\quad = k\,(\nabla\tilde{\Delta}_h^{-1}d_t\boldsymbol{\eta}^{j+1}, \nabla\mathbf{P}^c_{h,\mathbf{L}^2}\nabla\times[\tilde{\Delta}_h^{-1} - \tilde{\Delta}]\boldsymbol{\pi}^{j+1})\\
&\quad \leq k\,\|\,\nabla\tilde{\Delta}_h^{-1}d_t\boldsymbol{\eta}^{j+1}\,\|_{\mathbf{L}^2}\,\frac{1}{h}\,\|\,\mathbf{P}^c_{h,\mathbf{L}^2}\nabla\times[\tilde{\Delta}_h^{-1} - \tilde{\Delta}]\boldsymbol{\pi}^{j+1}\,\|_{\mathbf{L}^2} \qquad (5.109)\\
&\quad \leq Ck\,\|\,\nabla\tilde{\Delta}_h^{-1}d_t\boldsymbol{\eta}^{j+1}\,\|_{\mathbf{L}^2}\,\|\,\boldsymbol{\pi}^{j+1}\,\|_{\mathbf{L}^2}\,.
\end{aligned}$$

Hence, we conclude

$$\begin{aligned}
&\frac{1}{2}\,d_t\|\,\mathbf{P}^c_{h,\mathbf{L}^2}\mathbf{e}^{j+1}\,\|^2_{\mathbf{L}^2} + \frac{k}{2}\,\|\,\mathbf{P}^c_{h,\mathbf{L}^2}d_t\mathbf{e}^{j+1}\,\|^2_{\mathbf{L}^2}\\
&\quad +\frac{\kappa}{2}\,\{\|\,\nabla\mathbf{e}^{j+1}\,\|^2_{\mathbf{L}^2} + \|\,\boldsymbol{\eta}^{j+1}\,\|^2_{\mathbf{W}^{-1,2}_{\rm per}}\}\\
&\quad -(\mathbf{l}_k(\tilde{\mathbf{m}}^j)\tilde{\mathbf{m}}^j - \mathbf{l}_k(\tilde{\mathbf{m}}^j_h)\tilde{\mathbf{m}}^j_h, \mathbf{P}^c_{h,\mathbf{L}^2}\mathbf{e}^{j+1})_h \qquad (5.110)
\end{aligned}$$

$$
\begin{aligned}
&\leq |(\mathcal{N}^{j+1}, \mathbf{P}^c_{h,\mathbf{L}^2}\mathbf{e}^{j+1})_h| + \kappa \| \nabla[\tilde{\mathbf{m}}^{j+1} - \mathbf{P}^c_{h,\mathbf{L}^2}\tilde{\mathbf{m}}^{j+1}] \|^2_{\mathbf{L}^2} \qquad (5.111)\\
&+C_\kappa \| [\tilde{\Delta}^{-1} - \tilde{\Delta}_h^{-1}]\boldsymbol{\eta}^{j+1} \|^2_{\mathbf{W}^{1,2}}\\
&-(d_t\mathbf{e}^{j+1} - \mathbf{l}_k(\tilde{\mathbf{m}}^j)\tilde{\mathbf{m}}^j + \mathbf{l}_k(\tilde{\mathbf{m}}_h^j)\tilde{\mathbf{m}}_h^j, \tilde{\Delta}_h^{-1}\boldsymbol{\eta}^{j+1})_h,
\end{aligned}
$$

$$
\begin{aligned}
(\nabla \times \boldsymbol{\pi}^{j+1}, \tilde{\Delta}^{-1}\boldsymbol{\eta}^{j+1})_h &\geq \frac{1}{2} d_t \| \boldsymbol{\eta}^{j+1} \|^2_{\mathbf{W}^{-1,2}_{\rm per}} + \frac{k}{2} \| d_t\boldsymbol{\eta}^{j+1} \|^2_{\mathbf{W}^{-1,2}_{\rm per}}\\
&-(d_t\boldsymbol{\eta}^{j+1}, [\tilde{\Delta}_{\mathcal{T}}^{-1} - \tilde{\Delta}^{-1}]\boldsymbol{\eta}^{j+1}),\\
&-h\,(\nabla\boldsymbol{\eta}^{j+1}, \nabla\tilde{\Delta}_{\mathcal{T}}^{-1}\boldsymbol{\eta}^{j+1}) - h \| \mathbf{H}^{j+1} \|_{\mathbf{W}^{1,2}} \| \nabla\tilde{\Delta}_{\mathcal{T}}^{-1}\boldsymbol{\eta}^{j+1} \|_{\mathbf{L}^2}] \qquad (5.112)\\
&-\beta\,(d_t\mathbf{e}^{j+1} - \mathbf{l}_k(\tilde{\mathbf{m}}^j)\tilde{\mathbf{m}}^j + \mathbf{l}_k(\tilde{\mathbf{m}}_h^j)\tilde{\mathbf{m}}_h^j, \tilde{\Delta}_{\mathcal{T}}^{-1}\boldsymbol{\eta}^{j+1})\\
&-(\boldsymbol{\pi}^{j+1}, \nabla \times [\tilde{\Delta}^{-1} - \tilde{\Delta}_h^{-1}]\boldsymbol{\eta}^{j+1})_h,
\end{aligned}
$$

$$
\begin{aligned}
-(\nabla \times \boldsymbol{\eta}^{j+1}, \tilde{\Delta}^{-1}\boldsymbol{\pi}^{j+1})_h &= \frac{1}{2} d_t \| \boldsymbol{\pi}^{j+1} \|^2_{\mathbf{W}^{-1,2}_{\rm per}}\\
&+\frac{k}{2} \| d_t\boldsymbol{\pi}^{j+1} \|^2_{\mathbf{W}^{-1,2}_{\rm per}} + \sigma \| \boldsymbol{\pi}^{j+1} \|^2_{\mathbf{W}^{-1,2}_{\rm per}}\\
&-(d_t\boldsymbol{\pi}^{j+1}, [\tilde{\Delta}_{\mathcal{T}}^{-1} - \tilde{\Delta}^{-1}]\boldsymbol{\pi}^{j+1})_h + \sigma\,(\boldsymbol{\pi}^{j+1}, [\Delta_h^{-1} - \tilde{\Delta}^{-1}]\boldsymbol{\pi}^{j+1})_h\\
&+\delta k\,(\nabla\boldsymbol{\pi}^{j+1}, \nabla\tilde{\Delta}_h^{-1}\boldsymbol{\pi}^{j+1})_h \qquad (5.113)\\
&-k\,(d_t\boldsymbol{\eta}^{j+1}, \nabla \times \tilde{\Delta}_{\mathcal{T}}^{-1}\boldsymbol{\pi}^{j+1})_h - (\boldsymbol{\eta}^{j+1}, \nabla \times [\tilde{\Delta}^{-1} - \tilde{\Delta}_h^{-1}]\boldsymbol{\pi}^{j+1})_h.
\end{aligned}
$$

We add (5.110), (5.112), and $\beta$ times (5.113). The leading cross-product terms in (5.110), (5.112) cancel, as well as the last term in (5.110), (5.112) that arise from penalization. We make use of the observations (5.105)-(5.107).

$$
\begin{aligned}
&\frac{1}{2} d_t\{\| \mathbf{e}^{j+1} \|^2_{\mathbf{L}^2} + \| \boldsymbol{\pi}^{j+1} \|^2_{\mathbf{W}^{-1,2}_{\rm per}} + \| \boldsymbol{\eta}^{j+1} \|^2_{\mathbf{W}^{-1,2}_{\rm per}}\}\\
&+\frac{k}{2}\{\| d_t\mathbf{e}^{j+1} \|^2_{\mathbf{L}^2} + \| d_t\boldsymbol{\pi}^{j+1} \|^2_{\mathbf{W}^{-1,2}_{\rm per}}\\
&+\| d_t\boldsymbol{\eta}^{j+1} \|^2_{\mathbf{W}^{-1,2}_{\rm per}}\} + \frac{\kappa}{2}\{\| \nabla\mathbf{e}^{j+1} \|^2_{\mathbf{L}^2} + \| \boldsymbol{\eta}^{j+1} \|^2_{\mathbf{W}^{-1,2}}\} \qquad (5.114)\\
&+\sigma \| \boldsymbol{\pi}^{j+1} \|^2_{\mathbf{W}^{-1,2}_{\rm per}} + \frac{h}{2} \| \boldsymbol{\eta}^{j+1} \|^2_{\mathbf{L}^2} + \delta k \| \boldsymbol{\pi}^{j+1} \|^2_{\mathbf{L}^2}\\
&-(\mathbf{l}_k(\tilde{\mathbf{m}}^j)\tilde{\mathbf{m}}^j - \mathbf{l}_k(\tilde{\mathbf{m}}_h^j)\tilde{\mathbf{m}}_h^j, \mathbf{P}^c_{h,\mathbf{L}^2}\mathbf{e}^{j+1}) - |(\mathcal{N}^{j+1}, \mathbf{P}^c_{h,\mathbf{L}^2}\mathbf{e}^{j+1})|
\end{aligned}
$$

$$
\begin{aligned}
&\leq C \,\| \boldsymbol{\eta}^{j+1} \|^2_{\mathbf{W}^{-1,2}_{\text{per}}} + Ch\{h \| \nabla\{\mathbf{H}^{j+1} - \mathbf{P}_{h,\mathbf{L}^2}\mathbf{H}^{j+1}\} \|_{\mathbf{L}^2} \| \boldsymbol{\eta}^{j+1} \|_{\mathbf{W}^{-1,2}_{\text{per}}} \\
&\quad + \| \mathbf{H}^{j+1} \|_{\mathbf{W}^{1,2}} \| \boldsymbol{\eta}^{j+1} \|_{\mathbf{W}^{-1,2}_{\text{per}}}\} + C \,, \| d_t \boldsymbol{\eta}^{j+1} \|_{\mathbf{W}^{-1,2}_{\text{per}}} \| \nabla[\tilde{\Delta}_{\mathcal{T}}^{-1} - \tilde{\Delta}^{-1}]\boldsymbol{\eta}^{j+1} \|_{\mathbf{L}^2} \\
&\quad + C \,\| d_t \boldsymbol{\pi}^{j+1} \|_{\mathbf{W}^{-1,2}_{\text{per}}} \| \nabla[\tilde{\Delta}_{\mathcal{T}}^{-1} - \tilde{\Delta}^{-1}]\boldsymbol{\pi}^{j+1} \|_{\mathbf{L}^2} \\
&\quad + \sigma \,\| \boldsymbol{\pi}^{j+1} \|_{\mathbf{W}^{-1,2}_{\text{per}}} \| \nabla[\tilde{\Delta}_{\mathcal{T}}^{-1} - \tilde{\Delta}^{-1}]\boldsymbol{\pi}^{j+1} \|_{\mathbf{L}^2} \\
&\quad + Ck \,\| d_t \boldsymbol{\eta}^{j+1} \|_{\mathbf{W}^{-1,2}_{\text{per}}} \| \boldsymbol{\pi}^{j+1} \|_{\mathbf{L}^2} + C \| \mathbf{H}^{j+1} - \mathbf{P}^c_{h,\mathbf{L}^2}\mathbf{H}^{j+1} \|^2_{\mathbf{L}^2} \\
&\quad + C \left\{ \| \mathbf{H}^j - \mathbf{P}^c_{h,\mathbf{L}^2}\mathbf{H}^j \|_{\mathbf{L}^2} + \| \mathbf{P}^c_{h,\mathbf{L}^2}\boldsymbol{\eta}^{j+1} \|_{\mathbf{L}^2} \right\} \| \nabla[\tilde{\Delta}^{-1} - \tilde{\Delta}_h^{-1}]\boldsymbol{\pi}^{j+1} \|_{\mathbf{L}^2} \\
&\quad + Ck \left\{ \| \mathbf{E}^{j+1} - \mathbf{P}^c_{h,\mathbf{L}^2}\mathbf{E}^{j+1} \|_{\mathbf{W}^{1,2}} \| \nabla \tilde{\Delta}_h^{-1} \boldsymbol{\pi}^{j+1} \|_{\mathbf{L}^2} \right. \\
&\quad + \frac{1}{2} d_t \| \mathbf{m}^{j+1} - \mathbf{P}^c_{h,\mathbf{L}^2}\mathbf{m}^{j+1} \|^2_{\mathbf{L}^2} + \frac{k}{2} \| d_t \mathbf{m}^{j+1} - \mathbf{P}^c_{h,\mathbf{L}^2} d_t \mathbf{m}^{j+1} \|^2_{\mathbf{L}^2} \\
&\quad + Ck \,\| \mathbf{E}^{j+1} - \mathbf{P}^c_{h,\mathbf{L}^2}\mathbf{E}^{j+1} \|_{\mathbf{L}^2} \| \boldsymbol{\eta}^{j+1} \|_{\mathbf{L}^2} + Ch \,\| \boldsymbol{\eta}^{j+1} \|_{\mathbf{L}^2} \| \boldsymbol{\pi}^{j+1} \|_{\mathbf{L}^2} \,.
\end{aligned}
$$

The crucial last term can be controlled by the left-hand side for $h = \mathcal{O}(k)$, and $\delta$ sufficiently large. As to the remaining two terms in (5.114) in front of the inequality sign, we may conclude as in (4.120)-(4.126). Further terms in $\boldsymbol{\mathcal{N}}^{j+1}$ look similar to (most of) those in (5.46), and do not cause further problems, thanks to the a priori bounds in Theorem 5.6; see also (5.50)-(5.51). We can now run an inductive argument again to verify the statement in the theorem. □

## 5.5 Computational Experiments

We compare results to (MLLG) for electrically conductive ferromagnets to those from (LLG), see Example 4.5, using $(\text{MLLG})^{P_2}_{k,h}$. For $\Omega \Supset \omega$, and $\partial_{\mathbf{n}} \tilde{\mathbf{m}}_h^{j+1} = 0$ on $\partial\omega$, the scheme reads:

1. For given $\mathbf{H}_h^j \in \boldsymbol{\mathcal{S}}^1(\omega)$, and all $\varphi_h \in \boldsymbol{\mathcal{S}}^1(\omega)$, compute $\tilde{\mathbf{m}}_h^{j+1} \in [\mathcal{S}^1(\omega)]^3$ from

$$
\begin{aligned}
\frac{1}{k}(\tilde{\mathbf{m}}_h^{j+1} - \mathbf{m}_h^j, \boldsymbol{\varphi})_h + \kappa\, (\nabla \tilde{\mathbf{m}}_h^{j+1}, \nabla \boldsymbol{\varphi}_h)_h &= \kappa\, (| \nabla \tilde{\mathbf{m}}_h^j |^2 \tilde{\mathbf{m}}_h^{j+1}, \boldsymbol{\varphi}_h)_h \\
-(\nabla[\boldsymbol{\varphi}_h \times \tilde{\mathbf{m}}_h^j], \nabla \tilde{\mathbf{m}}_h^{j+1})_h &- \Big(\mathbf{m}_h^j \times \{D\phi(\mathbf{m}_h^j) - \mathbf{H}_h^j - \mathbf{f}_h^{j+1}\}, \boldsymbol{\varphi}_h)_h \\
-\kappa\, (D\phi(\mathbf{m}_h^j) - \mathbf{H}_h^j - \mathbf{f}_h^{j+1}, \boldsymbol{\varphi}_h\Big)_h & \\
+\kappa \Big(\langle D\phi(\mathbf{m}_h^j) - \mathbf{H}_h^j - \mathbf{f}_h^{j+1}\rangle_{\mathbb{R}^3} \mathbf{m}_h^j, \boldsymbol{\varphi}_h\Big)_h \,. & \qquad (5.115)
\end{aligned}
$$

2. For given $\mathbf{E}_h^j, \mathbf{H}_h^j \in \boldsymbol{\mathcal{S}}_0^1(\Omega)$, and all $\boldsymbol{\psi}_h \in \boldsymbol{\mathcal{S}}_0^1(\Omega)$, compute $\mathbf{E}_h^{j+1} \in \boldsymbol{\mathcal{S}}_0^1(\Omega)$ from

$$\begin{aligned} \frac{1}{k}(\mathbf{E}_h^{j+1} - \mathbf{E}_h^j, \boldsymbol{\psi}_h)_h + \delta k\,(\nabla \mathbf{E}_h^{j+1}, \nabla \boldsymbol{\psi}_h)_h \\ + \sigma\,(\mathbf{E}_h^{j+1}, \boldsymbol{\psi}_h)_h = (\nabla \times \mathbf{H}_h^j, \boldsymbol{\psi}_h)_h\,. \end{aligned} \tag{5.116}$$

3. For given $\mathbf{H}_h^j \in \boldsymbol{\mathcal{S}}^1(\omega)$, $\tilde{\mathbf{m}}_h^{j+1}, \mathbf{m}^j \in [\mathcal{S}^1(\omega)]^3$, and all $\boldsymbol{\psi}_h \in \boldsymbol{\mathcal{S}}_0^1(\Omega)$, solve

$$\begin{aligned} \frac{1}{k}(\mathbf{H}_h^{j+1} - \mathbf{H}_h^j, \boldsymbol{\psi}_h)_h + h\,(\nabla \mathbf{H}_h^{j+1}, \nabla \boldsymbol{\psi}_h)_h \\ = -(\nabla \times \mathbf{E}_h^{j+1}, \boldsymbol{\psi}_h)_h + \frac{\beta}{k}(\tilde{\mathbf{m}}_h^{j+1} - \mathbf{m}_h^{j+1}, \boldsymbol{\psi}_h)_h\,, \\ \mathbf{m}_h^{j+1} = \frac{\tilde{\mathbf{m}}_h^{j+1}}{|\,\tilde{\mathbf{m}}_h^{j+1}\,|}\,. \end{aligned} \tag{5.117}$$

**Example 5.1** *Consider data from Example 4.5, page 194, and* $(\kappa, \delta, \beta, \sigma) = (1, 1, 1, 100)$, $\mathbf{E}_0 = 0$. *In order to ensure* $(5.11)_2$, *we start with* $\mathbf{H}_0(\mathbf{x}) = (\nabla u_h(\mathbf{x}), 0)^\top$, $u \in \mathcal{S}_0^1(\Omega)$, *and* $u$ *solves (4.19), for a right-hand side that uses* $\mathbf{m} = \mathbf{m}_0$. *Magnetizations* $(\tilde{m}_{h,1}^j, \tilde{m}_{h,2}^j)^\top$ *show pretty the same dynamics as computed by (LLG) in Example 4.5, but change at larger times, see Figure 5.1 in comparison to those simulations in Figure 4.1 on page 196. Corresponding electric fields* $(E_{h,1}^j, E_{h,2}^j)^\top\big|_\omega$ *are displayed in Figure 5.2, showing eddy-like structures that may be the reason for a different magnetization pattern of (MLLG).*

*This computational comparison of magnetization patterns evidences that (LLG) is not sufficient to explain (dynamics of) microstructures in electrically conducting ferromagnets, where (MLLG) is considered the more accurate model.*

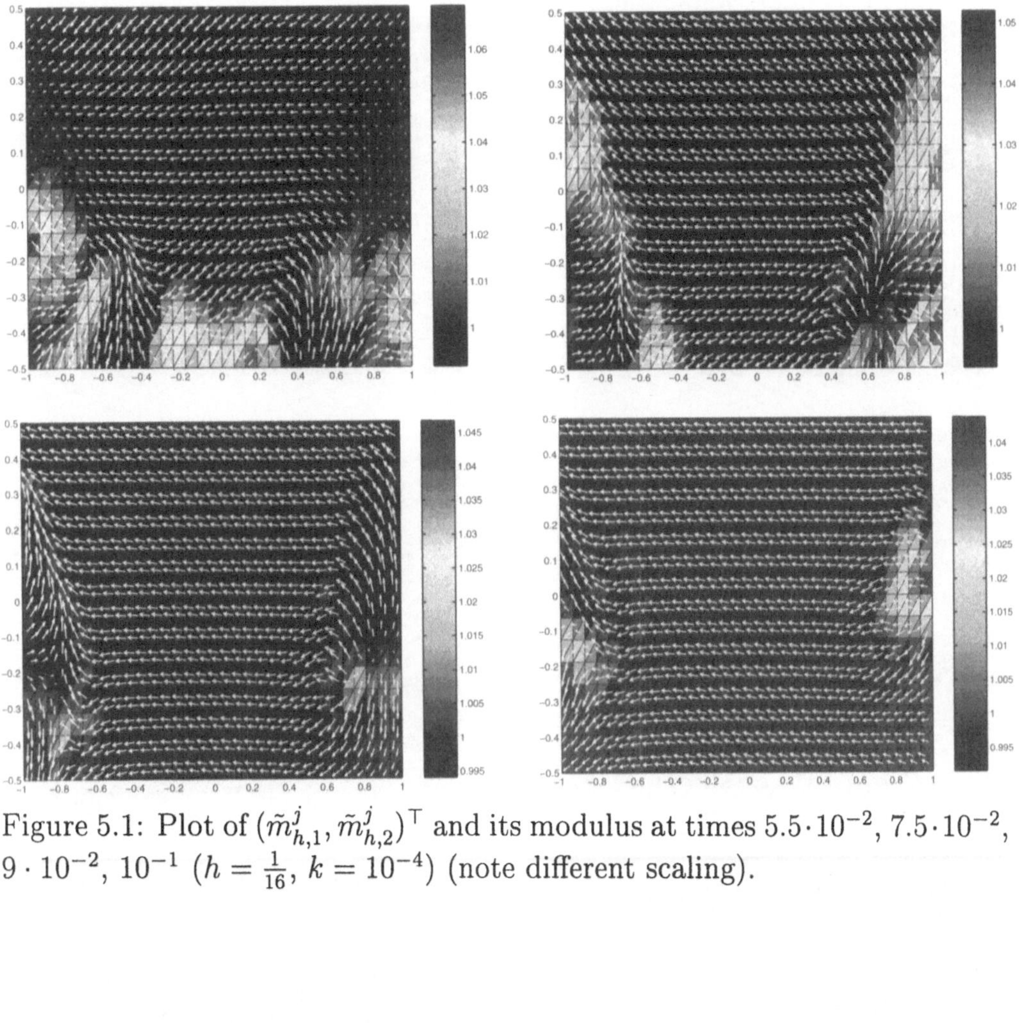

Figure 5.1: Plot of $(\tilde{m}^j_{h,1}, \tilde{m}^j_{h,2})^\top$ and its modulus at times $5.5 \cdot 10^{-2}$, $7.5 \cdot 10^{-2}$, $9 \cdot 10^{-2}$, $10^{-1}$ ($h = \frac{1}{16}$, $k = 10^{-4}$) (note different scaling).

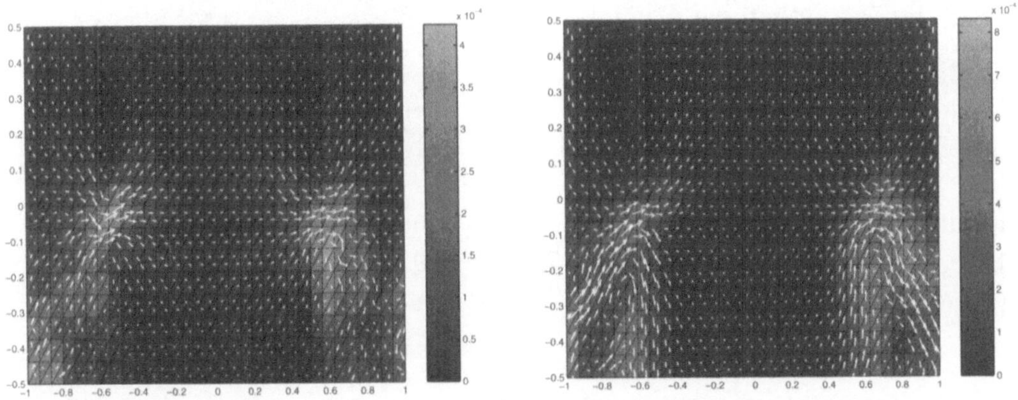

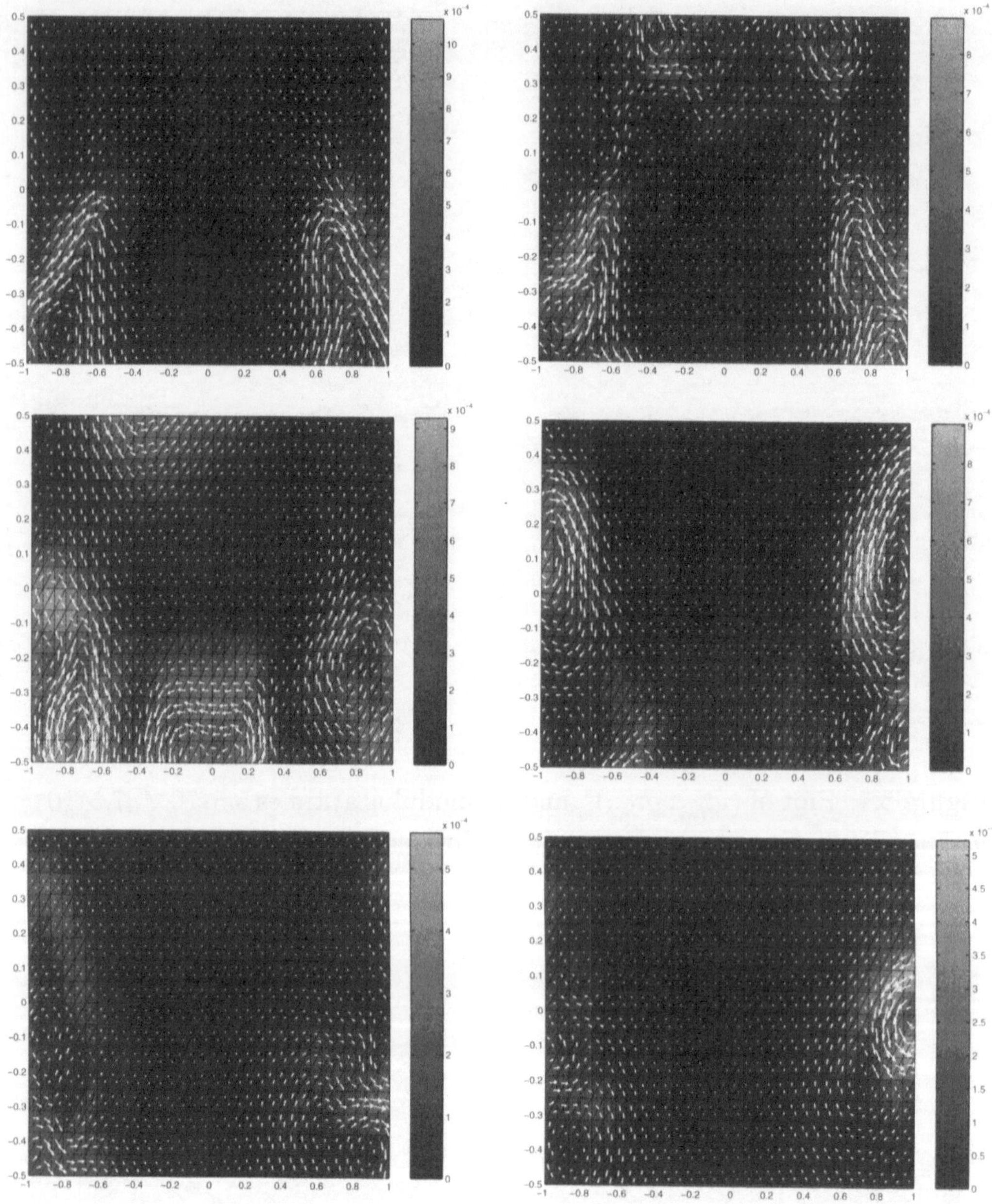

Figure 5.2: Plot of $(E^j_{h,1}, E^j_{h,2})^\top\big|_\omega$ (and its modulus) at times $t = 5 \cdot 10^{-3}$, $10^{-3}$, $1.5 \cdot 10^{-2}$, $2 \cdot 10^{-2}$, $5.5 \cdot 10^{-2}$, $7.5 \cdot 10^{-2}$, $9 \cdot 10^{-2}$, $10^{-1}$ ($h = \frac{1}{16}$, $k = 10^{-4}$) (note different scaling).

# Chapter 6

# Nematic Liquid Crystals

## 6.1 Introduction

The (simplified) Ericksen-Leslie equations are modified Navier-Stokes equations that take account of the liquid crystallinity. For $T > 0$, let $\mathbf{u} : \mathbb{R}^d \times \mathbb{R}^+ \supset \omega \times [0,T] \ni (\mathbf{x},t) \mapsto \mathbf{u}(\mathbf{x},t) \in \mathbb{R}^d$ and $\mathbf{d} : \mathbb{R}^d \times \mathbb{R}^+ \supset \omega \times [0,T] \ni (\mathbf{x},t) \mapsto \mathbf{d}(\mathbf{x},t) \in \mathbb{R}^d$ be the velocity of the liquid crystal flow, and the orientation of the liquid crystal molecules, resp., and $p : \mathbb{R}^d \times \mathbb{R}^+ \supset \omega \times [0,T] \ni (\mathbf{x},t) \mapsto p(\mathbf{x},t) \in \mathbb{R}$ denotes the pressure, for $d = 2, 3$. In this chapter, $\omega = (0, 2D)^2$, for $D > 0$. Let $\mathbf{u}_0, \mathbf{d}_0 \in \mathbf{W}^{2,2}_{\text{per}}(\omega, \mathbb{R}^2)$, and $\mathbf{u}_0 \in \mathbf{J}_1$. Then the equations (P) read, for a given force $\mathbf{f} \in W^{1,2}\big(I; \dot{\mathbf{W}}^{-1,2}_{\text{per}}(\omega, \mathbb{R}^2)\big)$,

$$\mathbf{u}_t - \nu \operatorname{div} \mathbf{D}(\mathbf{u}) + \mathbf{u} \cdot \nabla \mathbf{u} + \nabla p + \lambda \operatorname{div}(\nabla \mathbf{d} \odot \nabla \mathbf{d}) = \mathbf{f}\,, \tag{6.1}$$

$$\mathbf{d}_t + \mathbf{u} \cdot \nabla \mathbf{d} - \varrho \Delta \mathbf{d} = \varrho |\nabla \mathbf{d}|^2 \mathbf{d}\,, \tag{6.2}$$

$$\operatorname{div} \mathbf{u} = 0\,, \quad |\mathbf{d}| = 1\,, \tag{6.3}$$

$$\mathbf{u}\big|_{t=0} = \mathbf{u}_0 \in \mathbf{J}_1 \cap \dot{\mathbf{W}}^{2,2}_{\text{per}}\,, \quad \mathbf{d}\big|_{t=0} = \mathbf{d}_0 \in \mathbf{W}^{3,2}_{\text{per}}(\omega, \mathcal{S}^1)\,, \tag{6.4}$$

for space-periodic functions, $t \geq 0$,

$$\mathbf{u}(\mathbf{x} + 2D\mathbf{e}_i, t) = \mathbf{u}(\mathbf{x}, t)\,, \quad \mathbf{d}(\mathbf{x} + 2D\mathbf{e}_i, t) = \mathbf{d}(\mathbf{x}, t)\,, \tag{6.5}$$

where $\mathbf{x} + 2D\mathbf{e}_i = (x_1, ..., x_i + 2D, .., x_2)$, $i = 1, 2$. — Here, $\mathbf{D}(\mathbf{u}) = \frac{1}{2}\{\nabla \mathbf{u} + (\nabla \mathbf{u})^\top\}$ denotes the symmetric part of the velocity gradient, and $\nu, \lambda, \varrho$ are positive constants.

**Remark 6.1** *Consider* (6.1)-(6.4), *with* $|\mathbf{d}| = 1$ *deleted from* (6.3). *Suppose that the corresponding solution* $\{\mathbf{v}, p, \mathbf{m}\}$ *is smooth. Then* $|\mathbf{d}(\mathbf{x}, t| = 1$, *in*

$\omega \times [0,T]$. — *The verification of this observation uses an argument which is adapted from [60, 64]: We multiply* (6.2) *by* $\mathbf{d}$*, and set* $z(\mathbf{x},t) = |\,\mathbf{d}\,|^2$*. Because of*

$$z_t = 2\,\langle \mathbf{d}, \mathbf{d}_t\rangle_{\mathbb{R}^d}\,, \quad \nabla z = 2\,\langle \mathbf{d}, \nabla\mathbf{d}\rangle_{\mathbb{R}^d}\,, \quad \Delta z = 2\,\langle \mathbf{d}, \Delta\mathbf{m}\rangle_{\mathbb{R}^d} + 2\,|\,\nabla\mathbf{d}\,|^2\,,$$

*we have together with* $z(\mathbf{x},0) = 1$*, for all* $\mathbf{x} \in \omega$*,*

$$z_t + 2\,\langle \mathbf{v}\cdot\nabla\mathbf{d}, \mathbf{d}\rangle_{\mathbb{R}^d} - \varrho\,\Delta z = 2\varrho\,|\,\nabla\mathbf{d}\,|^2\big(|\,\mathbf{d}\,|^2 - 1\big)\,.$$

*Setting* $w = z - 1$*, we find* $w(\mathbf{x},0) = 0$*, and*

$$w_t - \varrho\,\Delta w + 2\,\langle \mathbf{v}\cdot\nabla\mathbf{d}, \mathbf{d}\rangle_{\mathbb{R}^d} = 2\varrho\,|\,\nabla\mathbf{d}\,|^2 w\,.$$

*We multiply by* $w$ *and integrate over* $\omega$*, together with* $\big(\mathbf{v}\cdot\nabla\mathbf{d}, \mathbf{d}\big) = 0$*, lead to*

$$\frac{1}{2}\frac{d}{dt}\int_\omega |\,w\,|^2\,d\mathbf{x} + \varrho\int_\omega |\,\nabla w\,|^2\,d\mathbf{x} \le 2\varrho\,\max_{\mathbf{x},t}|\,\nabla\mathbf{d}\,|^2\int_\omega |\,w\,|^2\,d\mathbf{x}\,.$$

*Application of Gronwall's Lemma then verifies the assertion.*

In literature, the constraint on the modulus of the director $\mathbf{d}$ is often replaced by a penalized formulation of (6.2) that exhibits potential structure, i.e., for a potential function $F_\varepsilon : \mathbf{W}^{1,2}(\omega,\mathbb{R}^2) \to \mathbb{R}$ we have

$$\mathbf{d}_{\varepsilon,t} + \mathbf{u}_\varepsilon\cdot\nabla\mathbf{d}_\varepsilon - \varrho\big(\Delta\mathbf{d}_\varepsilon - \mathbf{l}_\varepsilon(\mathbf{d}_\varepsilon)\big) = 0\,, \quad \text{where} \quad \mathbf{l}_\varepsilon(\mathbf{d}_\varepsilon) = \nabla F_\varepsilon(\mathbf{d}_\varepsilon)\,. \tag{6.6}$$

Typically, the Ginzburg-Landau approximation $\mathbf{l}_\varepsilon(\mathbf{d}_\varepsilon) = \frac{1}{\varepsilon}\big(|\,\mathbf{d}_\varepsilon\,|^2 - 1\big)\mathbf{d}_\varepsilon$ is used, for $\varepsilon > 0$ small. For this regularized version of (6.1)-(6.4), existence of weak solutions [89, 90] and local strong (even classical) solutions [91] is shown. In contrast, derivation of sharp a priori estimates independent of the penalty parameter for certain quantities of interest is a nontrivial matter. In addition, $\varepsilon$ has to be properly chosen in numerical modeling as regards the time-step $k$ and mesh-width $h$.

A first numerical analysis of (6.1), (6.6), (6.3), (6.4) (for $\omega \subset \mathbb{R}^3$ and Dirichlet boundary data) has been given in [93], where a stable Hermite-type discretization of the implicit Euler method is proposed; let $\boldsymbol{\mathcal{H}}_h := \boldsymbol{\mathcal{U}}_h \times \mathcal{P}_h \times \boldsymbol{\mathcal{R}}_h$ be a finite element subspace of $\boldsymbol{\mathcal{H}} := \mathbf{W}^{1,2}(\omega,\mathbb{R}^2) \times L^2(\omega)/\mathbb{R} \times$

$\mathbf{W}^{2,2}(\omega,\mathbb{R}^2)$, and $\mathbf{f}=0$. Given $\varepsilon>0$, then $\{\mathbf{u}_h^{j+1},\tilde{p}_h^{j+1},\mathbf{d}_h^{j+1}\}\in\boldsymbol{\mathcal{H}}_h$ solves for all $\{\boldsymbol{\varphi}_h,\chi_h,\boldsymbol{\phi}_h\}\in\boldsymbol{\mathcal{H}}_h$,

$$\begin{aligned}
&(d_t\mathbf{u}_h^{j+1},\boldsymbol{\varphi}_h)_h+\nu\,(\mathbf{D}(\mathbf{u}_h^{j+1}),\mathbf{D}(\boldsymbol{\varphi}_h))_h+(\mathbf{u}_h^{j+1}\cdot\nabla\mathbf{u}_h^{j+1},\boldsymbol{\varphi}_h)_h && (6.7)\\
&\quad-(\tilde{p}_h^{j+1},\operatorname{div}\boldsymbol{\varphi}_h)_h+\lambda\,(\Delta\mathbf{d}_h^{j+1},\nabla\mathbf{d}_h^{j+1}\boldsymbol{\varphi}_h)_h=0\,,\\
&(\operatorname{div}\mathbf{u}_h^{j+1},\chi_h)_h=0\,, && (6.8)\\
&(\nabla d_t\mathbf{d}_h^{j+1},\nabla\boldsymbol{\phi}_h)_h-(\Delta\boldsymbol{\phi}_h,\nabla\mathbf{d}_h^{j+1}\mathbf{u}_h^{j+1})_h && (6.9)\\
&\quad+(\varrho\,\Delta\mathbf{d}_h^{j+1}-\mathbf{l}_\varepsilon(\mathbf{d}_h^{j+1}),\Delta\boldsymbol{\phi}_h)_h=0\,.
\end{aligned}$$

Basically, the reason for using this conforming approach searching $\mathbf{d}_h^j\in\ell^2\big(I_k;\boldsymbol{\mathcal{R}}_h\big)\subset L^2\big(I;\mathbf{W}^{2,2}\big)$ is to mimic the following energy decay property for the original problem (6.1), (6.6), (6.3), (6.4) solved by $\{\mathbf{u}_\varepsilon,p_\varepsilon,\mathbf{d}_\varepsilon\}$,

$$\frac{dE}{dt}=-\Big(\nu\,\|\,\mathbf{D}(\mathbf{u}_\varepsilon)\,\|_{\mathbf{L}^2}^2+\lambda\varrho\,\|\,\Delta\mathbf{d}_\varepsilon-\mathbf{l}_\varepsilon(\mathbf{d}_\varepsilon)\,\|_{\mathbf{L}^2}^2\Big),\tag{6.10}$$

for

$$E=\frac{1}{2}\,\|\,\mathbf{u}_\varepsilon\,\|_{\mathbf{L}^2}^2+\frac{\lambda}{2}\,\|\,\nabla\mathbf{d}_\varepsilon\,\|_{\mathbf{L}^2}^2+\lambda\int_\omega F_\varepsilon(\mathbf{d}_\varepsilon)\,\mathrm{d}\mathbf{x}\,.\tag{6.11}$$

Alternatively, a mixed formulation of this problem using the additional function $\mathbf{W}=\nabla\mathbf{d}$ is proposed in [94]. We look for solutions $\{\mathbf{u},p,\mathbf{W},\mathbf{d}\}\in\boldsymbol{\mathcal{V}}\equiv\mathbf{W}^{1,2}_{\text{per}}(\omega,\mathbb{R}^2)\times L_0^2(\omega)\times\mathbf{H}(\omega;\operatorname{div})\times\mathbf{W}^{1,2}_{\text{per}}(\omega,\mathbb{R}^2)$, for almost every $t\in[0,T]$. A conforming finite element realization of Euler's time discretization scheme reads: Find a solution $\{\mathbf{u}_h^{j+1},\tilde{p}_h^{j+1},\mathbf{W}_h^{j+1},\mathbf{d}_h^{j+1}\}\in\boldsymbol{\mathcal{V}}_h\subset\boldsymbol{\mathcal{V}}$, such that for all $\{\boldsymbol{\varphi}_h,\chi_h,\mathbf{Z}_h,\boldsymbol{\phi}_h\}\in\boldsymbol{\mathcal{V}}_h$ holds

$$\begin{aligned}
&(d_t\mathbf{u}_h^{j+1},\boldsymbol{\varphi}_h)_h+\nu\,(\mathbf{D}(\mathbf{u}_h^{j+1}),\mathbf{D}(\boldsymbol{\varphi}_h))_h+(\mathbf{u}_h^{j+1}\cdot\nabla\mathbf{u}_h^{j+1},\boldsymbol{\varphi}_h)_h && (6.12)\\
&\quad-(\tilde{p}_h^{j+1},\operatorname{div}\boldsymbol{\varphi}_h)_h+\lambda\,(\operatorname{div}\mathbf{W}_h^{j+1},\nabla\mathbf{d}_h^{j+1}\boldsymbol{\varphi}_h)_h=0\,,\\
&(\operatorname{div}\mathbf{u}_h^{j+1},\chi_h)_h=0\,, && (6.13)\\
&(d_t\mathbf{W}_h^{j+1},\mathbf{Z}_h)_h-(\operatorname{div}\mathbf{Z}_h,\nabla\mathbf{d}_h^{j+1}\mathbf{u}_h^{j+1})_h && (6.14)\\
&\quad+\varrho\,(\operatorname{div}\mathbf{W}_h^{j+1}-\mathbf{l}_\varepsilon(\mathbf{d}_h^{j+1}),\operatorname{div}\mathbf{Z}_h)_h=0\,,\\
&(\nabla\mathbf{d}_h^{j+1},\nabla\boldsymbol{\phi}_h)_h-(\operatorname{div}\mathbf{W}_h^{j+1},\boldsymbol{\phi}_h)_h=0\,. && (6.15)
\end{aligned}$$

**Remark 6.2** *1. In [93, 94], the authors use the identity*

$$\operatorname{div}(\nabla\mathbf{d}\odot\nabla\mathbf{d})=\frac{1}{2}\nabla\langle\nabla\mathbf{d},\nabla\mathbf{d}\rangle_{\mathbb{R}^4}+(\nabla\mathbf{d})^\top\Delta\mathbf{d}\,.$$

*The first contribution is then gathered in* $\tilde{p}^{j+1}$ *together with the hydrostatic pressure* $p^{j+1}$.

2. $\mathbf{W}_h \subset \mathbf{H}(\omega; \mathrm{div})$ *is assembled from RT- or BDFM-elements, see [14].*

Optimal convergence behavior with respect to parameters $k, h$ for both methods is proven under the assumption of sufficient smoothness of solutions to (6.1), (6.6), (6.3), (6.4); however, in both works [93, 94], error analyses are presented for the *penalized* formulation of the problem, and given upper bounds for solutions to (6.1), (6.6), (6.3), (6.4) in strong norms show *exponential* increase in powers of $\varepsilon^{-1}$; it is argued there that the reason for this is mainly because of the discrete finite element spaces giving such a bound for $\{\mathbf{d}_h^j\}_{j=0}^J \in \ell^\infty(I_k; \mathcal{R}_h)$, jointly with the non-convex character of $\mathbf{l}_\varepsilon(\cdot)$, which requires a perturbation-type argument to deal with.

In this chapter, the key towards getting constants that are independent of $\varepsilon$ is its interpretation in terms of numerical parameters, i.e., $\varepsilon = \varepsilon(k, h)$. We address the following goals:

1. Analyze the **penalized implicit Euler scheme** in 2D for stable mixed finite element methods that satisfy the LBB-constraint (with respect to $\mathrm{div}\, \mathbf{u} = 0$), using in particular $\mathbf{d}_h^j|_{T\in\mathcal{T}_h} \in [\mathcal{S}^1(\omega)]^2$.

2. Construct an efficient **projection-based time-splitting scheme** to solve (6.1), (6.2), using element-wise affine finite elements for the unknowns $\mathbf{u}_h^j\big|_{K\in\mathcal{T}}, \mathbf{d}_h^j\big|_{K\in\mathcal{T}}, p_h^j\big|_{K\in\mathcal{T}}$, $0 \leq j \leq M$.

3. Verification of optimal convergence results in 2D in weaker norms compared to those in [93, 94] which apply to the general setting of regularities for all considered numerical schemes proposed here, involving constants that do *not* show exponential increase with respect to $\varepsilon^{-1}$; see, in particular, Remark 6.7, item 3. on page 256 and Example 6.2 on page 282 for sharpness of these results.

To reach 2., we propose a semidiscretization scheme in time $(\mathbf{P})_{k;\gamma}^P$, $\gamma \in \mathbb{N}_0$, that is based on a projection idea [28, 123, 43], for $\mathbf{u}^0 = \mathbf{u}_0$, $\mathbf{d}^0 = \tilde{\mathbf{d}}_0 = \mathbf{d}_0$. The $(j+1)$th iteration step reads as follows, for $j \geq 0$.

1. Given $\mathbf{u}^j, \mathbf{d}^j, \tilde{\mathbf{d}}^j$, compute $\tilde{\mathbf{d}}^{j+1} \in \mathbf{W}^{1,2}_{\mathrm{per}}(\omega, \mathbb{R}^2)$ from

$$\frac{1}{k}\{\tilde{\mathbf{d}}^{j+1} - \mathbf{d}^j\} + \mathbf{u}^j \cdot \nabla \tilde{\mathbf{d}}^{j+1} - \varrho \Delta \tilde{\mathbf{d}}^{j+1} = \varrho \,|\nabla \tilde{\mathbf{d}}^j|^2 \tilde{\mathbf{d}}^{j+1}. \tag{6.16}$$

2. Given $\mathbf{u}^j, \tilde{\mathbf{d}}^{j+1}$, determine $\tilde{\mathbf{u}}^{j+1} \in \dot{\mathbf{W}}^{1,2}_{\text{per}}(\omega, \mathbb{R}^2)$ from

$$\frac{1}{k}\{\tilde{\mathbf{u}}^{j+1} - \mathbf{u}^j\} - \nu\,\Delta\tilde{\mathbf{u}}^{j+1} + \mathbf{u}^j \cdot \nabla\tilde{\mathbf{u}}^{j+1} + \lambda\,\mathrm{div}\big(\nabla\tilde{\mathbf{d}}^{j+1} \odot \nabla\tilde{\mathbf{d}}^{j+1}\big) = \mathbf{f}^{j+1}. \tag{6.17}$$

3. Given $\tilde{\mathbf{u}}^{j+1}, \tilde{\mathbf{d}}^{j+1}$, compute $\{\mathbf{u}^{j+1}, \mathbf{d}^{j+1}\} \in \mathbf{J}_0(\omega) \times \mathbf{W}^{1,2}_{\text{per}}(\omega, \mathbb{R}^2)$ via

$$\frac{1}{k}\{\mathbf{u}^{j+1} - \tilde{\mathbf{u}}^{j+1}\} + \nabla p^{j+1} = 0, \quad \mathrm{div}\,\mathbf{u}^{j+1} = 0. \tag{6.18}$$

$$\mathbf{d}^{j+1} = \frac{\tilde{\mathbf{d}}^{j+1}}{|\,\tilde{\mathbf{d}}^{j+1}\,|^{2-\gamma}}, \quad \gamma \in \mathbb{N}_0. \tag{6.19}$$

For convergence analysis, this scheme can be reformulated as a semi-explicit penalized quasi-compressibility method:

$$\begin{aligned} d_t\tilde{\mathbf{d}}^{j+1} + \frac{1}{k}\{1 - \frac{1}{|\,\tilde{\mathbf{d}}^j\,|^{2-\gamma}}\}\tilde{\mathbf{d}}^j - \varrho\,\Delta\tilde{\mathbf{d}}^{j+1} & \qquad (6.20)\\ +\mathbf{P}_{\mathbf{J}_0}\tilde{\mathbf{u}}^j \cdot \nabla\tilde{\mathbf{d}}^{j+1} = \lambda\,|\,\nabla\tilde{\mathbf{d}}^{j+1}\,|^2\tilde{\mathbf{d}}^{j+1}, & \\ d_t\tilde{\mathbf{u}}^{j+1} - \nu\,\Delta\tilde{\mathbf{u}}^{j+1} + \mathbf{P}_{\mathbf{J}_0}\tilde{\mathbf{u}}^j \cdot \nabla\tilde{\mathbf{u}}^{j+1} + \nabla p^j & \qquad (6.21)\\ +\lambda\,\mathrm{div}\big(\nabla\tilde{\mathbf{d}}^{j+1} \odot \nabla\tilde{\mathbf{d}}^{j+1}\big) = \mathbf{f}^{j+1}, & \\ \mathrm{div}\,\tilde{\mathbf{u}}^{j+1} - k\Delta p^{j+1} = 0. & \qquad (6.22) \end{aligned}$$

Here, (6.20) is obtained if we insert (6.19) into (6.16), and (6.21) comes from substitution of (6.18) into (6.17). Note that the penalization parameter of the side constraint $|\,\mathbf{d}^j\,| = 1$ is $\varepsilon = k$, for a penalization function $\tilde{\mathbf{l}}_\varepsilon(\varphi) = \frac{1}{\varepsilon}\big(1 - \frac{1}{|\varphi|^{2-\gamma}}\big)\varphi$.

Numerical analysis for incompressible Navier-Stokes equations [68] is done in general under the following premises:

- *(A1), concerning regularity of the domain:* The unique solution $\mathbf{u} \in \mathbf{J}_1$ of the stationary, incompressible Stokes problem $\mathbf{A}\mathbf{u} = \mathbf{f}$ is already in $\mathbf{J}_1 \cap \mathbf{W}^{2,2}$, provided that $\mathbf{f} \in \mathbf{L}^2_0$, and satisfies $\|\,\mathbf{u}\,\|_{\mathbf{W}^{2,2}} \leq C\,\|\,\mathbf{A}\mathbf{u}\,\|_{\mathbf{L}^2}$.

- *(A2), concerning regularity of the given data:* Let $\mathbf{u}_0 \in \mathbf{J}_1 \cap \mathbf{W}^{2,2}$ and $\mathbf{f} \in W^{1,\infty}(I; \mathbf{L}^2_0)$.

Note that (A1) is satisfied for $\omega = [0, 2D]^2$. — The following result is verified in Section 6.4.

**Theorem 6.1** *Let $\{\mathbf{u},\mathbf{d},p\}$ solve (P) for $\omega = [0,2D]^2$, and $\{\mathbf{u}^j,\mathbf{d}^j,p^j\}_{j=0}^J$ solves $(\mathrm{P})_{k;\gamma}^P$, for $k \le k_0(t_J)$, $\gamma \in \mathbb{N}_0$, and $0 < t_J < T(\mathbf{u}_0,\mathbf{d}_0)$. Suppose that (A2) is satisfied. Then there exists a constant $C = C(\nu,\varrho,\gamma,\mathbf{f},\mathbf{u}_0,\mathbf{d}_0,\omega,t_J)$ independent of $k$ such that*

$$\max_{0\le j\le J}\Big\{\| \mathbf{u}(t_j) - \mathbf{u}^j \|_{\mathbf{W}^{-1,2}_{\mathrm{per}}} + \| \mathbf{d}(t_j) - \mathbf{d}^j \|_{\mathbf{L}^2}\Big\}$$

$$+\Big(k\sum_{j=0}^{J}\{\| \mathbf{u}(t_j) - \mathbf{u}^j \|^2_{\mathbf{L}^2} + \| \mathbf{d}(t_j) - \mathbf{d}^j \|^2_{\mathbf{W}^{1,2}}\} + \sqrt{k}\,\| p(t_j) - p^j \|_{L^2/\mathbb{R}}\Big)^{1/2}$$

$$+\Big(\sum_{j=0}^{J}\{\| \mathbf{d}(t_j) - \mathbf{d}^j \|^4_{\mathbf{L}^4} + \| \langle \mathbf{d}(t_j) - \mathbf{d}^j, \mathbf{d}(t_j)\rangle_{\mathbb{R}^2} \|^2_{L^2}\}\Big)^{1/2} \le C\,k\,.$$

**Remark 6.3** *1. In Section 6.3, we show corresponding error bounds for the implicit penalized Euler method*

$$\begin{aligned}
&d_t\mathbf{u}^{j+1} - \nu\,\Delta\mathbf{u}^{j+1} + \mathbf{u}^j\cdot\nabla\mathbf{u}^{j+1} + \lambda\,\mathrm{div}\big(\nabla\mathbf{d}^j \odot \mathbf{d}^{j+1}\big) + \nabla p^{j+1} = \mathbf{f}^{j+1}\,,\\
&d_t\mathbf{d}^{j+1} - \varrho\,\Delta\mathbf{d}^{j+1} + \mathbf{l}_\varepsilon(\mathbf{d}^{j+1}) + \mathbf{u}^j\cdot\nabla\mathbf{d}^{j+1} = \varrho\,\langle\nabla\mathbf{d}^j,\nabla\mathbf{d}^{j+1}\rangle_{\mathbb{R}^4}\mathbf{d}^{j+1}\,,\\
&\mathrm{div}\mathbf{u}^{j+1} = 0\,,\quad \mathbf{u}^0 = \mathbf{u}_0\,,\quad \mathbf{d}^0 = \mathbf{d}_0\,.
\end{aligned}$$

*2. Larger values $\gamma > 0$ enhance stability of the scheme, see Section 4.2.2 for a corresponding discussion. The analysis covers cases $\gamma \in \mathbb{N}_0$.*

*3. The smallness condition for the time-step ensures well-posedness of the algorithm $(\mathrm{P})_{k;\gamma}^P$.*

*4. As is well-known in CFD, Chorin's projection method suffers from boundary layers which limit the accuracy of the pressure close to the boundary, see [107, 108] and Section 6.5 for improved projection schemes that avoid this deficiency.*

When a mixed finite element method is used for the semidiscretization given in Remark 6.3, item 1., choices of pairings for discrete velocities and pressures need to be balanced to give a stable discretization. Actually, there are lots of stable (non-)conforming mixed methods available, see e.g. [14]. We abstract from special choices and require the following properties to be satisfied by the finite element pairing $\{\mathbf{V}_h, L^h\}$:

- *(B1), concerning continuity properties:* Each $\mathbf{v}_h \in \mathbf{V}_h$ satisfies

$$\int_\Gamma \{\mathbf{v}_h\big|_K - \mathbf{v}_h\big|_{K'}\}\,\mathrm{d}s = 0\,,$$

on inter-element faces $\Gamma = K \cap K'$, for $K, K' \in \mathcal{T}_h$.

- *(B2), concerning admissible class of triangulations $\mathcal{T}$: There exist interpolation operators $i_h : \mathbf{W}^{2,2} \to \mathbf{V}_h$, $j_h : L_0^2 \cap W^{1,2} \to L_h$ such that*

$$\text{(a)} \quad \| \mathbf{u} - i_h \mathbf{u} \|_{\mathbf{L}^2} + h \| \mathbf{u} - i_h \mathbf{u} \|_{\mathbf{W}^{1,2}} \leq C h^2 \| \Delta \mathbf{u} \|_{\mathbf{L}^2},$$
$$\text{(b)} \quad \| p - j_h p \|_{L^2} \leq C h \| \nabla p \|_{\mathbf{L}^2},$$

*with a constant $C = C(\mathcal{T})$ that is independent of $h$.*

- *(B3), concerning stability of the finite element pairing: The discrete LBB-condition is valid for the chosen pair of finite element spaces $\{\mathbf{V}_h, L^h\}$: For every $q_h \in L^h$ there exists a nontrivial function $\boldsymbol{\phi}_h \in \mathbf{V}_h$ such that*

$$|\left(q_h, \operatorname{div} \boldsymbol{\phi}_h\right)_h| \geq \kappa_0 \| \nabla \boldsymbol{\phi}_h \|_{\mathbf{L}^2} \| q_h \|_{\mathbf{L}^2},$$

*where $\kappa_0$ is a positive number that is independent from the mesh-size $h$.*

*(B1)* to *(B3)* provide the setup for a stable numerical discretization in space of the implicit Euler scheme $(\mathrm{P})_{k;h}^{E}$ that will be discussed in Section 6.3.

**Remark 6.4** *Here and in the sequel, we consider discretizations of $\Delta$ rather than* $\operatorname{div} \mathbf{D}(\mathbf{u})$ *in* (6.1)*; the second scenario may cause additional problems if non-conforming finite element realizations are used, due to violation of the discrete version of Korn's inequality. However, techniques are known to handle this deficiency, [79].*

Alternatively, numerical ingredients differ when $(\mathrm{P})_{k,h;\gamma}^{P}$ is considered, where (6.22) highlights the character of $(\mathrm{P})_{k}^{P}$ as a (semi-implicit) pressure stabilization method, which allows to use 'classically unstable finite element pairings' in a stable manner, [70]. We use pairings of piecewise affine, globally continuous finite element functions on a quasi-uniform mesh $\mathcal{T}$ for spatial discretization. The problem $(\mathrm{P})_{k,h;\gamma}^{P}$, $\gamma \in \mathbb{N}_0$ reads: Let $\{\mathbf{u}_h^j, \mathbf{d}_h^j\}_{j=0}^{J}$ be given, and $\{\mathbf{u}_h^0, \mathbf{d}_h^0\} = \{\mathbf{P}_{h;\mathbf{W}^{1,2}}\mathbf{u}_0, \mathbf{P}_{h;\mathbf{W}^{1,2}}\mathbf{d}_0\}$, for $\omega = [0, 2D]^2$. For $j \geq 0$, and all $\varphi_h, \boldsymbol{\phi}_h \in \left[\mathcal{S}^1(\omega)\right]^2$, and $\chi_h \in \mathcal{S}^1(\omega)/\mathbb{R}$:

1. Given $\mathbf{u}_h^j, \mathbf{d}_h^j, \tilde{\mathbf{d}}_h^j$, compute $\tilde{\mathbf{d}}_h^{j+1} \in \left[\mathcal{S}^1(\omega)\right]^2 \subset \mathbf{W}^{1,2}(\omega, \mathbb{R}^2)$ from

$$\begin{aligned} \frac{1}{k}\left(\tilde{\mathbf{d}}_h^{j+1} - \mathbf{d}_h^j, \varphi_h\right)_h + \varrho\left(\nabla \tilde{\mathbf{d}}_h^{j+1}, \nabla \varphi_h\right)_h& \\ +\left(\mathbf{u}_h^j \cdot \nabla \tilde{\mathbf{d}}_h^{j+1}, \varphi_h\right)_h = \varrho\left(|\nabla \tilde{\mathbf{d}}_h^j|^2 \tilde{\mathbf{d}}_h^{j+1}, \varphi_h\right)_h&. \end{aligned} \tag{6.23}$$

2. Given $\mathbf{u}_h^j, \tilde{\mathbf{d}}_h^{j+1}$, determine $\tilde{\mathbf{u}}_h^{j+1} \in \left[\mathcal{S}^1(\omega)\right]^2 \subset \mathbf{W}^{1,2}(\omega, \mathbb{R}^2)$ from

$$\frac{1}{k}\left(\tilde{\mathbf{u}}_h^{j+1} - \mathbf{u}_h^j, \boldsymbol{\phi}_h\right)_h + \nu\left(\nabla\tilde{\mathbf{u}}_h^{j+1}, \nabla\boldsymbol{\phi}_h\right)_h + \left(\mathbf{u}_h^j \cdot \nabla\tilde{\mathbf{u}}_h^{j+1}, \boldsymbol{\phi}_h\right)_h$$
$$-\lambda\left(\nabla\tilde{\mathbf{d}}_h^j \odot \nabla\tilde{\mathbf{d}}_h^j, \nabla\boldsymbol{\phi}_h\right)_h = \left(\mathbf{f}_h^{j+1}, \boldsymbol{\phi}_h\right)_h. \tag{6.24}$$

3. Given $\tilde{\mathbf{u}}_h^{j+1}, \tilde{\mathbf{d}}_h^{j+1}$, compute $\left\{\mathbf{u}_h^{j+1}, \mathbf{d}_h^{j+1}, p_h^{j+1}\right\} \in \left[\mathcal{S}^1(\omega)\right]^2 \times \mathcal{S}^1(\omega)/\mathbb{R}$ via

$$\left(\nabla p_h^{j+1}, \nabla\chi_h\right)_h = \frac{1}{k}\left(\tilde{\mathbf{u}}_h^{j+1}, \nabla\chi_h\right)_h, \tag{6.25}$$
$$\left(\mathbf{u}_h^{j+1}, \boldsymbol{\varphi}_h\right)_h = \left(\tilde{\mathbf{u}}_h^{j+1}, \boldsymbol{\varphi}_h\right)_h - k\left(\nabla p_h^{j+1}, \boldsymbol{\varphi}_h\right)_h, \tag{6.26}$$
$$\left(\mathbf{d}_h^{j+1}, \boldsymbol{\phi}_h\right)_h = \left(\frac{\tilde{\mathbf{d}}_h^{j+1}}{|\tilde{\mathbf{d}}_h^{j+1}|^{2-\gamma}}, \boldsymbol{\phi}_h\right)_h, \quad \gamma \in \mathbb{N}_0. \tag{6.27}$$

Section 6.4 gives the proof of the following result.

**Theorem 6.2** *Let $\{\mathbf{u}, \mathbf{d}, p\}$ solve* (P), *and $\{\mathbf{u}_h^j, \mathbf{d}_h^j, p_h^j\}_{j=0}^J$, solves* $(\mathrm{P})_{k,h;\gamma}^P$, *for $\gamma \in \mathbb{N}_0$, $0 < t_J < T(\mathbf{u}_0, \mathbf{d}_0)$. Let $\omega = [0, 2D]^2$, and $\{k, h\} \leq \{k_0(t_J), h_0(t_J)\}$. Suppose that the assumptions (A2),(B1),(B2),(B3) are satisfied. Then there exists a constant $C = C(\nu, \varrho, \gamma, \lambda, \mathbf{f}, \mathbf{u}_0, \mathbf{d}_0, \omega, t_J)$ that does not depend on $\{k, h\}$, such that*

$$\max_{0\leq j\leq M}\left\{\| \mathbf{A}^{-1/2}(\mathbf{u}(t_j) - \mathbf{u}_h^j) \|_{\mathbf{L}^2} + \| \mathbf{d}(t_j) - \mathbf{d}_h^j \|_{\mathbf{L}^2}\right\}$$
$$+\left(k\sum_{m=0}^{M}\left\{\| \mathbf{u}(t_j) - \mathbf{u}_h^j \|_{\mathbf{L}^2}^2 + \| \mathbf{d}(t_j) - \mathbf{d}_h^j \|_{\mathbf{W}^{1,2}}^2\right\} + \sqrt{k}\, \| p(t_j) - p_h^j \|_{L^2/\mathbb{R}}\right)^{1/2}$$
$$+\left(\sum_{j=0}^{J}\left\{\| \mathbf{d}(t_j) - \mathbf{d}_h^j \|_{\mathbf{L}^4}^4 + \| \langle \mathbf{d}(t_j) - \mathbf{d}_h^j, \mathbf{d}(t_j)\rangle_{\mathbb{R}^2} \|_{L^2}^2\right\}\right)^{1/2} \leq C\left\{k + h\right\}.$$

**Remark 6.5** *Corresponding results can be proved for a nonlinear variant of* (6.16), (6.17), *where $\varrho\left(|\nabla\tilde{\mathbf{d}}_h^j|^2\tilde{\mathbf{d}}_h^{j+1}, \boldsymbol{\varphi}_h\right)_h$ resp. $\lambda\left(\nabla\tilde{\mathbf{d}}_h^j \odot \nabla\tilde{\mathbf{d}}_h^j, \nabla\boldsymbol{\phi}_h\right)_h$ are replaced by $\varrho\left(|\nabla\tilde{\mathbf{d}}_h^{j+1}|^2\tilde{\mathbf{d}}_h^{j+1}, \boldsymbol{\varphi}_h\right)_h$ resp. $\lambda\left(\nabla\tilde{\mathbf{d}}_h^{j+1} \odot \nabla\tilde{\mathbf{d}}_h^{j+1}, \nabla\boldsymbol{\phi}_h\right)_h$.*

The remainder of this chapter is organized as follows. In Section 6.2, we show well-posedness of (P) and verify a priori bounds in strong norms. Section 6.3 deals with the penalized Euler method in semidiscretized and fully discrete form, i.e., $(\mathrm{P})_k^E$ resp. $(\mathrm{P})_{k;h}^E$. In Section 6.4, we analyze the projection-based schemes $(\mathrm{P})_k^P$ and $(\mathrm{P})_{k,h;\gamma}^P$. Computational experiments are reported in Section 6.5.

## 6.2 Analysis of (P)

In [89, 91], F. H. Lin & C. Liu show existence of global weak and local strong, classical solutions to (P) (for $\omega \subset \mathbb{R}^3$), where (6.2) is replaced by (6.6); however, a priori estimates for $\mathbf{d} \in L^2(I; \mathbf{W}^{2,2}_{\text{per}}(\omega, \mathbb{R}^2))$ depend on $\varepsilon$.

We deal with (P) directly in the setting of space-periodic situations in 2D; existence of local weak solutions follows from a combination of global and local (in space) energy estimates due to M. Struwe [120] and a fixed point argument.

We introduce the notion of weak solutions to (P).

**Definition 6.1** *Let* $\{\mathbf{u}_0, \mathbf{d}_0\} \in \mathbf{J}_1 \times \mathbf{W}^{1,2}_{\text{per}}(\omega, \mathbb{R}^2)$, $|\,\mathbf{d}_0\,| = 1$, *almost everywhere, and* $\mathbf{f} \in W^{1,\infty}(I; \mathbf{L}^2_0(\omega))$, *on* $\omega = (0, 2D)^2$. *We say that* $\{\mathbf{u}, \mathbf{d}, p\}$ *is a weak solution to* (6.1), (6.2) *if*

$$\begin{aligned}
&\mathbf{u} \in L^\infty\big(I; \mathbf{L}^2(\omega, \mathbb{R}^2)\big) \cap L^2\big(I; \mathbf{J}_1\big), \quad \mathbf{u}_t \in L^2\big(I; \mathbf{J}_1^\star(\omega)\big),\\
&\mathbf{d} \in L^\infty\big(I; \mathbf{W}^{1,2}_{\text{per}}(\omega, \mathcal{S}^1)\big) \cap L^2\big(I; \mathbf{W}^{2,2}_{\text{per}}((\omega, \mathcal{S}^1)\big), \quad \mathbf{d}_t \in L^2\big(I; \mathbf{L}^2(\omega, \mathbb{R}^2)\big),
\end{aligned}$$

*satisfies for every* $\{\boldsymbol{\phi}, \boldsymbol{\varphi}\} \in C^1\big(I; C^\infty_{\text{per}}(\omega, \mathbb{R}^2) \cap \mathbf{J}_1\big) \times C^1\big(I; C^\infty_{\text{per}}(\omega, \mathbb{R}^2)\big)$, *and* $\boldsymbol{\phi}(0) = \boldsymbol{\phi}(T) = 0$,

$$\begin{aligned}
&-\int_I \big(\mathbf{u}, \boldsymbol{\phi}_t\big)\,\mathrm{d}s + \nu \int_I \big(\nabla\mathbf{u}, \nabla\boldsymbol{\phi}\big)\,\mathrm{d}s + \int_I \big(\mathbf{u}\cdot\nabla\mathbf{u}, \boldsymbol{\phi}\big)\,\mathrm{d}s\\
&\quad = \lambda \int_I \big(\nabla\mathbf{d} \odot \nabla\mathbf{d}, \nabla\boldsymbol{\phi}\big)\,\mathrm{d}s + \int_I \big(\mathbf{f}, \boldsymbol{\phi}\big)\,\mathrm{d}s\,, \qquad (6.28)\\
&\int_I \big(\mathbf{d}_t, \boldsymbol{\varphi}\big)\,\mathrm{d}s + \varrho \int_I \big(\nabla\mathbf{d}, \nabla\boldsymbol{\varphi}\big)\,\mathrm{d}s + \int_I \big(\mathbf{u}\cdot\nabla\mathbf{d}, \boldsymbol{\varphi}\big)\,\mathrm{d}s \qquad (6.29)\\
&\quad = \varrho \int_I \big(|\,\nabla\mathbf{d}\,|^2\mathbf{d}, \boldsymbol{\varphi}\big)\,\mathrm{d}s\,,
\end{aligned}$$

*with* $\{\mathbf{u}(\mathbf{x}, 0), \mathbf{d}(\mathbf{x}, 0)\} = \{\mathbf{u}_0(\mathbf{x}), \mathbf{d}_0(\mathbf{x})\}$ *in the sense of traces, and for all* $T \geq 0$,

$$\begin{aligned}
&\|\,\mathbf{u}(T)\,\|^2_{\mathbf{L}^2} + \|\,\nabla\mathbf{d}(T)\,\|^2_{\mathbf{L}^2}\\
&\quad + \int_I \big\{\|\,\mathbf{u}_t\,\|^2_{\mathbf{W}^{-1,2}_{\text{per}}} + \|\,\nabla\mathbf{u}\,\|^2_{\mathbf{L}^2} + \|\,\mathbf{d}_t\,\|^2_{\mathbf{L}^2}\big\}\,\mathrm{d}s \leq C(\varrho, \lambda, \mathbf{f}, \mathbf{u}_0, \mathbf{d}_0)\,.
\end{aligned}$$

### 6.2.1 Existence of solutions to an auxiliary problem

Consider the following auxiliary problem on $\omega = [0, 2D]^2$: given $\{\mathbf{U}, \mathbf{f}\} \in L^\infty(I; \mathbf{J}_1 \cap \mathbf{W}^{2,2}_{\text{per}}) \cap W^{1,\infty}(I; \mathbf{L}^2_0) \cap W^{1,2}(I; \mathbf{W}^{1,2}_{\text{per}}) \times W^{1,2}(I; \mathbf{L}^2_0)$, such that $\mathbf{U}(0) = \mathbf{u}_0$, find weak solutions $\{\mathbf{u}, \mathbf{d}\}$ to

$$\mathbf{u}_t - \nu\,\Delta\mathbf{u} + \mathbf{u}\cdot\nabla\mathbf{u} + \nabla p + \lambda\,\mathrm{div}(\nabla\mathbf{d}\odot\nabla\mathbf{d}) = \mathbf{f}\,, \tag{6.30}$$

$$\mathbf{d}_t - \varrho\,\Delta\mathbf{d} + \mathbf{U}\cdot\nabla\mathbf{d} = \varrho\,|\,\nabla\mathbf{d}\,|^2\mathbf{d}\,, \tag{6.31}$$

$$\mathrm{div}\,\mathbf{u} = 0\,, \quad |\,\mathbf{d}\,| = 1\,, \tag{6.32}$$

$$\mathbf{u}(0) = \mathbf{u}_0 \in \mathbf{J}_1\,, \quad \mathbf{d}(0) = \mathbf{d}_0 \in \mathbf{W}^{1,2}(\omega, \mathcal{S}^1)\,. \tag{6.33}$$

The notion of weak solutions to (6.30)-(6.33) is immediate from Definition 6.1. In the sequel, we denote

$$E(\mathbf{U}; \varphi, \omega) = \int_\omega |\,\nabla\varphi\,|^2\,\mathrm{d}\mathbf{x}\,.$$

**Lemma 6.1** *Let $\{\mathbf{d}, \mathbf{u}\}$ be a weak solution to* (6.30)-(6.33). *There holds*

$$\int_I \|\,\mathbf{d}_t(s)\,\|^2_{\mathbf{L}^2}\,\mathrm{d}s + E(\mathbf{U}; \mathbf{d}(T), \omega) \le E(\mathbf{U}; \mathbf{d}(0), \omega)\,.$$

**Proof:**
Test (6.31) by $\mathbf{d}_t$, exploit $d_t|\,\mathbf{d}\,|^2 = 0$ almost everywhere, and use Gronwall's inequality for

$$\|\,d_t\mathbf{d}\,\|^2_{\mathbf{L}^2} + \frac{\varrho}{2}d_t\|\,\nabla\mathbf{d}\,\|^2_{\mathbf{L}^2} \le \|\,\mathbf{U}\,\|^2_{\mathbf{L}^\infty}\,\|\,\nabla\mathbf{d}\,\|^2_{\mathbf{L}^2} + \frac{1}{2}\,\|\,d_t\mathbf{d}\,\|^2_{\mathbf{L}^2}\,.$$

□

The following result reports on evolution of local energy portions in space-time.

**Lemma 6.2** *Let $\{\mathbf{d}, \mathbf{u}\}$ be a weak solution to* (6.30) *to* (6.33). *Let $\mathbf{x}_0 \in \omega$, and $B_R(\mathbf{x}_0)$ be a ball centered at $\mathbf{x}_0$, for $0 < R \le 2R_0$. There holds, for $R \le R_0$,*

$$E\big(\mathbf{U}; \mathbf{d}(T), B_R(\mathbf{x}_0)\big) \le 2(1+\varrho^2)E\big(\mathbf{U}; \mathbf{d}_0, B_{2R}(\mathbf{x}_0)\big) + C\frac{T}{R^2}E(\mathbf{U}; \mathbf{d}_0, \omega)\,.$$

**Proof:**
Let $\phi \in C_0^\infty\big(B_{2R}(\mathbf{x}_0)\big)$ satisfy $0 \le \phi \le 1$, $\phi \equiv 1$ on $B_r(\mathbf{x}_0)$, with $|\nabla \phi| \le \frac{2}{R}$, and test (6.31) with $\mathbf{d}_t\phi^2$. We find

$$\| \phi \mathbf{d}_t \|_{\mathbf{L}^2}^2 + \frac{\varrho}{2} d_t \| \phi \nabla \mathbf{d} \|_{\mathbf{L}^2}^2 \le 2\varrho \int_\omega |\nabla \mathbf{d}| |\mathbf{d}_t| |\nabla \phi| |\phi| \, \mathrm{d}\mathbf{x} + \| \phi \mathbf{U} \|_{\mathbf{L}^\infty} \| \phi \nabla \mathbf{d} \|_{\mathbf{L}^2} \| \phi \mathbf{d}_t \|_{\mathbf{L}^2}.$$

We employ Young's inequality to the first term on the right-hand side, together with Lemma 6.1. Gronwall's lemma serves to control the second term then. □

Similar to how we concluded from Lemma 4.7 in Chapter 4, we draw the following conclusion from this lemma: given $\varepsilon_1 > 0$, $\mathbf{d}_0 \in \mathbf{W}^{1,2}_{\text{per}}(\omega, \mathcal{S}^1)$, there exists a number $T_1 > 0$ depending only on a maximal number $R_1 > 0$ and $\alpha > 0$, such that

$$\sup_{\mathbf{x}_0 \in \omega} E\big(\mathbf{U}; \mathbf{d}_0, B_{2R_1}(\mathbf{x}_0)\big) < \varepsilon_1 ,$$

with the property that any smooth solution $\mathbf{d}$ to (6.30)-(6.33) satisfies

$$\sup_{\mathbf{x}_0 \in \omega,\, 0 \le t \le T_1} E\big(\mathbf{U}; \mathbf{d}(t), B_{R_1}(\mathbf{x}_0)\big) < 2\varepsilon_1 .$$

**Lemma 6.3** *Problem* (6.30) *to* (6.33) *has a local weak solution.*

**Proof:**
The subsequent analyses are formal and can be made rigorous on the level of Galerkin method, using eigenfunctions $\{\mathbf{w}_j\}_{j=1}^\infty$ resp. $\{r_j\}_{j=1}^\infty$ of $\mathbf{A}$ resp. $\Delta$: we look for approximate solutions $\mathbf{u}_m = \sum_{j=1}^m \alpha_{jm}(t)\mathbf{w}_j$ and $\mathbf{d}_m = \sum_{j=1}^m \beta_{jm}(t)\mathbf{r}_j$, such that the identities (6.28), (6.29) are satisfied on $\mathbf{W}^m := \{\mathbf{w}_1, .., \mathbf{w}_m\}$ resp. $\mathbf{R}^m := \{\mathbf{r}_1, .., \mathbf{r}_m\}$. Existence of solutions to this ordinary differential equation is then a consequence of (6.34).

(a) We multiply (6.30), (6.31) by $\{\mathbf{u}, -\Delta\mathbf{d}\}$. Rearranging terms leads to

$$\begin{aligned} &\frac{1}{2} d_t \{ \| \mathbf{u} \|_{\mathbf{L}^2}^2 + \| \nabla \mathbf{d} \|_{\mathbf{L}^2}^2 \} + \{ \nu \| \nabla \mathbf{u} \|_{\mathbf{L}^2}^2 + \varrho \| \Delta \mathbf{d} \|_{\mathbf{L}^2}^2 \} \\ &\quad = \lambda \big( \nabla \mathbf{d} \odot \nabla \mathbf{d}, \nabla \mathbf{u} \big) + (\mathbf{U} \cdot \nabla \mathbf{d}, \Delta \mathbf{d}) - \varrho \, (|\nabla \mathbf{d}|^2, \Delta \mathbf{d}) \qquad (6.34) \\ &\quad \le \Big\{ \frac{\lambda}{\nu} + \frac{1}{\varrho} + \varrho \Big\} \{ \| \nabla \mathbf{d} \|_{\mathbf{L}^4}^4 + \| \mathbf{U} \|_{\mathbf{L}^4}^4 \} + \frac{1}{4} \{ \nu \| \nabla \mathbf{u} \|_{\mathbf{L}^2}^2 + \varrho \| \Delta \mathbf{d} \|_{\mathbf{L}^2}^2 \} . \end{aligned}$$

We exploit Lemmata 6.2, 4.6 to control the first term on the right-hand side: Let $\varepsilon_1 > 0$ be given and $R_1 > 0$ be determined as above and $\{\phi_i\}$ be a set of smooth cut-off functions subordinate to a cover of $\omega$ by balls $\{B_{2R_1}(\mathbf{x}_i)\}$ with finite overlap and such that $0 \leq \phi_i \leq 1$, such that $|\nabla\phi| \leq \frac{2}{R_1}$, and $\sum_i \phi_i^2 = 1$. Then, we infer

$$\begin{aligned}\|\nabla\mathbf{d}\|_{\mathbf{L}^4}^4 &= \sum_i \int_\omega |\nabla\mathbf{d}|^4\phi_i^2\,\mathrm{dx} \\ &\leq C\sup_i E\big(\mathbf{U};\mathbf{d}(t),B_{2R_1}(\mathbf{x}_i)\big)\Big(\int_\omega |\nabla^2\mathbf{d}|^2\,\mathrm{dx} + \frac{1}{R_1^2}E(\mathbf{d}_0)\Big) \\ &\leq C\varepsilon_1\Big(\int_\omega |\nabla^2\mathbf{d}|^2\,\mathrm{dx} + \frac{1}{R_1^2}E(\mathbf{d}_0)\Big)\,.\end{aligned} \tag{6.35}$$

Consequently, the critical term in (6.34) can be controlled for small times, i.e., $T_1 > 0$ sufficiently small.

(b) We test (6.30) by $\mathbf{A}^{-1}\mathbf{u}_t$. Thanks to $(\mathbf{u}\cdot\nabla\varphi,\boldsymbol{\phi}) = -(\mathbf{u}\cdot\nabla\boldsymbol{\phi},\varphi)$, we find

$$\frac{1}{2}d_t\|\,d_t\mathbf{u}\,\|_{\dot{\mathbf{W}}_{\mathrm{per}}^{-1,2}}^2 + \frac{\nu}{2}d_t\|\,\mathbf{u}\,\|_{\mathbf{L}^2}^2 \leq C\left(\|\nabla\mathbf{d}\|_{\mathbf{L}^4}^4 + \|\,\mathbf{u}\,\|_{\mathbf{L}^4}^4\right) + \frac{1}{2}\|\,\mathbf{u}_t\,\|_{\dot{\mathbf{W}}_{\mathrm{per}}^{-1,2}}^2\,. \tag{6.36}$$

Hence, we obtain an upper bound $\|\,\mathbf{u}_t\,\|_{L^2(I;\dot{\mathbf{W}}_{\mathrm{per}}^{-1,2})} \leq C$ from (a), and Gronwall's inequality.

(c) From the uniform estimates in (a), (b) (valid on the level of Galerkin discretization) we may take a (relabeled) subsequence $\{\mathbf{u}_m,\mathbf{d}_m\}$, for $I = [0,T^\star]$ sufficiently small. Due to the compactness theorem of J. Lions, cf. [125], we may conclude

1. $\mathbf{u}_m \rightharpoonup \mathbf{u}$ in $L^2(I;\mathbf{J}_1)$, $\mathbf{u}_m \overset{\star}{\rightharpoonup} \mathbf{u}$ in $L^\infty(I;\mathbf{J}_0)$, and $\mathbf{u}_{m,t} \rightharpoonup \mathbf{u}_t$ in $L^2(I;\mathbf{J}_1^*)$,
2. $\mathbf{d}_m \rightharpoonup \mathbf{d}$ in $L^2(I;\mathbf{W}_{\mathrm{per}}^{2,2})$, $\mathbf{d}_m \overset{\star}{\rightharpoonup} \mathbf{d}$ in $L^\infty(I;\mathbf{W}_{\mathrm{per}}^{1,2})$, and $\mathbf{d}_{m,t} \rightharpoonup \mathbf{d}_t$ in $L^2(I;\mathbf{L}_0^2)$,
3. $\mathbf{u}_m \to \mathbf{u}$ in $L^2(I;\mathbf{J}_0)$, and $\mathbf{d}_m \to \mathbf{d}$ in $L^2(I;\mathbf{W}_{\mathrm{per}}^{1,2})$.

It is most crucial to check convergence of the nonlinear terms, i.e., for $m \to \infty$,

$$\int_I (\mathbf{u}_m \cdot \nabla \mathbf{u}_m, \boldsymbol{\varphi}_m)\, \mathrm{d}s \to \int_I (\mathbf{u} \cdot \nabla \mathbf{u}, \boldsymbol{\varphi})\, \mathrm{d}s\,, \tag{6.37}$$

$$\int_I (\nabla \mathbf{d}_m \odot \nabla \mathbf{d}_m, \nabla \boldsymbol{\phi}_m)\, \mathrm{d}s \to \int_I (\nabla \mathbf{d} \odot \nabla \mathbf{d}, \nabla \boldsymbol{\phi})\, \mathrm{d}s\,, \tag{6.38}$$

$$\int_I (|\nabla \mathbf{d}_m|^2 \mathbf{d}_m, \boldsymbol{\varphi}_m)\, \mathrm{d}s \to \int_I (|\nabla \mathbf{d}|^2 \mathbf{d}, \boldsymbol{\varphi})\, \mathrm{d}s\,. \tag{6.39}$$

where $\{\boldsymbol{\phi}_m, \boldsymbol{\varphi}_m\} \to \{\boldsymbol{\phi}, \boldsymbol{\varphi}\}$ uniformly in $C^1(\omega_T)$. — We start with (6.37).

$$\begin{aligned}\int_I (\mathbf{u}_m \cdot \nabla \mathbf{u}_m, \boldsymbol{\varphi}_m)\, \mathrm{d}s &= -\int_I (\mathbf{u}_m \cdot \nabla [\boldsymbol{\varphi}_m - \boldsymbol{\varphi}], \mathbf{u}_m)\, \mathrm{d}s \\ -\int_I ([\mathbf{u}_m - \mathbf{u}] \cdot \nabla \boldsymbol{\varphi}, \mathbf{u}_m)\, \mathrm{d}s &- \int_I (\mathbf{u} \cdot \nabla \boldsymbol{\varphi}, [\mathbf{u}_m - \mathbf{u}])\, \mathrm{d}s - \int_I (\mathbf{u} \cdot \nabla \boldsymbol{\varphi}, \mathbf{u})\, \mathrm{d}s\,,\end{aligned} \tag{6.40}$$

We confine to considering the second term on the right-hand side.

$$\begin{aligned} II &\le \int_I \| \mathbf{u}_m - \mathbf{u} \|_{\mathbf{L}^4} \| \mathbf{u}_m \|_{\mathbf{L}^4} \| \nabla \boldsymbol{\varphi} \|_{\mathbf{L}^2}\, \mathrm{d}s \\ &\le C \int_I \| \mathbf{u}_m - \mathbf{u} \|_{\mathbf{L}^2}^{1/2} \| \mathbf{u}_m - \mathbf{u} \|_{\mathbf{W}^{1,2}}^{1/2} \| \nabla \mathbf{u}_m \|_{\mathbf{L}^2}^{1/2}\, \mathrm{d}s, \\ &\le C \Big( \int_I \| \mathbf{u}_m - \mathbf{u} \|_{\mathbf{L}^2}\, \mathrm{d}s \Big)^{1/2} \Big( \int_I \| \mathbf{u}_m - \mathbf{u} \|_{\mathbf{W}^{1,2}} \| \nabla \mathbf{u}_m \|_{\mathbf{L}^2} \mathrm{d}s \Big)^{1/2} \to 0\,, \end{aligned} \tag{6.41}$$

thanks to interpolation of $\mathbf{L}^4$ between $\mathbf{L}^2$ and $\mathbf{W}^{1,2}$. We consider (6.38).

$$\begin{aligned} \int_I (\nabla \mathbf{d}_m \odot \nabla \mathbf{d}_m, \nabla \boldsymbol{\phi}_m)\, \mathrm{d}s &= \int_I (\nabla [\mathbf{d}_m - \mathbf{d}] \odot \nabla d_m, \nabla \boldsymbol{\phi}_m)\, \mathrm{d}s \\ + \int_I (\nabla \mathbf{d} \odot \nabla [\mathbf{d}_m - \mathbf{d}], \nabla \boldsymbol{\phi}_m)\, \mathrm{d}s &+ \int_I (\nabla \mathbf{d} \odot \nabla \mathbf{d}, \nabla [\boldsymbol{\phi}_m - \boldsymbol{\phi}])\, \mathrm{d}s \\ + \int_I (\nabla \mathbf{d} \odot \nabla \mathbf{d}, \nabla \boldsymbol{\phi})\, \mathrm{d}s\,. \end{aligned} \tag{6.42}$$

We consider the first term on the right-hand side; the other terms can be dealt with similarly.

$$\begin{aligned} I &\le \int_I \| \nabla[\mathbf{d}_m - \mathbf{d}] \|_{\mathbf{L}^4} \| \nabla \mathbf{d}_m \|_{\mathbf{L}^4} \| \nabla \phi_m \|_{\mathbf{L}^2} \, ds \\ &\le C \int_I \| \nabla[\mathbf{d}_m - \mathbf{d}] \|_{\mathbf{L}^4} \| \mathbf{d}_m \|_{\mathbf{W}^{2,2}}^{1/2} \, ds \\ &\le C \Big( \int_I \| \nabla[\mathbf{d}_m - \mathbf{d}] \|_{\mathbf{L}^2}^2 \, ds \Big)^{1/2} \Big( \int_I \| \mathbf{d}_m - \mathbf{d} \|_{\mathbf{W}^{2,2}}^2 \, ds \Big)^{1/2} \to 0 . \end{aligned} \tag{6.43}$$

Finally, we consider (6.39).

$$\begin{aligned} \int_I (| \nabla \mathbf{d}_m |^2 \mathbf{d}_m, \varphi_m) \, ds &= \int_I (\langle \nabla \mathbf{d}_m, \nabla[\mathbf{d}_m - \mathbf{d}] \rangle_{\mathbb{R}^4} \mathbf{d}_m, \varphi_m) \, ds \\ &+ \int_I (\langle \nabla[\mathbf{d}_m - \mathbf{d}], \nabla \mathbf{d} \rangle_{\mathbb{R}^4} \mathbf{d}_m, \varphi_m) \, ds + \int_I (| \nabla \mathbf{d} |^2 [\mathbf{d}_m - \mathbf{d}], \varphi_m) \, ds \\ &+ \int_I (| \nabla \mathbf{d} |^2 \mathbf{d}, \varphi_m) \, ds . \end{aligned} \tag{6.44}$$

We confine to an analysis of the first and third term on the right-hand side.

$$\begin{aligned} I &\le \Big( \int_I \| \nabla \mathbf{d}_m \|_{\mathbf{L}^2}^2 \| \mathbf{d}_m \|_{\mathbf{L}^\infty}^2 \, ds \Big)^{1/2} \Big( \int_I \| \nabla[\mathbf{d} - \mathbf{d}_m] \|_{\mathbf{L}^2}^2 \, ds \Big)^{1/2} \to 0 , \\ III &\le \int_I \| \nabla \mathbf{d} \|_{\mathbf{L}^2}^2 \| \mathbf{d}_m - \mathbf{d} \|_{\mathbf{L}^\infty} \| \varphi_m \|_{\mathbf{L}^2} \, ds \\ &\le C \int_I \| \nabla[\mathbf{d}_m - \mathbf{d}] \|_{\mathbf{W}^{1,q}} \, ds \to 0 \quad (q > 2) . \end{aligned}$$

This proves Lemma 6.3. □

We prove further a priori statements for the solution to (6.30)-(6.33). Arguments are again formal and can be repeated for Faedo-Galerkin approximates to make them rigorous.

**Lemma 6.4** *Let* $\{\mathbf{u}, \mathbf{d}, p\}$ *solve* (6.30)-(6.33), *for* $\omega = [0, 2D]^2$, $\mathbf{u}_0 \in \mathbf{J}_1 \cap \dot{\mathbf{W}}^{2,2}_{\text{per}}$, $\mathbf{d}_0 \in \mathbf{W}^{3,2}_{\text{per}}$, *such that* $|\,\mathbf{d}\,| = 1$, $\forall \mathbf{x} \in \omega$, $\mathbf{f} \in W^{1,\infty}(I; \mathbf{L}^2_0)$, $q > 2$, *and* $0 < t_J < T$. *Suppose*

$$\mathbf{U} \in L^\infty(I; \mathbf{J}_1 \cap \dot{\mathbf{W}}^{2,2}_{\text{per}}) \cap W^{1,\infty}(I; \mathbf{J}_0) \cap W^{1,2}(I; \mathbf{J}_1) .$$

*There exist constants* $\tilde{C} = \tilde{C}(\mathbf{u}_0, \mathbf{d}_0, \mathbf{f}, \omega, T; \mathbf{U})$, $\tilde{C}_q = \tilde{C}_q(\mathbf{u}_0, \mathbf{d}_0, \mathbf{f}, \omega, T; \mathbf{U}, q)$, *for* $2 < q < \infty$ *such that*

$$\text{(a)}\quad \operatorname*{ess\,sup}_{[0,T]}\{\| \mathbf{u} \|_{\mathbf{L}^2} + \| \mathbf{d} \|_{\mathbf{W}^{1,2}}\} + \Big(\int_I \{\| \nabla \mathbf{u}(s) \|^2_{\mathbf{L}^2} + \| \mathbf{d}_t(s) \|^2_{\mathbf{L}^2} + \| \Delta \mathbf{d}(s) \|^2_{\mathbf{L}^2}\}\,\mathrm{d}s\Big)^{1/2} \le \tilde{C},$$

$$\text{(b)}\quad \operatorname*{ess\,sup}_{[0,T]}\{\| \mathbf{u}_t \|_{\mathbf{L}^2} + \| \nabla \mathbf{d}_t \|_{\mathbf{L}^2} + \| \nabla \mathbf{u} \|_{\mathbf{L}^2} + \| \Delta \mathbf{d} \|_{\mathbf{L}^2}\} + \Big(\int_I \{\| \nabla \mathbf{u}_t(s) \|^2_{\mathbf{L}^2} + \| \Delta \mathbf{d}_t(s) \|^2_{\mathbf{L}^2}\}\,\mathrm{d}s\Big)^{1/2} \le \tilde{C},$$

$$\text{(c)}\quad \Big(\int_I \{\| \mathbf{u}_{tt}(s) \|^2_{\dot{\mathbf{W}}^{-1,2}_{\text{per}}} + \| \mathbf{d}_{tt}(s) \|^2_{\mathbf{L}^2} + \tau(s)\, \| \mathbf{u}_{tt}(s) \|^2_{\mathbf{L}^2}\}\,\mathrm{d}s\Big)^{1/2} \le \tilde{C},$$

$$\text{(d)}\quad \operatorname*{ess\,sup}_{[0,T]}\{\| \Delta \mathbf{d} \|_{\mathbf{L}^q} + \| \Delta \mathbf{u} \|_{\mathbf{L}^2} + \| \nabla p \|_{\mathbf{L}^2}\} + \int_I \{\| \Delta \mathbf{u}(s) \|^2_{\mathbf{L}^q} + \| \nabla p(s) \|^2_{\mathbf{L}^q}\}\,\mathrm{d}s \le \tilde{C}_q,$$

*for* $\tau(s) = \min\{1, s\}$. — *More precisely, there exists* $\alpha > 0$, *such that*

$$\tilde{C}(\mathbf{u}_0, \mathbf{d}_0, \mathbf{f}, \omega, T; \mathbf{U}) = \hat{C}(\mathbf{u}_0, \mathbf{d}_0, \mathbf{f}, \omega; \mathbf{U}) \exp\big(\overline{C}T^\alpha\big),$$
$$\tilde{C}_q(\mathbf{u}_0, \mathbf{d}_0, \mathbf{f}, \omega, T; \mathbf{U}, q) = \hat{C}_q(\mathbf{u}_0, \mathbf{d}_0, \mathbf{f}, \omega; \mathbf{U}, q) \exp\big(\overline{C}T^\alpha\big).$$

*In addition, for values* $q \in (2, \infty)$,

$$\mathbf{u} \in C(I; \mathbf{J}_1 \cap \dot{\mathbf{W}}^{2,2}_{\text{per}}) \cap L^2(I; \mathbf{J}_1 \cap \dot{\mathbf{W}}^{2,q}_{\text{per}}),$$
$$\mathbf{u}_t \in C(I; \mathbf{L}^2_0) \cap L^2(I; \dot{\mathbf{W}}^{1,2}_{\text{per}}), \quad \mathbf{u}_{tt} \in L^2(I; \dot{\mathbf{W}}^{-1,2}_{\text{per}}),$$
$$|\, \mathbf{d}(\mathbf{x}, t) \,| = 1, \quad \forall\, (\mathbf{x}, t) \in \omega_T, \quad \mathbf{d} \in C(I; \mathbf{W}^{1,2}_{\text{per}}) \cap L^2(I; \mathbf{W}^{2,q}_{\text{per}}),$$
$$\mathbf{d}_t \in C(I; \mathbf{W}^{1,2}_{\text{per}}) \cap L^2(I; \mathbf{W}^{2,2}_{\text{per}}), \quad \mathbf{d}_{tt} \in L^2(I; \mathbf{L}^2).$$

**Remark 6.6** *1. Existence of solutions in the given spaces follows again by identifying limits of Galerkin approximates solving a finite dimensional version of* (6.30)-(6.33) *with solutions to* (6.30)-(6.33), *using the a priori estimates derived before. We skip this part.*

*2. We employ the continuous embedding property* $L^q(I; \mathbf{X}) \cap W^{1,q'}(I; \mathbf{X}^\star) \hookrightarrow C(I; \mathbf{H})$, *for a Gelfand triple* $\mathbf{X} \hookrightarrow \mathbf{H} \simeq \mathbf{H}^\star \hookrightarrow \mathbf{X}^\star$.

**Proof:**
(b) We formally differentiate in time.

$$
\begin{aligned}
&\mathbf{u}_{tt} - \nu\,\Delta\mathbf{u}_t + \mathbf{u}_t\cdot\nabla\mathbf{u} + \mathbf{u}\cdot\nabla\mathbf{u}_t + \nabla p_t \\
&\quad + \lambda\Big(\operatorname{div}\big(\nabla\mathbf{d}_t\odot\nabla\mathbf{d} + \nabla\mathbf{d}\odot\nabla\mathbf{d}_t\big)\Big) = \mathbf{f}_t\,,
\end{aligned} \tag{6.45}
$$

$$
\begin{aligned}
&\mathbf{d}_{tt} - \varrho\,\Delta\mathbf{d}_t + \mathbf{U}_t\cdot\nabla\mathbf{d} + \mathbf{U}\cdot\nabla\mathbf{d}_t \\
&\quad = \varrho\,|\,\nabla\mathbf{d}\,|^2\mathbf{d}_t + 2\,\langle\nabla\mathbf{d},\nabla\mathbf{d}_t\rangle_{\mathbb{R}^4}\mathbf{d}\,.
\end{aligned} \tag{6.46}
$$

We multiply by $\{\mathbf{u}_t, -\Delta\mathbf{d}_t\}$.

$$
\begin{aligned}
&\frac{1}{2}\frac{d}{dt}\{\|\,\mathbf{u}_t\,\|^2_{\mathbf{L}^2} + \|\,\nabla\mathbf{d}_t\,\|^2_{\mathbf{L}^2}\} + \{\nu\,\|\,\nabla\mathbf{u}_t\,\|^2_{\mathbf{L}^2} + \varrho\,\|\,\Delta\mathbf{d}_t\,\|^2_{\mathbf{L}^2}\} \\
&\leq C\Big\{\|\,\mathbf{u}_t\,\|^2_{\mathbf{L}^4}\,\|\,\nabla\mathbf{u}\,\|_{\mathbf{L}^2} + \|\,\nabla\mathbf{d}_t\,\|_{\mathbf{L}^4}\,\|\,\nabla\mathbf{d}\,\|_{\mathbf{L}^4}\,\|\,\nabla\mathbf{u}_t\,\|_{\mathbf{L}^2} + \|\,\mathbf{f}_t\,\|^2_{\mathbf{L}^2}\Big\} \\
&\quad + C\Big\{\|\,\mathbf{U}_t\,\|_{\mathbf{L}^4}\,\|\,\nabla\mathbf{d}\,\|_{\mathbf{L}^4}\,\|\,\Delta\mathbf{d}_t\,\|_{\mathbf{L}^2} + \|\,\mathbf{U}\,\|_{\mathbf{L}^4}\,\|\,\nabla\mathbf{d}_t\,\|_{\mathbf{L}^4}\,\|\,\Delta\mathbf{d}_t\,\|_{\mathbf{L}^2} \\
&\quad + \|\,\nabla\mathbf{d}\,\|^2_{\mathbf{L}^4}\,\|\,\mathbf{d}_t\,\|_{\mathbf{L}^\infty}\,\|\,\Delta\mathbf{d}_t\,\|_{\mathbf{L}^2} + \|\,\nabla\mathbf{d}\,\|_{\mathbf{L}^4}\,\|\,\nabla\mathbf{d}_t\,\|_{\mathbf{L}^4}\,\|\,\Delta\mathbf{d}_t\,\|_{\mathbf{L}^2}\Big\}\,.
\end{aligned} \tag{6.47}
$$

Because of $\|\cdot\|_{\mathbf{L}^4} \leq C\,\|\cdot\|^{1/2}_{\mathbf{L}^2}\,\|\cdot\|^{1/2}_{\mathbf{W}^{1,2}}$, we can control terms on the right-hand side, using (a), (b). The remainder of (b) follows from fundamental theorem of calculus.

(c) We multiply (6.45), (6.46) by $\{\mathbf{A}^{-1}\mathbf{u}_{tt}, \mathbf{d}_{tt}\}$. Moreover, we make use of

$$
\int_I \big(|\,\nabla\mathbf{d}\,|^2\mathbf{d}_t, \mathbf{d}_{tt}\big)\,ds = \big(|\,\nabla\mathbf{d}(T)\,|^2, |\,\mathbf{d}_t(T)\,|^2\big)\Big|_{t=0}^{T} - 2\int_I \big(|\,\mathbf{d}_t\,|^2\nabla\mathbf{d}, \nabla\mathbf{d}_t\big)\,ds\,. \tag{6.48}
$$

$$
\begin{aligned}
&\{\|\,\mathbf{u}_{tt}\,\|^2_{\dot{\mathbf{W}}^{-1,2}_{\text{per}}} + \|\,\mathbf{d}_{tt}\,\|^2_{\mathbf{L}^2}\} + \frac{1}{2}\frac{d}{dt}\{\nu\,\|\,\mathbf{u}_t\,\|^2_{\mathbf{L}^2} + \varrho\,\|\,\nabla\mathbf{d}_t\,\|^2_{\mathbf{L}^2}\} \\
&\leq C\|\,\mathbf{f}_t\,\|^2_{\mathbf{L}^2} + C\Big\{\{\|\,\mathbf{u}_t\,\|_{\mathbf{L}^2}\,\|\,\nabla\mathbf{u}\,\|_{\mathbf{L}^2} + \|\,\mathbf{u}\,\|_{\mathbf{L}^8}\,\|\,\nabla\mathbf{u}_t\,\|_{\mathbf{L}^{\frac{4}{3}}}\}\|\,\mathbf{A}^{-1}\mathbf{u}_{tt}\,\|_{\mathbf{L}^8}\Big\} \\
&\quad + C\Big\{\{\|\,\mathbf{U}_t\,\|_{\mathbf{L}^4}\,\|\,\nabla\mathbf{d}\,\|_{\mathbf{L}^4} + \|\,\mathbf{U}\,\|_{\mathbf{L}^4}\,\|\,\nabla\mathbf{d}_t\,\|_{\mathbf{L}^4}\}\|\,\mathbf{d}_{tt}\,\|_{\mathbf{L}^2}\Big\} \\
&\quad + C\Big\{\|\,\nabla\mathbf{d}_t\,\|_{\mathbf{L}^2}\,\|\,\nabla\mathbf{d}\,\|_{\mathbf{L}^4}\,\|\,\nabla\mathbf{A}^{-1}\mathbf{u}_{tt}\,\|_{\mathbf{L}^4}\Big\} \\
&\quad + C\Big\{\|\,\nabla\mathbf{d}\,\|_{\mathbf{L}^4}\,\|\,\nabla\mathbf{d}_t\,\|_{\mathbf{L}^2}\,\|\,\mathbf{d}_t\,\|^2_{\mathbf{L}^4} + \frac{d}{dt}\|\,\nabla\mathbf{d}\,\|^2_{\mathbf{L}^4}\|\,\mathbf{d}_t\,\|^2_{\mathbf{L}^4}\Big\}
\end{aligned} \tag{6.49}
$$

Integration over $I$ together with (a), (b) then show (c).

(d) According to (6.2),

$$\varrho \| \Delta \mathbf{d} \|_{\mathbf{L}^q} \leq \| \mathbf{d}_t \|_{\mathbf{L}^q} + \| \mathbf{U} \cdot \nabla \mathbf{d} \|_{\mathbf{L}^q} + \varrho \| \nabla \mathbf{d} \|^2_{\mathbf{L}^{2q}} \leq C + \| \mathbf{u} \|_{\mathbf{L}^{2q}} \| \nabla \mathbf{d} \|_{\mathbf{L}^{2q}}, \tag{6.50}$$

which is bounded thanks to (b). Then, we may infer from Cattabriga's theorem ([58], Theorem 5.4) existence of solutions that satisfy the bound

$$\begin{aligned} \| \mathbf{u} \|_{\mathbf{W}^{2,2}} + \| p \|_{W^{1,2}} &\leq \| \mathbf{u}_t + \mathbf{u} \cdot \nabla \mathbf{u} + \operatorname{div} \left( \nabla \mathbf{d} \odot \nabla \mathbf{d} \right) \|_{\mathbf{L}^2} \qquad (6.51) \\ &\leq C + \| \mathbf{u} \|_{\mathbf{L}^4} \| \nabla \mathbf{u} \|_{\mathbf{L}^4} + \| \Delta \mathbf{d} \|_{\mathbf{L}^4} \| \nabla \mathbf{d} \|_{\mathbf{L}^4} \\ &\leq C + \| \mathbf{u} \|^2_{\mathbf{L}^4} \| \nabla \mathbf{u} \|_{\mathbf{L}^2} + \frac{1}{2} \| \Delta \mathbf{u} \|_{\mathbf{L}^2}. \end{aligned}$$

The bound for $\| \mathbf{u} \|_{L^2(I;\mathbf{W}^{2,p})} + \| p \|_{L^2(I;W^{1,p}/\mathbb{R})}$ follows from a similar argument.

(c) To show the remaining assertion there, we multiply (6.45) by $\mathbf{u}_{tt}$.

$$\begin{aligned} \| \mathbf{u}_{tt} \|^2_{\mathbf{L}^2} + \frac{\nu}{2} d_t \| \nabla \mathbf{u}_t \|^2_{\mathbf{L}^2} &\leq \| \mathbf{f}_t \|^2_{\mathbf{L}^2} + \| \mathbf{u}_t \|^2_{\mathbf{L}^4} \| \nabla \mathbf{u} \|^2_{\mathbf{L}^4} + \| \mathbf{u} \|^2_{\mathbf{L}^\infty} \| \nabla \mathbf{u}_t \|^2_{\mathbf{L}^2} \\ &\quad + \| \Delta \mathbf{d}_t \|^2_{\mathbf{L}^2} \| \nabla \mathbf{d} \|^2_{\mathbf{L}^\infty} + \| \Delta \mathbf{d} \|^2_{\mathbf{L}^2} \| \nabla \mathbf{d}_t \|^2_{\mathbf{L}^2}. \end{aligned}$$

After multiplication by $\tau(s)$ and integration over $I$, the previous results verify the assertion.

We can benefit from (d) in (6.47), (6.49) when dealing with $\mathbf{U}_t$ to obtain the specific structure of $\tilde{C}$ resp. $\tilde{C}_q$. □

In the next step, we address uniqueness of solutions to (6.30)-(6.33) which are in fact strong according to the given arguments.

**Lemma 6.5** *The solution* $\{\mathbf{u}, \mathbf{d}, p\}$ *to* (6.30)-(6.33) *is unique.*

**Proof:**
Let $\{\mathbf{u}_i, \mathbf{d}_i, p_i\}_{i=1,2}$ be two solutions to (6.30)-(6.33), such that $\{\mathbf{u}_1 - \mathbf{u}_2, \mathbf{d}_1 - \mathbf{d}_2\} = \{0, 0\}$. Set $\{\mathbf{e}, \boldsymbol{\eta}\} \equiv \{\mathbf{u}_1 - \mathbf{u}_2, \mathbf{d}_1 - \mathbf{d}_2\}$ and subtract the corresponding

equations (6.30), (6.33) then multiply by $\{\mathbf{e}, -\Delta\boldsymbol{\eta}\}$.

$$
\begin{aligned}
&\frac{1}{2}d_t\{\|\mathbf{e}\|_{\mathbf{L}^2}^2 + \|\nabla\boldsymbol{\eta}\|_{\mathbf{L}^2}^2\} + \{\nu\|\nabla\mathbf{e}\|_{\mathbf{L}^2}^2 + \varrho\|\Delta\boldsymbol{\eta}\|_{\mathbf{L}^2}^2\} \qquad (6.52)\\
&\le |(\mathbf{u}_1\cdot\nabla\mathbf{e} - \mathbf{e}\cdot\nabla\mathbf{u}_2, \mathbf{e})|\\
&\quad +|(\nabla\mathbf{d}_1\odot\nabla\boldsymbol{\eta} - \nabla\boldsymbol{\eta}\odot\nabla\mathbf{d}_2, \nabla\mathbf{e})| + |(\mathbf{U}\cdot\nabla\boldsymbol{\eta}, \Delta\boldsymbol{\eta})|\\
&\quad +|(\langle\nabla\boldsymbol{\eta}, \nabla[\mathbf{d}_1+\mathbf{d}_2]\rangle_{\mathbb{R}^4}\mathbf{d}_1, \Delta\boldsymbol{\eta})| + |(|\nabla\mathbf{d}_2|^2\boldsymbol{\eta}, \Delta\boldsymbol{\eta})|\\
&\le \|\nabla\mathbf{u}_2\|_{\mathbf{L}^4}\|\mathbf{e}\|_{\mathbf{L}^2}\|\mathbf{e}\|_{\mathbf{L}^4} + \{\|\nabla\mathbf{d}_1\|_{\mathbf{L}^\infty} + \|\nabla\mathbf{d}_2\|_{\mathbf{L}^\infty}\}\|\nabla\boldsymbol{\eta}\|_{\mathbf{L}^2}\|\nabla\mathbf{e}\|_{\mathbf{L}^2}\\
&\quad +\|\mathbf{U}\|_{\mathbf{L}^\infty}\|\nabla\boldsymbol{\eta}\|_{\mathbf{L}^2}\|\Delta\boldsymbol{\eta}\|_{\mathbf{L}^2} + \|\nabla\mathbf{d}_2\|_{\mathbf{L}^8}^2\|\boldsymbol{\eta}\|_{\mathbf{L}^4}\|\Delta\boldsymbol{\eta}\|_{\mathbf{L}^2}\\
&\quad +\|\nabla\boldsymbol{\eta}\|_{\mathbf{L}^2}\|\nabla[\mathbf{d}_1+\mathbf{d}_2]\|_{\mathbf{L}^\infty}\|\mathbf{d}_1\|_{\mathbf{L}^\infty}\|\Delta\boldsymbol{\eta}\|_{\mathbf{L}^2}.
\end{aligned}
$$

Application of Young's inequality and Gronwall's lemma settles the proof. □

Next, we quantify the difference in solutions to (6.30)-(6.33) for different $\mathbf{U}_i$.

**Lemma 6.6** *Let $\{\mathbf{u}^i, \mathbf{d}^i, p^i\}_{i=1,2}$ be solutions to (6.30)-(6.33), for $\mathbf{U}^i$, $i = 1, 2$, and $T > 0$. Then there holds*

$$
\begin{aligned}
\text{(a)}\quad &\operatorname*{ess\,sup}_{[0,T]}\{\|\mathbf{u}^1-\mathbf{u}^2\|_{\mathbf{W}^{1,2}} + \|\mathbf{d}^1-\mathbf{d}^2\|_{\mathbf{W}^{1,2}}\}\\
&+\Big(\int_0^T \{\|\mathbf{u}^1-\mathbf{u}^2\|_{\mathbf{W}^{2,2}} + \|\mathbf{d}^1-\mathbf{d}^2\|_{\mathbf{W}^{2,2}}\}^2\,\mathrm{d}s\Big)^{1/2}\\
&\le C\Big(\int_0^T \|\mathbf{U}^1-\mathbf{U}^2\|_{\mathbf{L}^2}^2\,\mathrm{d}s\Big)^{1/2},\\
\text{(b)}\quad &\operatorname*{ess\,sup}_{[0,T]}\Big\{\|[\mathbf{u}^1-\mathbf{u}^2]_t\|_{\mathbf{L}^2} + \|[\mathbf{d}^1-\mathbf{d}^2]_t\|_{\mathbf{W}^{1,2}}\Big\}\\
&+\Big(\int_0^T \{\|\nabla[\mathbf{u}^1-\mathbf{u}^2]_t\|_{\mathbf{L}^2}^2\,\mathrm{d}s\Big)^{1/2} + \Big(\int_0^T \|\Delta[\mathbf{d}^1-\mathbf{d}^2]_t\|_{\mathbf{L}^2}^2\,\mathrm{d}s\Big)^{1/2}\\
&\le C\Big(\int_0^T \|\mathbf{U}^1-\mathbf{U}^2\|_{\mathbf{L}^4}^2\,\mathrm{d}s\Big)^{1/2} + C\Big(\int_0^T \|[\mathbf{U}^1-\mathbf{U}^2]_t\|_{\mathbf{L}^2}^2\,\mathrm{d}s\Big)^{1/2},\\
\text{(c)}\quad &\operatorname*{ess\,sup}_{[0,T]}\{\|\mathbf{u}^1-\mathbf{u}^2\|_{\mathbf{W}^{2,2}} + \|\mathbf{d}^1-\mathbf{d}^2\|_{\mathbf{W}^{2,2}}\} \le \operatorname*{ess\,sup}_{[0,T]}\|\mathbf{U}^1-\mathbf{U}^2\|_{\mathbf{L}^2}\\
&+C\Big(\int_0^T \|\mathbf{U}^1-\mathbf{U}^2\|_{\mathbf{L}^4}^2\,\mathrm{d}s\Big)^{1/2} + C\Big(\int_0^T \|[\mathbf{U}^1-\mathbf{U}^2]_t\|_{\mathbf{L}^2}^2\,\mathrm{d}s\Big)^{1/2}.
\end{aligned}
$$

**Proof:**
(a) We proceed by bootstrapping: the first step is similar to the one in the proof of Lemma 6.5 (see there for notation) with the additional term

$$(\mathbf{U}^1 \cdot \nabla \mathbf{d}^1 - \mathbf{U}^2 \cdot \nabla \mathbf{d}^2, \Delta \boldsymbol{\eta}) = ([\mathbf{U}^1 - \mathbf{U}^2] \cdot \nabla \mathbf{d}^1, \Delta \boldsymbol{\eta}) + (\mathbf{U}^2 \cdot \nabla \boldsymbol{\eta}, \Delta \boldsymbol{\eta})$$
$$\leq C \| \mathbf{U}^1 - \mathbf{U}^2 \|_{\mathbf{L}^2} \| \nabla \mathbf{d}^1 \|_{\mathbf{L}^\infty} \| \Delta \boldsymbol{\eta} \|_{\mathbf{L}^2} + \| \mathbf{U}^2 \|_{\mathbf{L}^4}^4 \| \nabla \boldsymbol{\eta} \|_{\mathbf{L}^4} \| \Delta \boldsymbol{\eta} \|_{\mathbf{L}^2} .$$

Hence

$$\operatorname*{ess\,sup}_{[0,T]} \{ \| \mathbf{e} \|_{\mathbf{L}^2} + \| \nabla \boldsymbol{\eta} \|_{\mathbf{L}^2} \} + \Big( \int_0^T \{ \| \nabla \mathbf{e} \|_{\mathbf{L}^2}^2 + \| \Delta \boldsymbol{\eta} \|_{\mathbf{L}^2}^2 \} \, \mathrm{d}s \Big)^{1/2}$$
$$\leq C \int_0^T \| \mathbf{U}^1 - \mathbf{U}^2 \|_{\mathbf{L}^2}^2 \, \mathrm{d}s . \tag{6.53}$$

We consider the subtracted equations (6.30), multiply by $-\Delta \mathbf{e}$, and use (6.53),

$$\frac{1}{2} d_t \| \nabla \mathbf{e} \|_{\mathbf{L}^2}^2 + \nu \| \Delta \mathbf{e} \|_{\mathbf{L}^2}^2 \leq C \, | \, (\mathbf{u}^1 \cdot \nabla \mathbf{e} - \mathbf{e} \cdot \nabla \mathbf{u}^2, \Delta \mathbf{e}) \, |$$
$$+ | \, (\operatorname{div}(\nabla \mathbf{d}^1 \odot \nabla \boldsymbol{\eta} - \nabla \boldsymbol{\eta} \odot \nabla \mathbf{d}^2), \Delta \mathbf{e}) \, | .$$

Now Gronwall's inequality and (6.53) control the crucial term from the last contribution via $\| \Delta \boldsymbol{\eta} \|_{\mathbf{L}^2} \| \nabla \mathbf{d}^i \|_{\mathbf{L}^\infty} \| \Delta \mathbf{e} \|_{\mathbf{L}^2}$.

(b) We start from the error identities

$$\mathbf{e}_{tt} - \nu \Delta \mathbf{e}_t + \mathbf{e}_t \cdot \nabla \mathbf{u}^1 - \mathbf{u}_t^2 \cdot \nabla \mathbf{e} + \mathbf{e} \cdot \nabla \mathbf{u}_t^1 - \mathbf{u}^2 \cdot \nabla \mathbf{e}_t$$
$$+ \operatorname{div}\big( \nabla \boldsymbol{\eta}_t \odot \nabla \mathbf{d}^1 - \nabla \mathbf{d}_t^2 \odot \nabla \boldsymbol{\eta} + \nabla \boldsymbol{\eta} \odot \nabla \mathbf{d}_t^1 - \nabla \mathbf{d}^2 \odot \nabla \boldsymbol{\eta}_t \big) = 0 ,$$
$$\boldsymbol{\eta}_{tt} - \varrho \Delta \boldsymbol{\eta}_t + [\mathbf{U}^1 - \mathbf{U}^2]_t \cdot \nabla \mathbf{d}^1 - \mathbf{U}_t^2 \cdot \nabla \boldsymbol{\eta} + [\mathbf{U}^1 - \mathbf{U}^2] \cdot \nabla \mathbf{d}_t^1 - \mathbf{U}^2 \cdot \nabla \boldsymbol{\eta}_2$$
$$= \varrho \big[ \langle \nabla \boldsymbol{\eta}, \nabla [\mathbf{d}^1 + \mathbf{d}^2] \rangle_{\mathbb{R}^4} \mathbf{d}_t^1 + | \nabla \mathbf{d}^2 |^2 \boldsymbol{\eta}_t \big]$$
$$+ 2\varrho \big[ \langle \nabla \mathbf{d}^1, \nabla \boldsymbol{\eta}_t \rangle_{\mathbb{R}^4} \mathbf{d}^1 - \langle \nabla \boldsymbol{\eta}, \nabla \mathbf{d}_t^2 \rangle_{\mathbb{R}^4} \mathbf{d}^1 + \langle \nabla \mathbf{d}^2, \nabla \mathbf{d}_t^2 \rangle_{\mathbb{R}^4} \boldsymbol{\eta} \big] .$$

We multiply the system by $\{ \mathbf{e}_t, -\Delta \boldsymbol{\eta}_t \}$. The terms on the right-hand side of the subsequent inequality are dealt with independently.

$$\frac{1}{2} d_t \{ \| \mathbf{e}_t \|_{\mathbf{L}^2}^2 + \| \nabla \boldsymbol{\eta}_t \|_{\mathbf{L}^2}^2 \} + \{ \nu \| \nabla \mathbf{e}_t \|_{\mathbf{L}^2}^2 + \varrho \| \Delta \boldsymbol{\eta}_t \|_{\mathbf{L}^2}^2 \} \leq I + II + III + IV . \tag{6.54}$$

$$I \equiv | \, (\mathbf{e}_t \cdot \nabla \mathbf{u}^1, \mathbf{e}_t) \, | + | \, (\mathbf{u}_t^2 \cdot \nabla \mathbf{e}, \mathbf{e}_t) \, | + | \, (\mathbf{e} \cdot \nabla \mathbf{u}_t^1, \mathbf{e}_t) \, | \tag{6.55}$$
$$\leq \{ \| \mathbf{e}_t \|_{\mathbf{L}^2} \| \nabla \mathbf{u}^1 \|_{\mathbf{L}^2} + \| \mathbf{u}_t \|_{\mathbf{L}^4} \| \nabla \mathbf{e} \|_{\mathbf{L}^2} + \| \mathbf{e} \|_{\mathbf{L}^4} \| \nabla \mathbf{u}_t^1 \|_{\mathbf{L}^2} \} \| \nabla \mathbf{e}_t \|_{\mathbf{L}^2} .$$

$II$ collects terms that arise from the second line of the first error identity. We illustrate the treatment for two of them.

$$\begin{aligned} II_a &\equiv |(\nabla\boldsymbol{\eta}_t \odot \nabla\mathbf{d}^1, \nabla\mathbf{e}_t)| + |(\nabla\mathbf{d}_t^2 \odot \nabla\boldsymbol{\eta}, \nabla\mathbf{e}_t)| \\ &\leq C\|\nabla\boldsymbol{\eta}_t\|_{\mathbf{L}^2}\|\nabla\mathbf{d}^1\|_{\mathbf{L}^\infty}\|\nabla\mathbf{e}_t\|_{\mathbf{L}^2} \\ &\quad + \|\nabla\mathbf{d}_t^2\|_{\mathbf{L}^2}^{1/2}\|\Delta\mathbf{d}_t^2\|_{\mathbf{L}^2}^{1/2}\|\nabla\boldsymbol{\eta}\|_{\mathbf{L}^2}^{1/2}\|\Delta\boldsymbol{\eta}\|_{\mathbf{L}^2}^{1/2}\|\nabla\mathbf{e}_t\|_{\mathbf{L}^2}\,. \end{aligned} \tag{6.56}$$

$$\begin{aligned} III &\equiv |([\mathbf{U}^1-\mathbf{U}^2]_t\cdot\nabla\mathbf{d}^1, \Delta\boldsymbol{\eta}_t)| + |(\mathbf{U}_t^2\cdot\nabla\boldsymbol{\eta}, \Delta\boldsymbol{\eta}_t)| \\ &\quad + |([\mathbf{U}^1-\mathbf{U}^2]\cdot\nabla\mathbf{d}_t^1, \Delta\boldsymbol{\eta}_t)| + |(\mathbf{U}^2\cdot\nabla\boldsymbol{\eta}_t, \Delta\boldsymbol{\eta}_t)| \\ &\leq \|[\mathbf{U}^1-\mathbf{U}^2]_t\|_{\mathbf{L}^2}\|\nabla\mathbf{d}^1\|_{\mathbf{L}^\infty}\|\Delta\boldsymbol{\eta}_t\|_{\mathbf{L}^2} + \|\mathbf{U}_t^2\|_{\mathbf{L}^4}\|\nabla\boldsymbol{\eta}\|_{\mathbf{L}^4}\|\Delta\boldsymbol{\eta}_t\|_{\mathbf{L}^2} \\ &\quad + \|[\mathbf{U}^1-\mathbf{U}^2]\|_{\mathbf{L}^4}\|\nabla\mathbf{d}_t^1\|_{\mathbf{L}^4}\|\Delta\boldsymbol{\eta}_t\|_{\mathbf{L}^2} + \|\mathbf{U}^2\|_{\mathbf{L}^\infty}\|\nabla\boldsymbol{\eta}_t\|_{\mathbf{L}^2}\|\Delta\boldsymbol{\eta}_t\|_{\mathbf{L}^2}\,. \end{aligned} \tag{6.57}$$

Again, we confine to 'representative' terms that are collected in $IV$.

$$\begin{aligned} IV &\equiv |(\langle\nabla\boldsymbol{\eta}, \nabla[\mathbf{d}_1+\mathbf{d}_2]\rangle_{\mathbb{R}^4}\mathbf{d}_t^1, \Delta\boldsymbol{\eta}_t)| + |(|\nabla\mathbf{d}^2|^2\boldsymbol{\eta}_t, \Delta\boldsymbol{\eta}_t)| \\ &\quad + |(\langle\nabla\mathbf{d}^1, \nabla\boldsymbol{\eta}_t\rangle_{\mathbb{R}^4}\mathbf{d}^1, \Delta\boldsymbol{\eta}_t)| + |(\langle\nabla\mathbf{d}^2, \nabla\mathbf{d}_t^2\rangle_{\mathbb{R}^4}\boldsymbol{\eta}, \Delta\boldsymbol{\eta}_t)| \\ &\leq \|\nabla\boldsymbol{\eta}\|_{\mathbf{L}^4}\|\nabla[\mathbf{d}_1+\mathbf{d}_2]\|_{\mathbf{L}^\infty}\|\mathbf{d}_t^1\|_{\mathbf{L}^4}\|\Delta\boldsymbol{\eta}_t\|_{\mathbf{L}^2} \\ &\quad + \||\nabla\mathbf{d}^2|^2\|_{\mathbf{L}^4}\|\boldsymbol{\eta}_t\|_{\mathbf{L}^4}\|\Delta\boldsymbol{\eta}_t\|_{\mathbf{L}^2} \\ &\quad + \|\nabla\mathbf{d}^1\|_{\mathbf{L}^\infty}\|\nabla\boldsymbol{\eta}_t\|_{\mathbf{L}^2}\|\mathbf{d}^1\|_{\mathbf{L}^\infty}\|\Delta\boldsymbol{\eta}_t\|_{\mathbf{L}^2} \\ &\quad + \|\nabla\mathbf{d}^2\|_{\mathbf{L}^\infty}\|\nabla\mathbf{d}_t^2\|_{\mathbf{L}^4}\|\boldsymbol{\eta}\|_{\mathbf{L}^4}\|\Delta\boldsymbol{\eta}_t\|_{\mathbf{L}^2}\,. \end{aligned} \tag{6.58}$$

The result then follows from (a) and Lemma (6.4), by Gronwall's lemma.

(c) Consider the difference equation corresponding to (6.31) by $\Delta\boldsymbol{\eta}$.

$$\begin{aligned} \|\Delta\boldsymbol{\eta}\|_{\mathbf{L}^2} &\leq \|\boldsymbol{\eta}_t\|_{\mathbf{L}^2} + \|\mathbf{U}^1\|_{\mathbf{L}^4}\|\nabla\boldsymbol{\eta}\|_{\mathbf{L}^4} + \|\nabla\mathbf{d}_1\|_{\mathbf{L}^\infty}\|\mathbf{U}^1-\mathbf{U}^2\|_{\mathbf{L}^2} \\ &\quad + \|\nabla[\mathbf{d}^1+\mathbf{d}^2]\|_{\mathbf{L}^\infty}\|\nabla\boldsymbol{\eta}\|_{\mathbf{L}^2} + \|\nabla\mathbf{d}^2\|_{\mathbf{L}^8}^2\|\boldsymbol{\eta}\|_{\mathbf{L}^4} \end{aligned} \tag{6.59}$$

and interpolate $\mathbf{L}^4$ between $\mathbf{L}^2$ and $\mathbf{W}^{1,2}$. Secondly, we multiply the difference equation corresponding to (6.30) by $\Delta\mathbf{e}$.

$$\begin{aligned} \|\Delta\mathbf{e}\|_{\mathbf{L}^2} &\leq \|\mathbf{e}_t\|_{\mathbf{L}^2} + \|\mathbf{u}^1\cdot\nabla\mathbf{e}\|_{\mathbf{L}^2} + \|\mathbf{e}\|_{\mathbf{L}^4}\|\nabla\mathbf{u}^1\|_{\mathbf{L}^4} \\ &\quad + \|\Delta\boldsymbol{\eta}\|_{\mathbf{L}^2}\|\nabla\mathbf{d}^1\|_{\mathbf{L}^\infty} + \|\nabla\boldsymbol{\eta}\|_{\frac{2q}{q-2}}\|\Delta\mathbf{d}^1\|_{\mathbf{L}^q}\,. \end{aligned} \tag{6.60}$$

Then, (6.60) helps to control the last two contributions. □

## 6.2.2 Existence and regularity results for (P)

We transfer statements from Section 6.2.2 for (6.30)-(6.33) to (6.1)-(6.4), using contraction mapping principle. We stress that solutions are strong, according to the previous studies.

**Theorem 6.3** *Let* $\omega \in [0, 2D]^2$, $\{\mathbf{u}_0, \mathbf{d}_0\} \in \mathbf{J}_1 \cap \dot{\mathbf{W}}^{2,2}_{\text{per}} \times \mathbf{W}^{3,2}_{\text{per}}$, $|\,\mathbf{d}_0\,| = 1$, *and* $\mathbf{f} \in W^{1,\infty}(I; \mathbf{L}^2_0)$. *There exists a local strong unique solution, and the solution enjoys all properties stated in Lemma 6.4.*

**Proof:**
For given $T > 0$ such that $I = [0, T]$ consider

$$\boldsymbol{\mathcal{X}}_T \equiv L^\infty(I; \mathbf{J}_1 \cap \dot{\mathbf{W}}^{2,2}_{\text{per}}) \cap W^{1,\infty}(I; \mathbf{L}^2_0) \cap W^{1,2}(I; \dot{\mathbf{W}}^{1,2}_{\text{per}}),$$

equipped with the norm $\|\cdot\|_{\boldsymbol{\mathcal{X}}_T}$, where

$$\|\,\varphi\,\|_{\boldsymbol{\mathcal{X}}_T} = \Big(\sup_I\{\|\,\varphi\,\|_{\mathbf{W}^{2,2}} + \|\,\varphi_t\,\|_{\mathbf{L}^2}\}^2 + \int_0^T \|\,\varphi_t\,\|^2_{\mathbf{W}^{1,2}}\,ds\Big)^{1/2}.$$

Choose an appropriate constant $\hat{C}(\mathbf{u}_0, \mathbf{d}_0, \mathbf{f}, \omega) > 0$, and define

$$S_T = \Big\{\varphi \in \boldsymbol{\mathcal{X}}_T, \quad \varphi(0) = \mathbf{u}_0, \quad \|\,\varphi\,\|_{\boldsymbol{\mathcal{X}}_T} \le 2\,\hat{C}(\mathbf{u}_0, \mathbf{d}_0, \mathbf{f}, \omega)\Big\}.$$

For $\mathcal{F} : S_T \to \boldsymbol{\mathcal{X}}_T$, define the fixed-point iteration $\mathbf{u}^{\ell+1} = \mathcal{F}(\mathbf{u}^\ell)$, $\ell \in \mathbb{N}_0$, that solves (6.30)-(6.33), with $\mathbf{U} = \mathbf{u}^\ell$. Thanks to Lemmata 6.4, 6.6, there exists a $T = t_1$ such that $\mathcal{F}$ maps $S_T$ into itself and is a contraction. Moreover, again by Lemma 6.6, we can take $T = t_1$ small enough to make $\mathcal{F}$ a strict contraction from $S_T$ to $S_T$. Thus, the contraction mapping theorem gives existence of a local unique solution. Existence of solutions $\{\mathbf{d}, p\}$ is now immediate.

We have to carry out the same steps as in the proof of Lemma 6.4 to show results corresponding to (a)-(d) as well as existence results. In particular, the 'modification' in (6.46) does not affect the verification of (6.47), (6.49) crucially. □

### 6.2.3 Global existence of solutions to (P) for small initial data

For small initial data

$$\| \mathbf{u}_0 \|_{\mathbf{W}^{2,2}} + \| \mathbf{d}_0 \|_{\mathbf{W}^{3,2}} \le \delta \,, \tag{6.61}$$

we may follow the argumentation in Chapter 4, Section 4.1.1, using Amann's results [5]. A continuation argument then settles global existence of solutions.

**Theorem 6.4** *Let* $\omega \in [0, 2D]^2$, *and* $\{\mathbf{u}_0, \mathbf{d}_0\} \in \mathbf{J}_1 \cap \dot{\mathbf{W}}^{2,2}_{\text{per}} \times \mathbf{W}^{3,2}_{\text{per}}(\omega, \mathcal{S}^1)$ *satisfies* (6.61), *for* $\mathbf{f} = 0$. *There exists a global, strong, unique solution, and the solution enjoys all properties stated in Lemma 6.4.*

**Proof:**
(a) We deal with (6.1)-6.4 directly and formally multiply (6.1), (6.2) by $\{\mathbf{u}, -\Delta\mathbf{d}\}$, respectively. Thanks to the identity $\operatorname{div}(\nabla\mathbf{d} \odot \nabla\mathbf{d}) = \nabla\big[\frac{|\nabla \mathbf{d}|^2}{2}\big] + \Delta\mathbf{d} \cdot \nabla\mathbf{d}$, we obtain

$$\begin{aligned} &\frac{1}{2} d_t \{ \| \mathbf{u} \|^2_{\mathbf{L}^2} + \| \nabla\mathbf{d} \|^2_{\mathbf{L}^2} \} + \nu \| \nabla\mathbf{u} \|^2_{\mathbf{L}^2} + \varrho \| \Delta\mathbf{d} \|^2_{\mathbf{L}^2} \\ &\quad \le C_\varrho \| \nabla\mathbf{d} \|^4_{\mathbf{L}^4} + \frac{\varrho}{2} \| \Delta\mathbf{d} \|^2_{\mathbf{L}^2} \,. \end{aligned} \tag{6.62}$$

We use $\| \cdot \|_{\mathbf{L}^4} \le C \| \cdot \|_{\mathbf{L}^2} \| \cdot \|_{\mathbf{W}^{1,2}}$, together with Gronwall's inequality to verify the bound

$$\| \mathbf{u}(t) \|^2_{\mathbf{L}^2} + \| \mathbf{d}(t) \|^2_{\mathbf{L}^2} \le \{ \| \mathbf{u}(0) \|^2_{\mathbf{L}^2} + \| \mathbf{d}(0) \|^2_{\mathbf{L}^2} \} \exp(Ct) \,. \tag{6.63}$$

We argue by bootstrapping: coming back to (6.62), and using (6.61), (6.63), there exists $T_1 > 0$, such that terms on the right-hand side of (6.62) can be absorbed by the fourth one on the left-hand side. We find for times $t < T_1$

$$\begin{aligned} &\| \mathbf{u}(t) \|^2_{\mathbf{L}^2} + \| \nabla\mathbf{d}(t) \|^2_{\mathbf{L}^2} + \frac{1}{2} \int_0^t \{ \nu \| \nabla\mathbf{u} \|^2_{\mathbf{L}^2} + \varrho \| \Delta\mathbf{d} \|^2_{\mathbf{L}^2} \} \, \mathrm{d}s \\ &\quad \le \| \mathbf{u}(0) \|^2_{\mathbf{L}^2} + \| \nabla\mathbf{d}(0) \|^2_{\mathbf{L}^2} \,. \end{aligned} \tag{6.64}$$

(b) We proceed correspondingly to (a) for higher order derivatives, applying $D^2$ to 6.1, (6.2), and multiplying (6.1), (6.2) by $\{\mathbf{u}, -\Delta\mathbf{d}\}$. We interpolate

$\mathbf{L}^4, \mathbf{L}^6$ between $\mathbf{L}^2$ and $\mathbf{W}^{1,2}$ in the sequel.

$$\begin{aligned}
&\frac{1}{2} d_t\{\| \nabla \mathbf{u} \|_{\mathbf{L}^2}^2 + \| \Delta \mathbf{d} \|_{\mathbf{L}^2}^2\} + \nu \| \Delta \mathbf{u} \|_{\mathbf{L}^2}^2 + \varrho \| \nabla^3 \mathbf{d} \|_{\mathbf{L}^2}^2 \\
&\leq \| \mathrm{div}[\nabla \mathbf{d} \odot \nabla \mathbf{d}] \|_{\mathbf{L}^2}^2 + \| \mathbf{u} \cdot \nabla \mathbf{u} \|_{\mathbf{L}^2}^2 + \| \nabla[\mathbf{u} \cdot \nabla \mathbf{d}] \|_{\mathbf{L}^2}^2 + \| \nabla[| \nabla \mathbf{d} |^2 \mathbf{d}] \|_{\mathbf{L}^2}^2 \\
&\leq \| \Delta \mathbf{d} \|_{\mathbf{L}^4}^2 \| \nabla \mathbf{d} \|_{\mathbf{L}^4}^2 + \| \mathbf{u} \|_{\mathbf{L}^4}^2 \| \nabla \mathbf{u} \|_{\mathbf{L}^4}^2 + \| \nabla \mathbf{u} \|_{\mathbf{L}^4}^2 \| \nabla \mathbf{d} \|_{\mathbf{L}^4}^2 \\
&\quad + \| \mathbf{u} \|_{\mathbf{L}^4}^2 \| \Delta \mathbf{d} \|_{\mathbf{L}^4}^2 + \| \nabla \mathbf{d} \|_{\mathbf{L}^6}^6 + C \| \nabla^2 \mathbf{d} \|_{\mathbf{L}^4}^2 \| \nabla \mathbf{d} \|_{\mathbf{L}^4}^2 \\
&\leq C \| \nabla \mathbf{d} \|_{\mathbf{L}^2} \| \Delta \mathbf{d} \|_{\mathbf{L}^2}^2 \| \nabla^3 \mathbf{d} \|_{\mathbf{L}^2} + C \| \mathbf{u} \|_{\mathbf{L}^2} \| \nabla \mathbf{u} \|_{\mathbf{L}^2}^2 \| \Delta \mathbf{u} \|_{\mathbf{L}^2} \\
&\quad + C \| \nabla \mathbf{u} \|_{\mathbf{L}^2} \| \nabla^2 \mathbf{u} \|_{\mathbf{L}^2} \| \nabla \mathbf{d} \|_{\mathbf{L}^2} \| \nabla^2 \mathbf{d} \|_{\mathbf{L}^2} \\
&\quad + C \| \mathbf{u} \|_{\mathbf{L}^2} \| \nabla \mathbf{u} \|_{\mathbf{L}^2} \| \Delta \mathbf{d} \|_{\mathbf{L}^2} \| \nabla^3 \mathbf{d} \|_{\mathbf{L}^2} \\
&\quad + C \| \nabla \mathbf{d} \|_{\mathbf{L}^2}^2 \| \nabla^2 \mathbf{d} \|_{\mathbf{L}^2}^4 + C \| \nabla^2 \mathbf{d} \|_{\mathbf{L}^2} \| \nabla^3 \mathbf{d} \|_{\mathbf{L}^2} \| \nabla \mathbf{d} \|_{\mathbf{L}^2} \| \nabla^2 \mathbf{d} \|_{\mathbf{L}^2} .
\end{aligned}$$

In a first step, we may apply Gronwall's lemma on $[0, T_1]$, and use (6.64) to find for $t \in [0, T_1]$,

$$\| \nabla \mathbf{u}(t) \|_{\mathbf{L}^2}^2 + \| \Delta \mathbf{d}(t) \|_{\mathbf{L}^2}^2 \leq \{\| \nabla \mathbf{u}(0) \|_{\mathbf{L}^2}^2 + \| \Delta \mathbf{d}(0) \|_{\mathbf{L}^2}^2\} \exp(Ct) .$$

Then, we can again argue by bootstrapping, for $T_1 > 0$ appropriately chosen, to find the bound

$$\begin{aligned}
&\| \nabla \mathbf{u}(t) \|_{\mathbf{L}^2}^2 + \| \Delta \mathbf{d}(t) \|_{\mathbf{L}^2}^2 + \frac{1}{2} \int_0^t \{\nu \| \Delta \mathbf{u} \|_{\mathbf{L}^2}^2 + \varrho \| \nabla^3 \mathbf{d} \|_{\mathbf{L}^2}^2\} \, \mathrm{d}s \\
&\quad \leq \{\| \nabla \mathbf{u}(0) \|_{\mathbf{L}^2}^2 + \| \Delta \mathbf{d}(0) \|_{\mathbf{L}^2}^2\} .
\end{aligned}$$

(c) Verification of the following results is now immediate,

$$\| \Delta \mathbf{u}(t) \|_{\mathbf{L}^2}^2 + \| \nabla^3 \mathbf{d}(t) \|_{\mathbf{L}^2}^2 \leq \| \Delta \mathbf{u}(0) \|_{\mathbf{L}^2}^2 + \| \nabla^3 \mathbf{d}(0) \|_{\mathbf{L}^2}^2 .$$

Hence we may summarize results from (a)-(c) to find for $T_1 > 0$ properly chosen,

$$\| \mathbf{u}(t) \|_{\mathbf{W}^{2,2}} + \| \mathbf{d}(t) \|_{\mathbf{W}^{3,2}} \leq \| \mathbf{u}(0) \|_{\mathbf{W}^{2,2}} + \| \mathbf{d}(0) \|_{\mathbf{W}^{3,2}} , \quad t \in [0, T_1] . \tag{6.65}$$

We can now proceed by continuation, and apply (a)-(c) for $[T_1, 2T_1]$. □

From Theorems 6.3, 6.4, we obtain local and global existence of solutions to (6.1)-(6.4). This provides the mathematical setting to start numerical analysis on in the sequel.

## 6.3 The semi-implicit Euler method

### 6.3.1 Semi-Discretization in time

We study the following semidiscretization in time of (6.1)-(6.4), using a penalization approach of type (6.6). For $\omega = [0, 2D]^2$,

$$d_t\mathbf{u}^{j+1} - \nu\,\Delta\mathbf{u}^{j+1} + \mathbf{u}^j \cdot \nabla\mathbf{u}^{j+1} + \lambda\,\mathrm{div}\big(\nabla\mathbf{d}^j \odot \nabla\mathbf{d}^{j+1}\big) \qquad (6.66)$$
$$+\nabla p^{j+1} = \mathbf{f}^{j+1}\,,$$
$$d_t\mathbf{d}^{j+1} - \varrho\,\Delta\mathbf{d}^{j+1} + \mathbf{l}_\varepsilon(\mathbf{d}^{j+1}) + \mathbf{u}^j \cdot \nabla\mathbf{d}^{j+1} \qquad (6.67)$$
$$= \varrho\,\langle\nabla\mathbf{d}^j, \nabla\mathbf{d}^{j+1}\rangle_{\mathbb{R}^4}\mathbf{d}^{j+1}\,,$$
$$\mathrm{div}\mathbf{u}^{j+1} = 0\,, \quad \mathbf{u}^0 = \mathbf{u}_0\,, \quad \mathbf{d}^0 = \mathbf{d}_0\,. \qquad (6.68)$$

The choice of $\varepsilon = \varepsilon(k)$ is clarified in the subsequent theorem.

**Theorem 6.5** *Suppose $\omega = [0, 2D]^2$, and $0 < t_J < T(\mathbf{u}_0, \mathbf{d}_0)$. Let $\{\mathbf{u}, \mathbf{d}, p\}$ solve (6.1)-(6.4), and $\{\mathbf{u}^j, \mathbf{d}^j, p^j\}_{j=0}^J$ solves (6.66)-(6.68), and (A2) be valid. Then, for $\varepsilon^{-1} = o(\frac{1}{k})$, and $k \le k_0(t_J)$,*

$$\max_{0\le j\le M}\Big\{\|\,\mathbf{u}(t_j) - \mathbf{u}^j\,\|_{\dot{\mathbf{W}}^{-1,2}_{\mathrm{per}}} + \|\,\mathbf{d}(t_j) - \mathbf{d}^j\,\|_{\mathbf{L}^2} + \sqrt{k}\,\|\,p(t_j) - p^j\,\|_{W^{-1,2}_{\mathrm{per}}}\Big\} +$$
$$\Big(k\sum_{m=0}^{M}\{\|\,\mathbf{u}(t_j) - \mathbf{u}^j\,\|^2_{\mathbf{L}^2} + \|\,\mathbf{d}(t_j) - \mathbf{d}^j\,\|^2_{\mathbf{W}^{1,2}} + \sqrt{k}\,\|\,p(t_j) - p^j\,\|^2_{L^2}\}\Big)^{\frac{1}{2}} \le C\,k.$$

*This result remains valid if $\varrho\,\langle\nabla\mathbf{d}^j, \nabla\mathbf{d}^{j+1}\rangle_{\mathbb{R}^4}\mathbf{d}^{j+1}$ in (6.67) is replaced by $\varrho\,|\,\nabla\mathbf{d}^j\,|^2\mathbf{d}^{j+1}$.*

**Remark 6.7** *1. To show well-posedness of (6.66)-(6.68), we benefit from Remark 4.3, item 2., for $k \le k_0(t_J)$ sufficiently small, to show existence of solutions to (6.67). As to (6.66), an a priori bound from Corollary 6.1, (c) for the director field helps to show existence of solutions $\{\mathbf{u}^j, p^j\}$. — Note that the subsequent convergence analysis already holds for values $\varepsilon > 1.8k$.*

*2. In [93, 94], the authors prove (in adapted form)*

$$\max_{0\le j\le J}\{\|\,\mathbf{u}(t_j) - \mathbf{u}^j\,\|_{\mathbf{L}^2} + \|\,\mathbf{d}(t_j) - \mathbf{d}^j\,\|_{\mathbf{L}^2}\}$$
$$+\Big(k\sum_{j=0}^{J}\{\|\,\mathbf{u}(t_j) - \mathbf{u}^j\,\|^2_{\mathbf{W}^{1,2}} + \|\,\mathbf{d}(t_j) - \mathbf{d}^j\,\|^2_{\mathbf{W}^{2,2}}\}\Big)^{1/2} \le \tilde{C}\,k,$$

*with* $\tilde{C} = \tilde{C}\big(\exp\big(\frac{t_j}{\varepsilon}\big)\big)$.

*This result is proved by a perturbation argument to cope with the Ginzburg-Landau approximation term. — In the present scenario, we extract stability properties from this term. For this purpose, we use test functions* $\{\mathbf{A}^{-1}\mathbf{e}^{j+1}, \boldsymbol{\eta}^{j+1}\}$ *(rather than* $\{\mathbf{e}^{j+1}, \mathbf{A}\boldsymbol{\eta}^{j+1}\}$*) in the error analysis below. Note that* $C$ *in the theorem does not depend on* $\varepsilon, k$ *any more.*

*3. From above, we conclude*

$$\max_{0\le j\le J}\{\| \mathbf{u}(t_j) - \mathbf{u}^j \|_{\mathbf{L}^2} + \| \nabla\{\mathbf{d}(t_j) - \mathbf{d}^j\} \|_{\mathbf{L}^2}\} \le C\sqrt{k}\,.$$

*This implication is supported by computational experiments, see Section 6.5.*

*4. Replacing* $\varrho\,\langle\nabla\mathbf{d}^j, \nabla\mathbf{d}^{j+1}\rangle_{\mathbb{R}^4}\mathbf{d}^{j+1}$ *in* (6.67) *by* $\varrho\,|\,\nabla\mathbf{d}^j\,|^2\mathbf{d}^{j+1}$ *simplifies the analysis; in effect, this situation can be handled more easily by means of the induction argument given in the proof together with the strong regularity statements from Theorem 6.3. Technically, this would mean that the 'list of terms' assembled in '*$\mathrm{rest}_A$*' in the proof is essentially not modified, but the argumentation becomes easier for the last but one term there, i.e.,*

$$\begin{aligned}|\,(|\,\nabla\boldsymbol{\eta}^j\,|^2\boldsymbol{\eta}^{j+1}, \boldsymbol{\eta}^{j+1})\,| &\le \|\,\nabla\boldsymbol{\eta}^j\,\|^2_{\mathbf{L}^4}\,\|\,\boldsymbol{\eta}^{j+1}\,\|^2_{\mathbf{L}^4}\\ &\le \frac{\varrho}{16}\,\|\,\nabla\boldsymbol{\eta}^{j+1}\,\|^2_{\mathbf{L}^2} + C_\varrho\,\|\,\nabla\boldsymbol{\eta}^j\,\|^2_{\mathbf{L}^2}\,\|\,\boldsymbol{\eta}^j\,\|^2_{\mathbf{W}^{2,2}}\,\|\,\Delta\boldsymbol{\eta}^{j+1}\,\|^2_{\mathbf{L}^2}\,.\end{aligned}$$

*Correspondingly, the four last terms in '*$\mathrm{rest}_B$*' shift to a more explicit character. The most crucial contribution there is the last one which can now be controlled as follows,*

$$\begin{aligned}&\|\,\nabla\boldsymbol{\eta}^j\,\|^2_{\mathbf{L}^4}\,\|\,\boldsymbol{\eta}^{j+1}\,\|_{\mathbf{L}^4}\,\|\,\Delta\boldsymbol{\eta}^{j+1}\,\|_{\mathbf{L}^2}\\ &\le \frac{\varrho}{16}\,\|\,\Delta\boldsymbol{\eta}^{j+1}\,\|^2_{\mathbf{L}^2} + C\,\|\,\boldsymbol{\eta}^j\,\|^4_{\mathbf{L}^2}\,\|\,\nabla\boldsymbol{\eta}^j\,\|^4_{\mathbf{L}^2}\,\|\,\nabla\boldsymbol{\eta}^{j+1}\,\|^2_{\mathbf{L}^2}\,.\end{aligned}$$

*The reason to deal with both scenarios in Theorem 6.5 is that the second one is more 'attractive' if it comes to a finite element realization of* (6.66)-(6.68) *where this question becomes more subtle, Section 6.4.2. The proof below is given for* (6.66)-(6.68).

**Proof:**

We introduce the shorthand notations $\mathbf{e}^j := \mathbf{u}(t_j) - \mathbf{u}^j$ and $\boldsymbol{\eta}^j = \mathbf{d}(t_j) - \mathbf{d}^j$. Then the error identities read as follows, using

$$\boldsymbol{\mathcal{R}}^{j+1}(\boldsymbol{\phi}) := -\frac{1}{k}\int_{t_j}^{t_{j+1}} (s - t_j)\boldsymbol{\phi}_{tt}(s)\,\mathrm{d}s\,, \tag{6.69}$$

$$
\begin{aligned}
& d_t \mathbf{e}^{j+1} - \nu\, \Delta \mathbf{e}^{j+1} = \boldsymbol{\mathcal{R}}^{j+1}(\mathbf{u}) \qquad (6.70)\\
& \quad -\mathbf{e}^j \cdot \nabla \mathbf{u}(t_{j+1}) - \mathbf{e}^j \cdot \nabla \mathbf{e}^{j+1} + \mathbf{u}(t_j) \cdot \nabla \mathbf{e}^{j+1} \\
& \quad +k\, d_t \mathbf{u}(t_{j+1}) \cdot \nabla \mathbf{e}^{j+1} - \lambda \operatorname{div}\big(\nabla \boldsymbol{\eta}^j \odot \nabla \mathbf{d}(t_{j+1}) - \nabla \boldsymbol{\eta}^j \odot \nabla \boldsymbol{\eta}^{j+1} \\
& \quad -\nabla \mathbf{d}(t_j) \odot \nabla \boldsymbol{\eta}^{j+1} + k\, \nabla d_t \mathbf{d}(t_{j+1}) \odot \nabla \mathbf{d}(t_{j+1})\big)\,, \\
& d_t \boldsymbol{\eta}^{j+1} - \varrho\, \Delta \boldsymbol{\eta}^{j+1} - \mathbf{l}_\varepsilon(\mathbf{d}^{j+1}) = \boldsymbol{\mathcal{R}}^{j+1}(\mathbf{d}) - \mathbf{e}^j \cdot \nabla \mathbf{d}(t_{j+1}) \qquad (6.71)\\
& \quad -\mathbf{e}^j \cdot \nabla \boldsymbol{\eta}^{j+1} + \mathbf{u}(t_j) \cdot \nabla \boldsymbol{\eta}^{j+1} + \varrho \Big( \langle \nabla \mathbf{d}(t_j), \nabla \mathbf{d}(t_{j+1}) \rangle_{\mathbb{R}^4} \boldsymbol{\eta}^{j+1} \\
& \quad +k\, \langle \nabla d_t \mathbf{d}(t_{j+1}), \nabla \mathbf{d}(t_{j+1}) \rangle_{\mathbb{R}^4} \mathbf{d}(t_{j+1}) \\
& \quad + \big\langle \nabla \boldsymbol{\eta}^j, \nabla \big(2\mathbf{d}(t_{j+1}) - \boldsymbol{\eta}^{j+1}\big) \big\rangle_{\mathbb{R}^4} \big(\mathbf{d}(t_{j+1}) - \boldsymbol{\eta}^{j+1}\big) \Big) + k\, d_t \mathbf{u}(t_{j+1}) \cdot \nabla \boldsymbol{\eta}^{j+1}\,.
\end{aligned}
$$

In the sequel, we choose constants $\tilde{C}_1 = \tilde{C}_1(T; \mathbf{d}, \mathbf{u})$, $\tilde{C}_2 = \tilde{C}_2(T; \mathbf{d}, \mathbf{u}) > 0$, $0 < t_J < T$ sufficiently large, and only depending on the solution to (P). Let

$$
\begin{aligned}
\boldsymbol{\mathcal{E}}^J \;:=\; & \| \mathbf{e}^J \|^2_{\dot{\mathbf{W}}^{-1,2}_{\mathrm{per}}} + \| \boldsymbol{\eta}^J \|^2_{\mathbf{L}^2} + k \sum_{j=0}^{J} \big\{ \nu \| \mathbf{e}^j \|^2_{\mathbf{L}^2} + \varrho \| \nabla \boldsymbol{\eta}^j \|^2_{\mathbf{L}^2} \big\} \\
& + k^2 \sum_{j=1}^{J} \big\{ \| d_t \mathbf{e}^j \|^2_{\dot{\mathbf{W}}^{-1,2}_{\mathrm{per}}} + \| \mathbf{d}_t \boldsymbol{\eta}^j \|^2_{\mathbf{L}^2} \big\} \qquad (6.72) \\
& + \frac{k}{\varepsilon} \sum_{j=0}^{J} \big\{ \| \langle \boldsymbol{\eta}^j, \mathbf{d}(t_j) \rangle_{\mathbb{R}^2} \|^2_{L^2} + \| \boldsymbol{\eta}^j \|^4_{\mathbf{L}^4} \big\} \le \tilde{C}_1 k^2 \exp\big(\tilde{C}_2 t_J\big)\,.
\end{aligned}
$$

(a) Test (6.70), (6.71) by $\{\mathbf{A}^{-1}\mathbf{e}^{j+1}, \boldsymbol{\eta}^{j+1}\}$.

$$
\begin{aligned}
& \frac{1}{2} d_t \big\{ \| \mathbf{e}^{j+1} \|^2_{\dot{\mathbf{W}}^{-1,2}_{\mathrm{per}}} + \| \boldsymbol{\eta}^{j+1} \|^2_{\mathbf{L}^2} \big\} + \frac{k}{2} \big\{ \| d_t \mathbf{e}^{j+1} \|^2_{\dot{\mathbf{W}}^{-1,2}_{\mathrm{per}}} + \| d_t \boldsymbol{\eta}^{j+1} \|^2_{\mathbf{L}^2} \big\} \\
& \quad + \nu \| \mathbf{e}^{j+1} \|^2_{\mathbf{L}^2} + \varrho \| \nabla \boldsymbol{\eta}^{j+1} \|^2_{\mathbf{L}^2} - \big(\mathbf{l}_\varepsilon(\mathbf{d}^{j+1}), \boldsymbol{\eta}^{j+1}\big) \qquad (6.73) \\
& \le C_{t_J} \big\{ \| \mathbf{A}^{-1} \boldsymbol{\mathcal{R}}^{j+1}(\mathbf{u}) \|^2_{\mathbf{L}^2} + \| \boldsymbol{\mathcal{R}}^{j+1}(\mathbf{d}) \|^2_{\mathbf{L}^2} \big\} + \frac{1}{t_J} \| \boldsymbol{\eta}^{j+1} \|^2_{\mathbf{L}^2} + \mathrm{rest}_A\,.
\end{aligned}
$$

Note that we have

$$
\begin{aligned}
& k \sum_{j=0}^{J} \big\{ \| \boldsymbol{\mathcal{R}}^{j+1}(\mathbf{d}) \|_{\mathbf{L}^2} + \| \boldsymbol{\mathcal{R}}^{j+1}(\mathbf{u}) \|_{\dot{\mathbf{W}}^{-1,2}_{\mathrm{per}}} \big\}^2 \le k^{-1} \sum_{j=0}^{J} \Big( \int_{t_j}^{t_{j+1}} (s - t_j)^2 \,\mathrm{d}s \Big) \\
& \quad \times \Big( \int_{t_j}^{t_{j+1}} \big\{ \| \mathbf{u}_{tt}(s) \|_{\dot{\mathbf{W}}^{-1,2}_{\mathrm{per}}} + \| \mathbf{d}_{tt}(s) \|_{\mathbf{L}^2} \big\}^2 \,\mathrm{d}s \Big) \le C k^2\,,
\end{aligned}
$$

and the last term in front of the inequality sign can be rewritten as

$$\begin{aligned}
&\frac{1}{\varepsilon}\big(|\,\mathbf{d}(t_{j+1})\,|^2 - |\,\mathbf{d}^{j+1}\,|^2, |\,\boldsymbol{\eta}^{j+1}\,|^2\big) + \frac{1}{\varepsilon}\big(|\,\mathbf{d}(t_{j+1})\,|^2 - |\,\mathbf{d}^{j+1}\,|^2, \langle \mathbf{d}(t_{j+1}), \boldsymbol{\eta}^{j+1}\rangle_{\mathbb{R}^2}\big) \\
&\quad = \frac{1}{\varepsilon}\big(\langle \boldsymbol{\eta}^{j+1}, 2\mathbf{d}(t_{j+1}) - \boldsymbol{\eta}^{j+1}\rangle_{\mathbb{R}^2}, \langle \mathbf{d}(t_{j+1}), \boldsymbol{\eta}^{j+1}\rangle_{\mathbb{R}^2} - |\,\boldsymbol{\eta}^{j+1}\,|^2\big) \qquad (6.74) \\
&\quad = \frac{1}{\varepsilon}\big(2\,\|\,\langle \boldsymbol{\eta}^{j+1}, \mathbf{d}(t_{j+1})\rangle_{\mathbb{R}^2}\,\|_{L^2}^2 + \|\,\boldsymbol{\eta}^{j+1}\,\|_{\mathbf{L}^4}^4 - 3\,\big(\langle \mathbf{d}(t_{j+1}), \boldsymbol{\eta}^{j+1}\rangle_{\mathbb{R}^2}, |\,\boldsymbol{\eta}^{j+1}\,|^2\big)\big)\,.
\end{aligned}$$

We may argue as in (4.96)-(4.100) to control the last term in (6.74). — The term '$\text{rest}_A$' summarizes the following contributions, for $q > 2$:

$$\begin{aligned}
\|\,\mathbf{e}^j\,\|_{\mathbf{L}^2}\,\|\,\nabla \mathbf{u}(t_{j+1})\,\|_{\mathbf{L}^4}\,\|\,\mathbf{A}^{-1}\mathbf{e}^{j+1}\,\|_{\mathbf{L}^4} &\le \frac{\nu}{16}\,\|\,\mathbf{e}^j\,\|_{\mathbf{L}^2}^2 + \\
&\quad C\,\|\,\nabla \mathbf{u}(t_{j+1})\,\|_{\mathbf{L}^4}^2\,\|\,\mathbf{e}^{j+1}\,\|_{\dot{\mathbf{W}}_{\rm per}^{-1,2}}^2\,, \\
\|\,\mathbf{e}^j\,\|_{\mathbf{L}^4}\,\|\,\nabla \mathbf{A}^{-1}\mathbf{e}^{j+1}\,\|_{\mathbf{L}^4}\,\|\,\mathbf{e}^{j+1}\,\|_{\mathbf{L}^2} &\le \frac{\nu}{16}\,\|\,\mathbf{e}^{j+1}\,\|_{\mathbf{L}^2}^2 + \\
&\quad C\,\|\,\mathbf{e}^j\,\|_{\mathbf{L}^2}^2\,\|\,\nabla \mathbf{e}^j\,\|_{\mathbf{L}^2}^2\,\|\,\mathbf{e}^{j+1}\,\|_{\dot{\mathbf{W}}_{\rm per}^{-1,2}}^2\,, \\
\|\,\mathbf{u}(t_{j+1})\,\|_{\mathbf{L}^\infty} & \\
\times\|\,\mathbf{u}(t_j)\,\|_{\mathbf{L}^\infty}\,\|\,\nabla \mathbf{A}^{-1}\mathbf{e}^{j+1}\,\|_{\mathbf{L}^2}\,\|\,\mathbf{e}^{j+1}\,\|_{\mathbf{L}^2} &\le \frac{\nu}{16}\,\|\,\mathbf{e}^{j+1}\,\|_{\mathbf{L}^2}^2 + \\
&\quad C\,\|\,\mathbf{u}(t_j)\,\|_{\mathbf{L}^\infty}^2\,\|\,\mathbf{e}^{j+1}\,\|_{\dot{\mathbf{W}}_{\rm per}^{-1,2}}^2\,, \\
\|\,\nabla \boldsymbol{\eta}^j\,\|_{\mathbf{L}^2}\,\|\,\nabla \mathbf{d}(t_{j+1})\,\|_{\mathbf{L}^\infty}\,\|\,\nabla \mathbf{A}^{-1}\mathbf{e}^{j+1}\,\|_{\mathbf{L}^2} &\le \frac{\varrho}{16}\,\|\,\nabla \boldsymbol{\eta}^j\,\|_{\mathbf{L}^2}^2 + \\
&\quad C\,\|\,\nabla \mathbf{d}(t_{j+1})\,\|_{\mathbf{L}^\infty}^2\,\|\,\mathbf{e}^{j+1}\,\|_{\dot{\mathbf{W}}_{\rm per}^{-1,2}}^2\,, \\
\|\,\nabla \boldsymbol{\eta}^j\,\|_{\mathbf{L}^4}\,\|\,\nabla \boldsymbol{\eta}^{j+1}\,\|_{\mathbf{L}^2}\,\|\,\nabla \mathbf{A}^{-1}\mathbf{e}^{j+1}\,\|_{\mathbf{L}^4} &\le \frac{\nu}{16}\,\|\,\mathbf{e}^{j+1}\,\|_{\mathbf{L}^2}^2 + \\
&\quad C\,\|\,\nabla \boldsymbol{\eta}^j\,\|_{\mathbf{L}^2}\,\|\,\Delta \boldsymbol{\eta}^j\,\|_{\mathbf{L}^2}\,\|\,\nabla \boldsymbol{\eta}^{j+1}\,\|_{\mathbf{L}^2}^2\,, \\
\|\,\nabla \mathbf{d}(t_{j+1})\,\|_{\mathbf{L}^\infty}\,\|\,\nabla \boldsymbol{\eta}^{j+1}\,\|_{\mathbf{L}^2} & \\
\times\|\,\nabla \mathbf{A}^{-1}\mathbf{e}^{j+1}\,\|_{\mathbf{L}^2} &\le \frac{\varrho}{16}\,\|\,\nabla \boldsymbol{\eta}^j\,\|_{\mathbf{L}^2}^2 + C\,\|\,\mathbf{e}^{j+1}\,\|_{\dot{\mathbf{W}}_{\rm per}^{-1,2}}^2\,, \\
k\,\|\,\nabla d_t\mathbf{d}(t_{j+1})\,\|_{\mathbf{L}^2}\,\|\,\nabla \mathbf{d}(t_{j+1})\,\|_{\mathbf{L}^\infty} & \\
\times\|\,\nabla \mathbf{A}^{-1}\mathbf{e}^{j+1}\,\|_{\mathbf{L}^2} &\le C k^2\,\|\,\nabla d_t\mathbf{d}(t_{j+1})\,\|_{\mathbf{L}^2}^2\,\|\,\nabla \mathbf{d}(t_{j+1})\,\|_{\mathbf{L}^\infty}^2 \\
&\quad +\|\,\mathbf{e}^{j+1}\,\|_{\dot{\mathbf{W}}_{\rm per}^{-1,2}}^2\,,
\end{aligned}$$

$$
\begin{aligned}
\| \mathbf{e}^j \|_{\mathbf{L}^2} \| \nabla \mathbf{d}(t_{j+1}) \|_{\mathbf{L}^\infty} \| \boldsymbol{\eta}^{j+1} \|_{\mathbf{L}^2} &\leq \frac{\nu}{16} \| \mathbf{e}^{j+1} \|^2_{\mathbf{L}^2} + \\
& \quad C \| \nabla \mathbf{d}(t_{j+1}) \|^2_{\mathbf{L}^\infty} \| \boldsymbol{\eta}^{j+1} \|^2_{\mathbf{L}^2} , \\
| ( \langle \nabla \mathbf{d}(t_j), \nabla \mathbf{d}(t_{j+1}) \rangle_{\mathbb{R}^4} \boldsymbol{\eta}^{j+1}, \boldsymbol{\eta}^{j+1} ) | &\leq \| \nabla \mathbf{d}(t_j) \|_{\mathbf{L}^\infty} \| \nabla \mathbf{d}(t_{j+1}) \|_{\mathbf{L}^\infty} \| \boldsymbol{\eta}^{j+1} \|^2_{\mathbf{L}} \\
k \int_{t_j}^{t_{j+1}} \| \nabla \mathbf{d}_t(s) \|_{\mathbf{L}^4} \, \mathrm{d}s & \\
\times \| \nabla \mathbf{d}(t_{j+1}) \|_{\mathbf{L}^4} \| \mathbf{d}(t_{j+1}) \|_{\mathbf{L}^\infty} \| \boldsymbol{\eta}^{j+1} \|_{\mathbf{L}^2} &\leq k^3 \int_{t_j}^{t_{j+1}} \| \Delta \mathbf{d}_t(s) \|_{\mathbf{L}^2} \, \mathrm{d}s \\
& \quad + C \| \nabla \mathbf{d}(t_{j+1}) \|^2_{\mathbf{L}^4} \| \boldsymbol{\eta}^{j+1} \|^2_{\mathbf{L}^2} , \\
\| \nabla \boldsymbol{\eta}^j \|_{\mathbf{L}^2} \| \nabla \mathbf{d}(t_{j+1}) \|_{\mathbf{L}^\infty} & \\
\times \| \nabla \mathbf{d}(t_{j+1}) \|_{\mathbf{L}^\infty} \| \boldsymbol{\eta}^{j+1} \|_{\mathbf{L}^2} &\leq \frac{\varrho}{16} \| \nabla \boldsymbol{\eta}^j \|^2_{\mathbf{L}^2} + \\
& \quad C \| \nabla \mathbf{d}(t_{j+1}) \|^2_{\mathbf{L}^\infty} \| \boldsymbol{\eta}^{j+1} \|^2_{\mathbf{L}^2} , \\
\| \nabla \boldsymbol{\eta}^j \|_{\mathbf{L}^4} \| \nabla \boldsymbol{\eta}^{j+1} \|_{\mathbf{L}^2} & \\
\times \| \mathbf{d}(t_{j+1}) \|_{\mathbf{L}^\infty} \| \boldsymbol{\eta}^{j+1} \|_{\mathbf{L}^4} &\leq \frac{\varrho}{16} \| \nabla \boldsymbol{\eta}^{j+1} \|^2_{\mathbf{L}^2} + \\
& \quad C \| \nabla \boldsymbol{\eta}^j \|^2_{\mathbf{L}^2} \| \Delta \boldsymbol{\eta}^j \|^2_{\mathbf{L}^2} \| \boldsymbol{\eta}^{j+1} \|^2_{\mathbf{L}^2} , \\
\| \nabla \boldsymbol{\eta}^j \|_{\mathbf{L}^\infty} \| \nabla \boldsymbol{\eta}^{j+1} \|_{\mathbf{L}^2} \| \boldsymbol{\eta}^{j+1} \|^2_{\mathbf{L}^4} &\leq \frac{1}{4\varepsilon} \| \boldsymbol{\eta}^{j+1} \|^4_{\mathbf{L}^4} + \\
& \quad C_q \varepsilon \| \Delta \boldsymbol{\eta}^j \|^2_{\mathbf{L}^q} \| \nabla \boldsymbol{\eta}^{j+1} \|^2_{\mathbf{L}^2} , \\
\| \nabla \boldsymbol{\eta}^j \|_{\mathbf{L}^2} \| \nabla \mathbf{d}(t_{j+1}) \|_{\mathbf{L}^\infty} \| \boldsymbol{\eta}^{j+1} \|^2_{\mathbf{L}^4} &\leq \frac{\varrho}{16} \| \nabla \boldsymbol{\eta}^{j+1} \|^2_{\mathbf{L}^2} + C \| \nabla \boldsymbol{\eta}^j \|^2_{\mathbf{L}^2} \\
& \quad \times \| \nabla \mathbf{d}(t_{j+1}) \|^2_{\mathbf{L}^\infty} \| \boldsymbol{\eta}^{j+1} \|^2_{\mathbf{L}^2} .
\end{aligned}
$$

In a second step, we test (6.70), (6.71) by $\{\mathbf{e}^{j+1}, -\Delta \boldsymbol{\eta}^{j+1}\}$.

$$
\begin{aligned}
& \frac{1}{2} d_t \{ \| \mathbf{e}^{j+1} \|^2_{\mathbf{L}^2} + \| \nabla \boldsymbol{\eta}^{j+1} \|^2_{\mathbf{L}^2} \} + \frac{k}{2} \{ \| d_t \mathbf{e}^{j+1} \|^2_{\mathbf{L}^2} + \| \nabla d_t \boldsymbol{\eta}^{j+1} \|^2_{\mathbf{L}^2} \} \\
& \quad + \nu \| \nabla \mathbf{e}^{j+1} \|^2_{\mathbf{L}^2} + \varrho \| \Delta \boldsymbol{\eta}^{j+1} \|^2_{\mathbf{L}^2} \qquad (6.75) \\
& \quad \leq -(\mathbf{l}_\varepsilon(\mathbf{d}^{j+1}), \Delta \boldsymbol{\eta}^{j+1}) + C \left\{ \| \boldsymbol{\mathcal{R}}^{j+1}(\mathbf{u}) \|^2_{\mathbf{L}^2} + \| \boldsymbol{\mathcal{R}}^{j+1}(\mathbf{d}) \|^2_{\mathbf{L}^2} \right\} + \mathrm{rest}_B .
\end{aligned}
$$

'$\mathrm{rest}_B$' summarizes the following terms:

$$
\begin{aligned}
| (\mathbf{e}^j \cdot \nabla \mathbf{u}(t_{j+1}), \mathbf{e}^{j+1}) | &\leq \frac{\nu}{16} \| \mathbf{e}^j \|^2_{\mathbf{L}^2} + C \| \mathbf{e}^j \|^2_{\mathbf{L}^2} \| \nabla \mathbf{u}(t_{j+1}) \|^2_{\mathbf{L}^\infty} , \\
| (\nabla \boldsymbol{\eta}^j \odot \nabla \mathbf{d}(t_{j+1}), \nabla \mathbf{e}^{j+1}) | &\leq \frac{\nu}{16} \| \nabla \mathbf{e}^{j+1} \|^2_{\mathbf{L}^2} + \\
& \quad C \| \nabla \boldsymbol{\eta}^j \|^2_{\mathbf{L}^2} \| \nabla \mathbf{d}(t_{j+1}) \|^2_{\mathbf{L}^\infty} ,
\end{aligned}
$$

$$
\begin{aligned}
|\, \nabla\boldsymbol{\eta}^j \odot \nabla\boldsymbol{\eta}^{j+1}, \nabla\mathbf{e}^{j+1} \,| &\leq \frac{\nu}{16} \|\, \nabla\mathbf{e}^{j+1} \,\|_{\mathbf{L}^2}^2 + \frac{\varrho}{16} \|\, \Delta\boldsymbol{\eta}^{j+1} \,\|_{\mathbf{L}^2}^2 + \\
&\quad C \|\, \nabla\boldsymbol{\eta}^j \,\|_{\mathbf{L}^4}^4 \|\, \nabla\boldsymbol{\eta}^{j+1} \,\|_{\mathbf{L}^2}^2 , \\
k \,|\, (\nabla d_t\mathbf{d}(t_{j+1}) \odot \nabla\mathbf{d}(t_{j+1}), \nabla\mathbf{e}^{j+1}) \,| &\leq \frac{\nu}{16} \|\, \nabla\mathbf{e}^{j+1} \,\|_{\mathbf{L}^2}^2 + \\
&\quad Ck^3 \int_{t_j}^{t_{j+1}} \|\, \nabla\mathbf{d}_t(s) \,\|_{\mathbf{L}^2}^2 \,\mathrm{d}s \,\|\, \nabla\mathbf{d}(t_{j+1}) \,\|_{\mathbf{L}^\infty}^2 \\
|\, (\mathbf{e}^j \cdot \nabla\mathbf{d}(t_{j+1}), \Delta\boldsymbol{\eta}^{j+1}) \,| &\leq \frac{\varrho}{16} \|\, \Delta\boldsymbol{\eta}^{j+1} \,\|_{\mathbf{L}^2}^2 + \\
&\quad C \|\, \boldsymbol{\eta}^j \,\|_{\mathbf{L}^2}^2 \|\, \nabla\mathbf{d}(t_{j+1}) \,\|_{\mathbf{L}^\infty}^2 , \\
\|\, \mathbf{e}^j \,\|_{\mathbf{L}^4} \|\, \nabla\boldsymbol{\eta}^{j+1} \,\|_{\mathbf{L}^4} \|\, \Delta\boldsymbol{\eta}^{j+1} \,\|_{\mathbf{L}^2} &\leq \frac{\varrho}{16} \|\, \Delta\boldsymbol{\eta}^{j+1} \,\|_{\mathbf{L}^2}^2 + \\
&\quad C \|\, \mathbf{e}^j \,\|_{\mathbf{L}^2}^2 \|\, \nabla\mathbf{e}^j \,\|_{\mathbf{L}^2}^2 \|\, \nabla\boldsymbol{\eta}^{j+1} \,\|_{\mathbf{L}^2}^2 , \\
|\, (\mathbf{u}(t_j) \cdot \nabla\boldsymbol{\eta}^{j+1}, \Delta\boldsymbol{\eta}^{j+1}) \,| &\leq \frac{\varrho}{16} \|\, \Delta\boldsymbol{\eta}^{j+1} \,\|_{\mathbf{L}^2}^2 + \\
&\quad C \|\, \mathbf{u}(t_j) \,\|_{\mathbf{L}^\infty}^2 \|\, \nabla\boldsymbol{\eta}^{j+1} \,\|_{\mathbf{L}^2}^2 , \\
|\, (\langle \nabla\mathbf{d}(t_j), \nabla\mathbf{d}(t_{j+1}) \rangle_{\mathbb{R}^4} \boldsymbol{\eta}^{j+1}, \Delta\boldsymbol{\eta}^{j+1}) \,| &\leq \frac{\varrho}{16} \|\, \Delta\boldsymbol{\eta}^{j+1} \,\|_{\mathbf{L}^2}^2 + \|\, \nabla\mathbf{d}(t_j) \,\|_{\mathbf{L}^\infty}^2 \\
&\quad \|\, \nabla\mathbf{d}(t_{j+1}) \,\|_{\mathbf{L}^\infty}^2 \|\, \boldsymbol{\eta}^{j+1} \,\|_{\mathbf{L}^2}^2 , \\
k \,|\, (\langle \nabla d_t\mathbf{d}(t_{j+1}), & \\
\nabla\mathbf{d}(t_{j+1} \rangle_{\mathbb{R}^4} \mathbf{d}(t_{j+1}), \Delta\boldsymbol{\eta}^{j+1}) \,| &\leq \frac{\varrho}{16} \|\, \Delta\boldsymbol{\eta}^{j+1} \,\|_{\mathbf{L}^2}^2 + Ck^3 \int_{t_j}^{t_{j+1}} \|\, \nabla\mathbf{d}_t(s) \,\|_{\mathbf{L}^2}^2 \\
&\quad \times \|\, \nabla\mathbf{d}(t_{j+1}) \,\|_{\mathbf{L}^\infty}^2 , \\
k \,\|\, d_t\mathbf{u}(t_{j+1}) \,\|_{\mathbf{L}^4} \|\, \nabla\boldsymbol{\eta}^{j+1} \,\|_{\mathbf{L}^4} & \\
\times \|\, \Delta\boldsymbol{\eta}^{j+1} \,\|_{\mathbf{L}^2} &\leq \frac{\varrho}{16} \|\, \Delta\boldsymbol{\eta}^{j+1} \,\|_{\mathbf{L}^2}^2 + Ck^6 \|\, \nabla\boldsymbol{\eta}^{j+1} \,\|_{\mathbf{L}^2}^2 \\
&\quad \times \int_{t_j}^{t_{j+1}} \|\, \mathbf{u}_t(s) \,\|_{\mathbf{L}^4}^4 \,\mathrm{d}s , \\
|\, (\langle \nabla\boldsymbol{\eta}^j, \nabla\mathbf{d}(t_{j+1}) \rangle_{\mathbb{R}^4} \mathbf{d}(t_{j+1}), \Delta\boldsymbol{\eta}^{j+1}) \,| &\leq \frac{\varrho}{16} \|\, \Delta\boldsymbol{\eta}^{j+1} \,\|_{\mathbf{L}^2}^2 \\
&\quad + C \|\, \nabla\boldsymbol{\eta}^j \,\|_{\mathbf{L}^2}^2 \|\, \nabla\mathbf{d}(t_{j+1}) \,\|_{\mathbf{L}^\infty}^2 , \\
\|\, \nabla\boldsymbol{\eta}^j \,\|_{\mathbf{L}^4} \|\, \nabla\mathbf{d}(t_{j+1}) \,\|_{\mathbf{L}^\infty} & \\
\times \|\, \boldsymbol{\eta}^{j+1} \,\|_{\mathbf{L}^4} \|\, \Delta\boldsymbol{\eta}^{j+1} \,\|_{\mathbf{L}^2} &\leq \frac{\varrho}{16} \|\, \Delta\boldsymbol{\eta}^{j+1} \,\|_{\mathbf{L}^2}^2 \\
&\quad + C \|\, \nabla\mathbf{d}(t_{j+1}) \,\|_{\mathbf{L}^\infty}^2 \|\, \nabla\boldsymbol{\eta}^j \,\|_{\mathbf{L}^4}^2 \|\, \nabla\boldsymbol{\eta}^{j+1} \,\|_{\mathbf{L}^4}^2
\end{aligned}
$$

$$\begin{aligned} &\| \nabla \boldsymbol{\eta}^j \|_{\mathbf{L}^4} \| \nabla \boldsymbol{\eta}^{j+1} \|_{\mathbf{L}^4} \\ &\times \| \mathbf{d}^{j+1} \|_{\mathbf{L}^\infty} \| \Delta \boldsymbol{\eta}^{j+1} \|_{\mathbf{L}^2} \leq \frac{\varrho}{16} \| \Delta \boldsymbol{\eta}^{j+1} \|_{\mathbf{L}^2}^2 \\ &\qquad + C \| \nabla \boldsymbol{\eta}^j \|_{\mathbf{L}^2}^2 \| \Delta \boldsymbol{\eta}^j \|_{\mathbf{L}^2}^2 \| \nabla \boldsymbol{\eta}^{j+1} \|_{\mathbf{L}^2}^2 , \\ &\| \nabla \boldsymbol{\eta}^j \|_{\mathbf{L}^8} \| \nabla \boldsymbol{\eta}^{j+1} \|_{\mathbf{L}^{8/3}} \\ &\times \| \boldsymbol{\eta}^{j+1} \|_{\mathbf{L}^\infty} \| \Delta \boldsymbol{\eta}^{j+1} \|_{\mathbf{L}^2} \leq C \| \nabla \boldsymbol{\eta}^j \|_{\mathbf{L}^2}^{1/4} \| \Delta \boldsymbol{\eta}^j \|_{\mathbf{L}^2}^{3/4} \\ &\qquad \times \| \nabla \boldsymbol{\eta}^{j+1} \|_{\mathbf{L}^2}^{3/4} \| \boldsymbol{\eta}^{j+1} \|_{\mathbf{L}^2}^{1/4} \| \Delta \boldsymbol{\eta}^{j+1} \|_{\mathbf{L}^2}^2 . \end{aligned} \tag{6.76}$$

In the last estimate, we use the inequalities $\| \cdot \|_{\mathbf{L}^\infty} \leq C \| \cdot \|_{\mathbf{W}^{2,2}}^{3/4} \| \cdot \|_{\mathbf{L}^2}^{1/4}$ resp. $\| \cdot \|_{\mathbf{L}^{8/3}} \leq C \| \cdot \|_{\mathbf{L}^2}^{3/4} \| \cdot \|_{\mathbf{W}^{1,2}}^{1/4}$.

Next, we bound the terms on the right-hand side of (6.75) (in summarized form). For this purpose, we again employ Theorem 6.3 (resp. Lemma 6.4, (c), (b)).

$$\begin{aligned} k \sum_{j=0}^{J} \big\{ \| \boldsymbol{\mathcal{R}}^{j+1}(\mathbf{d}) \|_{\mathbf{L}^2} + \| \boldsymbol{\mathcal{R}}^{j+1}(\mathbf{u}) \|_{\mathbf{L}^2} \big\}^2 \leq k^{-1} \sum_{j=0}^{J} \Big( \int_{t_j}^{t_{j+1}} (s - t_j) \, \mathrm{d}s \Big) \\ \times \Big( \int_{t_j}^{t_{j+1}} s \big\{ \| \mathbf{u}_{tt}(s) \|_{\mathbf{L}^2} + \| \mathbf{d}_{tt}(s) \|_{\mathbf{L}^2} \big\}^2 \, \mathrm{d}s \Big) \leq Ck . \end{aligned} \tag{6.77}$$

For the third contribution, we find

$$\begin{aligned} k \sum_{j=0}^{J} \| \mathbf{l}_\varepsilon(\mathbf{d}^{j+1}) \|_{\mathbf{L}^2}^2 &\leq \frac{k}{\varepsilon^2} \sum_{j=0}^{J} \| \langle \boldsymbol{\eta}^{j+1}, \boldsymbol{\eta}^{j+1} + 2\,\mathbf{d}(t_{j+1}) \rangle_{\mathbb{R}^2} \|_{L^2}^2 \| \mathbf{d}(t_{j+1}) \|_{\mathbf{L}^\infty}^2 \\ &+ \frac{k}{\varepsilon^2} \sum_{j=0}^{J} \| \langle \boldsymbol{\eta}^{j+1}, \boldsymbol{\eta}^{j+1} + 2\,\mathbf{d}(t_{j+1}) \rangle_{\mathbb{R}^2} \|_{L^4}^2 \| \boldsymbol{\eta}^{j+1} \|_{\mathbf{L}^4}^2 \qquad (6.78) \\ &\leq \frac{k}{\varepsilon^2} \sum_{j=0}^{J} \big\{ \| \boldsymbol{\eta}^{j+1} \|_{\mathbf{L}^4}^4 + \| \langle \boldsymbol{\eta}^{j+1}, \mathbf{d}(t_{j+1}) \rangle_{\mathbb{R}^2} \|_{L^2}^2 + \| \boldsymbol{\eta}^{j+1} \|_{\mathbf{L}^2}^3 \| \nabla \boldsymbol{\eta}^{j+1} \|_{\mathbf{L}^2}^5 \big\} . \end{aligned}$$

In the last step, we used the interpolation result

$$\| \boldsymbol{\eta}^{j+1} \|_{\mathbf{L}^8}^4 \| \boldsymbol{\eta}^{j+1} \|_{\mathbf{L}^4}^4 \leq C \| \boldsymbol{\eta}^{j+1} \|_{\mathbf{L}^2}^3 \| \boldsymbol{\eta}^{j+1} \|_{\mathbf{W}^{1,2}}^5 .$$

(b) We proceed by induction, assuming (6.72) to be valid (the case '$J = 1$' is trivial): For $k \leq k_0(\tilde{C}_1, \tilde{C}_2, T)$ sufficiently small, we can use (6.72) to control terms on the right-hand side of 'rest$_B$'. Together with (6.77), (6.78), we find

$$\| \mathbf{e}^J \|_{\mathbf{L}^2}^2 + \| \nabla \boldsymbol{\eta}^J \|_{\mathbf{L}^2}^2 + \frac{k^2}{2} \sum_{j=0}^{J-1} \{ \| d_t \mathbf{e}^{j+1} \|_{\mathbf{L}^2}^2 + \| \nabla d_t \boldsymbol{\eta}^{j+1} \|_{\mathbf{L}^2}^2 \} \tag{6.79}$$
$$+ \nu k \sum_{j=0}^{J-1} \{ \| \nabla \mathbf{e}^{j+1} \|_{\mathbf{L}^2} + \varrho \| \Delta \boldsymbol{\eta}^{j+1} \|_{\mathbf{L}^2} \}^2 \leq \tilde{C}_1 k \exp(\tilde{C}_2 t_J) \equiv \overline{C}(t_J) k \,.$$

In particular, we have an upper bound for $\max_{0 \leq j \leq J} \{ \| \Delta \boldsymbol{\eta}^j \|_{\mathbf{L}^2} + \| \nabla \mathbf{e}^j \|_{\mathbf{L}^2} \}$ that is independent from $k$. This is sufficient to complete the induction argument: we can absorb terms that arise to bound 'rest$_A$' through corresponding ones on the left-hand side of (6.75), apart from one term arising from the last but one inequality, where $\| \Delta \boldsymbol{\eta}^j \|_{\mathbf{L}^q}$, $q > 2$ enters the picture. We now sketch the verification of the following bound for $q \searrow 2$,

$$\max_{0 \leq j \leq J-1} \| \Delta \boldsymbol{\eta}^{j+1} \|_{\mathbf{L}^q} \leq \overline{C}_q(t_J) \left\{ 1 + k^{-\alpha(q)} \right\}, \quad \text{and } 0 < \alpha(q) \searrow 0, \, \overline{C}_q \to \infty. \tag{6.80}$$

We copy the argument (6.50) for (6.71). The most crucial term to be handled is then

$$\| d_t \boldsymbol{\eta}^{j+1} \|_{\mathbf{L}^q} + \| \mathbf{l}_\varepsilon (\mathbf{d}^{j+1}) \|_{\mathbf{L}^q}, \quad q > 2.$$

We use the interpolation result $\| \cdot \|_{\mathbf{L}^q} \leq C \| \cdot \|_{\mathbf{L}^2}^{2/q} \| \cdot \|_{\mathbf{W}^{1,2}}^{1-2/q}$ to deal with both terms: to control the first one, we make use of (6.72) together with (6.79) (where negative orders of $k$ enter to the estimate (6.80).

For $\| \mathbf{l}_\varepsilon (\mathbf{d}^{j+1}) \|_{\mathbf{L}^q}$, we follow the argumentation of (6.78): the terms in the last row then appear in slightly stronger norms; however, an interpolation argument together with (6.72), (6.79) then shows (6.80).

We can now come back to the term in question in 'rest$_A$'; choosing $\varepsilon = \mathcal{O}(k^{3\alpha(q)})$ then completes the inductive argument,

$$\| \mathbf{e}^{J+1} \|_{\dot{\mathbf{W}}_{\text{per}}^{-1,2}}^2 + \| \boldsymbol{\eta}^{J+1} \|_{\mathbf{L}^2}^2 + \frac{k^2}{2} \sum_{j=1}^{J+1} \{ \| d_t \mathbf{e}^j \|_{\dot{\mathbf{W}}_{\text{per}}^{-1,2}}^2 + \| d_t \boldsymbol{\eta}^j \|_{\mathbf{L}^2}^2 \} \tag{6.81}$$
$$+ \nu (1 - C(\overline{C}(t_J)) \sqrt{k}) k \sum_{j=1}^{J+1} \{ \| \mathbf{e}^j \|_{\mathbf{L}^2} + \varrho \| \nabla \boldsymbol{\eta}^j \|_{\mathbf{L}^2} \}^2 \leq C(t_{J+1}, \tilde{C}_1, \tilde{C}_2) k^2 \,.$$

By a bootstrap argument, we can again go through the upper bounds of '$\text{rest}_A$' (and '$\text{rest}_B$' as well) to see that multiplied errors in one term disappear of higher order; this may occur by interpolating the error stated in the $\mathbf{L}^4$-norm in the last but one contribution of $\text{rest}_A$ between $\mathbf{L}^2$ and $\mathbf{W}^{1,2}$, and exploit know available bounds for these quantities. Thus, we may recover (6.72) at $J+1$, for $\tilde{C}_1, \tilde{C}_2$ chosen sufficiently large. □

**Corollary 6.1** *Let the assumptions of Theorem 6.5 be valid. Then the solution to* $(\mathrm{P})_k$ *satisfies*

$$\text{(a)} \quad \Big(k\sum_{j=0}^{J} \| 1 - |\mathbf{d}^j|^2 \|_{L^2}^2\Big)^{1/2} \leq C\, k^{3/2}\,,$$

$$\text{(b)} \quad \max_{0\leq j\leq J}\Big\{\| d_t\mathbf{u}^j \|_{\mathbf{L}^2} + \| \nabla d_t\mathbf{d}^j \|_{\mathbf{L}^2} + \| \nabla\mathbf{u}^j \|_{\mathbf{L}^2} + \| \Delta\mathbf{d}^j \|_{\mathbf{L}^2}\Big\} \leq C\,,$$

$$\text{(c)} \quad \max_{1\leq j\leq J}\big\{\| \Delta\mathbf{d}^j \|_{\mathbf{L}^q} + \| \Delta\mathbf{u}^j \|_{\mathbf{L}^2} + \| \nabla p^j \|_{\mathbf{L}^2}\big\} \leq C\,.$$

*where* $C = C(t_J, \omega, \mathbf{d}_0, \mathbf{u}_0, \mathbf{f})$ *does not depend on* $k, \varepsilon$, *and* $q > 2$.

**Proof:**
We start with a sketch of proof for (c): At first, we go through the argument (6.78) (now for $\mathbf{L}^q$, $q>2$); together with part (b) and Cattabriga's theorem ([58], Theorem 5.4), we sharpen (b) to $\max_{1\leq j\leq J} \| \Delta\mathbf{d}^j \|_{\mathbf{L}^q} \leq C_q$, for $q>2$. We may now benefit from this result and (b) when testing (6.66) by $\mathbf{A}\mathbf{u}^{j+1}$.

Item (a) immediately follows from Theorem 6.5 and the relation

$$\begin{aligned} \| \,|\mathbf{d}(t_{j+1})|^2 - |\mathbf{d}^{j+1}|^2 \|_{L^2}^2 &= \| \langle \mathbf{e}^{j+1}, \mathbf{e}^{j+1} + 2\,\mathbf{d}(t_{j+1})\rangle_{\mathbb{R}^2} \|_{L^2}^2 \qquad (6.82)\\ &\leq 2\Big\{\| \mathbf{e}^{j+1} \|_{\mathbf{L}^4}^4 + \| \langle \mathbf{e}^{j+1}, \mathbf{d}(t_{j+1})\rangle_{\mathbb{R}^2} \|_{L^2}^2\Big\}. \end{aligned}$$

The results of (b) are implications of (6.79), (6.81) and Lemma 6.4. □

### 6.3.2 The fully discrete case

Let

$$\begin{aligned}
\mathbf{V}_h &\equiv \big\{\mathbf{v}_h \in \mathbf{L}^2_0(\omega,\mathbb{R}^2) : \mathbf{v}_h\big|_K \in \boldsymbol{\mathcal{P}}_A(K), \text{ for every } K\in\mathcal{T}\big\}, \\
&\quad \text{with norm } \big(\|\cdot\|^2_{\mathbf{L}^2} + \|\nabla_{\mathcal{T}}\cdot\|^2_{\mathbf{L}^2}\big)^{1/2}, \\
L^h &\equiv \big\{q_h \in L^2_0(\omega) : q_h\big|_K \in \omega(K), \text{ for every } K\in\mathcal{T}\big\}, \text{ with norm } \|\cdot\|_{L^2}, \\
\mathbf{W}_h &\equiv \big\{\mathbf{v}_h \in \mathbf{L}^2(\omega,\mathbb{R}^2) : \mathbf{v}_h\big|_K \in \boldsymbol{\mathcal{P}}_B(K), \text{ for every } K\in\mathcal{T}\big\},
\end{aligned}$$

such that *(B1)-(B3)* are valid. $\mathbf{V}_h$ and $\mathbf{W}_h$ may differ by their piecewise polynomial degree $\boldsymbol{\mathcal{P}}_i(K)$, $i\in\{A,B\}$ of basis functions; however, we suppose *(B1)-(B3)* to be valid for both spaces of finite element functions. An appropriate choice is for instance $\mathbf{V}_h = \boldsymbol{\mathcal{S}}^{1,NC}(\omega)$, $\mathbf{W}_h = \boldsymbol{\mathcal{S}}^1(\omega)$, $L^h = \mathcal{S}^0(\omega)/\mathbb{R}$. — We look for triple $\{\mathbf{u}^j_h, p^j_h, \mathbf{d}^j_h\}\in \mathbf{V}_h\times L^h\times\mathbf{W}_h$, for every $1\le j\le J$, such that for every $\{\varphi_h,\chi_h,\boldsymbol{\phi}_h\}\in\mathbf{V}_h\times L^h\times\mathbf{W}_h$,

$$\begin{aligned}
&\big(d_t\mathbf{u}^{j+1}_h,\varphi_h\big)_h + \nu\big(\nabla\mathbf{u}^{j+1}_h,\nabla\varphi_h\big)_h + b_h(\mathbf{u}^j_h,\mathbf{u}^{j+1}_h,\varphi_h) && (6.83)\\
&\quad -\lambda\big(\nabla\mathbf{d}^j_h\odot\nabla\mathbf{d}^{j+1}_h,\nabla\varphi_h\big)_h - \big(p^{j+1}_h,\operatorname{div}\varphi_h\big)_h = \big(\mathbf{f}^{j+1},\varphi_h\big)_h, \\
&\big(d_t\mathbf{d}^{j+1}_h,\boldsymbol{\phi}_h\big)_h + \varrho\big(\nabla\mathbf{d}^{j+1}_h,\nabla\boldsymbol{\phi}_h\big)_h + \big(\mathbf{l}_\varepsilon(\mathbf{d}^{j+1}_h),\boldsymbol{\phi}_h\big)_h && (6.84)\\
&\quad + b(\mathbf{u}^j_h,\mathbf{d}^{j+1}_h,\boldsymbol{\phi}_h) = \varrho\big(|\nabla\mathbf{d}^j_h|^2\mathbf{d}^{j+1}_h,\boldsymbol{\phi}_h\big)_h, \\
&\big(\operatorname{div}\mathbf{u}^{j+1}_h,\chi_h\big)_h = 0, \quad \mathbf{u}^0_h = \mathbf{P}^i_{h,\mathbf{W}^{1,2}}\mathbf{u}^0, \quad \mathbf{d}^0_h = \mathbf{P}^c_{h,\mathbf{W}^{1,2}}\mathbf{d}^0, && (6.85)
\end{aligned}$$

where $\mathbf{P}^i_{h,\mathbf{W}^{1,2}}$, $i\in\{c,nc\}$ refers to conforming or non-conforming discretization; for $\mathbf{u}\in\mathbf{V}_h$, and $\mathbf{v},\mathbf{w}\in\mathbf{X}_h$, for $\mathbf{X}_h\in\{\mathbf{V}_h,\mathbf{W}_h\}$,

$$b(\mathbf{u},\mathbf{v},\mathbf{w}) = (\mathbf{u}\cdot\nabla\mathbf{v},\mathbf{w}), \quad b_h(\mathbf{u},\mathbf{v},\mathbf{w}) \equiv \frac{1}{2}\big\{(\mathbf{u}\cdot\nabla\mathbf{v},\mathbf{w})_h - (\mathbf{u}\cdot\nabla\mathbf{w},\mathbf{v})_h\big\}.$$

Note that $b_h(\mathbf{u},\mathbf{v},\mathbf{v}) = 0$. — We recall some standard results on finite elements for (Navier-) Stokes equations: Let $-\Delta_h^{-1}:\mathbf{V}^*_h\to\mathbf{V}_h$, such that $\mathbf{v}_h = -\Delta_h^{-1}\boldsymbol{\psi}$ satisfies $(\nabla\mathbf{v}_h,\nabla\boldsymbol{\phi}_h)_h = (\boldsymbol{\psi},\boldsymbol{\phi}_h)$, for all $\boldsymbol{\phi}_h\in\mathbf{V}_h$. We denote

$$\begin{aligned}
\mathbf{J}_h &\equiv \big\{\mathbf{v}_h\in\mathbf{V}_h : (\chi_h,\operatorname{div}\mathbf{v}_h)_h = 0, \text{ for all } \chi_h\in L^h\big\}, \text{ with norm } \|\cdot\|_{\mathbf{L}^2}, \\
\mathbf{K}_h &\equiv \big\{\mathbf{v}_h\in\mathbf{V}_h : (\chi_h,\operatorname{div}\mathbf{v}_h)_h = 0, \text{ for all } \chi_h\in L^h\big\}, \\
&\quad \text{with norm } \big(\|\cdot\|^2_{\mathbf{L}^2} + \|\nabla_{\mathcal{T}}\cdot\|^2_{\mathbf{L}^2}\big)^{1/2}.
\end{aligned}$$

We define the projection $\mathbf{P}_{\mathbf{J}_h} : \mathbf{L}^2(\omega) \to \mathbf{J}_h$ through $(\mathbf{u} - \mathbf{P}_{\mathbf{J}_h}\mathbf{u}, \boldsymbol{\phi})_h = 0$, for all $\boldsymbol{\phi}_h \in \mathbf{J}_h$. In case that *(B1)-(B3)* are satisfied, we have [68]

$$\begin{aligned} \| \boldsymbol{\phi} - \mathbf{P}_{\mathbf{J}_h}\boldsymbol{\phi} \|_{\mathbf{L}^2} + h \| \nabla_{\mathcal{T}}(\boldsymbol{\phi} - \mathbf{P}_{\mathbf{J}_h}\boldsymbol{\phi}) \|_{\mathbf{L}^2} & \qquad (6.86) \\ \leq Ch^2 \| \Delta\boldsymbol{\phi} \|_{\mathbf{L}^2} \quad \forall \boldsymbol{\phi} \in \mathbf{J}_1 \cap \mathbf{W}^{2,2}, & \\ \| \boldsymbol{\phi} - \mathbf{P}_{\mathbf{J}_h}\boldsymbol{\phi} \|_{\mathbf{L}^2} \leq Ch \| \nabla\boldsymbol{\phi} \|_{\mathbf{L}^2} \quad \forall \boldsymbol{\phi} \in \mathbf{J}_1 . & \qquad (6.87) \end{aligned}$$

We need a discrete version of Sobolev's inequalities. The following result is similar to Lemma 4.4 in [68], and we adopt its proof to the present case.

**Lemma 6.7** *Suppose that (B1),(B2),(B3) are valid, with* $\omega = [0, 2D]^2$. *There exists a constant* $C(\mathcal{T})$ *that is independent of* $h$, *such that the following inequalities hold true, for all* $\boldsymbol{\phi} \in \dot{\mathbf{W}}^{1,2}_{\text{per}} + \mathbf{V}_h$,

$$\| \boldsymbol{\phi} \|_{\mathbf{L}^4} \leq C \| \nabla_{\mathcal{T}}\boldsymbol{\phi} \|^{1/2}_{\mathbf{L}^2} \| \boldsymbol{\phi} \|^{1/2}_{\mathbf{L}^2} . \qquad (6.88)$$

**Proof:**
Let $\boldsymbol{\phi}_h \in \dot{\mathbf{W}}^{1,2}_{\text{per}} + \mathbf{V}_h$, and define $\mathbf{w} \in \dot{\mathbf{W}}^{1,2}_{\text{per}}$ that solves $\Delta\mathbf{w} = \Delta_h\boldsymbol{\phi}_h$. By standard approximation results on quasi-uniform meshes $\mathcal{T}$ (see also (B2)), there holds

$$\| \mathbf{w} - \boldsymbol{\phi}_h \|_{\mathbf{L}^2} + h \| \nabla_{\mathcal{T}}[\mathbf{w} - \boldsymbol{\phi}_h] \|_{\mathbf{L}^2} \leq Ch \| \nabla_{\mathcal{T}}\boldsymbol{\phi}_h \|_{\mathbf{L}^2} . \qquad (6.89)$$

Hence, we find $\| \nabla\mathbf{w} \|_{\mathbf{L}^2} \leq C \| \nabla_{\mathcal{T}}\boldsymbol{\phi}_h \|_{\mathbf{L}^2}$, in particular. In the sequel, we use the piecewise constant interpolant $\overline{\boldsymbol{\phi}_h} = | K |^{-1} \int_K \boldsymbol{\phi}_h \, \mathrm{d}\mathbf{x}$. Then, an elementwise application of (B2) (generalized to arbitrary $\mathbf{L}^p$-spaces, see Theorem 16.1 in [29]) is possible, and we find

$$\begin{aligned} \| \boldsymbol{\phi}_h \|_{\mathbf{L}^4} &\leq \| \boldsymbol{\phi}_h - \overline{\boldsymbol{\phi}_h} \|_{\mathbf{L}^4} + \| \overline{\boldsymbol{\phi}_h - \mathbf{w}} \|_{\mathbf{L}^4} + \| \overline{\mathbf{w}} - \mathbf{w} \|_{\mathbf{L}^4} + \| \mathbf{w} \|_{\mathbf{L}^4} \\ &\leq C \big\{ h^{1/2} \| \nabla_{\mathcal{T}}\boldsymbol{\phi}_h \|_{\mathbf{L}^2} + \| \overline{\boldsymbol{\phi}_h - \mathbf{w}} \|_{\mathbf{L}^4} \\ &\qquad + h^{1/2} \| \nabla\mathbf{w} \|_{\mathbf{L}^2} + \| \mathbf{w} \|^{1/2}_{\mathbf{L}^2} \| \nabla\mathbf{w} \|^{1/2}_{\mathbf{L}^2} \big\} \\ &\leq C \big\{ \| \boldsymbol{\phi}_h \|^{1/2}_{\mathbf{L}^2} \| \nabla_{\mathcal{T}}\boldsymbol{\phi}_h \|^{1/2}_{\mathbf{L}^2} + \| \overline{\boldsymbol{\phi}_h - \mathbf{w}} \|_{\mathbf{L}^4} \big\} , \end{aligned}$$

using the decomposition $\| \mathbf{w} - \boldsymbol{\phi}_h \|_{\mathbf{L}^2} + \| \boldsymbol{\phi}_h \|_{\mathbf{L}^2}$, (6.89), and inverse inequality. Finally, by inverse inequality, (6.89), and $\| \nabla\mathbf{w} \|_{\mathbf{L}^2} \leq C \| \nabla_{\mathcal{T}}\boldsymbol{\phi}_h \|_{\mathbf{L}^2}$, we obtain

$$\| \overline{\boldsymbol{\phi}_h - \mathbf{w}} \|_{\mathbf{L}^4} \leq Ch^{-1/2} \| \boldsymbol{\phi}_h - \mathbf{w} \|_{\mathbf{L}^2} \leq C \| \boldsymbol{\phi}_h \|^{1/2}_{\mathbf{L}^2} \| \nabla_{\mathcal{T}}\boldsymbol{\phi}_h \|^{1/2}_{\mathbf{L}^2} .$$

□

**Remark 6.8** *There holds, for all* $\boldsymbol{\phi} \in \mathbf{W}^{1,2}_{\text{per}} + \mathbf{W}_h$,

$$\| \boldsymbol{\phi} \|_{\mathbf{L}^4} \le C \left\{ \| \boldsymbol{\phi} \|^{1/2}_{\mathbf{L}^2} + \| \nabla_{\mathcal{T}} \boldsymbol{\phi} \|^{1/2}_{\mathbf{L}^2} \right\} \| \boldsymbol{\phi} \|^{1/2}_{\mathbf{L}^2} .$$

*This result can be verified with* $\tilde{\Delta}$, $\tilde{\Delta}_{\mathcal{T}}$ *instead of* $\Delta$, $\Delta_{\mathcal{T}}$ *in the previous proof.*

We consider different projections onto finite element spaces, for $i \in \{c, nc\}$: the $\mathbf{L}^2$-projection, defined by $(\boldsymbol{\phi} - \mathbf{P}^i_{h,\mathbf{L}^2}\boldsymbol{\phi}, \boldsymbol{\varphi}_h) = 0$, the $\mathbf{W}^{1,2}$-projection, defined by $(\boldsymbol{\phi} - \mathbf{P}^i_{h,\mathbf{W}^{1,2}}\boldsymbol{\phi}, \boldsymbol{\varphi}_h) + (\nabla_{\mathcal{T}}[\boldsymbol{\phi} - \mathbf{P}^i_{h,\mathbf{W}^{1,2}}\boldsymbol{\phi}], \nabla_{\mathcal{T}}\boldsymbol{\varphi}_h) = 0$.

In the sequel, it is crucial to have control on the following terms. $\Gamma^1_h, \Gamma^2_h, \Gamma^3_h$ are analyzed in [68]. A face of a triangle K is denoted by $\Gamma \subset \partial K$, with normal $\mathbf{n}_\Gamma$ pointing to the exterior.

$$\Gamma^1_h(\mathbf{v}, \boldsymbol{\varphi}) = \sum_{K \in \mathcal{T}} \int_{\partial K} \langle \partial_{\mathbf{n}_\Gamma} \mathbf{v}, \boldsymbol{\varphi} \rangle_{\mathbb{R}^2} \, ds , \tag{6.90}$$

$$\Gamma^2_h(\mathbf{v}, \boldsymbol{\psi}, \boldsymbol{\varphi}) = \sum_{K \in \mathcal{T}} \int_{\partial K} \langle \mathbf{v}, \mathbf{n}_\Gamma \rangle_{\mathbb{R}^2} \langle \boldsymbol{\psi}, \boldsymbol{\varphi} \rangle_{\mathbb{R}^2} \, ds , \tag{6.91}$$

$$\Gamma^3_h(q, \boldsymbol{\varphi}) = \sum_{K \in \mathcal{T}} \int_{\partial K} q \langle \boldsymbol{\varphi}, \mathbf{n}_\Gamma \rangle_{\mathbb{R}^2} \, ds , \tag{6.92}$$

$$\Gamma^4_h(\mathbf{v}, \mathbf{w}, \boldsymbol{\varphi}) = \sum_{K \in \mathcal{T}} \int_{\partial K} \langle (\nabla \mathbf{v} \odot \nabla \mathbf{w}) \mathbf{n}_\Gamma, \boldsymbol{\varphi} \rangle_{\mathbb{R}^2} \, ds . \tag{6.93}$$

**Lemma 6.8** *(partly from [68]) Suppose that (B1),(B2),(B3) are satisfied. Then, for all* $\mathbf{v}, \boldsymbol{\psi} \in \mathbf{J}_1 \cap \dot{\mathbf{W}}^{2,2}_{\text{per}}$, $\rho \in W^{1,2}$, $\mathbf{a}, \mathbf{b} \in \mathbf{W}^{2,2}_{\text{per}}$, *and* $\boldsymbol{\varphi} \in (\mathbf{J}_1 + \mathbf{K}_h)$, *there exists a constant* $C = C(\mathcal{T})$ *that does not depend on* $h$, *such that*

(a) $|\Gamma^1_h(\mathbf{v}, \boldsymbol{\varphi})| \le Ch \| \nabla_{\mathcal{T}} \boldsymbol{\varphi} \|_{\mathbf{L}^2} \| \Delta \mathbf{v} \|_{\mathbf{L}^2}$,

(b) $|\Gamma^2_h(\mathbf{v}, \boldsymbol{\psi}, \boldsymbol{\varphi})| + |\Gamma^2_h(\boldsymbol{\psi}, \mathbf{v}, \boldsymbol{\varphi})|$
$\le Ch \| \nabla_{\mathcal{T}} \boldsymbol{\varphi} \|_{\mathbf{L}^2} \{ \| \nabla \mathbf{v} \|_{\mathbf{L}^2} \| \boldsymbol{\psi} \|_{\mathbf{L}^\infty} + \| \nabla \boldsymbol{\psi} \|_{\mathbf{L}^2} \| \mathbf{v} \|_{\mathbf{L}^\infty} \}$,

(c) $|\Gamma^3_h(\rho, \boldsymbol{\varphi})| \le Ch \| \nabla_{\mathcal{T}} \boldsymbol{\varphi} \|_{\mathbf{L}^2} \| \nabla \rho \|_{L^2}$,

(d) $|\Gamma^4_h(\mathbf{a}, \mathbf{b}, \boldsymbol{\varphi})| \le Ch \| \nabla_{\mathcal{T}} \boldsymbol{\varphi} \|_{\mathbf{L}^2} \{ \| \Delta \mathbf{a} \|_{\mathbf{L}^2} \| \nabla \mathbf{b} \|_{\mathbf{L}^\infty} + \| \Delta \mathbf{b} \|_{\mathbf{L}^2} \| \nabla \mathbf{a} \|_{\mathbf{L}^\infty} \}$.

**Proof:**
(Sketch) Items (a) to (c) can be found in [68]. We can follow the same strategy as in [68], Lemma 4.1, to verify (d): we may assume $\boldsymbol{\varphi} \in \mathbf{K}_h$.

$$\begin{aligned} |\Gamma^4_h(\mathbf{a}, \mathbf{b}, \boldsymbol{\varphi})| &\le Ch \| \nabla_{\mathcal{T}} \boldsymbol{\varphi}_h \|_{\mathbf{L}^2} \| \nabla [\nabla \mathbf{a} \odot \nabla \mathbf{b}] \|_{\mathbf{L}^2} \\ &\le Ch \| \nabla_{\mathcal{T}} \boldsymbol{\varphi}_h \|_{\mathbf{L}^2} \{ \| \Delta \mathbf{a} \|_{\mathbf{L}^2} \| \nabla \mathbf{b} \|_{\mathbf{L}^\infty} + \| \Delta \mathbf{b} \|_{\mathbf{L}^2} \| \nabla \mathbf{a} \|_{\mathbf{L}^\infty} \} . \end{aligned}$$

□

The main result in this section is formulated in the following theorem.

**Theorem 6.6** *Suppose that* $\{\mathbf{u}, p, \mathbf{d}\}$ *solves* (P), $\{\mathbf{u}_h^j, p_h^j, \mathbf{d}_h^j\}_{j=0}^J$ *solves* (6.83)-(6.85) *for* $\varepsilon^{-1} = o(\frac{1}{k})$, $k^{-1/2} = o(\frac{1}{h})$, *and (A2),(B1),(B2),(B3) are satisfied. Let* $\{k, h\} \leq \{k_0(t_J), h_0(t_J)\}$. *here exists a constant* $C = C(\omega, t_J, \mathbf{u}_0, \mathbf{d}_0; \mathcal{T})$ *that does not depend on* $k, \varepsilon, h$, *such that*

$$\max_{0\leq j\leq J}\Big\{\| \mathbf{A}^{-1/2}(\mathbf{u}(t_j) - \mathbf{u}_h^j) \|_{\mathbf{L}^2} + \| \mathbf{d}(t_j) - \mathbf{d}_h^j \|_{\mathbf{L}^2}\Big\} + \Big(k\sum_{j=0}^{J}\{\| \mathbf{u}(t_j) - \mathbf{u}_h^j \|_{\mathbf{L}^2}^2$$
$$+\| \mathbf{d}(t_j) - \mathbf{d}_h^j \|_{\mathbf{W}^{1,2}}^2\} + \frac{1}{\varepsilon}\| \mathbf{d}(t_j) - \mathbf{d}_h^j \|_{\mathbf{L}^4}^4\}\Big)^{1/2} \leq C\{k + h\}.$$

*Moreover, there holds*

$$\max_{0\leq j\leq J} \| \, | \mathbf{d}_h^j |^2 - 1 \, \|_{L^2} \leq C\,\{k + h\}.$$

**Proof:**
*1st step:* If we take into account the non-conformity of finite elements choices, we have for all $\boldsymbol{\varphi}_h \in \mathbf{V}_h$, and $\boldsymbol{\phi}_h \in \mathbf{W}_h$,

$$\begin{aligned}
&\left(d_t\mathbf{u}^{j+1}, \boldsymbol{\varphi}_h\right)_h + \nu\left(\nabla\mathbf{u}^{j+1}, \nabla\boldsymbol{\varphi}_h\right)_h + b_h(\mathbf{u}^j, \mathbf{u}^{j+1}, \boldsymbol{\varphi}_h) && (6.94)\\
&\quad -\lambda\left(\nabla\mathbf{d}^j \odot \nabla\mathbf{d}^{j+1}, \nabla\boldsymbol{\varphi}_h\right)_h - \left(p^{j+1}, \operatorname{div}\boldsymbol{\varphi}_h\right)_h\\
&\qquad = \left(\mathbf{f}^{j+1}, \boldsymbol{\varphi}_h\right)_h + \Gamma_h\left(\mathbf{u}^{j+1}, p^{j+1}, \mathbf{d}^{j+1}, \boldsymbol{\varphi}_h\right),\\
&\left(d_t\mathbf{d}^{j+1}, \boldsymbol{\phi}_h\right)_h + \varrho\left(\nabla\mathbf{d}^{j+1}, \nabla\boldsymbol{\phi}_h\right)_h + \left(\mathbf{l}_\varepsilon(\mathbf{d}^{j+1}), \boldsymbol{\phi}_h\right)_h && (6.95)\\
&\quad +b_h(\mathbf{u}^j, \mathbf{d}^{j+1}, \boldsymbol{\phi}_h) = \varrho\left(| \nabla\mathbf{d}^j |^2\mathbf{d}_h^{j+1}, \boldsymbol{\phi}_h\right)_h + \Gamma_h^2(\mathbf{u}^j, \mathbf{d}^{j+1}, \boldsymbol{\phi}_h),
\end{aligned}$$

where

$$\begin{aligned}
\Gamma_h\left(\mathbf{u}^{j+1}, p^{j+1}, \mathbf{d}^{j+1}, \boldsymbol{\varphi}_h\right) &= \Gamma_h^1\left(\mathbf{u}^{j+1}, \boldsymbol{\varphi}_h\right) + \Gamma_h^2\left(\mathbf{u}^j, \mathbf{u}^{j+1}, \boldsymbol{\varphi}_h\right) && (6.96)\\
&\quad +\Gamma_h^3\left(p^{j+1}, \boldsymbol{\varphi}_h\right) + \Gamma_h^4\left(\mathbf{d}^j, \mathbf{d}^{j+1}, \boldsymbol{\varphi}_h\right).
\end{aligned}$$

Let $\{\mathbf{e}^j, \boldsymbol{\eta}^j\} = \{\mathbf{u}^j - \mathbf{u}_h^j, \mathbf{d}^j - \mathbf{d}_h^j\}$. We subtract equations (6.94), (6.95) from (6.83), (6.84), and choose the test functions $\{\mathbf{A}_h^{-1}\mathbf{e}^{j+1}, \mathbf{P}_{h;\mathbf{L}^2}^c\boldsymbol{\eta}^{j+1}\}$, for $\mathbf{A}_h : \mathbf{V}_h \to \mathbf{V}_h^*$ the discretized Stokes operator. Note that

$$\begin{aligned}
(\nabla_{\mathcal{T}}\mathbf{e}^{j+1}, \nabla_{\mathcal{T}}\mathbf{A}_h^{-1}\mathbf{e}^{j+1})_h &= (\nabla_{\mathcal{T}}[\mathbf{e}^{j+1} \pm \mathbf{P}_{\mathbf{J}_h}\mathbf{u}^{j+1}], \nabla_{\mathcal{T}}\mathbf{A}_h^{-1}\mathbf{e}^{j+1})_h && (6.97)\\
&= \| \mathbf{P}_{\mathbf{J}_h}\mathbf{e}^{j+1} \|_{\mathbf{L}^2}^2 + (\nabla_{\mathcal{T}}[\mathbf{u}^{j+1} - \mathbf{P}_{\mathbf{J}_h}\mathbf{u}^{j+1}], \nabla_{\mathcal{T}}\mathbf{A}_h^{-1}\mathbf{e}^{j+1})_h,
\end{aligned}$$

$$\begin{aligned} |\,(p^{j+1}, \operatorname{div}\varphi_h)_h\,| &= |\,(p^{j+1} - j_h p^{j+1}, \operatorname{div}\varphi_h)_h\,| \\ &\leq Ch \,\|\, \nabla p^{j+1} \,\|_{\mathbf{L}^2} \,\|\, \nabla_{\mathcal{T}} \varphi_h \,\|_{\mathbf{L}^2}\,, \end{aligned} \tag{6.98}$$

$$\begin{aligned} &|\,(\Gamma_h(\mathbf{u}^{j+1}, p^{j+1}, \mathbf{d}^{j+1}, \varphi_h)\,| \\ &\leq C_q h \,\|\, \nabla_{\mathcal{T}} \varphi_h \,\|_{\mathbf{L}^2} \Big\{ \|\, \mathbf{u}^{j+1} \,\|_{\mathbf{W}^{2,2}} \|\, \mathbf{u}^j \,\|_{\mathbf{W}^{2,2}}^{3/4} \,\|\, \mathbf{u}^j \,\|_{\mathbf{L}^2}^{1/4} \,\|\, \nabla \mathbf{u}^{j+1} \,\|_{\mathbf{L}^2}\,, \\ &\quad + \|\, \mathbf{u}^{j+1} \,\|_{\mathbf{W}^{2,2}}^{3/4} \,\|\, \mathbf{u}^{j+1} \,\|_{\mathbf{L}^2}^{1/4} \,\|\, \mathbf{u}^j \,\|_{\mathbf{W}^{1,2}} + \|\, \nabla p^{j+1} \,\|_{\mathbf{L}^2} \\ &\quad + \|\, \mathbf{d}^j \,\|_{\mathbf{W}^{2,2}} \,\|\, \mathbf{d}^{j+1} \,\|_{\mathbf{W}^{2,q}} + \|\, \mathbf{d}^{j+1} \,\|_{\mathbf{W}^{2,2}} \,\|\, \mathbf{d}^j \,\|_{\mathbf{W}^{2,q}} \Big\}, \quad q > 2\,, \end{aligned} \tag{6.99}$$

$$\begin{aligned} |\,\Gamma_h^2(\mathbf{u}^j, \mathbf{d}^{j+1}, \varphi_h)\,| \;\leq\; & Ch \,\|\, \nabla_{\mathcal{T}} \varphi_h \,\|_{\mathbf{L}^2} \Big\{ \|\, \mathbf{u}^j \,\|_{\mathbf{L}^2}^{1/2} \,\|\, \mathbf{u}^j \,\|_{\mathbf{W}^{2,2}}^{1/2} \,\|\, \nabla \mathbf{d}^{j+1} \,\|_{\mathbf{L}^2} \\ & + \|\, \mathbf{d}^{j+1} \,\|_{\mathbf{L}^2}^{1/2} \,\|\, \mathbf{d}^{j+1} \,\|_{\mathbf{W}^{2,2}}^{1/2} \,\|\, \nabla \mathbf{u}^j \,\|_{\mathbf{L}^2} \Big\}, \end{aligned} \tag{6.100}$$

$$\begin{aligned} &\varepsilon \Big( \mathbf{l}_\varepsilon(\mathbf{d}^{j+1}) - \mathbf{l}_\varepsilon(\mathbf{d}_h^{j+1}), \boldsymbol{\eta}^{j+1} \Big)_h \\ &= \Big( \big[|\, \mathbf{d}^{j+1} \,|^2 - 1\big] \mathbf{d}^{j+1} - \big[|\, \mathbf{d}_h^{j+1} \,|^2 - 1\big] \mathbf{d}_h^{j+1}, \boldsymbol{\eta}^{j+1} \Big)_h \\ &= \Big( \langle \boldsymbol{\eta}^{j+1}, 2\mathbf{d}^{j+1} - \boldsymbol{\eta}^{j+1} \rangle_{\mathbb{R}^2} \mathbf{d}^{j+1}, \boldsymbol{\eta}^{j+1} \Big)_h + \Big( \big[|\, \mathbf{d}_h^{j+1} \,|^2 - 1\big], |\, \boldsymbol{\eta}^{j+1} \,|^2 \Big)_h \\ &= 2 \,\|\, \langle \boldsymbol{\eta}^{j+1}, \mathbf{d}^{j+1} \rangle_{\mathbb{R}^2} \,\|_{\mathbf{L}^2}^2 - \Big( |\, \boldsymbol{\eta}^{j+1} \,|^2, \langle \boldsymbol{\eta}^{j+1}, \mathbf{d}^{j+1} \rangle_{\mathbb{R}^2} \Big)_h \\ &\quad - \Big( |\, \boldsymbol{\eta}^{j+1} \,|^2, \langle \boldsymbol{\eta}^{j+1}, 2\mathbf{d}^{j+1} - \boldsymbol{\eta}^{j+1} \rangle_{\mathbb{R}^2} \Big)_h + \Big( |\, \boldsymbol{\eta}^{j+1} \,|^2, \big[|\, \mathbf{d}^{j+1} \,|^2 - 1\big] \Big)_h, \end{aligned} \tag{6.101}$$

$$\begin{aligned} &\varepsilon \Big( \mathbf{l}_\varepsilon(\mathbf{d}^j) - \mathbf{l}_\varepsilon(\mathbf{d}_h^j), \mathbf{d}^j - \mathbf{P}_{h;\mathbf{L}^2}^c \mathbf{d}^j \Big)_h \\ &= \Big( \big[|\, \mathbf{d}^j \,|^2 - 1\big] \mathbf{d}^j - \big[|\, \mathbf{d}^j - \boldsymbol{\eta}^j \,|^2 - 1\big] \, [\mathbf{d}^j - \boldsymbol{\eta}^j], \mathbf{d}^j - \mathbf{P}_{h;\mathbf{L}^2}^c \mathbf{d}^j \Big)_h \\ &= \Big( 2 \,(\langle \mathbf{d}^j, \boldsymbol{\eta}^j \rangle_{\mathbb{R}^2} - |\, \boldsymbol{\eta}^j \,|^2) [\mathbf{d}^j - \boldsymbol{\eta}^j], \mathbf{d}^j - \mathbf{P}_{h;\mathbf{L}^2}^c \mathbf{d}^j \Big)_h \\ &\quad + \Big( \big[|\, \mathbf{d}^j \,|^2 - 1\big] \boldsymbol{\eta}^j, \mathbf{d}^j - \mathbf{P}_{h;\mathbf{L}^2}^c \mathbf{d}^j \Big)_h \\ &\leq \Big\{ \|\, \langle \mathbf{d}^j, \boldsymbol{\eta}^j \rangle_{\mathbb{R}^2} \,\|_{\mathbf{L}^2} + \|\, \boldsymbol{\eta}^j \,\|_{\mathbf{L}^4}^2 \Big\} \Big\{ \|\, \mathbf{d}^j \,\|_{\mathbf{L}^\infty} \,\|\, \mathbf{d}^j - \mathbf{P}_{h;\mathbf{L}^2}^c \mathbf{d}^j \,\|_{\mathbf{L}^2} \\ &\quad + \|\, \boldsymbol{\eta}^j \,\|_{\mathbf{L}^4} \,\|\, \mathbf{d}^j - \mathbf{P}_{h;\mathbf{L}^2}^c \mathbf{d}^j \,\|_{\mathbf{L}^4} \Big\} \\ &\quad + \|\, |\, \mathbf{d}^j \,|^2 - 1 \,\|_{\mathbf{L}^2} \,\|\, \boldsymbol{\eta}^j \,\|_{\mathbf{L}^4} \,\|\, \mathbf{d}^j - \mathbf{P}_{h;\mathbf{L}^2}^c \mathbf{d}^j \,\|_{\mathbf{L}^4}\,. \end{aligned} \tag{6.102}$$

Together with these results, we obtain

$$\begin{aligned}
&\frac{1}{2} d_t\Big\{\| \mathbf{A}_h^{-1/2}\mathbf{e}^{j+1} \|_{\mathbf{L}^2}^2 + \| \boldsymbol{\eta}^{j+1} \|_{\mathbf{L}^2}^2\Big\} + \nu \| \mathbf{e}^{j+1} \|_{\mathbf{L}^2}^2 + \frac{\varrho}{2} \| \nabla\boldsymbol{\eta}^{j+1} \|_{\mathbf{L}^2}^2 \\
&\quad + \frac{k}{2}\Big\{\| \mathbf{A}_h^{-1/2} d_t\mathbf{e}^{j+1} \|_{\mathbf{L}^2}^2 + \| d_t\boldsymbol{\eta}^{j+1} \|_{\mathbf{L}^2}^2\Big\} \qquad (6.103)\\
&\quad + \frac{1}{8\varepsilon}\Big\{2 \| \langle\boldsymbol{\eta}^{j+1}, \mathbf{d}^{j+1}\rangle_{\mathbb{R}^2} \|_{\mathbf{L}^2}^2 + \| \boldsymbol{\eta}^{j+1} \|_{\mathbf{L}^4}^4\Big\} \\
&\leq \nu \| \mathbf{u}^{j+1} - \mathbf{P}_{\mathbf{J}_h}\mathbf{u}^{j+1} \|_{\mathbf{L}^2}^2 + \frac{\varrho}{2} \| \nabla_{\mathcal{T}}\big[\mathbf{d}^{j+1} - \mathbf{P}^c_{h;\mathbf{L}^2}\mathbf{d}^{j+1}\big] \|_{\mathbf{L}^2}^2 \\
&\quad + \big| \big(| \boldsymbol{\eta}^{j+1} |^2, \big[| \mathbf{d}^{j+1} |^2 - 1\big]\big) \big| \\
&\quad + Ch^2\Big\{\| \nabla p^{j+1} \|_{\mathbf{L}^2}^2 + \| \Delta\mathbf{u}^{j+1} \|_{\mathbf{L}^2}^2\Big\} + \| \mathbf{A}_h^{-1/2}\mathbf{e}^{j+1} \|_{\mathbf{L}^2}^2 + \mathrm{rest}_C\,.
\end{aligned}$$

The term '$\mathrm{rest}_C$' gathers terms together that correspond to (most of) those in '$\mathrm{rest}_A$' in (6.73); it is by means of Lemma 6.7 that terms can be treated similarly. Norms $\| \cdot \|_{\mathbf{W}^{2,q}}$, $q > 2$ are avoided for finite element functions by using inverse inequalities. Moreover, $\| \varphi_h \|_{\mathbf{L}^\infty} \leq C\{\| \varphi_h \|_{\mathbf{L}^2} + \| \nabla_{\mathcal{T}}\varphi_h \|_{\mathbf{L}^2}\}$.

$$\begin{aligned}
\| \mathbf{e}^j \|_{\mathbf{L}^2} \| \nabla\mathbf{u}^{j+1} \|_{\mathbf{L}^2} \| \mathbf{A}_h^{-1}\mathbf{e}^{j+1} \|_{\mathbf{L}^\infty} &\leq \frac{\nu}{16} \| \mathbf{e}^j \|_{\mathbf{L}^2}^2 + \\
&\quad C \| \nabla\mathbf{u}^{j+1} \|_{\mathbf{L}^2}^2 \| \nabla_{\mathcal{T}}\mathbf{A}_h^{-1}\mathbf{e}^{j+1} \|_{\mathbf{L}^2}^2 , \\
\| \mathbf{e}^j \|_{\mathbf{L}^\infty} \| \nabla_{\mathcal{T}}\mathbf{A}_h^{-1}\mathbf{e}^{j+1} \|_{\mathbf{L}^2} \| \mathbf{e}^{j+1} \|_{\mathbf{L}^2} &\leq \frac{\nu}{16} \| \mathbf{e}^{j+1} \|_{\mathbf{L}^2}^2 + \\
&\quad C \| \mathbf{e}^j \|_{\mathbf{L}^\infty}^2 \| \nabla_{\mathcal{T}}\mathbf{A}_h^{-1}\mathbf{e}^{j+1} \|_{\mathbf{L}^2}^2 , \\
\| \mathbf{u}^j \|_{\mathbf{L}^\infty} \| \nabla_{\mathcal{T}}\mathbf{A}_h^{-1}\mathbf{e}^{j+1} \|_{\mathbf{L}^2} \| \mathbf{e}^{j+1} \|_{\mathbf{L}^2} &\leq \frac{\nu}{16} \| \mathbf{e}^{j+1} \|_{\mathbf{L}^2}^2 + \\
&\quad C \| \mathbf{u}^j \|_{\mathbf{L}^\infty}^2 \| \nabla_{\mathcal{T}}\mathbf{A}_h^{-1}\mathbf{e}^{j+1} \|_{\mathbf{L}^2}^2 , \\
\| \nabla\boldsymbol{\eta}^j \|_{\mathbf{L}^2} \| \nabla\mathbf{d}^{j+1} \|_{\mathbf{L}^\infty} \| \nabla_{\mathcal{T}}\mathbf{A}_h^{-1}\mathbf{e}^{j+1} \|_{\mathbf{L}^2} &\leq \frac{\varrho}{16} \| \nabla\boldsymbol{\eta}^j \|_{\mathbf{L}^2}^2 + \\
&\quad C \| \nabla\mathbf{d}^{j+1} \|_{\mathbf{L}^\infty}^2 \| \mathbf{A}_h^{-1/2}\mathbf{e}^{j+1} \|_{\mathbf{L}^2}^2 , \\
\| \nabla\mathbf{P}^c_{h;\mathbf{L}^2}\boldsymbol{\eta}^j \|_{\mathbf{L}^\infty} \| \nabla\boldsymbol{\eta}^{j+1} \|_{\mathbf{L}^2} & \\
\times\big\{\| \nabla_{\mathcal{T}}\big[\mathbf{A}^{-1} - \mathbf{A}_h^{-1}\big]\mathbf{e}^{j+1} \|_{\mathbf{L}^2} & \\
+\| \nabla\mathbf{A}^{-1}\mathbf{e}^{j+1} \|_{\mathbf{L}^2}\big\} &\leq Ch^{-1} \| \nabla\mathbf{P}^c_{h;\mathbf{L}^2}\boldsymbol{\eta}^j \|_{\mathbf{L}^2} \| \nabla\boldsymbol{\eta}^{j+1} \|_{\mathbf{L}^2} \\
&\quad \times\big\{h \| \mathbf{e}^{j+1} \|_{\mathbf{L}^2} + \| \mathbf{A}^{-1/2}\mathbf{e}^{j+1} \|_{\mathbf{L}^2}\big\} \\
\| \nabla\mathbf{d}^{j+1} \|_{\mathbf{L}^\infty} \| \nabla\boldsymbol{\eta}^{j+1} \|_{\mathbf{L}^2} \| \nabla_{\mathcal{T}}\mathbf{A}_h^{-1}\mathbf{e}^{j+1} \|_{\mathbf{L}^2} &\leq \frac{\varrho}{16}\| \nabla\boldsymbol{\eta}^{j+1} \|_{\mathbf{L}^2}^2 + C\| \mathbf{A}_h^{-1/2}\mathbf{e}^{j+1} \|_{\mathbf{L}^2}^2 \\
\| \mathbf{e}^{j-1} \|_{\mathbf{L}^2} \| \nabla\mathbf{d}^j \|_{\mathbf{L}^\infty} \| \boldsymbol{\eta}^j \|_{\mathbf{L}^2} &\leq \frac{\nu}{16}\| \mathbf{e}^{j-1} \|_{\mathbf{L}^2}^2 + C\| \nabla\mathbf{d}^j \|_{\mathbf{L}^\infty}^2 \| \boldsymbol{\eta}^j \|_{\mathbf{L}^2}^2
\end{aligned}$$

$$\begin{aligned}
|(|\nabla \mathbf{d}^j|^2\boldsymbol{\eta}^{j+1}, \boldsymbol{\eta}^{j+1})| &\le \|\nabla \mathbf{d}^j\|_{\mathbf{L}^\infty}^2 \|\boldsymbol{\eta}^{j+1}\|_{\mathbf{L}^2}^2\,, \\
\|\nabla\boldsymbol{\eta}^j\|_{\mathbf{L}^2} \|\nabla \mathbf{d}^j\|_{\mathbf{L}^\infty}^2 \|\boldsymbol{\eta}^{j+1}\|_{\mathbf{L}^2} &\le \frac{\varrho}{16}\|\nabla\boldsymbol{\eta}^j\|_{\mathbf{L}^2}^2 + C\|\nabla \mathbf{d}^j\|_{\mathbf{L}^\infty}^2 \|\boldsymbol{\eta}^{j+1}\|_{\mathbf{L}^2}^2\,.
\end{aligned}$$

Next, we only list the most crucial terms; we also refer to (4.125) for a similar study: let $q > 2$,

$$\begin{aligned}
\|\nabla \mathbf{P}^c_{h;\mathbf{L}^2}\boldsymbol{\eta}^j\|_{\mathbf{L}^2}^2 \|\mathbf{d}^{j+1}\|_{\mathbf{L}^\infty} \|\mathbf{P}^c_{h;\mathbf{L}^2}\boldsymbol{\eta}^{j+1}\|_{\mathbf{L}^\infty} &\le \frac{\varrho}{16}\|\nabla\boldsymbol{\eta}^{j+1}\|_{\mathbf{L}^2}^2 + \\
&\quad \|\nabla \mathbf{P}^c_{h;\mathbf{L}^2}\boldsymbol{\eta}^j\|_{\mathbf{L}^2}^4 \|\mathbf{d}^{j+1}\|_{\mathbf{L}^\infty}^2\,, \\
\|\nabla \mathbf{P}^c_{h;\mathbf{L}^2}\boldsymbol{\eta}^j\|_{\mathbf{L}^2} \|\nabla \mathbf{P}^c_{h;\mathbf{L}^2}\boldsymbol{\eta}^j\|_{\mathbf{L}^q} \|\boldsymbol{\eta}^{j+1}\|^2_{\mathbf{L}^{\frac{4q}{q-2}}} &\le Ch^{\frac{2-q}{q}}\|\nabla \mathbf{P}^c_{h;\mathbf{L}^2}\boldsymbol{\eta}^j\|_{\mathbf{L}^2}^2 \|\nabla\boldsymbol{\eta}^{j+1}\|^2_{\mathbf{L}^{\frac{4q}{q-2}}}\,, \\
\|\nabla\boldsymbol{\eta}^j\|_{\mathbf{L}^2} \|\nabla \mathbf{d}^j\|_{\mathbf{L}^\infty} \|\boldsymbol{\eta}^{j+1}\|_{\mathbf{L}^4}^2 &\le \frac{\varrho}{16}\|\nabla\boldsymbol{\eta}^{j+1}\|_{\mathbf{L}^4}^2 + C_\varrho\|\nabla\boldsymbol{\eta}^j\|_{\mathbf{L}^2}^2 \\
&\quad \times\|\nabla \mathbf{d}^{j+1}\|_{\mathbf{L}^\infty}^2 \|\boldsymbol{\eta}^{j+1}\|_{\mathbf{L}^2}^2\,.
\end{aligned}$$

*2nd step:* We proceed by induction and assume that the following inequality is valid, for constants $\tilde{C}_1 = \tilde{C}_1(T;\mathbf{d},\mathbf{u})$, $\tilde{C}_2 = \tilde{C}_1(T;\mathbf{d},\mathbf{u})$,

$$\begin{aligned}
\boldsymbol{\mathcal{F}}^J &:= \|\mathbf{A}_h^{-1/2}\mathbf{e}^J\|_{\mathbf{L}^2}^2 + \|\boldsymbol{\eta}^J\|_{\mathbf{L}^2}^2 + k\sum_{j=0}^J\Big\{\nu\|\mathbf{e}^j\|_{\mathbf{L}^2}^2 + \varrho\|\nabla\boldsymbol{\eta}^j\|_{\mathbf{L}^2}^2\Big\} \\
&\quad + k^2\sum_{j=1}^J\Big\{\|d_t\mathbf{A}_h^{-1/2}\mathbf{e}^j\|_{\mathbf{L}^2}^2 + \|\mathbf{d}_t\boldsymbol{\eta}^j\|_{\mathbf{L}^2}^2\Big\} \qquad (6.104) \\
&\quad + \frac{k}{\varepsilon}\sum_{j=0}^J\Big\{\|\langle\boldsymbol{\eta}^j, \mathbf{d}(t_j)\rangle_{\mathbb{R}^2}\|_{\mathbf{L}^2}^2 + \|\boldsymbol{\eta}^j\|_{\mathbf{L}^4}^4\Big\} \le \tilde{C}_1\{h+k\}^2\exp(\tilde{C}_2 t_J)\,.
\end{aligned}$$

The case '$J = 1$' is again trivial. We sum up in (6.103), employ the results in (a) and the induction assumption to find

$$\boldsymbol{\mathcal{F}}^{J+1} \le C(t_{J+1}, \tilde{C}_1)\,\{h+k\}^2\,. \qquad (6.105)$$

Thus, the last but one inequality in part (a) that involves $q$ can be handled now for values $\frac{1}{\sqrt{k}} = o(\frac{1}{h})$, and $k \le k_0(t_{J+1}, \tilde{C}_1)$, $h \le h_0(t_{J+1}, \tilde{C}_1)$ sufficiently small, we recover (6.104) at $J+1$.

The assertion now follows from a standard result for the error $\|(\mathbf{A}_h^{-1} - \mathbf{A}^{-1})(\mathbf{u}(t_j) - \mathbf{u}_h^j)\|_{\mathbf{L}^2}$. □

The penalized semi-implicit Euler method computes iterates $\{\mathbf{d}^j\}_{j=0}^J$ of modulus close to one, see Corollary 6.1. Moreover, the LBB-condition restricts the choices of finite element spaces to find a stable discretization of $(\mathbf{P})_k^E$. The advantage of projection schemes as sophisticated splitting schemes for incompressible fluid flow problems is to reduce computational effort and increase flexibility of the semidiscrete problem with respect to stable finite element pairings.

## 6.4 The Projection Schemes $(\mathrm{P})_k^P$ and $(\mathrm{P})_{k;h}^P$

In 1968, A. Chorin [28] and R. Temam [123] independently proposed a projection scheme to solve the incompressible Navier-Stokes equations, for $\omega \subset \mathbb{R}^2$ a bounded domain,

$$\begin{aligned}
&\mathbf{u}_t - \nu\,\Delta\mathbf{u} + \mathbf{u}\cdot\nabla\mathbf{u} + \nabla p = \mathbf{f}\,, && (6.106)\\
&\operatorname{div}\mathbf{u} = 0\,, \quad \mathbf{u}\big|_{\partial\omega} = 0\,, \qquad \mathbf{u}(0) = \mathbf{u}_0 \in \tilde{\mathbf{J}}_1 \cap \mathbf{W}^{2,2}\,.
\end{aligned}$$

The idea is to split computation of iterates for velocity and pressure in each step by first solving a Burgers' type equation. The guess $\tilde{\mathbf{u}}^{j+1} \in \mathbf{W}_0^{1,2}$ is then projected onto the space of (weakly) incompressible functions, $\tilde{\mathbf{J}}_0 \ni \mathbf{u}^{j+1} = \mathbf{P}_{\tilde{\mathbf{J}}_0}\tilde{\mathbf{u}}^{j+1}$. — The scheme reads: Given $\mathbf{u}^j$, for $0 \le j \le J$.

1. Compute $\tilde{\mathbf{u}}^{j+1} \in \mathbf{W}_0^{1,2}$ from

$$\frac{1}{k}\{\tilde{\mathbf{u}}^{j+1} - \mathbf{u}^j\} - \nu\,\Delta\tilde{\mathbf{u}}^{j+1} + \mathbf{u}^j\cdot\nabla\tilde{\mathbf{u}}^{j+1} = \mathbf{f}^{j+1}\,. \tag{6.107}$$

2. Determine $\{\mathbf{u}^{j+1}, p^{j+1}\} \in \tilde{\mathbf{J}}_0 \times W^{1,2}/\mathbb{R}$ from

$$\frac{1}{k}\{\mathbf{u}^{j+1} - \tilde{\mathbf{u}}^{j+1}\} + \nabla p^{j+1} = 0\,. \tag{6.108}$$

Note that the projection step amounts to solving a Poisson problem for the pressure function,

$$-\Delta p^{j+1} = -\frac{1}{k}\operatorname{div}\tilde{\mathbf{u}}^{j+1}\,, \quad \partial_{\mathbf{n}}p^{j+1}\big|_{\partial\omega} = 0\,, \tag{6.109}$$

and afterwards we determine $\mathbf{u}^{j+1} \in \tilde{\mathbf{J}}_0$ from (6.18) by means of a simple algebraic update.

Despite of their attractivity from a computational viewpoint, (standard) projection schemes suffer from marked boundary layers that are caused by the non-physical boundary condition stated in (6.109). Moreover, Chorin's projection method can be re-interpreted as a semi-implicit pressure stabilization method [111, 107], see (6.109), which allows for finite element pairings that do not meet the LBB-constraint — like $\mathbf{P1} - P1$ for velocity and pressure space. Modified projection schemes leading to improved convergence behavior are proposed and analyzed in [107, 108], see also Section 6.5. However, it is because of the 'popularity' of Chorin's projection method as well as its stabilizing features in a finite element realization that we propose $(\mathrm{P})_k^P$ in this form, rather than using those improved modified projection schemes.

Optimal convergence behavior of Chorin's scheme for general fluid flows is proved in [107], and we refer to this work for further discussions of this and related projection schemes of first and second order.

In Section 6.4.1, we analyze $(\mathrm{P})_k^P$, restated in the form (6.20)-(6.22). It is mainly due to the strongly nonlinear term $\mathrm{div}\,\big(\nabla \mathbf{d}^{j+1} \odot \nabla \mathbf{d}^{j+1}\big)$ that we restrict our analysis to the case $\omega = [0, 2D]^2$. We did not succeed in clarifying the case of Dirichlet boundary data satisfactorily.

The analysis is accomplished be first quantifying the perturbation that pressure stabilization imposes onto the velocity iterate; this error portion is already well-understood from [107]. In a second step, we investigate the penalization effect in (6.20). Communication between both equations is mainly given through the velocity term in (6.20), which is considered in the last part of the error analysis.

In Section 6.4.2, we turn to a stable spatial discretization $(\mathrm{P})_{k,h;\gamma}^P$, using piecewise affine, globally continuous functions for iterates of velocity, pressure, and director field.

## 6.4.1 Error analysis for $(\mathrm{P})_{k;\gamma}^P$

The main goal of this section is to verify the subsequent theorem. The error analysis for $(\mathrm{P})_{k,\gamma}^P$ is done via reformulation of the scheme in the form (6.20)-(6.22).

**Theorem 6.7** *Let* $\{\mathbf{u}, p, \mathbf{d}\}$ *solve* (6.1)-(6.4), *whereas* $\{\tilde{\mathbf{u}}^j, p^j, \tilde{\mathbf{d}}^j\}_{j=0}^J$ *solves* (6.16)-(6.19), *for* $\gamma \in \mathbb{N}_0$, *and* $0 < t_J < T(\mathbf{u}_0, \mathbf{d}_0)$. *Suppose that (A2) is valid, for* $\omega = [0, 2D]^2$. *There exists a constant* $C = C(t_J; \mathbf{u}_0, \mathbf{d}_0, \omega, \mathbf{f})$ *such* *that for $k \leq k_0(t_J)$,*

$$\max_{0\leq j\leq J}\{\| \mathbf{u}(t_j) - \tilde{\mathbf{u}}^j \|_{\mathbf{W}^{-1,2}_{\rm per}} + \| \mathbf{d}(t_j) - \tilde{\mathbf{d}}^j \|_{\mathbf{L}^2}\} + \Big(k\sum_{j=0}^{J}\{\| \mathbf{u}(t_j) - \tilde{\mathbf{u}}^j \|^2_{\mathbf{L}^2}$$
$$+\| \mathbf{d}(t_j) - \tilde{\mathbf{d}}^j \|^2_{\mathbf{W}^{1,2}} + k \| p(t_j) - p^j \|^2_{L^2/\mathbb{R}}\}\Big)^{1/2} \leq C\,k\,.$$

The proof of this theorem relies on the following result for the projection scheme (6.107), (6.108) that is proved in [107].

**Theorem 6.8** *(from [107]) Let $\{\tilde{\mathbf{u}}^j, p^j\}$ be the solution to* (6.107)-(6.109), *for $\omega \subset \mathbb{R}^2$ bounded, whereas $\{\mathbf{u}, p\}$ solves* (6.106), *for $\mathbf{f} \in L^\infty(I, \mathbf{L}^2)$, and $0 \leq t_j \leq t_J$. Assume that (A2) is valid. Then, for sufficiently small $k \leq k_0(t_J)$, there exists a constant $C = C(t_J; \mathbf{u}, \mathbf{f}, \omega)$, such that the following*

1. *convergence estimates hold true, for $\tau_j := \min\{1, t_j\}$:*

$$\max_{0\leq j\leq J}\Big\{\| \mathbf{u}(t_j) - \tilde{\mathbf{u}}^j \|_{\mathbf{L}^2} + \tau_j \| p(t_j) - p^j \|_{W^{-1,2}} + \sqrt{\tau_j}\, \| p(t_j) - p^j \|_{L^2/\mathbb{R}}\Big\}$$
$$+\Big(k\sum_{j=0}^{J}\{\| \mathbf{u}(t_j) - \mathbf{u}^j \|_{\mathbf{W}^{1,2}} + \sqrt{k}\, \| \nabla\big[p(t_j) - p^j\big] \|_{L^2}\}^2\Big)^{1/2} \leq C\,k\,.$$

2. *a priori results are valid:*

$$\max_{1\leq j\leq J}\Big\{\| d_t\tilde{\mathbf{u}}^j \|_{\mathbf{L}^2} + \| \Delta\tilde{\mathbf{u}}^j \|_{\mathbf{L}^2} + \| \nabla p^j \|_{\mathbf{L}^2}\Big\} + \Big(k\sum_{j=0}^{J} \| d_t\tilde{\mathbf{u}}^j \|^2_{\mathbf{W}^{1,2}}\Big)^{1/2} \leq C\,.$$

We can now start with the verification of Theorem 6.7, restricting ourselves to the case $\gamma = 0$ in the whole section. The proof is split into several steps. In the proof, we omit the tilde notation for both the velocity and director iterates.

**Proof:**

*1st step:* Consider the auxiliary problem: determine solutions $\{\mathbf{v}^j, \pi^j, \mathbf{b}^j\}$ to

$$d_t\mathbf{v}^{j+1} - \nu\,\Delta\mathbf{v}^{j+1} + \mathbf{v}^j\cdot\nabla\mathbf{v}^{j+1} + \nabla\pi^j \tag{6.110}$$
$$= \mathbf{f}^{j+1} - \lambda\,\mathrm{div}\big(\nabla\mathbf{d}(t_{j+1}) \odot \nabla\mathbf{d}(t_{j+1})\big)\,,$$
$$\mathrm{div}\,\mathbf{v}^{j+1} - k\Delta\pi^{j+1} = 0\,, \tag{6.111}$$
$$d_t\mathbf{b}^{j+1} + \frac{1}{k}\{1 - \frac{1}{|\,\mathbf{b}^j\,|}\}\mathbf{b}^{j+1} - \varrho\,\Delta\mathbf{b}^{j+1} \tag{6.112}$$
$$+\mathbf{u}(t_j)\cdot\nabla\mathbf{b}^{j+1} = \varrho\,|\,\nabla\mathbf{b}^j\,|^2\mathbf{b}^{j+1}\,.$$

According to Lemma 6.4, the right-hand side of (6.110) is in $L^\infty(I;\mathbf{L}^2)$, so we can apply Theorem 6.8.

$$\max_{0\le j\le J}\Big\{\|\mathbf{u}(t_j)-\mathbf{v}^j\|_{\mathbf{L}^2}+\tau_j\,\|p(t_j)-\pi^j\|_{\dot W^{-1,2}_{\mathrm{per}}}\Big\} \tag{6.113}$$
$$+\Big(k\sum_{j=0}^{J}\big\{\|\mathbf{u}(t_j)-\mathbf{v}^j\|_{\mathbf{W}^{1,2}}+\sqrt{k}\,\|\nabla\big[p(t_j)-\pi^j\big]\|_{\mathbf{L}^2}\big\}^2\Big)^{1/2}\le C\,k\,.$$

Approximation behavior of iterates $\{\mathbf{b}^j\}_{j=0}^J$ from (6.112) has been investigated in Theorem 4.7 for the Landau-Lifshitz equation. The additional convection term does not involve further problems here, thanks to regularity properties stated in Lemma 6.4.

$$\max_{0\le j\le J}\|\mathbf{d}(t_j)-\mathbf{b}^j\|_{\mathbf{L}^2}+\Big(k\sum_{j=0}^{J}\big\{\|\mathbf{d}(t_j)-\mathbf{b}^j\|^2_{\mathbf{W}^{1,2}} \tag{6.114}$$
$$+\frac{1}{k}\big[\|\mathbf{d}(t_j)-\mathbf{b}^j\|^4_{\mathbf{L}^4}+\|\langle\mathbf{d}(t_j)-\mathbf{b}^j,\mathbf{d}(t_j)\rangle_{\mathbb{R}^2}\|^2_{\mathbf{L}^2}\big]\big\}\Big)^{1/2}\le C\,k\,.$$

In addition, it is owing to Theorem 6.8, item 2., and Corollary 4.4, (b) that we have for sufficiently small $k\le k_0(t_J)$,

$$\max_{1\le j\le J}\Big\{\|d_t\mathbf{v}^j\|_{\mathbf{L}^2}+\|\Delta\mathbf{v}^j\|_{\mathbf{L}^2}+\|\nabla\pi^j\|_{\mathbf{L}^2}+\|d_t\mathbf{b}^j\|_{\mathbf{L}^2}+\|\Delta\mathbf{b}^j\|_{\mathbf{L}^2}\Big\}\le C\,. \tag{6.115}$$

*2nd step:* We need to control $\big\{\mathbf{e}^j,\lambda^j,\boldsymbol{\eta}^j\big\}:=\big\{\mathbf{v}^j-\mathbf{u}^j,\pi^j-p^j,\mathbf{b}^j-\mathbf{d}^j\big\}$ that solves

$$d_t\mathbf{e}^{j+1}-\nu\,\Delta\mathbf{e}^{j+1}+\nabla\lambda^j=\mathbf{e}^j\cdot\nabla\mathbf{v}^{j+1}-\mathbf{e}^j\cdot\nabla\mathbf{e}^{j+1}+\mathbf{v}^j\cdot\nabla\mathbf{e}^{j+1} \tag{6.116}$$
$$+k\,d_t\mathbf{v}^{j+1}\cdot\nabla\mathbf{e}^{j+1}-\lambda\,\mathrm{div}\,\big(\nabla\boldsymbol{\eta}^{j+1}\odot\nabla\mathbf{b}^{j+1}$$
$$-\nabla\boldsymbol{\eta}^{j+1}\odot\nabla\boldsymbol{\eta}^{j+1}-\nabla\mathbf{b}^{j+1}\odot\nabla\boldsymbol{\eta}^{j+1}\big)\,,$$
$$\mathrm{div}\,\mathbf{e}^{j+1}-k\,\Delta\lambda^{j+1}=0\,, \tag{6.117}$$
$$d_t\boldsymbol{\eta}^{j+1}-\varrho\,\Delta\boldsymbol{\eta}^{j+1}+\frac{1}{k}\Big\{\big\{1-\frac{1}{|\mathbf{b}^j|}\big\}\mathbf{b}^{j+1}-\big\{1-\frac{1}{|\mathbf{d}^j|}\big\}\mathbf{d}^j\Big\} \tag{6.118}$$
$$=-\mathbf{e}^j\cdot\nabla\mathbf{b}^{j+1}-\mathbf{e}^j\cdot\nabla\boldsymbol{\eta}^{j+1}$$
$$+\mathbf{v}^j\cdot\nabla\boldsymbol{\eta}^{j+1}+k\,d_t\mathbf{v}^{j+1}\cdot\nabla\boldsymbol{\eta}^{j+1}+\varrho\big(|\nabla\mathbf{b}^j|^2\mathbf{b}^{j+1}-|\nabla\mathbf{d}^j|^2\mathbf{d}^{j+1}\big)\,.$$

We test (6.116), (6.118) by $\{-\Delta^{-1}\mathbf{e}^{j+1}, \boldsymbol{\eta}^{j+1}\}$, respectively, for $\Delta^{-1}: \dot{\mathbf{W}}^{-1,2}_{\text{per}} \to \dot{\mathbf{W}}^{1,2}_{\text{per}}$. To deal with the pressure, we infer from (6.117)

$$\begin{aligned} k\left(\lambda^j, \lambda^{j+1}\right) &= k \,\| \lambda^{j+1} \|^2 - k^2 \left(\lambda^{j+1}, d_t\lambda^{j+1}\right) \\ &= k \,\| \lambda^{j+1} \|^2 - k\left(\Delta^{-1}\mathbf{e}^{j+1}, \nabla\lambda^{j+1}\right) \\ &\geq \frac{k}{2}\{\| \lambda^{j+1} \|^2_{\mathbf{L}^2} - \| \mathrm{div}\Delta^{-1} d_t\mathbf{e}^{j+1} \|^2_{\mathbf{L}^2}\}, \end{aligned} \tag{6.119}$$

where we used the identity $\mathrm{div}\Delta^{-1} = \Delta^{-1}\mathrm{div}$ on $\dot{\mathbf{W}}^{-1,2}_{\text{per}}$.

The time shift in the penalization term in (6.118) is considered next. The difference can be re-written in the form

$$\begin{aligned} &\frac{1}{k}\Big\{\frac{1}{|\mathbf{b}^j|} - \frac{1}{|\mathbf{d}^j|}\Big\}\{\boldsymbol{\eta}^j - \mathbf{b}^j\} + \frac{1}{k}\Big\{1 - \frac{1}{|\mathbf{b}^j|}\Big\}\{\boldsymbol{\eta}^j + k\, d_t\mathbf{b}^{j+1}\} \\ &= \frac{|\boldsymbol{\eta}^j|^2 - 2\,\langle\boldsymbol{\eta}^j, \mathbf{b}^j\rangle_{\mathbb{R}^2}}{k(|\mathbf{d}^j| + |\mathbf{b}^j|^2)|\mathbf{b}^j||\mathbf{d}^j|}\{\boldsymbol{\eta}^j - \mathbf{b}^j\} \\ &\quad + \frac{1}{k}\Big\{1 - \frac{1}{|\mathbf{b}^j|}\Big\}\{\boldsymbol{\eta}^j + k\, d_t\mathbf{b}^{j+1}\}, \end{aligned} \tag{6.120}$$

similar to (4.110), (4.111). We test this term by $\boldsymbol{\eta}^{j+1}$ to obtain

$$\begin{aligned} &\frac{1}{k}\int_\Omega \frac{|\boldsymbol{\eta}^j|^4 + 2|\langle\boldsymbol{\eta}^j, \mathbf{b}^j\rangle_{\mathbb{R}^2}|^2 - 3\,\langle\boldsymbol{\eta}^j, \mathbf{b}^j\rangle_{\mathbb{R}^2}|\boldsymbol{\eta}^j|^2}{(|\mathbf{d}^j| + |\mathbf{b}^j|^2)|\mathbf{b}^j||\mathbf{d}^j|}\,\mathrm{d}\mathbf{x} \\ &+ \Big(\frac{|\boldsymbol{\eta}^{j+1}|^2 - 2\langle\boldsymbol{\eta}^j, \mathbf{b}^j\rangle_{\mathbb{R}^2}}{(|\mathbf{d}^j| + |\mathbf{b}^j|^2)|\mathbf{b}^j||\mathbf{d}^j|}\{\boldsymbol{\eta}^j - \mathbf{b}^j\}, d_t\boldsymbol{\eta}^{j+1}\Big). \end{aligned}$$

We may deal with this term like in the proof of Theorem 4.9. Hence, we obtain for values $k \leq k_0(t_J)$

$$\begin{aligned} &\frac{1}{2}\{d_t\| \nabla\Delta^{-1}\mathbf{e}^{j+1} \|^2_{\mathbf{L}^2} + d_t\| \boldsymbol{\eta}^{j+1} \|^2_{\mathbf{L}^2} + k\,\| d_t\boldsymbol{\eta}^{j+1} \|^2_{\mathbf{L}^2}\} + \frac{k}{8}\| \nabla\Delta^{-1} d_t\mathbf{e}^{j+1} \|^2_{\mathbf{L}^2} \\ &+\nu\,\| \mathbf{e}^{j+1} \|^2_{\mathbf{L}^2} + \varrho\,\| \nabla\boldsymbol{\eta}^{j+1} \|^2_{\mathbf{L}^2} + k\,\| \lambda^{j+1} \|^2_{L^2} \\ &+\frac{1}{8k}\,\{\| \mathbf{e}^j \|^4_{\mathbf{L}^4} + \| \langle\mathbf{e}^j, \mathbf{b}^j\rangle_{\mathbb{R}^2} \|^2_{\mathbf{L}^2}\} \\ &\leq |\left(\mathbf{l}_\varepsilon(\mathbf{b}^{j+1}) - \mathbf{l}_\varepsilon(\mathbf{d}^{j+1}), \boldsymbol{\eta}^{j+1}\right)| + \mathcal{H}_1 + \mathcal{H}_2, \end{aligned} \tag{6.121}$$

where $\mathcal{H}_1$ collects the terms on the right-hand side of (6.116), whereas $\mathcal{H}_2$ gathers those of (6.118). — Dealing with $\mathcal{H}_2$ efficiently again requires an inductive argument which works like in the proof of Theorem 6.6. This settles the proof of the theorem. □

### 6.4.2 Error analysis for $(\mathrm{P})_{k,h;\gamma}^P$

It has already been pointed out in the introduction that the projection scheme allows for stable application of $P1$ conforming finite elements for *all* involved quantities. It is possible to prove optimal convergence behavior for $(\mathrm{P})_{k,h;\gamma}^P$ as it is given in Theorem 6.2 because of the following a priori bounds for the solution to (6.16)-(6.19).

**Lemma 6.9** *Suppose that (A2) is valid for* (6.16)-(6.19), *for* $\omega = [0, 2D]^2$, $\gamma \in \mathbb{N}_0$, *and* $k \leq k_0(t_J)$. *There exists a constant* $C = C(t_J; \mathbf{u}, \mathbf{f}, \omega)$ *such that*

$$\max_{0\leq j\leq J}\Big\{\| d_t\tilde{\mathbf{d}}^j \|_{\mathbf{L}^2} + \| \nabla\tilde{\mathbf{d}}^j \|_{\mathbf{L}^2} + \| \mathbf{u}^j \|_{\mathbf{L}^2}\Big\}$$
$$+\Big(k\sum_{j=0}^{J+1} \| \nabla p^j \|_{L^2}^2 + \| \Delta\tilde{\mathbf{d}}^j \|_{\mathbf{L}^2}^2 + \| \nabla\tilde{\mathbf{u}}^j \|_{\mathbf{L}^2}^2\Big)^{1/2} \leq C\,.$$

**Proof:**
Most of the results follow from Theorem 6.7. The result for the pressure follows from (6.117), and the upper bound for $\{\mathbf{d}^j\}_{j=0}^J \in \ell^2(I_k; \mathbf{W}^{2,2})$ is obtained from multiplication of (6.118) by $-\Delta\eta^{j+1}$ and using (6.120), in particular. This result allows to apply $\mathbf{e}^{j+1}$ as test function in (6.116). □

The main result of this section is formulated in the subsequent lemma which implies Theorem 6.2.

**Lemma 6.10** *Suppose that* $\omega = [0, 2D]^2$ *and (A2),(B1),(B2),(B3) for* (6.16)-(6.19), *and* (6.23)-(6.27), *resp., and* $\{k, h\} \leq \{k_0(t_J), h_0(t_J)\}$. *There exists a constant* $C = C(\nu, \gamma, \lambda, \mathbf{f}, \mathbf{u}_0, \mathbf{d}_0, \omega, t_J)$ *that does not depend on* $k, h$ *such that*

$$\max_{0\leq j\leq M}\Big\{\| \Delta^{-1/2}(\mathbf{u}^j - \mathbf{u}_h^j) \|_{\mathbf{L}^2} + \| \mathbf{d}^j - \mathbf{d}_h^j \|_{\mathbf{L}^2}\Big\}$$
$$+\Big(k\sum_{m=0}^{M}\{\| \mathbf{u}^j - \mathbf{u}_h^j \|_{\mathbf{L}^2}^2 + \| \mathbf{d}^j - \mathbf{d}_h^j \|_{\mathbf{W}^{1,2}}^2\} + \sqrt{k}\,\| p^j - p_h^j \|_{L^2/\mathbb{R}}\Big)^{1/2}$$
$$+\Big(\sum_{j=0}^{J}\{\| \mathbf{d}^j - \mathbf{d}_h^j \|_{\mathbf{L}^4}^4 + \| \langle \mathbf{d}^j - \mathbf{d}_h^j, \mathbf{d}^j\rangle_{\mathbb{R}^2} \|_{\mathbf{L}^2}^2\}\Big)^{1/2} \leq C\,\{k + h\}\,,$$

**Proof:**
(Sketch) The technique is similar to the verification of Theorem 6.8, step 2.: for a corresponding error notation, only the first penalty term in (6.118) is shifted towards being fully explicit, which deletes the last contribution in the last term in brackets prior to (6.120). We test the error identities (6.116), (6.118) by $\{-\Delta_h^{-1}\mathbf{e}^{j+1}, \mathbf{P}^c_{h;\mathbf{L}^2}\boldsymbol{\eta}^{j+1}\}$ and argue similarly to (6.97), (6.101). We exploit best approximation properties from Lemma 6.9 to verify the result.

□

# 6.5 Computational Experiments

Convergence results for $(\mathrm{P})^P_{k;h}$ in Theorem 6.2 are obtained for periodic boundary data; this setting is suitable to control the strong nonlinear effects inherent to the problem, with no interference from contributions that are caused by boundary effects. As is known from Navier-Stokes equations [107], the projection scheme of Chorin gives significant pollution to pressure approximates close to the boundary, see Remark 6.9 below. Hence, we address a possible pollution effect to iterates of the director field in $(\mathrm{P})^P_{k;h}$ in Examples 6.1 and 6.2; here, we also study a modified scheme which is exempted from producing boundary layers in the approximation of the pressure, based on the Chorin-Uzawa method, [107]. Error estimates for the original scheme are given above, and we report on convergence behavior of both methods for iterates of velocity field, pressure, and director field. We conclude this section with the physically relevant Example 6.3 taken from [93, 94] of a liquid crystal fluid flow involving singularities (see also [87] and the literature cited therein).

**Remark 6.9** *(Dirichlet boundary value problem, $\omega \subset \mathbb{R}^d$) Chorin's projection method suffers from generating pressure iterates with marked boundary layers due to the prescription of the non-physical boundary condition $\partial_{\mathbf{n}} p^j = 0\big|_{\partial\omega}$. In Chapter 8 of [107], a variant called Chorin-Uzawa scheme is introduced to avoid this deficiency (for Dirichlet boundary data $\mathbf{u}\big|_{\partial\omega} = 0$). If applied to* (6.106)*, the method reads as follows: Start with a triple $\{\mathbf{u}^0, p^0, \tilde{p}^0\}$ such that*

$$\| \mathbf{u}^0 - \mathbf{u}_0 \|_{\mathbf{L}^2} + \sqrt{k}\, \| p^0 - p(0) \|_{L^2/\mathbb{R}} \leq C\, k\,, \quad \tilde{p}^0 = 0\,, \quad p^0 \in L^2_0(\omega)\,.$$

*Given $\{\tilde{\mathbf{u}}^j, \mathbf{u}^j, \tilde{p}^j, p^j\}$, for $j \geq 0$, compute new iterates from:*

$$1. \quad \frac{1}{k}\{\tilde{\mathbf{u}}^{j+1} - \mathbf{u}^j\} - \nu\,\Delta\tilde{\mathbf{u}}^{j+1} + \mathbf{u}^j \cdot \nabla\tilde{\mathbf{u}}^{j+1} \qquad (6.122)$$

$$+\nabla\{p^j - \tilde{p}^j\} = \mathbf{f}^{j+1}\,, \quad \mathbf{u}^{j+1}\big|_{\partial\omega} = 0\,.$$

$$2. \quad \frac{1}{k}\{\mathbf{u}^{j+1} - \tilde{\mathbf{u}}^{j+1}\} + \nabla\tilde{p}^{j+1} = 0\,, \qquad (6.123)$$

$$\operatorname{div}\mathbf{u}^{j+1} = 0\,, \quad \mathbf{u}^{j+1}\big|_{\partial\omega}\cdot\mathbf{n} = 0\,.$$

$$3. \quad p^{j+1} = p^j - \alpha\operatorname{div}\tilde{\mathbf{u}}^{j+1}\,, \quad \alpha < 1\,. \qquad (6.124)$$

*Hence, we distinguish between the Lagrange multiplier $\tilde{p}^j$ and $p^j$ as physically meaningful quantity; see [107] for an analysis of this scheme. In particular, the following error bounds are shown there, for solutions $\{\mathbf{u}, p\} \in W^{1,\infty}(I; \mathbf{W}^{1,2}) \times W^{1,\infty}(I; L^2/\mathbb{R})$,*

$$\max_{0\leq j\leq M}\Big\{\|\,\tilde{\mathbf{u}}^j - \mathbf{u}(t_j)\,\|_{\mathbf{L}^2} + \sqrt{k}\,\|\,p^j - p(t_j)\,\|_{L^2/\mathbb{R}}\Big\} \qquad (6.125)$$

$$+\Big(k\sum_{j=0}^{J}\|\,\mathbf{u}^j - \mathbf{u}(t_j)\,\|^2_{\mathbf{W}^{1,2}}\Big)^{1/2} \leq C\,\big(1 + \log\frac{1}{k}\big)\,k\,.$$

*The crucial improvement over a corresponding result for Chorin's scheme is the error statement for $\mathbf{u}^j \in \ell^2(I_k; \mathbf{W}^{1,2}(\omega, \mathbb{R}^2))$ which reflects absence of boundary layers in space; this result can be transferred to a corresponding statement for $\{p^j\}_{j=0}^J \in \ell^2(I_k; L^2/\mathbb{R})$. For further analysis and computational studies of the Chorin-Uzawa method, we refer to [107]. If it comes to a finite element realization, we observe a loss of stabilization effect of the Chorin-Uzawa scheme for the pairing $\mathbf{P}1 - P1$. This can be theoretically understood from its reformulation as a semi-explicit artificial compressibility method, i.e.,*

$$\operatorname{div}\mathbf{u}^\varepsilon + \varepsilon\,p^\varepsilon_t = 0\,, \quad p^\varepsilon(0) = p_0\,.$$

*To make sure stable usage of $\mathbf{P}1 - P1$ elements, we follow the scenario proposed in [107] where (6.124) is replaced by its stabilization, for $\delta > 0$,*

$$(\mathrm{Id} - \delta\,h^2\,\Delta_h p_h^{j+1})\,p_h^{j+1} = -\alpha\nu\operatorname{div}_{\mathcal{T}}\tilde{\mathbf{u}}_h^{j+1} + p_h^j\,, \quad \partial_{\mathbf{n},h}p_h^{j+1}\big|_{\partial\omega} = 0\,. \qquad (6.126)$$

*Figure 6.1 shows existence resp. absence of marked boundary layers in case of the Chorin- (left) and the Chorin-Uzawa scheme in original (middle; using*

$\mathbf{P1}-P1$) and stabilized form, respectively. Computations are done for an academic example, with $\omega=(0,1)^2$, $\nu=1$, and

$$\mathbf{u}(x,y,t)=\begin{pmatrix} x^2(1-x)^2(2y-6y^2+4y^3) \\ -y^2(1-y)^2(2x-6x^2+4x^3) \end{pmatrix}, \quad p(x,y,t)=(x^2-\frac{1}{3}). \tag{6.127}$$

*For these computations, the convection part in (6.106) and in the projection schemes is omitted.*

*We close with some additional comments on (the analysis of) the Chorin-Uzawa scheme:*

*1. The prescription of accurate initial data for the pressure is a severe restriction of this algorithm. However, a modification of Chorin-Uzawa using special time-grids is constructed in [107] that avoids this deficiency. Since we mainly focus on (non-)occurrence of boundary layers in the structure of errors in the pressure, we consider the basic method here.*

*2. The regularity assumptions that preceed (6.125) are not valid in general. However, it is again possible by means of using adopted time-grids to free from this constraint; see Chapter 10 in [107] for details.*

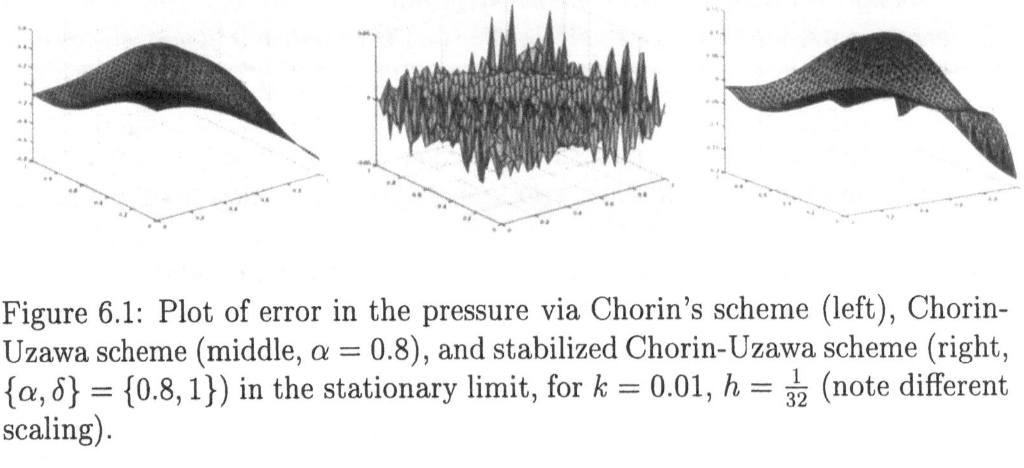

Figure 6.1: Plot of error in the pressure via Chorin's scheme (left), Chorin-Uzawa scheme (middle, $\alpha=0.8$), and stabilized Chorin-Uzawa scheme (right, $\{\alpha,\delta\}=\{0.8,1\}$) in the stationary limit, for $k=0.01$, $h=\frac{1}{32}$ (note different scaling).

In the next two examples, we study both variants of projection methods in the context of the Ericksen-Leslie equations. In particular, we investigate possible pollution effects to iterates of the director field. For this purpose, we introduce $\{\mathbf{g}^j\}_{j=0}^J$ to (6.128) to study an academic test case with known solution.

**Example 6.1** *Similar to* $(\mathrm{P})_k^P$, *we propose* $(\tilde{\mathrm{P}})_k^P$ *by using Chorin-Uzawa method. The method reads, for given data* $\{\mathbf{f}^j\}_{j=0}^J, \{\mathbf{g}^j\}_{j=0}^J \in \ell^2\big(I_k; \mathbf{L}^2(\omega, \mathbb{R}^2)\big)$ *and initial data* $\{\tilde{\mathbf{d}}^0, \mathbf{u}^0, \tilde{\mathbf{u}}^0, p^0, \tilde{p}^0\}$, *such that* $p^0 \in L_0^2(\omega)$, *and*

$$\| \mathbf{u}^0 - \mathbf{u}_0 \|_{\mathbf{L}^2} + \| \mathbf{d}^0 - \mathbf{d}_0 \|_{\mathbf{L}^2} + \sqrt{k} \, \| p^0 - p(0) \|_{L^2/\mathbb{R}} \leq C\,k\,, \quad p^0 \in L_0^2(\omega)\,.$$

*Let* $\{\tilde{\mathbf{d}}^j, \mathbf{d}^j, \tilde{\mathbf{u}}^j, \mathbf{u}^j, \tilde{p}^j, p^j\}$ *be given, for* $j \geq 0$.

1. *Compute* $\tilde{\mathbf{d}}^{j+1}$ *from*

$$\begin{aligned} &\frac{1}{k}\{\tilde{\mathbf{d}}^{j+1} - \mathbf{d}^j\} + \mathbf{u}^j \cdot \nabla \tilde{\mathbf{d}}^{j+1} - \varrho \, \Delta \tilde{\mathbf{d}}^{j+1} \\ &= \varrho \, | \, \nabla \tilde{\mathbf{d}}^{j+1} \, |^2 \tilde{\mathbf{d}}^{j+1} + \mathbf{g}^{j+1}, \quad \tilde{\mathbf{d}}^{j+1}\big|_{\partial\omega} = \mathbf{F}^{j+1}\,. \end{aligned} \tag{6.128}$$

2. *Determine* $\tilde{\mathbf{u}}^{j+1}$, *with* $\tilde{\mathbf{u}}^{j+1}\big|_{\partial\omega} = 0$, *from*

$$\begin{aligned} &\frac{1}{k}\{\tilde{\mathbf{u}}^{j+1} - \mathbf{u}^j\} - \nu \, \Delta \tilde{\mathbf{u}}^{j+1} + \mathbf{u}^j \cdot \nabla \tilde{\mathbf{u}}^{j+1} \\ &+ \lambda \, \mathrm{div}\big(\nabla \tilde{\mathbf{d}}^{j+1} \odot \nabla \tilde{\mathbf{d}}^{j+1}\big) + \nabla\{p^j - \tilde{p}^j\} = \mathbf{f}^{j+1}\,. \end{aligned} \tag{6.129}$$

3. *Compute* $\{\mathbf{u}^{j+1}, \mathbf{d}^{j+1}, p^{j+1}\}$ *via*

$$\frac{1}{k}\{\mathbf{u}^{j+1} - \tilde{\mathbf{u}}^{j+1}\} + \nabla \tilde{p}^{j+1} = 0 \tag{6.130}$$

$$\mathrm{div}\, \mathbf{u}^{j+1} = 0\,, \quad \mathbf{u}^{j+1}\big|_{\partial\omega} = 0\,,$$

$$p^{j+1} = p^j - \alpha \, \mathrm{div}\, \tilde{\mathbf{u}}^{j+1}, \quad \alpha < 1\,, \tag{6.131}$$

$$\mathbf{d}^{j+1} = \frac{\tilde{\mathbf{d}}^{j+1}}{| \, \tilde{\mathbf{d}}^{j+1} \, |}\,. \tag{6.132}$$

*We compute* $\{\mathbf{f}^j, \mathbf{g}^j\}_{j=0}^J$ *from* $\{\mathbf{u}, p\}$ *given in* (6.127), *and the director field* $\mathbf{d}(x, y, t) = (\frac{1}{2}x, \sqrt{1 - \frac{1}{4}x^2})^\top$. *Figure 6.2 displays profiles of the error of the velocity field* $(\lim_{j\to\infty} \| \mathbf{u} - \mathbf{u}_h^j \|_{\mathbf{L}^2} \approx 6.9-3$, *resp.* $2.9-3)$ *as well as the error of the pressure (top) and the profile of the error of the director field* $(\lim_{j\to\infty} \| \mathbf{d} - \mathbf{d}_h^j \|_{\mathbf{L}^2} \approx 3.1-3$, *resp.* $6.4-4)$ *as well as the modulus of the computed director field (bottom). For these experiments, the convective part* $\mathbf{u} \cdot \nabla \mathbf{u}$ *in* (6.1) *is deleted from the problem.*

*For $(\tilde{\mathrm{P}})_k^P$, we observe marked boundary layers for the pressure (of thickness $\mathcal{O}(h)$), rather than small layers (of magnitude $\mathcal{O}(\sqrt{k})$) for $(\mathrm{P})_k^P$ (see [42, 111, 107, 11] for a theoretical justification). Surprisingly, the quality of the director field is not affected by boundary layers or instabilities in all considered schemes.*

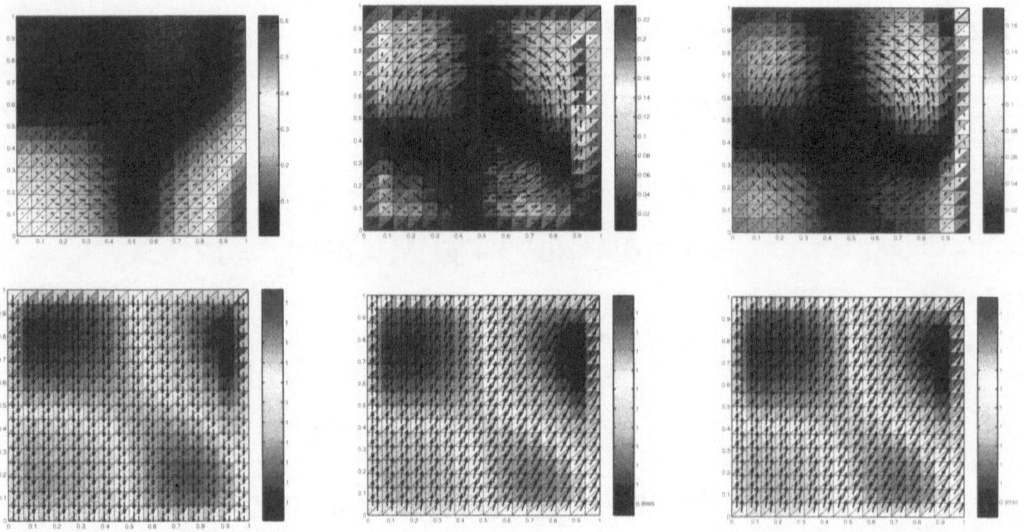

Figure 6.2: Plot of the error of velocity field and pressure (colormap) (top) and error of the computed director field and modulus of it (bottom) in the stationary limit per column; Chorin's scheme (left), Chorin-Uzawa (middle, $\alpha = 0.8$), and the stabilized Chorin-Uzawa scheme (right), $\{\alpha, \delta\} = \{0.8, 0.25\}$), for $k = 0.01$, $h = \frac{1}{16}$ (note different scaling).

The next example provides data of measured errors.

**Example 6.2** *Theorem 6.7 is valid for periodic boundary data. Here, we present computational results to study effects due to existing boundaries, using Dirichlet boundary data. For $\omega = (-1, 1)^2$, let*

$$
\begin{aligned}
\mathbf{u}(x,y,t) &= (1+t^3)\begin{pmatrix} x^2(1-x)^2(2y-6y^2+4y^3) \\ -y^2(1-y)^2(2x-6x^2+4x^3) \end{pmatrix}, \\
\mathbf{d}(x,y,t) &= \begin{pmatrix} xt \\ \sqrt{1-x^2t^2} \end{pmatrix}, \\
p(x,y,t) &= (x^2 - \frac{1}{3})t
\end{aligned}
$$

*be the exact solution to* (6.1)-(6.4), *with additional term* $\mathbf{g} \in L^2(I; \mathbf{L}^2(\omega, \mathbb{R}^2))$ *on the right-hand side of* (6.2). *Both,* $\mathbf{f}$ *and* $\mathbf{g}$ *are then computed from* $\{\mathbf{u}, \mathbf{d}, p\}$ *given above.*

*Tables 6.1 resp. 6.2 display* $L^2$*-errors at* $t = 0.5$ *for different quantities, using* $(\mathrm{P})_k^P$ *resp.* $(\tilde{\mathrm{P}})_k^P$. *We observe first order of convergence for the director field in both cases, and reduced rate of convergence for* $\nabla \mathbf{d}^{j+1}$. *No pollution with respect to the director field due to Chorin's method as part of* $(\mathrm{P})_k^P$ *is observed. In Theorem 6.7, optimal convergence behavior is stated for velocity iterates of* $(\mathrm{P})_k^P$ *in* $\ell^\infty(I_k; \mathbf{W}^{-1,2}(\omega, \mathbb{R}^2))$. *In the present setting, we observe* 0.5 *as convergence rate in* $\mathbf{L}^2$ *at time* $t = 0.5$, *and approximately* 0.25 *for* $\nabla \mathbf{u}$, *with no significant difference in the accuracy for both methods,* $(\mathrm{P})_k^P$ *and* $(\tilde{\mathrm{P}})_k^P$. *Theorem 6.7 does not provide convergence statements for pressure iterates in* $\ell^\infty(I_k; L^2/\mathbb{R})$; *however, experimentally we found the order of convergence close to* 0.25. *We observe smaller errors for pressure iterates computed by* $(\tilde{\mathrm{P}})_k^P$ *but do not observe an improved convergence rate in our experiments.*

| $\ell$ | $\tilde{u}_1^{j+1}$ | $\nabla \tilde{u}_1^{j+1}$ | $\operatorname{div} \tilde{\mathbf{u}}^{j+1}$ | $\operatorname{div} \mathbf{u}^{j+1}$ | $\tilde{d}_2^{j+1}$ | $\nabla \tilde{d}_2^{j+1}$ | $p^{j+1}$ |
|---|---|---|---|---|---|---|---|
| 1 | 1.43 − 2 | 6.89 − 2 | 5.87 − 2 | 4.93 − 2 | 3.05 − 3 | 1.52 − 2 | 2.10 − 1 |
| 2 | 1.35 − 2 | 6.66 − 2 | 5.63 − 2 | 4.74 − 2 | 1.29 − 3 | 7.65 − 3 | 1.98 − 1 |
| 3 | 1.19 − 2 | 5.85 − 2 | 5.15 − 2 | 4.34 − 2 | 5.39 − 4 | 4.50 − 3 | 1.79 − 1 |
| 4 | 9.66 − 3 | 4.84 − 2 | 4.44 − 2 | 3.75 − 2 | 2.55 − 4 | 3.33 − 3 | 1.51 − 1 |
| 5 | 6.97 − 3 | 3.68 − 2 | 3.58 − 2 | 3.03 − 2 | 1.29 − 4 | 2.63 − 3 | 1.21 − 1 |
| 6 | 4.61 − 3 | 2.75 − 2 | 2.71 − 2 | 2.30 − 2 | 7.30 − 5 | 2.17 − 3 | 9.69 − 2 |
| 7 | 3.32 − 3 | 2.38 − 2 | 1.96 − 2 | 1.69 − 2 | 5.43 − 5 | 1.64 − 3 | 8.45 − 2 |
| order | 0.46 | 0.28 | 0.3 | 0.27 | 0.92 | 0.44 | 0.25 |

Table 6.1: $L^2$-errors at $t = 0.5$ for $(\mathrm{P})_{k,h}^P$, and time-steps $k = 0.1 \cdot 2^{-\ell+1}$ ($h = \frac{1}{32}$).

We study the motion of two point defects of degrees $\pm 1$ by means of $(\mathrm{P})_k^P$ in the last example.

| $\ell$ | $\tilde{u}_1^{j+1}$ | $\nabla\tilde{u}_1^{j+1}$ | $\operatorname{div}\tilde{\mathbf{u}}^{j+1}$ | $\operatorname{div}\mathbf{u}^{j+1}$ | $\tilde{d}_2^{j+1}$ | $\nabla\tilde{d}_2^{j+1}$ | $p^{j+1}$ |
|---|---|---|---|---|---|---|---|
| 1 | 1.41 − 2 | 6.82 − 2 | 5.81 − 2 | 4.89 − 2 | 3.03 − 3 | 1.52 − 2 | 1.26 − 1 |
| 2 | 1.32 − 2 | 6.40 − 2 | 5.50 − 2 | 4.63 − 2 | 1.28 − 3 | 7.60 − 3 | 1.11 − 1 |
| 3 | 1.15 − 2 | 5.63 − 2 | 4.96 − 2 | 4.18 − 2 | 5.31 − 4 | 4.44 − 3 | 9.43 − 2 |
| 4 | 9.14 − 3 | 4.62 − 2 | 4.23 − 2 | 3.57 − 2 | 2.47 − 4 | 3.26 − 3 | 7.89 − 2 |
| 5 | 6.67 − 3 | 3.56 − 2 | 3.44 − 2 | 2.91 − 2 | 1.25 − 4 | 2.58 − 3 | 6.49 − 2 |
| 6 | 4.61 − 3 | 2.75 − 2 | 2.67 − 2 | 2.29 − 2 | 7.26 − 5 | 2.16 − 3 | 5.37 − 2 |
| 7 | 3.40 − 3 | 2.40 − 2 | 2.03 − 2 | 1.74 − 2 | 5.47 − 5 | 1.69 − 3 | 4.43 − 2 |
| order | 0.42 | 0.29 | 0.35 | 0.28 | 0.91 | 0.43 | 0.25 |

Table 6.2: $L^2$-errors at $t = 0.5$ for $(\tilde{\mathrm{P}})_{k,h}^P$, and time-steps $k = 0.1 \cdot 2^{-\ell+1}$ ($h = \frac{1}{32}$).

**Example 6.3** *(Annihilation of Singularities) Let* $\omega = (-1,1)^2$, $\mathbf{f} = 0$, $\mathbf{u}|_{\partial\omega} = 0$, *and* $\mathbf{d}_0 = \hat{\mathbf{d}}/\sqrt{|\,\hat{\mathbf{d}}\,|^2 + \eta^2}$, *where* $\hat{\mathbf{d}}(x,y) = (x^2 + y^2 - \frac{1}{4}, y)^\top$; *cf. [90, 93]. We fix constants* $\{\nu, \lambda, \varrho\} = \{1, \frac{1}{5}, 1\}$ *in* (6.1)-(6.2). *For the given mesh-size* $h = \frac{1}{16}$, *we choose* $\eta = 0.2$ *to regularize the singularities, following the idea of [93]. Figure 6.3 displays* $\mathbf{d}^{j+1}$ *and its modulus at different times. The solution is computed from* (6.23)-(6.27), *for* $k = 0.01$.

*We observe annihilation of two singularities at time* $t \approx 0.28$; *see also [93] for computational studies of the same problem.*

As a summary, we draw the following conclusions from our experiments:

1. Using $P1$-elements for all quantities in the context of (stabilized) projection methods to solve Ericksen-Leslie equations is a good alternative towards using higher order [93] or mixed methods [94].

2. Dealing with the non-convexity in problem (6.1)-(6.4) only allows for sharp convergence statements in weak norms, see Theorem 6.7, with deteriorate convergence rates in higher norms.

3. No pollution effect of director fields is found for $(\mathrm{P})_{k,h}^P$; its variant $(\tilde{\mathrm{P}})_{k,h}^P$ does not show improved convergence behavior.

4. The scheme $(\mathrm{P})_{k,h}^P$ works well in situations with (point) defects.

5. At each time-step, we solve linear problems in our computational experiments; however, we expect nonlinear strategies to be more appropriate for complex flows.

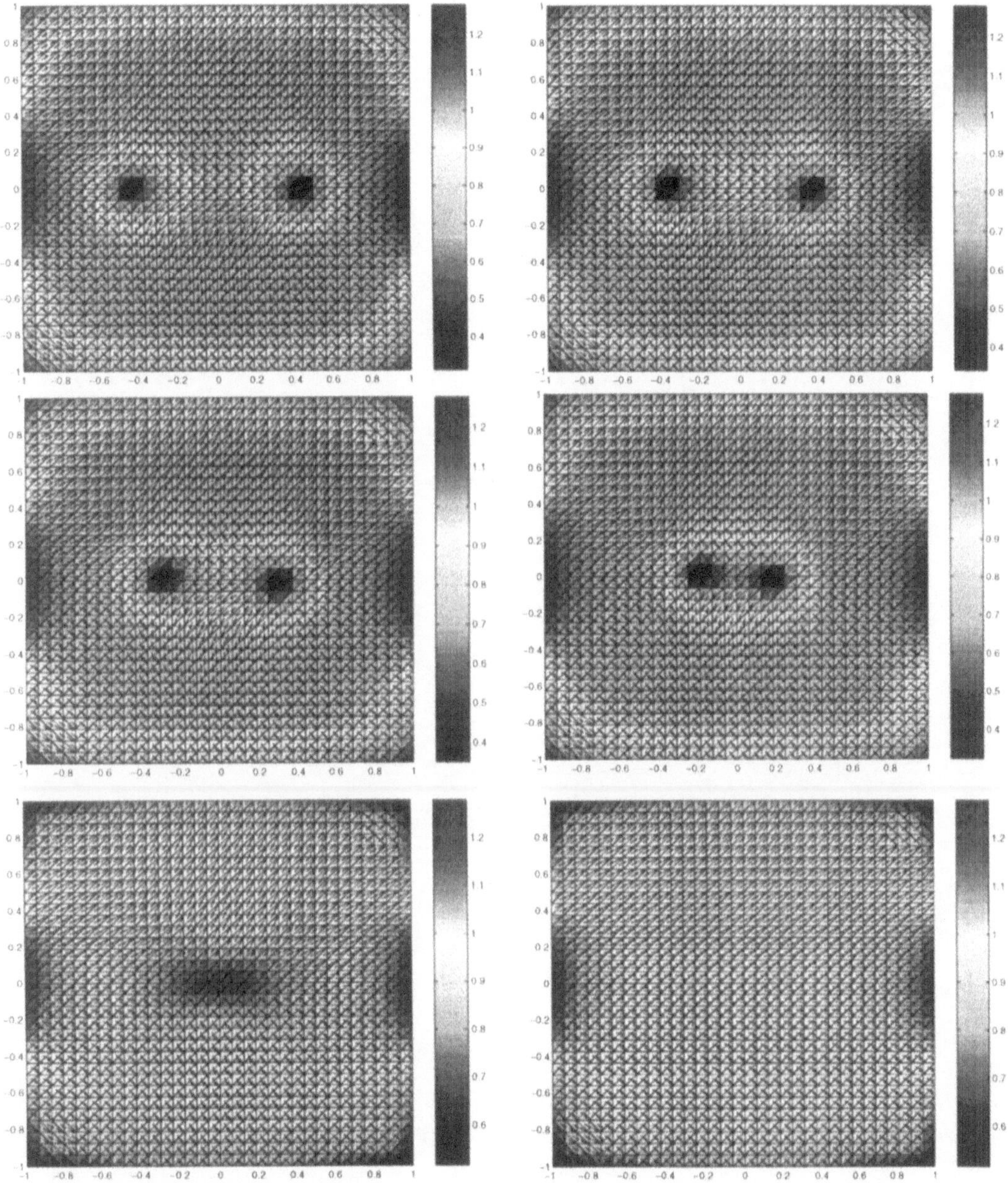

Figure 6.3: Plot of director field and its modulus at $t = 5i \cdot 10^{-2}$, $1 \le i \le 5$, and $t = 0.4$ ($h = \frac{1}{16}$, $k = 0.01$ in $(\mathbf{P})^{P}_{k;h}$) (note different scaling).

# Chapter 7

# Summary and Outlook

Ferromagnetic materials are widely used as recording media; advanced techniques in designing and analyzing such materials, for example writing heads in magnetic recording, can have an enormous impact on future technologies. Their mathematical theory started with the introduction of the Landau-Lifshitz free energy; a numerical analysis of existing strategies to solve the corresponding minimization problem is presented in part I of this monograph. Dynamics of the magnetization (e.g., switching processes of (averaged) magnetizations) is given by a torque balance, which leads to the (LLG) equation; electromagnetic coupling effects are incorporated in the (MLLG) equation. Part II closes with the numerical analysis for the nematic liquid crystal problem which also imposes a non-convex constraint onto its solution.

The goal of prime importance in this work is to investigate recently constructed numerical schemes for the given problems on the basis of existing regularities, and to construct new, more efficient schemes. Its realization requires the development of new numerical tools that are adopted to properties of the specific problems. We summarize these achievements in the following list.

- (Local) minimizers of the discretized non-convex Landau-Lifshitz energy functional exhibit mesh-dependent multi-scale structures (laminates, domain branching, closure domains), which makes their computation costly (Section 1.1).
- Increased symmetry properties (e.g., cubic compared to uniaxial ferromagnets) improve convergence properties of discrete direct minimization methods (Section 1.2).

- Convexification leads to computing macroscopic magnetizations. Special care is necessary to numerically solve the mixed problem: nonconforming functions (Section 2.1, [20]) or stabilized conforming methods (Section 2.2, [52]) help to attain a stable method.

- Stabilized finite element methods lead to strongly converging finite element magnetizations as minimizers of the discretized (degenerated) convexified problem for (locally) smooth minimizers (Section 2.3).

- An active set based strategy turns out to be efficient to compute Young-measure valued solutions of the unmodified Landau-Lifshitz energy density in a conforming discretization approach (Chapter 3, [82]).

- Projection schemes to solve (LLG) can be reformulated as semi-explicit penalization methods. More stable 'scaled projection' methods are developed on the basis of new penalization concepts that perform no projection onto the sphere any more. The Ginzburg-Landau penalization is a well-established strategy to verify existence of solutions for the limiting problem. Numerically, this strategy turns out less attractive, if compared to the new variants that are proposed in Chapter 4.

- The dynamics of electrically conducting ferromagnets requires to take account of Maxwell's equations. Splitting strategies are proposed and stretched time-grids that allow for practical convergence statements (in non-negative norms). A simple finite element discretization requires a further stabilization strategy. This material is presented in Chapter 5.

- (Nematic) Liquid Crystal problems share the non-convex constraint for the director field. A projection scheme is proposed and analyzed that efficiently copes with this constraint as well as incompressibility of the fluid; the scheme combines ideas of W. E & X. P. Wang [43] and A. Chorin [28]. An analysis is presented in Chapter 6 for low order Lagrange-type finite elements which shows optimal convergence (in weak norms).

Analyses of proposed schemes in part II are presented for strong solutions in two dimensions that exist at least locally in time for the considered nonstationary problems; in this framework, we can identify optimal penalization parameters in terms of numerical parameters and deal with nonlinear phenomena in an efficient way. However, there remain challenging open questions to be answered concerning numerical schemes that try to resolve weak solutions, allowing for (point) defects, etc. In order to handle such problems reliably, we expect schemes which only allow for restricted free choices of numerical parameters (i.e., stability constraints $F(\varepsilon, k, h) > 0$ will be more restrictive) and special grid geometries (i.e., locally refined meshes according to local energies of solutions[1]); the situation will even become more involved for three-dimensional applications. It is our hope that this work will stimulate further research in this exciting and challenging field of numerical analysis for microstructure evolution models.

---

[1] By evidence, local mesh refinement is also used to realize adaptive concepts, but this is not what we have in mind here.

# Bibliography


[1] A. AHARONI, *Introduction to the theory of ferromagnetism*, Oxford University Press (1996).

[2] F. ALOUGES, *A new algorithm for computing liquid crystal stable configurations: The harmonic case*, SIAM J. Num. Anal. **34**, pp. 1708-1726 (1997).

[3] F. ALOUGES, A. SOYEUR, *On global weak solutions for Landau-Lifshitz equations: existence and nonuniqueness*, Nonlin. Anal., Theory, Meth. & Appl. **18**, pp. 1071-1084 (1992).

[4] H. W. ALT, *Lineare Funktionalanalysis*, 2nd edition, Springer (1992).

[5] H. AMANN, *Quasilinear parabolic systems under nonlinear boundary conditions*, Arch. Rat. Mech. Anal. **92**, pp. 153-192 (1986).

[6] D. N. ARNOLD, R. S. FALK, *A uniformly accurate finite element method for the Mindlin-Reissner plate*, SIAM J. Numer. Anal. **26**, pp. 1276-1290 (1989).

[7] J. M. BALL, *A version of the fundamental theorem for Young measures*, in: PDEs and Continuum Models of Phase Transition. M. Rascle, D. Serre, M. Slemrod (eds.), Lecture Notes in Physics **344**, pp. 207-215, Springer (1989).

[8] R. E. BANK, *PLTMG: A Software Package for Solving Elliptic Partial Differential Equations. User's Guide 6.0.*, SIAM Philadelphia (1990).

[9] S. Baraket, L. Lassoued, *Bifurcation analysis of solutions to a Landau-Lifshitz problem with external fields*, Houston Journal of Mathematics, University of Houston **23**, pp. 669-683 (1997).

[10] G. Bertotti, *Hysteresis in magnetism*, Academic Press (1998).

[11] H. Blum, *Asymptotic error expansion and defect correction in the finite element method*, habilitation thesis, Universität Heidelberg (1990).

[12] J. H. Bramble, J. E. Pasciak, O. Steinbach, *On the stability of the $L^2$-projection in $H^1(\Omega)$*. Preprint available at `http://www.math.tamu.edu/pasciak/` (1999).

[13] S.C. Brenner, L.R. Scott, *The mathematical theory of finite element methods*, Texts in Applied Mathematics **15**, Springer (1994).

[14] F. Brezzi, M. Fortin, *Mixed and hybrid finite element methods*, Springer (1991).

[15] F. Brezzi, K.-J. Bathe, M. Fortin, *Mixed-interpolated elements for Reissner-Mindlin plates*, Int. J. Num. Meth. Eng. **28** (1989), pp. 1787-1801.

[16] W. F. Brown, Jr., *Magnetostatic interactions*, North-Holland, Amsterdam, 1962.

[17] W. F. Brown, *Micromagnetics*, Interscience (1963).

[18] W. F. Brown, Jr., *Magnetostatic principles in ferromagnetism*, Springer, New York, (1966).

[19] C. Carstensen, *Merging the Bramble-Pasciak-Steinbach and the Crouzeix-Thomée criterion for $H^1$-stability of the $L^2$-projection onto finite element spaces*, Math. Comp. [in print].

[20] C. Carstensen, A. Prohl, *Numerical Analysis of Relaxed Micromagnetics by Penalised finite elements*, Numer. Math. (published online) (2001).

[21] C. CARSTENSEN, T. ROUBÍČEK, *Numerical approximation of Young measures in non-convex variational problems*, Num. Math. **84**, pp. 395-415 (2000).

[22] S. CHANDRASEKHAR, *Liquid Crystals*, Cambridge, 2nd ed. (1992).

[23] K.-C. CHANG, W.-Y. DING, R. YE, *Finite-time blow-up of the heat flow of harmonic maps from surfaces*, J. Diff. Geom. **36**, pp. 507-515 (1992).

[24] Y. CHEN, *A Remark on the Regularity for the Landau-Lifshitz equation*, Appl. Anal. **63**, pp. 207-221 (1996).

[25] Y. CHEN, B. GUO, *Two dimensional Landau-Lifshitz equation*, J. Partial Diff. Eqs. **9**, pp. 313-322 (1996).

[26] R. CHOKSI, R. V. KOHN, *Bounds on the micromagnetic energy of a uniaxial ferromagnet*, Commun. Pure Appl. Math. **51**, no. **3**, pp. 259-289 (1998).

[27] R. CHOKSI, R. V. KOHN, F. OTTO, *Domain branching in uniaxial ferromagnets: a scaling law for the minimum energy*, Comm. Math. Phys. **201**, pp. 61-79 (1999).

[28] A. J. CHORIN, *Numerical solution of the Navier-Stokes equations*, Math. Comp. **22**, pp. 745-762 (1968).

[29] P. G. CIARLET, *Basic Error estimates for elliptic problems*, in: Handbook of Numerical Analysis, vol. **II**, Finite Element Methods (Part 1), North-Holland (1991).

[30] P. CLÉMENT, *Approximation by finite element functions using local regularization*, Sér. Rouge Anal. Numér. (RAIRO) **R-2**, pp. 77-84 (1975).

[31] R. COHEN, R. HARDT, D. KINDERLEHRER, S.-Y. LIN, M. LUSKIN, *Minimum energy configurations for liquid crystals*, in: IMA Volumes in Mathematics and its Applications (J. Ericksen, D. Kinderlehrer eds.), vol. **5**, pp. 99-121, Springer (1987).

[32] R. COHEN, S.-Y. LIN, M. LUSKIN, *Relaxation and gradient methods for molecular orientation in liquid crystals*, Comput. Phys. Commun. **53**, pp. 455-465 (1989).

[33] D. J. CRAIK, R. S. TEBBLE, *Ferromagnetism and Ferromagnetic Domains*, North-Holland (1965).

[34] M. CROUZEIX, V. THOMÉE, *The stability in $L^p$ and $W^{1,p}$ of the $L^2$-projection onto finite element function spaces*, Math. Comp. **48**, pp. 521-532 (1987).

[35] B. DACOROGNA, *Direct Methods in the Calculus of Variations*, Applied Math. Sciences **78**, Springer (1989).

[36] T. DAVIS, E. GARTLAND, *Finite element analysis of the Landau-De Gennes minimization problem for liquid crystals*, SIAM J. Numer. Anal. **35**, pp. 336-362 (1998).

[37] E. DEAN, R. GLOWINSKI, C. H. LI, *Applications of operator splitting methods to the numerical solution of nonlinear problems in continuum mechanics and physics*, in: Mathematics Applied to Science, Academic press, pp. 13-64 (1988).

[38] D. DEMUS, J. GOODBY, G. W. GRAY, H.-W. SPIESS, V. VILL (eds.), *Physical Properties of Liquid Crystals*, Wiley-VCH (1999).

[39] A. DE SIMONE, *Energy Minimizers for Large Ferromagnetic Bodies*, Arch. Rat. Math. Anal. **125**, pp. 99-143 (1993).

[40] M. J. DONAHUE, `http://math.nist/gov/~MDonahue/`.

[41] W. E, *Selected Problems in Material Science*, in: *World Mathematics 2000*, Springer (2000).

[42] W. E, J. G. LIU, *Projection method I: Convergence and numerical boundary layers*, SIAM J. Numer. Anal. **32**, pp. 1017-1057 (1995).

[43] W. E, X.-P. WANG, *Numerical Methods for the Landau-Lifshitz equation*, SIAM J. Numer. Anal. **38**, pp. 1647-1665 (2000).

[44] I. EKELAND, R. TEMAM, *Convex Analysis and Variational Problems*, North-Holland (1976).

[45] J. EELLS, L. LEMAIRE, *A report on harmonic maps*, Bull. London Math. Soc. **10**, pp. 1-68 (1978).

[46] J. EELLS, L. LEMAIRE, *Another report on harmonic maps*, Bull. London Math. Soc. **20**, pp. 385-524 (1988).

[47] J. EELLS, J. H. SAMPSON, *Harmonic mappings of Riemannian manifolds*, Amer. J. Math. **86**, pp. 109-160 (1964).

[48] J. EELLS, J. C. WOOD, *Restrictions on harmonic maps of surfaces*. Topology **15**, pp. 263-266 (1976).

[49] J. FIDLER, T. SCHREFL, *Micromagnetic modelling — the current state of the art*, J. Phys. D: Appl. Phys. **33**, pp. R135-R156 (2000).

[50] H.C. FOGEDBY, *Theoretical aspects of mainly low dimensional magnetic systems*, Lecture Notes in Phys. **131**, Springer (1980).

[51] D. R. FREDKIN, T. R. KOEHLER, *Hybrid method for computing demagnetizing fields*, IEEE Trans. Magn. **26**, pp. 415-417 (1990).

[52] S. A. FUNKEN, A. PROHL, *On stabilized finite element methods in relaxed micromagnetism*, Preprint **99-18**, Universität Kiel (1999).

[53] P. G. DE GENNES, *The Physics of Liquid Crystals*, Oxford (1974).

[54] M. GIAQUINTA, *Introduction to regularity theory for nonlinear elliptic systems*, Birkhäuser (1993).

[55] M. GIAQUINTA, *Multiple integrals in the calculus of variations and nonlinear elliptic systems*, Ann. Math. Studies **105**, Princeton University Press (1983).

[56] D. GILBARG, N. S. TRUDINGER, *Elliptic Partial Differential Equations of the Second Order*, 2nd ed., Springer (1983).

[57] T. L. GILBERT, *A Lagrangian formulation of gyromagnetic equation of the magnetization field*, Phys. Rev. **100**, pp. 1243*ff* (1955).

[58] V. GIRAULT, P. A. RAVIART, *Finite Element Methods for Navier-Stokes equations*, Springer (1986).

[59] A. P. GUIMARAES, *Magnetism and Magnetic Resonance in Solids*, Wiley (1998).

[60] B. GUO, M.-C. HONG, *The Landau-Lifshitz equation of the ferromagnetic spin chain and harmonic maps*, Calc. Var. **1**, pp. 311-334 (1993).

[61] B. GUO, M.-C. HONG, F. SU, *The global attractors for the Landau-Lifshitz equation of the ferromagnetic spin chain on compact manifolds*, AMS/IP vol. **3**, pp. 213-227 (1997).

[62] B. GUO, F. SU, *Global weak solution for the Landau-Lifshitz-Maxwell equation in three space dimensions*, J. Math. Anal. Appl. **211**, pp. 326-346 (1997).

[63] B. GUO, F. SU, *The global smooth solution for Landau-Lifshitz-Maxwell equation without dissipation*, J. Partial Diff. Eqs. **11**, pp. 193-208 (1998).

[64] B. GUO, Y. WANG, *Generalized Landau-Lifshitz systems and harmonic maps*, Science in China (Ser. A), vol. **39**, no. **12**, pp. 1242-1257 (1996).

[65] R. HARDT, D. KINDERLEHRER, AND F. H. LIN, *Existence and partial regularity of static liquid crystal configurations*, Comm. Math. Phys. **105**, pp. 547-570 (1986).

[66] R. HARDT, F. H. LIN, *Stability of singularities of minimizing harmonic maps*, J. Differential Geom. **29**, pp. 113-123 (1989).

[67] R. HERTEL, H. KRONMÜLLER, *Adaptive finite element mesh refinement techniques in three-dimensional micromagnetic modeling*, IEEE Trans. Magn. **34**, pp. 3922-3930 (1998).

[68] J. G. HEYWOOD, R. RANNACHER, *Finite element approximation of the nonstationary Navier-Stokes problem. I. Regularity of solutions and second order error estimates for spatial discretization*, SIAM J. Numer. Anal. **19**, pp. 275-311 (1982).

[69] A. HUBERT, R. SCHÄFER, *Magnetic Domains*, Springer (1998).

[70] T. J. R. Hughes, L. P. Franca, M. Balestra, *A new finite element formulation for computational fluid mechanics: V. Circumventing the Babuška-Brezzi condition: A stable Petrov-Galerkin formulation of the Stokes problem accommodating equal order interpolation*, Comp. Meth. Appl. Mech. Eng. **59**, pp. 85-99 (1986).

[71] R. D. James, D. Kinderlehrer, *An example of frustration in a ferromagnetic material*, in: Defects, Singularities, and Patterns in Nematic Liquid Crystals: Mathematical and Physical Aspects, NATO Meeting Series, J. M. Coron, F. Helein, J. M. Ghidaglia, eds., Kluwer Academic Publishers, Dordrecht, Netherlands, pp. 201-221 (1990).

[72] R. D. James, D. Kinderlehrer, *Frustration in ferromagnetic materials*, Contin. Mech. Thermodyn. **2**, pp. 215-239 (1990).

[73] P. Joly, O. Vacus, *Mathematical and numerical studies of non linear ferromagnetic materials*, Math. Mod. and Num. Anal. **33**, pp. 593-626 (1999).

[74] P. Keast, *Moderate-degree tetrahedral quadrature formulas*, Comp. Meth. Appl. Mech. Engrg. **55**, pp. 339-348 (1986).

[75] K.J. Kirk, *Nanomagnets for Sensors and data storage*, Contemp. Phys. **41**, pp. 61-78 (2000).

[76] K. J. Kirk, J. N. Chapman, and C. D. Wilkinson, Appl. Phys. Lett. **71**, pp. 539*ff* (1997).

[77] S. Klainerman, *Global existence for nonlinear wave equations*, Comm. Pure Appl. Math. **33**, pp. 43-101 (1980).

[78] T. R. Koehler, *Hybrid FEM-BEM method for fast micromagnetic calculations*, Physica B **233**, pp. 302-307 (1997).

[79] R. Kouhia, R. Stenberg, *A linear nonconforming finite element method for nearly incompressible elasticity and Stokes flow*, Comp. Meth. Appl. Mech. Engrg. **124**, pp. 195-212 (1995).

[80] M. Kružík, *Numerical solution to relaxed problems in micromagnetics*, Preprint (1997).

[81] M. KRUŽÍK, *Maximum principle based algorithm for hysteresis in micromagnetics*, Caesar-Preprint **8** (2001).

[82] M. KRUŽÍK, A. PROHL, *Young measure approximation in micromagnetics*, Numer. Math. (published online) (2001).

[83] M. KRUŽÍK, T. ROUBÍČEK, *Weierstrass-type maximum principle for microstructure in micromagnetics*, Preprintreihe des Max-Planck-Instituts für Mathematik in den Naturwissenschaften Leipzig **99-40** (1999).

[84] L. D. LANDAU, E. M. LIFSHITZ, *On the theory of the dispersion of magnetic permeability in ferromagnetic bodies*, Phys. Z. Sowje. **8** (1935), reproduced in: Collected papers of L. D. Landau (D. ter. Haar eds.), pp. 101-114, Pergamon (1965).

[85] F. LESLIE, *Some constitutive equations for liquid crystals*, Arch. Rat. Mech. Anal. **28**, pp. 265-283 (1968).

[86] E. LIFSHITZ, *On the magnetic structure of iron*, J. Phys. USSR **8**, pp. 337-346 (1944).

[87] F. H. LIN, *Nonlinear theory of defects in nematic liquid crystals; phase transition and flow phenomena*, Comm. Pure Appl. Math. **42**, pp. 789-814 (1989).

[88] F. H. LIN, *Some dynamic properties of Ginzburg-Landau vortices*, Comm. Pure Appl. Math. **49**, pp. 323-359 (1996).

[89] F. H. LIN, C. LIU, *Nonparabolic dissipative systems, modeling the flow of liquid crystals*, Comm. Pure Appl. Math. **48**, pp. 501-537 (1995).

[90] F. H. LIN, C. LIU, *Global existence of solutions for the Ericksen-Leslie system*, Arch. Rat. Mech. Appl. [accepted].

[91] F. H. LIN, C. LIU, *Existence of Solutions for the Ericksen-Leslie System*, manuscript (2000).

[92] S.-Y. LIN, M. LUSKIN, *Relaxation methods for liquid crystal problems*, SIAM J. Numer. Anal. **26**, pp. 1310-1324 (1989).

[93] C. LIU, N. J. WALKINGTON, *Approximation of Liquid Crystal Flows*, SIAM J. Numer. Anal. **37**, pp. 725-741 (2000).

[94] C. LIU, N. J. WALKINGTON, *Mixed methods for the approximation of liquid crystal flows*, CNA-Preprint **99-23**, Carnegie Mellon University, Pittsburgh (1999).

[95] M. LUSKIN, L. MA, *Analysis of the finite element approximation of microstructure in micromagnetics*, SIAM J. Numer. Anal. **29**, pp. 320-331 (1992).

[96] L. MA, *Analysis and computation for a variational problem in micromagnetics*, Ph.D. thesis, University of Minnesota, Minneapolis, MN, (1991).

[97] P. MONK, *A finite element method for approximating the time-harmonic Maxwell equations*, Numer. Math. **63**, pp. 243-261 (1992).

[98] P. MONK, O. VACUS, *Error estimates for a numerical scheme for ferromagnetic problems*, SIAM J. Numer. Anal. **36**, pp. 696-718 (1999).

[99] J. C. NÉDÉLEC, *Mixed finite elements in* $\mathbb{R}^3$, Numer. Math. **35**, pp. 315-341 (1980).

[100] J. C. NÉDÉLEC, *A new family of mixed finite elements in* $\mathbb{R}^3$, Numer. Math. **50**, pp. 57-81 (1986).

[101] P. PEDREGAL, *Parametrized Measures and Variational Principles*, Birkhäuser, Basel (1997).

[102] P. PEDREGAL, *Numerical computation of parametrized measures*, Numer. Funct. Anal. Opt. **16**, pp. 1049-1066 (1995).

[103] P. PEDREGAL, *On the numerical analysis of nonconvex variational problems*, Numer. Math. **74**, pp. 325-336 (1996).

[104] P. PEDREGAL, *Relaxation in ferromagnetism: the rigid case*, J. Nonl. Sci. **4**, pp. 105-125 (1994).

[105] P. PEDREGAL, *Parametrized Measures and Variational Principles*, Birkhäuser, Basel (1997).

[106] F. Pistella, V. Valente, *Numerical Stability for a Discrete Model in the Dynamics of Ferromagnetic Bodies*, manuscript (1997).

[107] A. Prohl, *Projection and Quasi-Compressibility methods for solving the incompressible Navier-Stokes equations*, Teubner (1997).

[108] A. Prohl, *On pressure approximation via projection methods in computational fluid dynamics*, Preprint **99-17**, Universität Kiel (1999).

[109] A. Prohl, *An adaptive finite element method for solving a double well problem describing crystalline microstructure*, $M^2AN$ **33**, pp. 781-796 (1999).

[110] A. Prohl, *Numerical Analysis in Nonstationary Micromagnetism and Nematic Liquid Crystals*, habilitation thesis, Universität Kiel (2001).

[111] R. Rannacher, *On Chorin's projection method for the incompressible Navier-Stokes equations*, in: LNM **1530**, eds.: J. G. Heywood, K. Masuda, R. Rautmann, S. A. Solonnikov: The Navier-Stokes equations II — Theory and Numerical Methods, pp. 167-183, Proc. Oberwolfach (1991).

[112] T. Roubíček, *Relaxation in Optimization Theory and Variational Calculus*, Walter de Gruyter (1997).

[113] T. Roubíček, M. Kružík, *Microstructure evolution model in micromagnetics*, Caesar-Preprint **3** (2000).

[114] K. Schittkowski, *NLPQL: A Fortran subroutine solving constrained nonlinear programming problems*, Annals of Operation Research **5**, pp. 485-500 (1985-6).

[115] R. M. Schoen, K. Uhlenbeck, *Boundary regularity and miscellaneous results on harmonic maps*, J. Diff. Geom. **18**, pp. 253-268 (1983).

[116] W. Scholz, *Micromagnetic Simulation of Thermally Activated Switching in Fine Particles*, diploma thesis, TU Wien (1999).

[117] T. SCHREFL, private communication.

[118] T. SCHREFL, J. FIDLER, K. J. KIRK, J. N. CHAPMAN, *Domain structures and switching mechanisms in patterned magnetic elements*, J. Magn. Mag. Mat. **175**, pp. 193-204 (1997).

[119] T. SCHREFL, J. FIDLER, K. J. KIRK, J. N. CHAPMAN, *A higher order FEM-BEM method for the calculation of domain processes in magnetic nano-elements*, IEEE Trans. Magn. **33**, pp. 4182-4184 (1997).

[120] M. STRUWE, *On the evolution of harmonic mappings of Riemannian surfaces*, Comment. Math. Helv. **60**, pp. 558-581 (1985).

[121] M. STRUWE, *Variational Methods*, Springer (1990).

[122] M. STRUWE, *Geometric Evolution Problems*, IAS/Park City Math. Series, vol. **2**, pp. 259-339 (1996).

[123] L. TARTAR, *Beyond Young Measures*, Meccanica **30** pp. 505-526 (1995).

[124] R. TEMAM, *Sur l'approximation de la solution des equations de Navier-Stokes par la methode de pas fractionnaires II*, Arch. Rat. Mech. Anal **33**, pp. 377-385 (1969).

[125] R. TEMAM, *Navier-Stokes equations*, rev. ed., Studies in Mathematics and its Applications **2**, North-Holland (1977).

[126] R. VERFÜRTH, *A review of a posteriori error estimation and adaptive mesh-refinement techniques*, Wiley-Teubner (1996).

[127] E. G. VIRGA, *Variational Theories for Liquid Crystals*, Chapman& Hall, series: Appl. Math. and Math. Comp. **8** (1994).

[128] A. VISINTIN, *On Landau-Lifchitz' equations for ferromagnetism*, Japan J. Appl. Math. **2**, pp. 49-84 (1985).

[129] A. VISINTIN, *Models of phase transitions*, Birkhäuser (1996).

[130] A. VISINTIN, *Modified Landau-Lifshitz equation in ferromagnetism*, Physica B **233**, pp. 365-369 (1997).

[131] A. VISINTIN, *On some models of ferromagnetism*, in: Free Boundary Problems I, Chiba, 1999 (N. Kenmochi, ed.) Gakuto Int. Series in Math. Sci. Appl. **13**, pp. 411-428 (2000).

[132] L. C. YOUNG, *Generalized curves and existence of an attained absolute minimum in the calculus of variations*, Comptes Rendus de la Société et des Lettres de Varsovie, Classe III **30**, pp. 212-234 (1937).

# Index